Fisheries Biology, Assessme
and Management

Fisheries Biology, Assessment and Management

Second edition

Michael King
Fisheries Consultant
Toogoom, Queensland
Australia

Blackwell
Publishing

Blackwell Publishing editorial offices:
Blackwell Publishing Ltd, 9600 Garsington Road, Oxford OX4 2DQ, UK
Tel: +44 (0)1865 776868
Blackwell Publishing Professional, 2121 State Avenue, Ames, Iowa 50014-8300, USA
Tel: +1 515 292 0140
Blackwell Publishing Asia Pty Ltd, 550 Swanston Street, Carlton, Victoria 3053, Australia
Tel: +61 (0)3 8359 1011

First published 1995 by Fishing News Books, a division of Blackwell Science Ltd
Reprinted 1996, 1997, 1998, 1999, 2001, 2003, 2006
Second edition published 2007 by Blackwell Publishing Ltd

6   2013

Library of Congress Cataloging-in-Publication Data
King, M. G.
   Fisheries biology, assessment, and management / Michael King. — 2nd ed.
       p. cm.
   Includes bibliographical references and index.
   ISBN: 978-1-4051-5831-2 (pbk. : alk. paper) 1. Fishery management.   2. Fish
populations.   I. Title.

   SH328.K55 2007
   639.2—dc22

                                                                  2007011457

A catalogue record for this title is available from the British Library

Set in 9.75/12.5pt Sabon
by Graphicraft Limited, Hong Kong

For further information on Blackwell Publishing, visit our website:
www.blackwellpublishing.com

Contents

Preface

Total catches from fisheries of the world have been decreasing at least over the time since the first edition of this book was published in 1995. Many stocks of fish are under threat or worse. According to the United Nations' Food and Agriculture Organization, about 60% of the world's 200 most valuable fish species are in trouble. Some species, such as the Atlantic halibut and Atlantic bluefin tuna, have been fished to the brink of extinction and populations of many other exploited species are threatened.

In the same way that excessive hunting on land has threatened terrestrial species, excessive fishing has reduced stocks of marine species to dangerously low levels. Not only are catches becoming less for a given amount of fishing, the sizes of individual fishes in catches are becoming smaller. Large and valuable predatory species have been removed from ecosystems and fishers are catching smaller and usually less desirable species. Often this means that fishers are 'fishing down marine food webs' by targeting seafood species that are at increasingly lower trophic levels. And within populations of particular target species, the large females that are responsible for producing a disproportionately large number of eggs are disappearing.

In the face of remorseless and heavy fishing by many nations on earth, a reasonable assumption would be that many marine species have been driven to extinction. But there is very little evidence to suggest that this is so. The fact is that if a furry, charismatic land animal went missing many people would be aware of it. But there are just too few people studying and counting marine species, and survey efforts are biased towards the ones of economic importance. Nevertheless, several species ranging from a skate in the Irish Sea to giant clams in Pacific islands have been driven to local extinction (for which the weasel word 'extirpation' is sometimes used). Pacific salmon have been fished out of many river systems and are unlikely to return. And there are a few species, such as the Californian white abalone, in which extinction has been averted at the last minute by human intervention. Although there are very few verified examples of overfishing leading to the extinction of a species, many fisheries have collapsed and target populations have been depleted to undesirably low levels. One, perhaps pessimistic, suggestion is that all major fisheries will be based on stock sizes of less than 10% of their original size by the middle of the present century. There is a renewed interest in the old principle that many populations require a relatively large minimum number of individuals to ensure that reproduction is successful; this may be particularly so in the sea where, in the case of widely spaced individuals, the chances of drifting sperm and eggs meeting must be very low.

Although heavy fishing is the major cause of depleted fish stocks, the situation is aggravated in many coastal fish stocks by damage to the marine environment. Marine ecosystems are suffering from habitat destruction, coastal development, pollution and eutrophication in addition to the removal of particular species by fishing. If a population, already depleted by excessive fishing, suffers additional impediments such as the environmental degradation of its key habitats, extinction is highly likely. If this occurs, the question of whether overfishing or environmental effects caused the loss of the species is merely academic.

The demand for seafood and for participation

in fishing is high. In the face of this demand, fisheries managers appear to have done a poor job. Perhaps politicians have yielded to this strong demand and allowed levels of fishing beyond that recommended by fisheries scientists. Or perhaps fisheries managers have concentrated too much on fish stocks to the exclusion of the effects of fishing on marine ecosystems. Or it may be that some managers are out of touch with the human element of fisheries. The science involved in fisheries management may be highly developed, but fisheries management is about managing people, not fish stocks.

The global crisis in fisheries has resulted from the open access and fishing subsidization policies of governments, both of which have led to over-exploitation and overcapitalization. Incidental catches of large quantities of unmarketable species (bycatches) are disrupting marine food webs and damaging ecosystems. In addition, illegal and unreported fishing are frustrating efforts to manage fisheries on a sustainable basis. Some positive trends include actions being taken to decrease the incidental catches of marine species, the ecolabelling of seafood sourced from sustainable fisheries and international cooperation in improving monitoring, control and surveillance. The number of areas that are effectively closed to fishing is increasing and more fishers and fishing communities are becoming involved in the cooperative management of fisheries.

In this new edition of *Fisheries Biology, Assessment and Management*, a fishery is considered as a component of a marine ecosystem that must also be managed. Consequently, a new chapter has been added to provide an introduction to marine ecology. The chapter on exploited species has been updated and, in line with the first edition, does not neglect the fisheries resources of warmer waters where a much larger range of species provides food security for local people. The chapter on fishing gear has been expanded to cover fishers and the effects of fishing on target species, non-target species and the environment. Although aquaculture is providing an increasing contribution to world seafood production, it is not covered explicitly in this book. In many cases,

aquaculture exacerbates the problems of wild fisheries in that it is associated with environmental damage (from the construction of ponds and the pollution of water by food and faeces), the need to collect juveniles from the wild (in over 20% of the farmed species) and the requirement for food containing fishmeal (produced from wild fisheries).

The chapters on parameter estimation and stock assessment have taken advantage of the wide availability of computers. Previous graphically-based examples are reworked, and new ones have been added, with step-by-step instructions on building computer spreadsheet models. Although computer models are more conveniently developed by writing programs in one of the common languages, spreadsheet programs are used in this text because of their wide use and transparency. In such transparent models, intermediate values and meaningful steps can be seen in cells, columns and rows. In simulation models in particular, transparency is not an inconsequential matter, particularly when demonstrating the working and assumptions behind models to colleagues, fisheries managers and members of the fishing industry. Descriptions in the appendices allow readers to build programs that search for optimum parameter values and to construct models with random variations that more realistically represent the vagaries of nature.

The final chapter on fisheries management has been expanded to include the setting of broader management objectives to protect ecosystems as well as fish stocks. Conventional management measures on inputs (the amount of fishing) and outputs (the amount of catch) are also discussed. The section on the co-management and community-based management of fisheries has been expanded to match the degree of increased interest in these areas. For the same reasons, the section on the use of marine protected areas in fisheries management has been updated and enlarged.

Finally, it must be said that virtually all of the world's environmental ills, including the over-exploitation of marine life, can be ascribed to the fact that the planet has too many people. The world's population has passed the six billion mark and is still growing rapidly. Each additional person is another source of pollution and adds to the demand for both living and non-living natural resources. Seafood is in danger of becoming a scarce and very expensive commodity for which there is an insatiable demand. There is a dire need to contain population growth and protect the environment. Not to do so will result in future generations being heirs to an abused planet with polluted seas and exhausted natural resources.

Michael King
Fisheries Consultant
Toogoom, Queensland
Australia

Acknowledgements

I thank Jennifer Kallie and Bryan Frankham for comments on drafts of parts of this second edition. I also thank students, friends and colleagues who kindly provided comments on, or information for, the first edition of the book; these include Ian Cartwright, Alistair McIlgorm, Marc Wilson, David Campbell, Melita Samoilys, Jeremy King, Neil Sarti, Alby Steffans, John Wakeford and Carol Scott. For a much wider range of comments and suggestions, I particularly thank Scoresby Shepherd, Alf Carver and John Wallace. I also thank the publishers, Nigel Balmforth and Kate Nuttall, for their support as well as the publisher's reviewer, Dr Ian Cowx, for many helpful suggestions.

Graphs produced using Microsoft Excel and Solver.

1 Ecology and ecosystems

1.1 Introduction

Studies in biology, the science of living things, can be directed at increasing levels of biological organization from molecules, cells, organs, to organisms (or species) and beyond to populations and ecosystems. This chapter is concerned with higher levels of biological organization – populations, communities and ecosystems. A population is a group of individuals belonging to the same species and a community is a collection of populations inhabiting a particular area. An ecosystem is a functional and relatively self-contained system that includes communities and their non-living environment.

Studies in fisheries biology have been directed at particular species targeted by fishing operations. Now, particularly in the light of decreasing catches and threatened marine ecosystems, there is a need for fisheries managers to take a broader view, one that includes the interdependence of target species, other species and the marine environment. The ecosystem is the basic unit of ecology, and can be defined as the study of the interactions between groups of organisms and their environment. The environment of an organism includes all external entities and factors that affect it, and therefore includes physical factors such as light, temperature and oxygen as well as other living things such as competitors, mates, predators, and parasites.

Although it is common, and often useful, to apply the term 'ecosystem' to particular entities, such as coral reef ecosystems or estuarine ecosystems, it must be realized that these are not isolated units. Ecosystems are linked to one another by biological and physical processes. In marine ecosystems, these linking processes include biological factors, such as migration and food chains, as well as physical ones, such as ocean currents and tides. Pursuing these linking factors, it becomes apparent that the entire planet can be regarded as an ecosystem, and is sometimes referred to as a biosphere. However, a more restricted view of an ecosystem – as the plants, animals and environment of a particular type of habitat, such as a coral reef – provides a more manageable entity for study and management.

1.2 Distribution and abundance

Populations are groups of individuals belonging to the same species. In fisheries biology, the word 'stock' is sometimes used interchangeably (and loosely) with 'population'. In the strict sense, a stock is a distinct, reproductively isolated population which exists within a defined spatial range.

1.2.1 Unit stocks

Fisheries studies and management are concerned with a unit stock, which may be defined as a discrete group of individuals that has the same gene pool, is self-perpetuating, and has little connection with adjacent groups of the same species. Although this definition may not satisfy biogeographers, it does describe a unit which, because it has similar biological characteristics, may be studied, assessed and managed as a discrete entity.

Some species exist within a wide geographic range as a collection of unit stocks. The cod, *Gadus morhua*, for example, is distributed across the North Atlantic (Fig. 1.1) and within this relatively large range, exists in more or less isolated

1

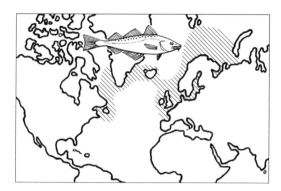

Fig. 1.1 The distribution of the cod, *Gadus morhua*, in the North Atlantic.

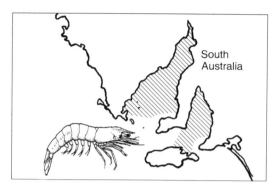

Fig. 1.2 Two separate unit stocks (shaded areas) of the penaeid prawn, *Penaeus latisulcatus*, in two gulfs in South Australia.

subpopulations or races. In such cases, fishing on one subpopulation appears to have no effect on others.

The boundaries of a unit stock are often difficult to determine, and many seemingly isolated populations may receive new recruits from other distant reproducing populations. Even in stocks of fishes on isolated reefs, the ability of larvae to drift and survive for a considerable time in the plankton allows them to reach other reefs some distance away. For example, the snapper, *Lutjanus kasmira*, which was deliberately introduced into Hawaii, spread throughout all the reefs and islands of the archipelago within a period of ten years (Oda & Parrish, 1981). Non-migratory species that live in widely separated areas, such as seamounts, must either rely on larval drift to replenish their populations or, if their larval lifespan is short, be self-sustaining.

From a fisheries assessment and management viewpoint, it is important to determine whether two adjacent stocks are either sufficiently inter-active to be regarded as a single unit stock, or independent enough to be treated as separate unit stocks. In most cases, several criteria are used to confirm or refute a stock's unit status. The penaeid prawn, *Penaeus latisulcatus*, for example, is caught by trawlers in two adjacent gulfs in South Australia (Fig. 1.2) but not in the area of sea between the two gulfs. As each gulf contains mature individuals and has coastal mangrove areas where juveniles are found, the stock in each gulf has the ability for self-replenishment. In addition, as

tagging studies have not revealed any migration of prawns between the two areas, the stocks in each of the two gulfs are regarded as two separate units for research and management purposes.

1.2.2 Spacing of organisms

Within a unit stock, individuals may be distributed uniformly, randomly or in aggregations (Fig. 1.3). A uniform or even distribution rarely occurs in nature, mainly because the environment is rarely uniform. Even if the environment is relatively even, such as on a sandy sea floor, the

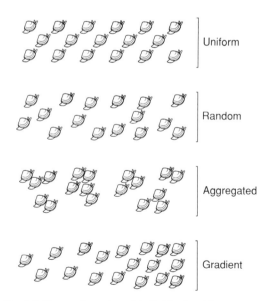

Fig. 1.3 Types of spacing of individuals within an area.

distribution of sedentary species is likely to be non-uniform as a result of the uneven settlement of larvae from the plankton. However, a uniform distribution may be approximated in species where there is competition, territoriality or aggression between individuals. Territorial reef fishes, for example, often exclude others of the same species from a range around a home base on the reef. Random distributions are also rare in nature, if only because the aspects of the environment on which the species depend, such as food and shelter, are not randomly distributed.

Although widely-spaced individuals avoid intraspecific competition, this is often at the expense of advantages which may accrue to those living in aggregations, groups, or schools, the most common type of distribution. The advantages of living in aggregations in mating species may include better access to sexual partners, and in broadcast spawners, an enhanced confluence of eggs and sperm. Aggregations may also provide an increased ability to locate food and a degree of protection from predators. For example, among aggregating sea urchins, fertilization success is high, the trapping of drift algae for food is enhanced, and the spine canopy of the aggregation is a formidable deterrent to predators. Whatever the spacing, the overall distribution of individuals or clumps will be influenced by differences or gradients in the environment. In all marine organisms, a differential distribution with depth is to be expected, and most species occur in maximum numbers over a relatively narrow optimal depth range.

In fisheries studies the estimation of abundance, or at least relative abundance (the number of individuals at one time relative to the number present at another time), is important in determining the effects of fishing and environmental disturbances. Methods of estimating abundance are presented in Section 4.2.

1.3 Population growth and regulation

Populations of all organisms fluctuate around a mean level as long as deaths are balanced by births. In cases of populations, such as many fish stocks, which are overexploited or threatened by environmental degradation, deaths will exceed births and numbers will decrease. When an exotic species is introduced into a 'new' and suitable environment (Section 1.5.4 'Species invasions, introductions and translocations') its population will increase, often in the absence of predators, until it is contained by the lack of food or living space. This section provides an introduction to populations as well as the factors that affect and regulate them.

1.3.1 Population growth

In the absence of limiting factors in conditions of unlimited food and living space, the increase of numbers in a population would be immense. For example, if a female shrimp produced 50 thousand female larvae, her descendants could number 2500 million females after just one additional annual spawning event as long as all the resulting larvae survived to reproduce. If N is the number of individuals in the population at a particular time, b is the birth rate, and d is the death rate, the population growth rate (I) is:

$$I = N(b - d) \tag{1.1}$$

It is the value of $(b - d)$, referred to as the intrinsic rate of population increase, that determines whether a population will decrease, remain stable (at zero population growth) or increase. As long as the average birth rate, b, exceeds the average death rate, d, the population (N) will increase. In addition, if populations are increasing, N will become larger with each generation causing the rate of increase to rise further. This multiplying rate of increase causes population numbers to increase as shown in the left-hand curve in Fig. 1.4: the curve becomes steeper and steeper until the population is expanding at an infinite rate. All organisms have such potential for exponential growth in the absence of any limiting environmental factors. But, as the world is not packed full of shrimps or any other organism, it is obvious that the increase in numbers in real populations is being held in check, or regulated, by one or more factors.

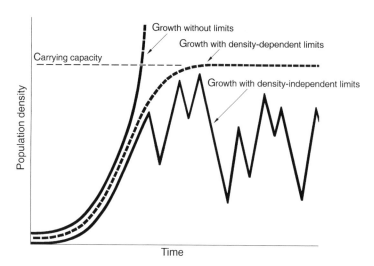

Fig. 1.4 Population growth curves, (from the left) without limits, with density-dependent limits and with density-independent limits.

1.3.2 Population regulation

All populations are limited in abundance by their requirement for resources – for essentials such as food and living space. Competition for these resources, and predation, cause the rate of population growth to decrease at high densities. These limiting factors are regarded as density-dependent because their effects increase as population density increases – for example, the effects of shortages of food and living space (starvation and crowding) increase with population size. Over time, population numbers follow an S-shaped curve in which an initial increasing rate of growth is followed by a decreasing rate as the curve approaches an asymptote (at what is known as the carrying capacity of the environment) imposed by one or more limiting factors (middle curve in Fig. 1.4).

Although populations of all organisms have limits to their growth, the human population is increasing, seemingly without limit. The carrying capacity of the planet in relation to human food has been increased by highly productive agricultural systems (although the distribution of food is inequitable) and more stringent social controls have allowed people to live in more crowded living spaces. How long this high rate of human increase can continue is not known. But leaving quality of life and aesthetic considerations aside, each additional person contributes to the planet's environmental ills and the demand for natural resources. The overexploitation of fisheries resources and degradation of the environment are attributable to the rapidly increasing human population.

The numbers in some populations vary around the carrying capacity of the environment but in others, numbers fluctuate widely in response to factors that are unrelated to population density. The classic terrestrial example of this is the fluctuating population of some insects that appear in large numbers in warm humid conditions and disappear as the weather becomes cooler and drier. In this case, the weather is imposing a density-independent limitation on population size – that is, the weather is unrelated to (is independent of) population density.

Some species of shrimps or prawns, often regarded as the insects of the sea, also have populations that vary greatly from year to year depending on conditions in shallow nursery areas. Populations of some species of fish thrive in brackish-water estuaries during ideal conditions but die in large numbers with influxes of freshwater during heavy rain and floods. Populations of molluscs living in shallow tidal pools may suffer high mortalities during extended periods of low tides and hot weather when water temperatures climb and the amount of dissolved oxygen decreases. Storms and human interference, including shoreline building projects, produce silt that reduces the amount of sunlight reaching light-dependent organisms, such as algae, giant clams and coral. In these examples, salinity, temperature, dissolved oxygen and subsurface light are imposing density-independent limits on populations.

1.3.3 Life history patterns

The evolution of a particular life-history pattern in a species, including specific growth rates, mortality rates, and reproductive strategies, depends on a complex array of selective forces imposed on a species by its environment. Stock numbers are a result of the recruitment rate, birth rate and mortality rate of the species and are under the control of density-dependent and density-independent effects.

In stable or predictable environments, species are more likely to be under the control of density-dependent effects, such as competition for food and space, and stock sizes may be relatively constant over time. In more variable environments, density-independent effects, such as extreme water temperatures, storms, and adverse currents, are likely to result in stock numbers fluctuating over time; any deviation away from optimal conditions will result in a decrease in numbers, and any subsequent improvement in conditions will be followed by an increase in numbers.

Many fish stocks appear to exist in numbers less than the carrying capacity of their environment and occasional large recruitments produce unusually large cohorts in many commercial fish stocks. The major part of the catch in some fisheries on long-lived species is often based on one or two large year classes which dominate and progress through length–frequency distributions over several years. The fact that these occasional strong year classes persist in adult fish stocks over many years suggests that the additional production is within the carrying capacity of the environment.

Classical theory relating the life histories of species to the physical and biological nature of their environment has been based on MacArthur and Wilson's concept of r- and K-selection (MacArthur & Wilson, 1967). In this concept, r denotes the intrinsic (unlimited) rate of increase in population size, and K is the environmental carrying capacity, or the upper limit of numbers which can be supported by the environment. Species may be classified by the life-history characteristics which control their populations. According to one theory (Pianka, 1970), r-selected species have the ability to increase rapidly in number to take advantage of temporarily favourable environments, and K-selected species inhabit stable environments where the ability to compete with rival species is more important (Table 1.1).

If a species lives in an unpredictable or variable environment, where the chances of survival are uncertain, evolution is likely to favour early maturity and a single massive reproductive event (semelparity). In this case, spawning early in life, in the face of poor survival prospects ensures the success of the species. Success, from a species' rather than an individual's viewpoint, is measured in terms of its ability to perpetuate itself – i.e. reproductive success. Growth in such species is likely to be rapid, mortality is often density-independent and high, and the lifespan will be short (see Box 1.1 'Live fast, die young').

If a species lives in a more constant or benign environment, however, the ability to reproduce many times (iteroparity) over an extended lifespan may represent the most successful life-history strategy. Reproductive success, therefore, is the result of a balance between the energy and matter devoted to reproduction, and that devoted to the maintenance and growth of the parent. Whether a greater or lesser proportion of available resources is devoted to reproduction or the well-being of the parent in a particular species, depends on the species' environment.

The attractiveness and oversimplification of the r–K theory has led many researchers to label a given species as either r-selected or K-selected, when, at best, a species may be placed somewhere along an r–K continuum based on its particular life-history characteristics. In addition,

Table 1.1 Correlates of *r*- and *K*-selection. Adapted from Pianka (1970).

	r-selection	*K*-selection
Environment	Fluctuating Unpredictable	Stable Predictable
Mortality	Density-independent High, often catastrophic	Density-dependent Constant and low
Population size	Variable in time Below carrying capacity	Constant in time At carrying capacity
Growth	Fast Short lifespan Small body size	Slow Long lifespan Large body size
Reproduction	Early maturity Semelparity High fecundity High reproductive effort	Delayed maturity Iteroparity Low fecundity Low reproductive effort
Selection for . . .	Productivity High rate of increase	Efficiency Competitive ability

---

Box 1.1

Live fast, die young

The vertebrate with the shortest recorded lifespan may be the coral reef pygmy goby, *Eviota sigillata*, which lives for just eight weeks and grows to less than 20 mm in length (Depczynski & Bellwood, 2005). Coral reef pygmy gobies spend their first three weeks as larvae in the open ocean before undergoing metamorphosis and returning to settle on the reef, where they mature within one to two weeks and have just three weeks in which to reproduce and contribute to the next generation. Although larval mortality in reef fishes is typically high, and the small body size of the goby limits the number of eggs produced, pygmy gobies are very successful and are found across the Indian and Pacific oceans. The key to the species' success may be parental care: adult males fan and guard the eggs until hatching.

When the chances of survival are small, evolution often favours a 'live fast, die young' strategy (*r*-selection in Table 1.1) in which rapid growth and maturation offset the reduced chances of survival. However, rather than producing a single and massive quantity of small eggs, the pygmy goby produces a smaller number of eggs (about 400 eggs in three clutches) and invests energy in protecting them. This reproductive strategy, which enhances the survival of offspring by egg protection, is found in a few other fish species and is common in gastropod molluscs, some of which produce elaborate cases to protect a few large eggs (see Chapter 2).

---

the uncritical application of competition-based theory is hazardous because little is known of the intensity of competition in different environments. An example of a study of various related species distributed at different depths (and therefore subject to different environmental conditions) is given in Box 1.2 'Life history patterns and depth'.

Box 1.2
Life-history patterns and depth

Caridean shrimps (Fig. B1.2.1) are distributed in different but often overlapping depths on the outer reef slopes of tropical Pacific islands and have been the subject of a study on the relationship of life-history patterns to depth (King & Butler, 1985). Species of caridean shrimps in shallower depths are exposed to a greater number of predators, and a more fluctuating environment than those in deeper water. Consistent with the predictions of r–K theory, species in shallower depths have higher growth rates, smaller and more eggs, earlier maturity, and shorter lifespans than deeper-water species. However, some of the findings were counter to competition-based ecological theory, and it was proposed that the decreased predation rates in deeper water allow deeper-water species to have an extended lifespan, an increased degree of iteroparity, and hence a corresponding increase in reproductive effort summed over the lifespan. As the planktonic larvae of caridean shrimps migrate through the height of the water column, the probability of larval survival is likely to decrease with increasing distance traversed, and with increasing time spent in the water body. This decrease in the probability of larval survival with increasing depth may be offset by the production of larger eggs by deeper-water species. Larvae hatching from larger eggs are often thought to have better survival prospects than those from small eggs; this may be due to an increased ability of larger larvae to withstand starvation, and to escape predation, or a shorter larval life and hence a shorter period of predation by planktivores.

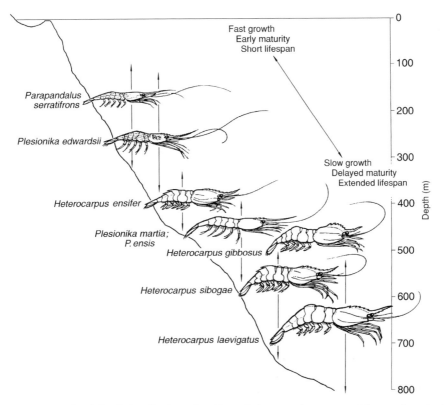

Fig. B1.2.1 The generalized depth distribution of caridean shrimps on the outer reef slopes of tropical Pacific islands. From shallow to deep water, species increasingly have slower growth, later maturity and longer lifespans. From King (1987).

An alternative theory involves 'bet-hedging' (Stearns, 1976), where the benefits of high fecundity are believed to be offset by an increased adult mortality. This assumes that the energetic costs of reproduction are high, and any increase in fecundity is accompanied by an increased parental mortality. The production of offspring in any one event is therefore a 'trade-off' against the expectation of future offspring production. When the survival of pre-recruits is considered rather than that of adults, the traits predicted for fluctuating and stable environments may oppose those predicted by the r–K theory. For example, in environments where larval survival is variable, selection should favour reproductive patterns featuring an extended lifespan and iteroparity (Murphy, 1968). In this case, the repeated production of smaller broods decreases the chances of failure during a single reproductive event.

Considering the complex array of life-history characteristics that contribute to a species' fitness in terms of survival, it is unlikely that any single ecological theory will adequately account for the evolved life-history patterns of all species. Nevertheless, concepts such as r–K theory provide useful clues to the likely range of life-history characteristics of species encountered in a particular environment.

Fisheries resource species may be classified according to their position on a continuum from those with a high natural mortality rate and a high reproductive capacity, to those with a low natural mortality rate and a low reproductive capacity. Short-lived species with high fecundity include shrimps, prawns, squid, and clupeoids, and long-lived species with low fecundity include sharks and many deeper-water and cold-water fish.

Short-lived species with high fecundity have the capacity to produce large numbers of pre-recruits when environmental conditions are suitable. Sardines, for example, may take advantage of the increased productivity in an intermittent upwelling, and juvenile penaeid prawns may benefit from occasional ideal conditions in a nursery area. Such species are subject to large fluctuations in recruitment, and therefore stock size from year to year; a corollary of this is that the relationship between recruitment and stock size is usually poor. If short-lived species with high fecundity are overfished, recruitment levels may remain high and variable even at low stock levels. As recruitment varies greatly due to environmental effects, decreases related to low stock size may be hard to detect and recruitment may fail without warning.

Long-lived species with low fecundity, on the other hand, produce a smaller and more constant number of pre-recruits, which have relatively high probabilities of survival. Deeper-water fish, for example, live in a relatively constant environment, and unexploited species are likely to produce similar numbers of pre-recruits from year to year. Here there is likely to be a much stronger relationship between stock size and recruitment, and, if the species is fished, recruitment and catch rates will decrease immediately.

Life histories can be affected quite rapidly by exploitation. Phenotypic and genetic life-history responses to fishing can include increased individual growth rates, reductions in age and size at maturity, and genetic changes due to the high mortality rates on particular parts of the population. These changes can be rapid (in less than one generation) or slower in the case of selection for individuals that are less susceptible to fishing (Hutchings, 2002).

Life-history patterns and diversity also vary along environmental gradients other than depth (see Box 1.2). The most obvious trends in the number of species are the decreasing diversity with increasing latitude (from the equator to the poles) and with increasing distance from what are believed to centres of origin and speciation.

In fish species there is a significant negative trend from high diversity at the equator to lower diversity in higher latitudes but such is the variation within various fish assemblages in all latitudes that latitude change accounts for only 15% of the variance in available data (Martha et al., 2002). Diversity is known to decrease with increasing distances away from centres of high diversity but the reasons for this are unclear. Centre of origin hypotheses have been subject to

Table 1.2 Maximum number of reef-building coral genera by area in the South Pacific. Veron (1986).

Genera	Area
70	Papua New Guinea
60	Solomon Islands, Vanuatu, New Caledonia, Fiji
50	Gilbert Islands, Guam, Saipan Islands, Mariana Islands
40	Marshall Islands, Samoa, Cook Islands
30	Tahiti, Society Islands, McKean Islands, Kanton Islands
20	Pitcairn, Oeno Atoll
10	Wake Islands, Midway Islands, Hawaii, Line Islands, Tonga, Marquesas

criticism and differences in species diversity have been attributed to allopatric speciation models that include plate tectonics, sea-level fluctuations and island age (Mooi & Gill, 2002). The trend of decreasing diversity of scleractinian corals from the Western to the Eastern Pacific with increasing distance from the centre of biological diversity for the Indo-Pacific in Indonesia is shown in Table 1.2.

With the decrease in diversity away from the Western Pacific there is a general increase in endemism. In spite of a low diversity of corals, Hawaii has a greater number of endemic species than other areas to the west (Veron, 1986).

1.4 Marine ecosystems

A small intertidal rock pool and a large body of water such as the entire Mediterranean Sea could each be described as an ecosystem – a relatively self-contained, functional system that includes communities and their non-living environment.

However, from a practical management viewpoint, an ecosystem is usually defined as some unit that can be studied, monitored and managed within the limits of available financial and human resources. Whereas managing a small rock pool is hardly worthwhile, managing an entire ocean is usually impractical. In most cases, managers will concentrate their efforts on a defined ecosystem that is large enough to encapsulate as many trophically dependent species and linked habitats as possible, but not so large that management (including monitoring and enforcement) is ineffective. Whether small or large, however, ecosystems are not independent of each other and boundaries placed between them are largely artificial.

Although, in most cases, ecosystems will be defined in terms of smaller manageable areas, some large geographical areas have been designated as ecosystems for management purposes. This has been done on the grounds that, although the areas contain a variety of habitat types, these are linked by ecological processes including currents and food chains. The Great Barrier Reef, for example, which stretches along 2000 km of the Australian coast, is managed by a single national authority.

Many larger marine ecosystems straddle political borders, necessitating the involvement of international and regional agencies in the management of shared resources and environments. Thus, for the purposes of monitoring and management, extensive regions of the sea have been identified as large marine ecosystems, or LMEs. These are large regions of ocean (of 2 000 000 km^2 or more) that share similar hydrographic characteristics and contain trophically dependent populations of aquatic species. LMEs can include semi-enclosed areas such as the Mediterranean and the Black Sea, continental shelf areas such as the Northwest Australian Shelf and coastal current systems such as the Benguela Current off the coast of south-western Africa. A FAO project on LMEs, launched after the 1992 Earth Summit (UNCED, Rio de Janeiro, 1992) has the primary objective of improving the long-term sustainability of international and coastal aquatic resources and environments. The project promotes the

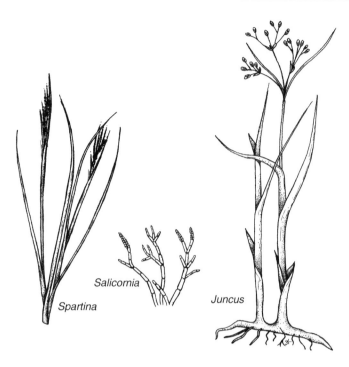

Fig. 1.5 Common saltmarsh plants. Cord grasses, *Spartina*; glassworts, *Salicornia*; and rushes, *Juncus*. Drawing courtesy of Jeremy King.

integrated management of coastal areas and the marine environment in order to halt or even reverse their deterioration.

A basic knowledge of the ecosystems on which marine species depend is required by those responsible for managing fish stocks and their habitats. This section provides an introduction under the three broad headings of: coastal waters; coral reefs and lagoons; and continental shelves and the open sea. These groupings could be said to include many separate ecosystems, and certainly many different habitats. More detailed ecological information can be gained from a good standard text such as Nybakken and Bertness (2005).

1.4.1 Coastal waters

Coastal zones are the interface between terrestrial and marine environments. At this interface, marine ecosystems receive nutrients from terrestrial sources via freshwater runoff, rivers and the scouring effect of high tides. The high productivity and accessibility of many coastal areas result in them being the most heavily fished areas of the sea.

In many coastal areas, freshwater and marine environments meet in wetlands and estuaries. In temperate areas, typical wetlands include salt marshes dominated by plants such as the cord grasses (*Spartina*), rushes (*Juncus*) and samphires or glassworts (*Salicornia*) shown in Fig. 1.5. A few organisms, including species of molluscs, crabs and smaller crustaceans, can tolerate the anoxic and highly saline substrates in salt marshes but much of the productivity is unused. Nutrients not used by the few resident saltmarsh organisms are transported via tides and runoff to enrich nearby estuaries and coastal areas.

Estuaries, areas at the mouths of creeks or rivers where freshwater and seawater meet and mix, are particularly rich in minerals and organic material. These semi-enclosed areas of brackish water with connections to the open sea are accessible to marine organisms that can tolerate varying conditions of salinity, temperature and turbidity. In much of an estuary, however, restricted light penetration and muddy substrates inhibit the presence of attached aquatic plants.

The usual widening of estuaries as they approach the sea results in the flow of water slowing down and releasing its load of lighter particles. These fine particles settle out of the water to form large banks of silt and mud where only a few species of algae can get a foothold, usually by attaching to shells and grit. However, diatoms and bacteria benefit from the high nutrient levels and make estuaries some of the most productive of all marine ecosystems. Estuaries often support large numbers of euryhaline, or salt tolerant species including many molluscs and worms that feed on the nutrient-rich deposits and, in turn, provide food for larger animals including fish. Nevertheless, estuaries usually have a lower species diversity than adjacent coasts as fewer organisms can tolerate the fluctuating environmental conditions.

The juveniles of many species, including shrimps, menhaden, anchovies, and mullets, grow in estuaries before migrating out to sea to breed. Although relatively few fish species are permanent residents of estuaries, many larger species including snappers, trevallies (jacks) and sharks periodically move into the lower reaches of estuaries to feed.

Some migratory fishes pass through estuaries to breeding grounds either in salt water or freshwater. Those that spend most of their lives at sea but move into rivers to release their eggs are known as anadromous species and the best known examples are the various species of salmon (family Salmonidae). Those that grow in rivers and move out to sea to breed are catadromous species and include freshwater eels of the genus *Anguilla*, that migrate out to places such as the Sargasso sea to spawn (see Section 2.3 'Fishes').

In subtropical and tropical regions the most notable salt tolerant plants of estuaries are the 80 or so species of specialized, but often unrelated, trees collectively known as mangroves. Most mangroves grow where the water is brackish – either in estuaries or on coasts where seawater is diluted by freshwater soaks or runoff from the land. Mangroves have several special adaptations to equip them for living in silty, waterlogged soils. Many species have evolved exposed (or aerial) root systems that support the tree mass and absorb oxygen. The orange mangrove has exposed knee roots, the red mangrove has long prop roots which grow down from the trunk, and the black mangrove has cable roots which extend over a large surface area and send up small peg-like aerial roots or pneumatophores (Fig. 1.6).

Several families of fish are commonly associated with mangroves including gobies, gudgeons, silver biddies, sardines, snappers, slip-mouths, puffers and mullets. But a much larger number of smaller and microscopic species, including diatoms and bacteria, takes advantage of the large amount of organic material associated with mangroves. Nutrients from terrestrial runoff are taken up by the mangroves, which, through leaf drop and decay, contribute to detrital food chains. It has been estimated that from 2–18.7

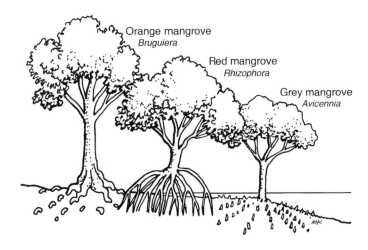

Fig. 1.6 Mangroves. Examples shown are the orange mangrove, *Bruguiera*, with knee roots, the red mangrove, *Rhizophora*, with prop roots, and the grey or black mangrove, *Avicennia*, with pneumatophores.

tonnes of leaf litter can be produced each year by 1 hectare of mangroves (various researchers in Saenger, 1994).

Although mangroves cover less than 0.1% of the global land surface, they appear to account for as much as a tenth of the dissolved organic carbon (DOC) that flows from land to the ocean. Studies in Brazil (Dittmar et al., 2006) suggest that the plants are one of the main sources of dissolved organic matter in the ocean. Net-like mangrove roots trap carbon-rich leaf litter from which dissolved organic matter is leached by tidal flow into coastal waters. However, mangrove foliage has declined by nearly half over the last few decades because of increasing coastal development and habitat damage. Organic matter dissolved in the world oceans contains a similar amount of carbon as atmospheric carbon dioxide and the balance between the two is part of the global carbon cycle that regulates atmospheric carbon dioxide and climate. It has been speculated that reductions in mangroves and dissolved organic carbon may threaten this delicate balance, with potential consequences for atmospheric composition and climate.

Mangroves also protect and extend shorelines. Their exposed roots dissipate the energy of waves and cause currents to slow down to release suspended material. As the siltation continues, there is an ecological succession in which the mangrove front advances towards the sea and plants that are less salt tolerant become established in the area behind the front. In many parts of the world, mangroves and wetlands are protected because of their recognized role in marine ecosystems, particularly as nursery areas for many marine species. Nevertheless, such areas are under constant threat from pollution and land reclamation (see Section 1.5.1 'Habitat modification and loss').

In temperate and tropical areas, the lowest tide levels of estuaries, lagoons and sheltered coastal areas are habitats for aquatic plants collectively called seagrasses (see Box 1.3 'The invasion of the sea by flowering plants'). Seagrasses are not true grasses, but have some grass-like features including leaves attached to short erect stems, creeping horizontal rhizomes and root systems. In addition to reproducing vegetatively via their network of rhizomes, which spread beneath the sand to form vast beds resembling underwater meadows, seagrasses produce small flowers that are fertilized by pollen floating in the sea.

In contrast with marine algae, relatively few animals consume seagrasses directly as the cellulose that constitutes the bulk of the plant makes it indigestible to most herbivores. Large animals that do feed on seagrasses include the green turtle as well as the Indo-Pacific dugong and its Caribbean relative, the manatee. Many more species, however, graze on the more easily digested fine mat of algae growing as epiphytes on the seagrass blades. Unlike marine algae that take nutrients from the surrounding water, seagrasses absorb nutrients via their roots and therefore recycle material that would otherwise be trapped in the substrate. As seagrass leaves eventually decay, the detritus formed is used as a source of food by a much wider range of marine species. Besides being highly productive areas, seagrass beds provide sheltered nursery areas in which the larvae and juveniles of many marine species live and grow before moving elsewhere as adults.

Away from the mouths of rivers, and without the input of terrestrially-derived nutrients, coastal areas are generally less productive but have a greater species diversity. About 40% of the world's coastline is fringed by sandy beaches. Sand is supplied to beaches from the erosion of cliffs and rocks, by rivers, from offshore sediments and is lost by being blown inland and washed offshore during storms. Sand is also moved from one beach to another by longshore currents created when waves strike the coast at an angle (Fig. 1.7). In many parts of the world, groynes have been built out at right angles to shorelines to obstruct the longshore transport of sand and encourage the build-up of sand; although sand is deposited on the up-current side of the groyne, it may be scoured away from the down-current side.

Beaches are made up of particles ranging in size from the fine powder of silica sand to the large,

Box 1.3
The invasion of the sea by flowering plants

Flowering plants first appeared on this earth about 100 million years ago, when the dinosaurs were wandering on dusty, grass-less soil, and became dominant over all terrestrial plants. Presumably competition for space and nutrients provided the driving force for some flowering plants to invade the sea. This invasion resulted in the successful establishment of a small group of flowering plants, the seagrasses, in the shallow margins of the sea. There are only about 50 different species of seagrasses, but these have been so successful that they are found in shallow coastal areas and estuaries in temperate and tropical regions around the world (Fig. B1.3.1).

Seagrasses play an important role in the coastal processes of many tropical and temperate shores by stabilizing drifting sand and contributing to marine food chains, particularly by leaf loss. Seagrasses shed their leaves seasonally and during storms these may be washed up on beaches in massive quantities. Seagrass leaves used to be collected to make alkaline potassium compounds, such as potassium hydroxide. The old name for these compounds was potash, and the name came from the way it was originally obtained – by burning seagrasses, leaching the ashes, and then evaporating the solution in large iron pots.

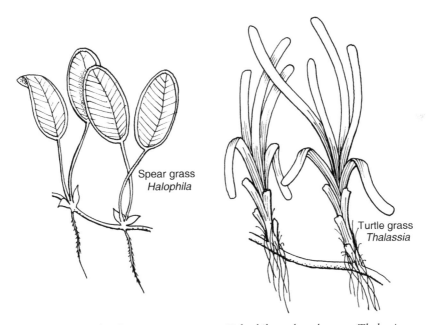

Spear grass
Halophila

Turtle grass
Thalassia

Fig. B1.3.1 Seagrasses. Examples shown are spear grass, *Halophila*, and turtle grass, *Thalassia*.

well-rounded pebbles and cobbles of shingle beaches. Many beaches in temperate climates have yellow sand derived from quartz (a mineral form of silica that crystallizes as hexagonal prisms). Darker sands may display colours due to small proportions of different minerals such as mica (layered silica), rutile (titanium dioxide), feldspar (aluminosilicates), and magnetite (iron oxide). Some coasts have striking black sand beaches formed by the wearing away of dark volcanic basalt. Besides minerals ground from rocks and terrigenous material brought to the ocean from rivers, sand often contains material of a biological origin including shell fragments

Fig. 1.7 The build-up of sand against a solid jetty in a longshore current that moves from right to left.

Box 1.4
Bioerosion

Bioerosion is the breaking down of substrates, usually coral, by the actions of various organisms (referred to as bioeroders). Some sponges, bivalve molluscs, and worms are internal bioeroders and bore into, and live in, the coral structure. External bioeroders, including some fish and sea urchins, feed on the surface of the coral. Two important bioeroders of corals are the colourful parrotfishes (family Scaridae) and sea urchins (Fig. B1.4.1).

Parrotfishes have massive fused teeth with which they scrape coral to feed on algae and symbiotic zooxanthellae within the coral polyps. The fishes have to graze large quantities of coral to gain a small amount of organic material and they appear to be continually evacuating clouds of fine coral particles. As each adult parrotfish can produce about one tonne of particulate matter each year, their contribution to the sand of lagoons and tropical beaches is considerable. Some sea urchins, such as various species of *Diadema*, are among the most important invertebrate bioeroders on coral reefs around the world (Glynn, 1997).

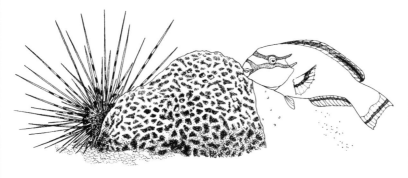

Fig. B1.4.1 Common bioeroders of coral. Left, the sea urchin, *Diadema*, and right, a parrotfish (family Scaridae).

and the white sand of tropical beaches may be entirely derived from calcareous algae and corals (see Box 1.4 'Bioerosion'). About one-third of all sea floor sediments and, in some places, beach sand consists of the remains of foraminiferans (see Section 1.6.2).

In many parts of the world, sandy beaches are under threat from sand mining and foreshore constructions which change the natural paths of sand replenishment. Sand is mined for use in the construction industry (in the making of concrete) and in the manufacture of glass, which is made by melting together silica sand (SiO_2), limestone ($CaCO_3$), and sodium salts at very high temperatures.

The number of intertidal organisms living on sandy beaches is related to sand particle size and wave action. Beaches with strong wave action usually have coarse sand, as lighter particles which remain in suspension, are carried away by the backwash of waves. Conversely, beaches with smaller waves and a less powerful backwash are likely to have finer sand. In general, beaches of coarse sand are less suitable habitats for intertidal species. Finer sand retains more interstitial water during low tides and many intertidal animals are dependent on this water to obtain dissolved oxygen and to prevent desiccation.

Substrate mobility, the movement of sand due to currents and wave action, also has a marked effect on the number of species inhabiting beaches. Shifting sand makes it difficult for plants to gain a foothold and for animals to burrow deep enough to avoid dislodgement. In addition, beaches with heavy surf generally have lower quantities of nutrients in the substrate as the backwash of waves tends to remove most organic material. These difficulties are exemplified to the extreme on a cobble beach, the formation of which depends on at least moderate wave action to prevent cobbles being covered by sediment. The impossibility of burrowing and attachment as well as the risk of being crushed by the rolling cobbles, makes a cobble beach one of the most formidable coastal habitats for marine organisms to colonize.

Although sheltered environments generally have a greater diversity of species, high energy coasts such as surf beaches support a smaller number of specially-adapted organisms. There are no plants on the mobile substrate and the beaches are populated by herbivores that can utilize phytoplankton, which is often concentrated by wave action. Users of the phytoplankton include suspension feeders typified by surf clams

of the genus *Donax*, which are found on high energy beaches around the world. These clams have wedge-shaped shells and large spade-like feet that allow them to burrow quickly and deeply to avoid repeated dislodgement by waves. Small mole crabs of the family Hippidae have rounded bodies with short strong limbs which provide them with the same ability. The colonization of such harsh environments suggests that the hazards involved are offset by some advantages, which may include living in a habitat where there are fewer predators and competitors.

Even though beach slope, sand particle size and wave action are interrelated, coasts are often classified on the basis of their exposure to waves as either, low, medium or high energy. The classification is useful even though there is obviously a continuum from low energy coasts such as sheltered bays and lagoons to high energy coasts such as exposed surf beaches. In general, it is found that the diversity and abundance of species decreases from low energy to high energy coasts. In a study of 105 beaches around the world, Bally (in McLachlan, 1983) found that both the abundance and mean number of macrofauna species decreased with increasing exposure to waves (Table 1.3).

Above the high tide mark, the backshore of a beach is submerged only during storms and may contain one or more berms – ridges of sand and debris running parallel to the beach. The foreshore extends from the low tide mark to the high tide mark. This littoral or intertidal zone, which lies between the extremes of high and low tides,

Table 1.3 Mean number of macrofauna species and abundance on beaches with different wave exposure. Mean data from 105 beaches worldwide. From Bally in McLachlan (1983).

Exposure	Number of species	Abundance (per m²)
Low energy beaches	30	1710
Medium energy beaches	17	752
High energy beaches	11	400

is the transitional area between terrestrial and marine ecosystems.

Tides are influenced mainly by the moon and the sun and, in spite of variations caused by weather patterns, are more or less regular (see Box 1.5 'The sun, the moon and the tides'). Most coasts of the world have one or two high and low tides each 24 hours and this regularity has allowed many aquatic animals to evolve for life in the intertidal zone. Intertidal organisms are exposed to the air for some part of each day and have to contend with fluctuating temperatures, the possibility of desiccation and the inability to gain oxygen and food when not covered with water. In spite of these difficulties, many different species inhabit the intertidal zone. The advantages of doing so may include escaping swimming predators, at least for some of each day, and gaining food that is less abundant or not available below the low tide mark.

A large number of marine animals, including those that live below low tide, display rhythms in spawning and feeding behaviours that are related to tides and the lunar cycle. A well-documented example of a species in which spawning is influenced by the tides is the grunion, *Leuresthes tenuis*, a fish that spawns on beaches on the Pacific coast of the United States. On certain nights of the highest spring tides, grunions move to the upper beach where they lay eggs that hatch about one lunar month later, when the fry can use the high tides to swim out to sea.

Tides are also responsible for the passive movements of marine animals during their life cycles and some species with a limited ability to swim appear to use tides to migrate in and out of inshore nursery areas. The juveniles of some penaeid shrimps or prawns, for example, live in estuarine nursery areas where they burrow during the day and emerge to feed at night as they drift with the tide. At times of the year when flood tides are more common during the night, juvenile prawns make a net movement up the estuaries each night before burrowing during the day. However, at the time of the year when ebb tides begin to predominate over flood tides each night, juvenile prawns make a net movement sea-

wards during each night and this time marks the period in which juvenile prawns move offshore to join adult stocks (Penn, 1975).

Many predatory fishes actively move inshore on flood tides to prey on burrowing animals that, after a prolonged period of exposure, leave their burrows to gain oxygen and food from the inflowing water. Bottom-feeding fishes including rays and skates move in with the tide to feed on clams, crabs and other small invertebrates that emerge from the substrate.

The ability to withstand the effects of exposure to air varies between species and this accounts for the marked patterns in the distribution of organisms at different intertidal levels on all coasts. This zonation in the intertidal zone is particularly noticeable on rocky shores, which are more densely populated and have a greater diversity of species than sandy shores. Animals that need to live on, or even attached to hard substrates include barnacles, bivalve molluscs such as oysters and mussels, gastropod molluscs such as limpets, and some soft-bodied species including anemones and sea squirts.

The littoral or intertidal zone lies between the extremes of low and high water at spring tides and is further divided into other zones based on a universal scheme proposed originally for temperate rocky shores (Stephenson & Stephenson, 1972). The scheme, which is a description of (rather than an explanation for) the zonation of organisms on rocky shores, appears essentially applicable to tropical shores as well as temperate ones. The midlittoral zone can be thought of as lying between the low and high tide marks of average tides – that is, half way between neap and spring tides. This zone is regularly exposed and submerged on most tidal cycles. Above the midlittoral zone is the supralittoral fringe, into which the tide reaches only during spring tides and above this a spray zone that is dampened by the spray of waves. Below the midlittoral zone is a infralittoral (or sublittoral) fringe that is exposed only during spring tides.

A generalized distribution of organisms on a rocky shore is shown in Fig. 1.8. Above the spring tide high water mark, where rocks are moistened

Box 1.5
The sun, the moon and the tides

The sea covering the earth is under the attractive forces of both the sun and the moon. The strength of these forces is related to the mass of the body and its distance from earth – force is directly proportional to mass but inversely proportional to the square of distance. In spite of the moon's smaller mass, its proximity to earth results in its attractive forces being much stronger than the sun's, and so provides the major influence on tides. In Fig. B1.5.1, the fine lines above the earth's surface show the tidal bulges of water caused by the gravitational pull of both the moon (the lunar tide) and the sun (the solar tide). The effect of these bodies is to create a bulge of water on opposite sides of the earth. Thus two areas on the earth's surface are directly under a bulge of water, and are experiencing high tides, at any one time. As the earth is rotating, each point on the earth's surface should (and many places do) have twice-daily or semidiurnal tides – that is, two high tides and two low tides each lunar day of 24 h 50 min.

Fig. B1.5.1a represents the full moon (when the moon, illuminated by the sun, is visible as a full sphere from earth) and the new moon (when the moon lies between the earth and the sun). In both cases, the pull of the moon and sun is working together and result in two large tidal bulges on opposite sides of the earth. These are the times of spring tides when the tidal range (the difference between the heights of high and low tide) is greatest. Fig. B1.5.1b represents the first and third quarters of the moon, when only half the moon is illuminated by the sun. In these cases, the pull of the moon and sun is working at right angles to each other. These are the times of neap tides when the tidal range is least. As the moon circles the earth each lunar month of about 29.5 days, there are two neap tides and two spring tides during this period.

Tides are modified by other factors such as the declination of the moon and the depth and size of ocean basins. This results in three basic tidal patterns in various parts of the world. Semidiurnal tides (two highs and two lows per day) occur in the Atlantic Ocean and the Indian Ocean. Diurnal tides (one high and one low per day) are common in shallow semi-enclosed seas such as the Gulf of Mexico and along the coast of Southeast Asia. Mixed tides, in which there are two high and two low tides of unequal height per day, are the most common type (shown at the bottom of Fig. B1.5.1). Tides and tide ranges vary due to the influence of coastal formations including headlands, bays and offshore islands. The world's largest tide range is at the head of the Bay of Fundy where the spring tide range reaches 15 m. In some estuaries with such large tidal ranges, the flood (rising) tide is held back by the restricting effect of decreasing depths and narrowing rivers, so that it rushes in on top of the water flowing downstream as a tidal bore. These turbulent masses of water have an almost vertical wave front which may reach heights of over 3 m and rush in at the 'speed of a galloping horse' – about 22 km/hr.

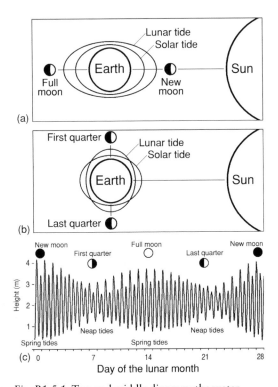

Fig. B1.5.1 Top and middle diagram: the water covering the earth (indicated by the fine lines above the earth's surface) is under the attractive forces of both the sun and the moon which result in (a) spring tides and (b) neap tides. Bottom diagram (c) shows mixed tides on the Pacific coast of Australia over a lunar month.

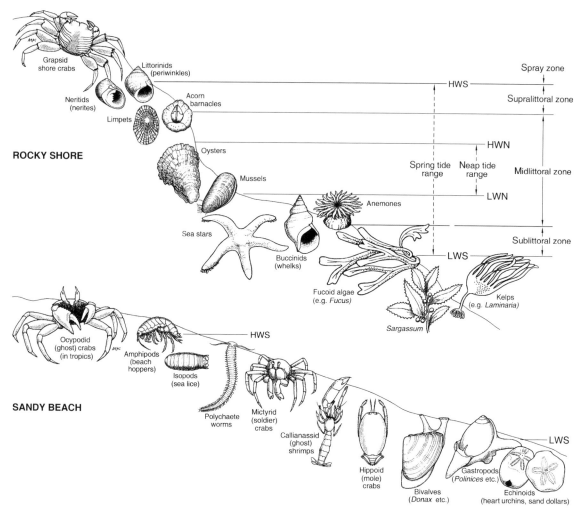

Fig. 1.8 The distribution of some characteristic organisms on a rocky shore (top) and a sandy beach (bottom) in the intertidal zone. Not all organisms shown are necessarily found in the same area. HWS = high water spring tides; LWS = low water spring tides; HWN = high water neap tides; LWN = low water neap tides.

by sea spray, grey or orange lichens and tar-like patches of cyanobacteria may be found and rock crabs (such as those of the family Grapsidae) dwell in damp crevices. The level of high water spring tides is often marked by bands of dark periwinkles (family Littorinidae) or, more commonly on tropical shores, the lighter coloured nerites (Neritidae). Almost universally, there is a light coloured band of acorn barnacles (balanomorph crustaceans) and limpets (molluscs of the family Acmaeidae) close to the average high tide mark and the upper limit of barnacles marks

the upper limit of the midlittoral zone. Species of oysters that can tightly close their shells to avoid desiccation when exposed are cemented to rocks around the mid-tide mark. Mussels, with their thinner shells, do less well at dealing with exposure and are often restricted to the lower tide zone where they form dense dark bands.

At the lower part of the midlittoral zone, animals such as sea squirts and anemones may be found with several species of algae, including *Porphyra* and *Fucus*, that can withstand exposure for limited periods. In temperate and cold waters, large kelps extend up just above the lowest of low tides (to the upper limit of the sublittoral fringe): these include the leathery oarweed, *Laminaria* in the northern hemisphere and similar species such as the strap-weed, *Ecklonia*, in the southern hemisphere. In tropical waters, where brown algae are less diverse, *Sargassum* may be present.

Many intertidal animals have adaptations in form or behaviour that assist in preventing desiccation when exposed at low tide. Some limpets, for example, form circular grooves or 'home scars' to which they return at low tide and these grooves help to seal the shells against water loss. Periwinkles live in crevices or in dense groups to reduce water loss by evaporation and small mussels, attached to rocks by their byssal threads, often form a dense black mat that retains water at low tide.

Although the upper limit of organisms in the intertidal zone is determined by physical factors such as desiccation, the lower limit is affected by biological factors, particularly competition and predation. Competition for attachment space between barnacles and mussels, for example, restricts the incursion of barnacles into the mussel zone below. Some predators, such as shore crabs, can tolerate exposure and pose a threat to periwinkles and limpets uncovered at low tide. Other predators that cannot tolerate exposure as well as shore crabs have to move in with the tide to feed on organisms living in the intertidal zone. The higher an organism lives in the intertidal zone the less time it is exposed to such predators. The lower limit of mussels is often determined by predatory sea stars, which, because they can withstand exposure for only a short time, have to feed on mussels lower down in the intertidal area. In a similar way the lower limit of barnacles is often determined by predatory gastropods such as whelks (Buccinidae), which bore into the barnacles at the lower end of their distribution.

Some predators, on rocky shores and elsewhere, have a beneficial effect in maintaining the diversity of species in a particular habitat. Such predators, referred to as keystone species, reduce the number of individuals of dominant species

that would otherwise out-compete other species in the same area. An example is the case of predatory sea stars that limit the size of mussel populations on rocky shores, allowing space for other organisms to settle and exist on the rock surface.

The zonation of species on sandy beaches is less distinct than on rocky shores as many animals migrate up and down the intertidal zone with the tides. Nevertheless some patterns exist (e.g. McLachlan, 1983) and a generalized distribution of species is shown in Fig. 1.8. The highest parts of beaches are often occupied by sand fleas or beach hoppers (amphipods) and, in tropical areas, by stalk-eyed ghost crabs (ocypodids). In the damp sand of the midlittoral zone, sea lice (isopods), polychaete worms and some molluscs may be found. The wave saturated surf zone at the lowest part of the beach is inhabited by surf clams (of the genera *Donax*, *Tivela*, and *Spisula*), burrowing urchins and sand dollars, carnivorous gastropod molluscs including moon snails (*Polinices* and *Natica*) and burrowing mole crabs of the family Hippidae.

Shallow coastal areas, particularly those subject to the input of nutrients from terrestrial sources, are some of the most productive areas in the sea and are important habitats and nursery areas for a wide range of marine species. Sheltered coastal areas, including bays and estuaries, are easily accessible to fishers from coastal communities and are important fishing areas. In developed countries, most recreational fishing takes place on the coast and in many cases this accounts for large collective catches of seafood. Many marine species in inshore areas are also targeted by commercial fleets and this is often a cause of conflict between artisanal and industrial fisheries.

Besides receiving beneficial nutrients from the land, coastal ecosystems may receive and be under threat from the runoff and discharge of pollutants from terrestrial activities including agriculture, industry and human habitation. These threats, including discharges that result in eutrophic and polluted waters, are discussed in Section 1.5. Coastal areas contain many habitats on which marine species depend and all fisheries management plans must address the protection of such areas.

1.4.2 Coral reefs and lagoons

Coral reefs are some of the most productive ecosystems in the world. In a short swim across a coral reef, a greater variety of living things can be seen than in most other places on the planet. On islands and tropical coasts without the input of nutrients from rivers and upwellings, coral reef ecosystems provide the productivity that supports food chains and species that are exploited as food by coastal communities. For many coastal people the seafood collected from coral reefs and lagoons provides the main source of dietary protein; this is particularly so on atolls and low-lying islands where agriculture is impossible.

Corals, with anemones, hydroids and jellyfishes, are members of the animal phylum Cnidaria (previously Coelenterata) which has over 9000 species distributed in all oceans. Cnidarians exist both as free-floating medusae (as jellyfishes) and sedentary polyps, and stages in the life cycles of many alter between these two forms. Polyps may be solitary (as in anemones) or colonial (as in most corals). The medusa form has tentacles hanging down from a bell-shaped body and the sedentary polyp form has an upward facing mouth surrounded by tentacles. In all cnidarians, the tentacles are armed with stinging threads, or nematocysts that are used to trap or kill their prey and a few warm-water species are capable of killing human beings.

The phylum Cnidaria is divided into three classes – the Scyphozoa, the Hydrozoa and the Anthozoa (Fig. 1.9). The scyphozoans are the non-colonial organisms commonly known as jellyfishes, most of which drift in the sea with the plankton. The related cubozoans include the square-bodied highly venomous jellyfishes such as the sea wasp (see Box 1.6 'Stinging cnidarians').

The hydrozoans exhibit both polyp and medusa forms. Free-floating forms, such as the familiar Portuguese man-of-war, *Physalia physalia*, are colonies of specialized individuals that appear superficially similar to true scyphozoan jellyfishes.

Other hydrozoan colonies are sessile and include both plant-like and coral-like forms. Plant-like forms may be branching or fern-like such as the stinging hydroid commonly called fire weed by divers. Some hydrozoans incorporate calcium in their skeletons to become hard and coral-like in form, such as in the fire coral, *Millepora*.

Anthozoans have only polyp forms and are divided into subclasses based on their symmetry around the polyp's central axis. Those with eight tentacles make up the subclass Octocorallia (or Alcyonaria) which includes gorgonians and soft corals. In gorgonians (Gorgonacea) the colonies have erect branched or fan-like skeletons of horn-like material. The soft corals (Alcyonacea) form colonies that are fleshy and sometimes tree-like. Other octocorals build skeletons that are stiffened with calcium and these include the blue coral, *Heliopora*, and the red organ-pipe coral, *Tubipora musica*.

Those with six tentacles (or multiples of six) belong to the subclass Hexacorallia (or Zoantharia) and include the anemones, black corals and hard or stony corals. Anemones (Actiniaria) are solitary polyps that are, in most cases, sessile and attached at their base; some coral reef species grow to a large size and are often home to several clown fish (*Amphiprion* spp.). The black corals of the order Antipatharia have coloured bushy shapes with a black skeleton of very dense horn-like material that can be highly polished to make black coral jewellery. The hard or stony corals of the order Scleractinia with their skeletons of calcium carbonate are the main contributors to the building of a coral reef, but other organisms including hydrozoan corals and calcareous algae also contribute to the reef mass.

Coral reefs are among the world's largest natural structures, and are built by the collective efforts of polyps that are generally less than 3 mm in diameter (Fig. 1.10). Each polyp has a cylindrical body with an upper opening that functions as both mouth and anus. In the case of hard corals, the lower half of the polyp produces a cup-like corallite of calcium carbonate around its base. When observed at night, the surfaces of many

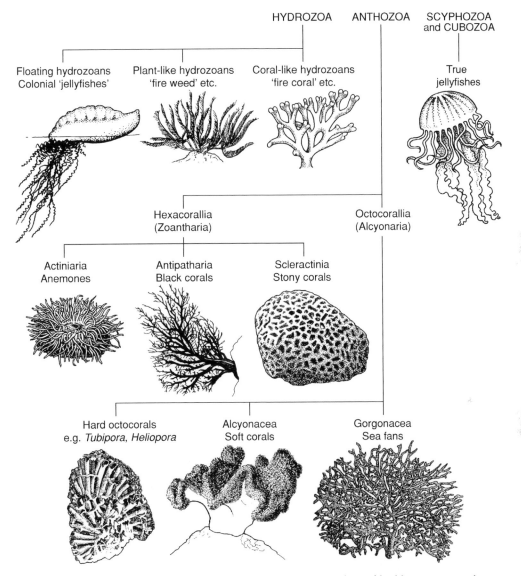

HYDROZOA   ANTHOZOA   SCYPHOZOA and CUBOZOA

Floating hydrozoans Colonial 'jellyfishes'

Plant-like hydrozoans 'fire weed' etc.

Coral-like hydrozoans 'fire coral' etc.

True jellyfishes

Hexacorallia (Zoantharia)

Octocorallia (Alcyonaria)

Actiniaria Anemones

Antipatharia Black corals

Scleractinia Stony corals

Hard octocorals e.g. *Tubipora*, *Heliopora*

Alcyonacea Soft corals

Gorgonacea Sea fans

Fig. 1.9 An abbreviated overview of the phylum Cnidaria, which contains the reef-building, stony corals.

coral colonies appear to be covered with a carpet of feeding polyps which retract into their protective corallites by day.

The tentacles of coral polyps possess specialized stinging or thread cells each of which contains a fluid-filled sac, called a nematocyst (see inset in Fig. 1.10). When a small sensory trigger on the surface of the cell is touched, a long thread coiled within the nematocyst is released with some force. The thread either clings to, or coils around its prey, usually drifting plankton, before the tentacle bearing the nematocyst transfers the food to the polyp's mouth. Nematocysts, sometimes containing powerful toxins, are found in all cnidarians.

Besides feeding on plankton that drift within reach of their tentacles, most corals are also able to feed indirectly from sunlight. The tissue of autotrophic corals contains single-celled plants called zooxanthellae, which, in the presence of sunlight, synthesize organic material that is passed on to the coral. As the food caught by the

Box 1.6
Stinging cnidarians

Although nematocysts are possessed by all cnidarians, the stings of most are harmless to humans. Most cnidarians either do not have nematocysts that are capable of penetrating human skin or have toxins that are insufficiently potent to cause injury. However, there are exceptions and these include some siphonophores (colonial hydrozoans), fire corals (coral-like hydrozoans) and jellyfishes (scyphozoans). The best-known siphonophore is the Portuguese man-of-war (*Physalia*) which has a large float and tentacles that extend for many metres. Their stings, which leave painful lines and welts across the skin, are not fatal but may cause allergic reactions and difficulty in breathing. Fire corals are superficially similar to corals although most are yellow or ochre in colour and have many very small pores (hence the generic name of *Millepora*). Various species grow in thickets with strong vertical flanges or slender branches that often divide at their ends into two parts (Fig. B1.6.1). These coral-like hydrozoans can deliver a strong sting and cause an itchy rash that persists for several days.

The most dangerous cnidarians are the cubozoan or box jellyfishes, so-called because of their four-sided bells. The deadly sea wasp, *Chironex fleckeri*, is large for a cubozoan with its bell reaching 25 cm in width and its tentacles extending over 4 m. Its powerful toxins cause extreme pain and even death from heart failure. The species occurs in sheltered bays and at the mouths of estuaries in tropical waters from Southeast Asia to northern Australia, particularly during the hot summer months.

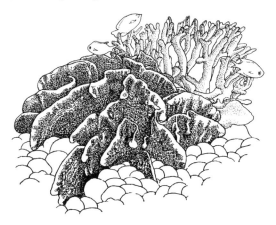

Fig. B1.6.1 Fire coral. At left the leaf-like fire coral, *Millepora platyphilla*, and at right, the branching fire coral, *Millepora dichotoma*. Both are mustard coloured with lighter margins or tips.

coral may provide nitrogen and phosphorus for the zooxanthellae as well as itself, the relationship is symbiotic and essentially the same as found in giant clams (see Section 2.2.1).

Because of their reliance on zooxanthellae, most corals, including those responsible for building reefs, can only live in clear, shallow, brightly lit water. However, there are some corals that do not rely on zooxanthellae and can live at depths where they do not have to compete for food and space with light-requiring stony corals. Corals of deeper water include the valuable black corals (Antipatharia) that are collected for use in the jewellery trade. In life, black corals (Fig. 1.9) are brown or yellow and the dense black skeleton underneath is the material suitable for cutting

and polishing. Large branches of black coral are gathered by scuba divers in depths below 50 m, where the time that a diver can spend collecting is quite short.

Most stony corals are colonial and living polyps are found only in a thin film on the surface of the coral. Individual polyps are connected to each other laterally by extensions of their tissue and the colony grows by asexual budding from the bases or the oral discs of the polyps. As polyps die their calcareous skeletons remain to form part of the coral base and new polyps grow on the skeletons left behind by their predecessors. The coral colony thus continues to grow outwards and upwards with each generation of polyps. In the case of massive reef-building corals, the major

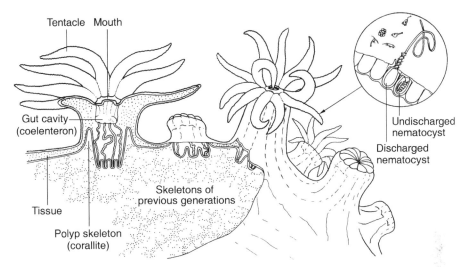

Fig. 1.10 Coral polyps. Three polyps are shown extended from their corallites. The circle shows an enlarged part of a tentacle with two stinging cells containing nematocysts, one of which is discharged.

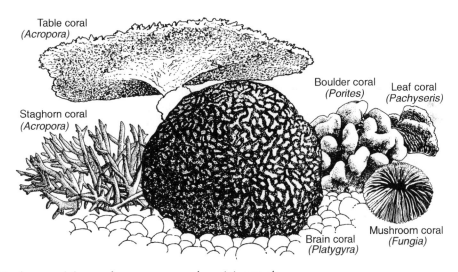

Fig. 1.11 The forms and shapes of some common scleractinian corals.

part, below the outer growing layer of living polyps, consists of the skeletons of countless millions of polyps – 1 kg of coral rock may contain over 80 000 polyp skeletons.

Most coral polyps form a colony that has a particular size and shape that makes it recognizable as a particular species of coral (Fig. 1.11). The family Acroporidae contains many species, all with small corallites but many different shapes. Table corals form broad plates, often in step-like formations down reef slopes, which like the broad canopy of a tropical rain forest, capture the maximum amount of sunlight. The more delicate branching species of *Acropora*, the staghorn corals, fill in gaps between the larger corals and grow in more sheltered lagoons.

Massive corals include the brain corals and boulder corals and these often form the bulk of coral reefs. Brain corals (*Favia*, *Platygyra* and others) have large corallites and are either dome-shaped or encrusting. Boulder corals (*Porites* spp.) have very small corallites and form large solid domes or clumps. All of the corals shown in Fig. 1.11 are colonial except the mushroom coral (*Fungia*) which exists as a solitary huge polyp, up to 20 cm in diameter, and is able to move short distances using its lateral tentacles.

On some areas of coral reefs, octocorals, which are all colonial, are more conspicuous than stony corals. Soft corals (of the order Alyconacea) form fleshy colonies that may be very abundant in some reef areas in the Indo-Pacific. Brightly coloured gorgonians or sea fans (Fig. 1.12), which are particularly common on coral reefs in the Atlantic, are one of the most spectacular of coral formations. Their polyps are distributed over horny and flexible skeletons which form fan-like branches with intricate meshes that are often set at right angles to the water movement to allow maximum exposure to drifting food. Hydrozoans such as the fire coral, *Millepora*, are also a major component of some areas on both Atlantic and Indo-Pacific reefs.

Existing coral colonies increase in size by asexual reproduction, in which new individuals bud off from parent polyps. However, the formation of new coral colonies relies on sexual reproduction, in which the larvae produced drift in the sea to settle in places distant from the parent corals.

Although some species brood fertilized eggs, most are broadcast spawners and release eggs and sperm into the water, often synchronously during a brief time each year. For many corals on the Great Barrier Reef there is a mass spawning event during a few nights after a full moon in late spring or early summer. Corals release millions of eggs and sperm bundles so that, to an underwater viewer, the mass spawning resembles an upside-down snowstorm. As they reach the surface, the bundles break apart and eggs, fertilized by sperm, develop and hatch into planulae larvae. During the mass spawning event, the developing eggs and larvae form dense reddish slicks on the sea's surface and these are referred to as coral dust by fishers. Planulae drift on the surface of the sea for several days and those that survive predation and unfavourable currents to reach shallow water, may settle on suitable hard surfaces and grow into polyps.

In spite of the potential for the wide distribution of corals, coral reefs are sparsely distributed and are especially uncommon in some areas such as the Eastern Atlantic. The requirements for reef-building or hermatypic corals to grow are quite specific. First, the water must be shallow and clear enough for sufficient sunlight to reach the zooxanthellae. Second, the temperature of the water must be above 20°C with optimal temperatures for reef development from 23–25°C. If these requirements are met, drifting coral larvae will settle on suitable hard surfaces and form colonies that have the potential to become coral reefs (see Box 1.7 'Types of coral reefs').

In contrast to the rocky coasts of cooler areas, the zonation of species on coral reefs is often ambiguous and studies on individual reefs may not be widely applicable. The tide range in many areas of extensive coral reefs is often small and the distribution of organisms appears more related to wave exposure and light penetration. Nevertheless, some generalizations can be made and some features are shown in the profile of a barrier reef and lagoon in Fig. 1.13.

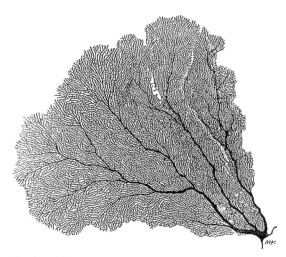

Fig. 1.12 The gorgonian sea fan, *Subergorgia*. Coloured red, yellow, violet or brown, gorgonians grow to over 3 m in width.

Box 1.7
Types of coral reefs

There are three basic types of coral reefs – fringing reefs, barrier reefs and atolls. The subsidence theory proposed by Charles Darwin after his voyage on the Beagle describes the formation of atolls by the gradual sinking of a newly-formed volcanic island. As long as the newly-formed island is in tropical waters and accessible to drifting coral larvae, it would soon acquire a fringing reef of coral. As the island slowly sinks, the reef front of the fringing reef around the island will grow upwards to form a barrier reef with a lagoon between the reef

and the island. As the island continues to sink and drop beneath the sea surface, the barrier reef continues to grow upwards to form a circular atoll. This geological sequence of events is shown in Fig. B1.7.1.

There are, however, coral reefs on the edges of continents and islands that are not volcanic and not subject to subsidence. In these cases, coral may have become established on suitable substrates such as rocky outcrops and shallow offshore banks to eventually form fringing and barrier reefs respectively and atoll formation is not an end result of these processes.

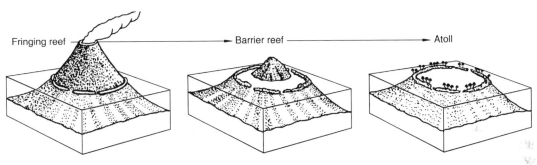

Fig. B1.7.1 The subsidence theory on the evolution of atolls from fringing reefs.

From seawards, as the barrier reef is approached the sea floor rises steeply, sometimes from great depths. Many islands of volcanic origin have no continental shelves and depths of over 1 km may be found within the same distance from the reef. As the reef slopes upwards, autotrophic corals start to become abundant above depths of about 50 m where light is about 4% of that at the surface. There is often a terrace at depths of about 15 m and above this a rich area of coral growth, mainly of massive species that can withstand the turbulence caused by breaking waves.

Towards the surface, where wave turbulence is the strongest, reefs may have a series of surge channels that often extend far into the reef itself, sometimes creating blowholes and dramatic geysers that send spray many metres into the air. The

surge channels, and the coral walls or buttresses that separate them, run approximately at right angles to the reef front or crest. Beyond this buttress zone, encrusting coralline algae forms an algal (or *Lithothamnion*) ridge in a turbulent area that contains very few corals. Towards the shore from the algal ridge, there are many corals and other organisms spread over a shallow reef flat and larger corals such as *Porites* may form boulders in which the tops are bleached and dead from exposure at low tides.

The breaking up of corals creates the areas of rubble often found at the back of the reef flat. As the reef flat gives way to the lagoon, smaller and more delicate corals, including branching species, thrive in the calmer conditions. Lagoons vary in extent from small moats to expanses of water many kilometres in width. In the case of islands

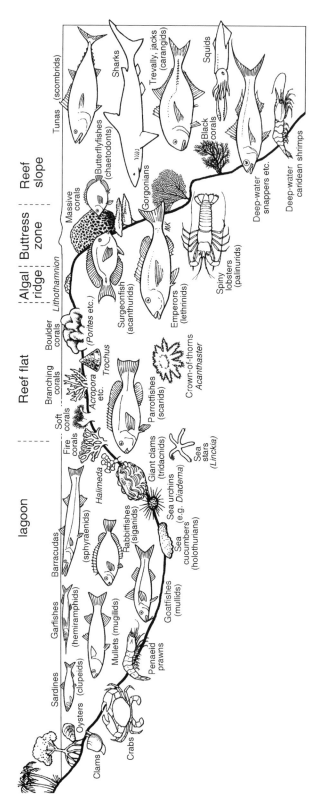

Fig. 1.13 A profile of a lagoon, barrier reef and beyond with some of the more conspicuous organisms. Although stony corals are the main reef builders, hydrocorals, calcareous green algae (*Halimeda*) and coralline red algae (*Lithothamnion*) contribute calcium carbonate to the reef mass. Coral predators include butterflyfishes, parrotfishes and crown-of-thorns, *Acanthaster planci.*

progressing towards atolls, the subsidence theory suggests that the size of a lagoon increases with the age of the island. The floors of lagoons are usually composed of sand derived from corals and calcareous algae and may support large beds of seagrass. Where waters are not affected by turbidity and low salinity, the floors of lagoons may be interspersed with patch reefs and pinnacles of coral.

Although scleractinian corals are the main reef builders, other organisms contribute calcium carbonate to the process. Most importantly, these include calcareous green algae of the genus *Halimeda* and coralline red algae of the genus *Lithothamnion* and others. Hydrocorals, molluscs, echinoderms and sponges all produce calcium carbonate although the relative importance of each differs between individual reefs and oceans. On Atlantic reefs, common calcium-producing organisms include the hydrocoral *Millepora* which is less dominant in the Indo-Pacific. Reefs in the Indo-Pacific, on the other hand, have a greater number of molluscs, including the giant clams (*Tridacna* and *Hippopus*), which contribute to calcium carbonate formation.

Many species, collectively called bioeroders, break down coral and other calcareous material to create fine coral particles that contribute to sand formation and limestone deposition. Some sponges, worms, and bivalves bore into corals, and other species, including some fish and sea urchins, feed on the surface of the coral (see Box 1.4 'Bioerosion'). Of the vertebrate predators of coral, parrotfishes (scarids) have fused teeth to scrape corals and butterflyfishes (chaetodonts) have pointed mouths to consume coral polyps as well as small crustaceans.

The most publicized predator of corals is the crown-of-thorns sea star, *Acanthaster planci*, which uses its extendable mouth to suck up coral polyps and has been accused of laying waste huge areas of coral reefs. There is controversy over what causes the destructive plagues of crown-of-thorns which have destroyed reefs at rates of 5 km a month. Perhaps there are natural fluctuations in population sizes, or perhaps surges in their numbers have been caused by pollution

from nearby coastal areas. The overcollecting of some of their molluscan predators, such as the large trumpet triton, *Charonia tritonis*, has also been blamed (see Fig. 2.5 in Chapter 2).

Corals in many areas have been affected by bleaching, a condition in which the corals' zooxanthellae are lost due to stress. Without the pigmentation of the zooxanthellae, the corals appear white and eventually die if the factors causing the stress are prolonged for more than a few weeks. Bleaching appears to be occurring on individual reefs (localized bleaching events) as well as on reefs over large geographic areas of the world (mass bleaching events). Although various abnormal environmental conditions will cause coral bleaching, elevated sea surface temperatures have been implicated in its present widespread occurrence (Glynn, 1993; Brown, 1997). Bleaching has been observed when the mean summer maximum temperature is raised by as little as 1 or 2°C. One of the first effects of elevated temperatures and a precursor to bleaching appears to be the impairment of the carbon dioxide fixation mechanisms in zooxanthellae (Jones et al., 1998).

Although global increases in sea temperatures can cause corals to bleach by losing their photosynthetic microalgal symbionts, there is a suggestion that some corals associate with microalgae that are more resistant to thermal stress (Sotka & Thacker, 2005). If so, coral reef resilience will depend on whether changes in host–symbiont associations can match the rates and amounts of changes in temperature, as well as on our ability to ameliorate continuing human impacts. Although reefs are under threat by human impacts, it is likely that reefs will change rather than disappear entirely, with some species already showing far greater tolerance to climate change and coral bleaching than others (Hughes et al., 2003).

Although the degradation of coral reefs began well before the present threats of bleaching were recorded, it appears that the rate of degradation may be increasing and coral reef ecosystems will not survive without immediate protection. Whether or not the destruction of coral is related to the global greenhouse effect (Section 1.5.5), the activities of people in local areas continue to be a threat to coral reefs. One common cause of localized reef destruction is the increasing amount of silt entering reef waters. Silt either smothers coral colonies or, even at lower levels, decreases the amount of sunlight reaching zooxanthellae. Activities resulting in high silt loads include harbour dredging, coastal building projects and tree clearing, particularly around rivers. In many countries in which coral collecting is not prohibited by law (or the laws are poorly enforced), corals are collected for sale as souvenirs and blocks of coral are used for the construction of buildings and walls.

Fishing on coral reefs or otherwise removing species selectively is also likely to affect a coral reef ecosystem. Parrotfishes that keep coral surfaces clear of algae, for example, are often speared at night by fishers using underwater torches. The overfishing of these species and other algal grazers leaves the coral exposed to excessive algal growth and eventual smothering. On the other hand, excessive erosion of corals can be caused by the removal of predators that keep coral bioeroders in check. Populations of urchins that erode corals, for example, are controlled by predators including triggerfishes (balistids), emperors (lethrinids) and wrasses (labrids). The overfishing of these species may allow urchin populations to increase dramatically and destroy coral formations (McClanahan et al., 1996).

In some areas, corals are being destroyed by the use of poisonous chemicals and explosives for fishing (see Chapter 3). Corals that have been destroyed by the use of explosives may not recover for many years. Dynamited reefs in the Philippines, for example, have taken an average of 38 years to recover after dynamiting has been stopped (Alcala and Gomez, 1979).

Coral reefs support the highest diversity, the highest standing stock biomass and the highest level of photosynthesis (Birkeland, 1997). Threats to coral reefs, therefore, can reduce marine biodiversity and, because of the role of corals in capturing carbon dioxide, perhaps even affect the planet itself. From a localized viewpoint, coral reefs support species that are necessary and customary food items for many coastal communities. In addition, reefs protect coastlines and villages from large ocean waves and are the source of the sand that builds up on beaches.

1.4.3 Continental shelves and the open sea

Extending away from the shore of continents there is often a flat and relatively shallow area of seabed called the continental shelf. In some areas, the continental shelf extends out from the shore for many kilometres while in others, the shelf is very narrow. At depths of about 200 m, the sea floor falls away steeply on the continental slope down to the continental rise, which slopes more gently to a relatively flat abyssal plain at depths of about 3–4 km. The vastness and monotony of the abyssal plain is broken up by deep ocean trenches and high mounts and ridges.

The crust or outer layer of the earth is not a continuous layer, but is made up of a number of lithospheric plates, somewhat like a cracked eggshell in which pieces float on and cover the fluid beneath. The separate plates that make up the earth's surface are in constant but irregular motion and their movements over geological time are referred to as continental drift. Continental drift has caused continents to periodically fuse and break apart throughout geological history. Over 250 million years ago, the world was an unrecognizable one, with all present-day continents bunched up as a single super-continent, Pangaea. At present, the lithospheric plates are moving at rates of up to 7 cm per year.

Movements of plates both towards and away from each other are shown diagrammatically in Fig. 1.14. At divergent boundaries, the continental plates move apart releasing hot, liquid magma from deep in the earth. The magma flows upwards and solidifies as basaltic rock, which forms a submarine ridge and sometimes islands. The best-known and most extensive submarine ridge is the Mid-Atlantic Ridge, formed between the African and South American plates. This ridge breaks through the sea's surface to form mid-ocean islands that stretch from the Azores in

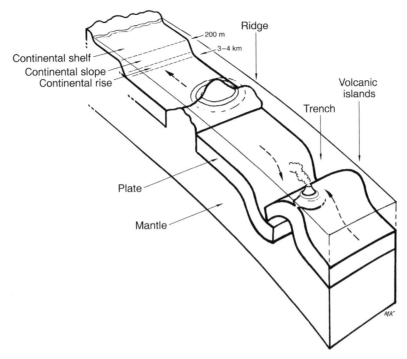

Fig. 1.14 An ocean profile showing the continental shelf, slope and rise and the movement of lithospheric plates. Arrows with broken lines show the direction of plate movement either apart to form ridges and islands or together in a subduction zone to form deep oceanic trenches and arcs of volcanic islands.

the North Atlantic to Tristan da Cunha in the South Atlantic (Fig. 1.15). With further spreading of continental plates, there is a sequence from ridges to rift valleys and ocean basins over geological time.

At convergent boundaries, where the plates move towards each other, one plate is forced below the other to form a deep oceanic trench in a process called subduction. The Pacific Ocean contains the largest number of trenches, including the Kermadec, Tonga, Bougainville, Mariana, Philippine, Japan and Aleutian trenches in the east and the Peru-Chile, Middle America and Cascadia trenches in the west. The ocean's greatest depth is in the Mariana Trench in the North Pacific Ocean, and is the result of the older and denser Pacific Plate to the east subducting beneath the younger and less dense Philippine Plate to the west. Destruction and melting of subducted crusts results in arcs of volcanoes running parallel to trenches and many of these break through the sea to form arcs of islands. The

Mariana Islands, for example, are formed on the western margin of the Mariana Trench and other volcanic arcs include the Aleutian Islands in the North Pacific and the Leeward and Windward Islands in the Caribbean, all of which lie adjacent to deep ocean trenches.

The movement of lithospheric plates creates unstable areas at their boundaries (Fig. 1.15) where earthquakes are common. Earthquakes occur at all plate boundaries and those created at convergent boundaries are often the most powerful. Convergent boundaries that surround the Pacific Ocean result in a dangerous earthquake-prone chain known as the Pacific Ring of Fire, which has boundaries that run from New Zealand through Papua New Guinea, Southeast Asia and Japan to the Aleutian Islands in the west and from the Andes Mountains in South America to Mexico in the east. The vertical movement of lithospheric plates beneath the sea is also responsible for the giant waves called tsunami (see Box 1.8 'Tsunami').

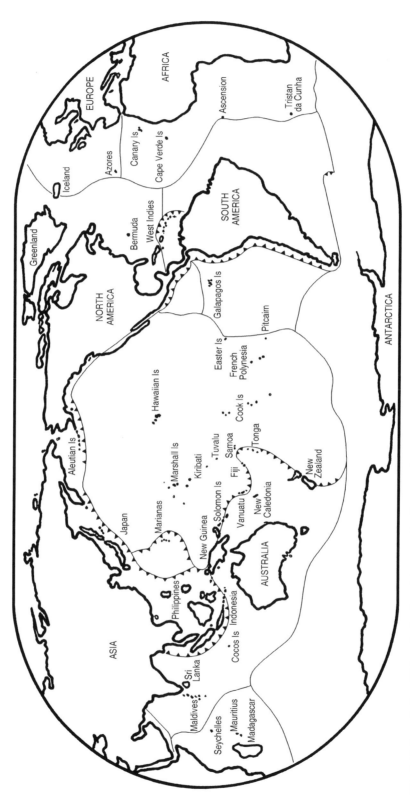

Fig. 1.15 Boundaries of lithospheric plates (shown as fine lines) on the ocean floor. Convergent boundaries are superimposed with black triangles.

Box 1.8
Tsunami

Sometimes the vertical movement of continental plates will result in shock waves, or surges, which travel away as a series of waves that are often called tidal waves but, because they have nothing to do with tides, are better referred to by their Japanese name, tsunami. Tsunami, although containing many millions of tonnes of water, travel as a series of low swells across the surface of the sea. At this stage, because the waves are flat bulges, perhaps less than a metre high and over 150 km apart, they are not usually observed by ships at sea. However, when these low but voluminous waves reach shallow water, friction with the sea floor causes their lower parts to slow down and their upper parts to rise up to form giant waves, sometimes over 30 m high.

An early and well-documented example of a volcanic eruption causing a tsunami is that of Krakatoa, which lies near the Java Trench and the buckled arc of the Indonesian islands. In 1883, Krakatoa erupted with such an explosion that the noise could be heard in Australia, 2500 km away. The resulting tsunami had waves over 20 m high that killed over 36 000 people along the Sunda Strait. The waves swept up a gunboat and deposited it over 3 km inland and their energy was recorded as far away as the English Channel. But, in terms of devastation, this tsunami was dwarfed by the one that killed over 220 000 people in more than a dozen countries around the Indian Ocean in December 2004. This tsunami followed an earthquake off the north-western tip of Sumatra in Aceh province which was hit by a wall of water within half an hour of the earthquake.

There are very few indications of the approach of a tsunami except perhaps the rapid withdrawal of water away from the shore as the trough of the wave reaches land before the first wave hits. There is a belief in some coastal communities that the approach of a tsunami can be predicted by the erratic behaviour of animals. More technical early warning systems involve seabed sensors that send signals via surface buoys and satellites to authorities. An alternative system depends on recording seismic waves that pass through the earth at speeds greater than those of tsunami-generated waves. However, because the speeds of tsunami waves can reach 700 km/hr, it is inevitable that coastal communities will receive warnings little in advance of approaching waves.

In addition to divergent and convergent boundaries a third type, called a transform boundary, occurs when two plates slide alongside each other in opposite directions; a well-studied example is the San Andreas Fault that runs through California. Volcanoes and volcanic islands also occur well away from plate boundaries at hotspots, places in which hot molten rock rises in plumes from within the earth's mantle and manages to breach the lithosphere plate. The well-known volcano and lava fields in Hawaii are at a hotspot situated near the centre of one of the world's largest lithosphere plates. Undersea volcanic peaks or seamounts result from volcanic activity associated with either hotspots or divergences and are particularly common in the Pacific Ocean. When continuing sea floor spreading moves the seamount away from its source of magma, it can no longer grow and may become eroded by waves to become a flat-topped guyot (or tablemount).

Continental shelves vary in their productivity depending on the quantity of nutrients that they receive from terrestrial sources and the most productive are the soft-bottom continental shelves that receive sediments and organic material from major rivers and deltas. Such rich areas are often centres of commercial fishing activity and, because the sea floor consists of sediments that are generally clear of obstructions such as rocks, the use of towed nets is common. As artisanal fishers exploit resources nearer the shore, soft-bottom shelf areas are often sites of conflict between non-commercial and commercial fishers.

Beyond the continental shelves, the open ocean is the least productive of all marine ecosystems with some tropical waters being the equivalent of terrestrial deserts. Nevertheless, surface layers of the open sea are exploited for pelagic fish, the most important of which are the tunas. Seamounts which rise close to the surface attract other targeted fishes including orange roughy and deepwater snappers. In the open sea, nutrients in surface waters are generally used up quite quickly by phytoplankton in the photic zone and, in the absence of mechanisms to replace these nutrients, the waters would be devoid of life. One such replacement mechanism involves winds and waves causing the vertical mixing of waters at different depths and the transport of nutrients up from below the photic zone.

Like most other substances, the mass per unit volume, or density, of water increases with decreasing temperature. Water, in fact becomes more dense with decreasing temperatures but only until a temperature of just under 4°C is reached; as water begins to freeze, its molecular structure expands resulting in a lower density (this property, incidentally, allows ice to form *over* a body of water rather than below it as do the solid forms of other liquids).

If the temperatures, and therefore densities, of surface waters and deeper waters are similar, wave-induced vertical mixing is possible. Convection currents involve the downward movement of surface water that is cool and dense followed by the upward movement of deeper less dense water. However, when the water column is highly stratified, with warm, low density water lying above cold, high density water, no vertical mixing is possible. This situation is typical of tropical oceans in which very warm water floats above a region of rapidly decreasing temperatures, called the thermocline. The thermocline prevents vertical mixing and effectively locks nutrient-rich water below it. Temperature profiles of water columns in polar, temperate and tropical seas are shown in Fig. 1.16.

In polar seas, the water column is not strongly stratified except for a weak thermocline that appears briefly in midsummer and nutrients in

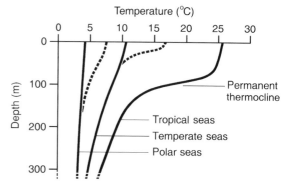

Fig. 1.16 Temperature changes with depth in tropical, temperate and polar seas. The broken lines represent temporary summer thermoclines.

the photic zone are generally not limited. In such high latitudes, autotroph activity is limited by available light and productivity is restricted to one or two months in midsummer. Even so, this single burst of productivity in Antarctica is sufficient to support a large biomass of krill and through the food chains, large populations of fish, whales, seals, penguins and albatrosses.

In temperate seas, a temporary thermocline occurs in the summer when surface waters are warmed by the sun, and this results in a period of low productivity. The onset of winter causes this thermocline to disappear and rough weather results in the water column becoming well mixed. Nutrients bought to the surface during winter result in a spring-induced bloom in phytoplankton abundance and high productivity.

In tropical seas, a permanent thermocline prevents vertical mixing from taking place. Although heterotrophic bacteria may have a role in the regeneration of nutrients in the photic zone, oceanic waters in the tropics are generally low in productivity. The clear blue water of tropical seas is that way because there are few phytoplankton floating in it.

Wind-driven surface currents may also move nutrients from one place to another and, more importantly, result in upwellings that bring deep, nutrient-rich water to the surface. Surface currents are driven by winds, which blow in response to areas of high and low atmospheric pressure. Near the equator, the sun warms the air

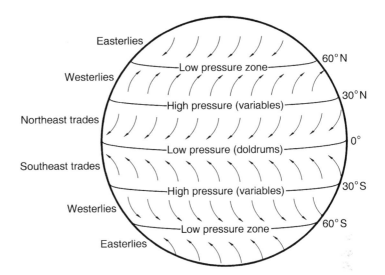

Easterlies
Westerlies
Northeast trades
Southeast trades
Westerlies
Easterlies

Low pressure zone — 60° N
High pressure (variables) — 30° N
Low pressure (doldrums) — 0°
High pressure (variables) — 30° S
Low pressure zone — 60° S

Fig. 1.17 High and low pressure bands on the earth's surface. Sinking air causes high pressure belts and rising air causes low pressure belts. The winds blow from high to low pressure areas but are deflected due to forces generated by the earth spinning from west to east.

and this rises to create a low-pressure zone. The hot air rising from the equator moves north and south and, as it cools, descends to create high pressure zones in latitudes of about 30 degrees to the north and the south. Cells (called Hadley cells) created by rising and sinking air masses result in the relatively constant bands of high and low pressure shown in Fig. 1.17. Winds blow from high pressure to low pressure areas, but not directly, as their direction is influenced by the deflecting effect of the spinning earth. The rotation of the earth from west to east imparts an easterly momentum to all objects moving on the earth's surface (see Box 1.9 'The Coriolis effect').

Areas of calms or variable winds are associated with bands of low and high pressure on the earth's surface. Sailing vessels crossing the equator, for example, are often becalmed in the low pressure zone called the doldrums. Thirty degrees away to the north and south, high pressure belts are characterized by winds that are variable in both direction and strength and known to sailors as the variables or horse latitudes. The latter name comes from the days when sailing ships bound for the West Indies commonly carried horses, and if the ships were becalmed for weeks on end in the variables, the horses were cast overboard to save on dwindling supplies of freshwater.

The generally steady trade winds that blow towards the equatorial low pressure zone are named for the assistance they gave to trading ships under sail. These blow from the north-east in the northern tropics and from the southeast in the southern tropics. Moving towards the poles from 30 degrees north and south, the winds blowing from the highs of the variables are deflected to blow from the west towards the east. In southern seas, the area called the Roaring Forties, south of 40 degrees, the strong westerly winds were used by sailing ships to make quick passages around the globe.

On the surface of the sea, winds drive currents that are diverted by land masses as well as by the Coriolis effect. This effect causes surface currents to be deflected to the right of the wind direction in the northern hemisphere, and to the left in the southern hemisphere. This results in similar surface current patterns in the Atlantic, Indian and Pacific oceans with a clockwise circulation in the northern hemisphere and an anticlockwise circulation in the southern hemisphere (Fig. 1.18). Currents circulate in large subtropical gyres around centres which often have very little water movement. The most renowned of these is the Sargasso Sea, well known for its rafts of the floating alga, *Sargassum*, and sailors' tales of fabulous sea creatures.

In the northern hemisphere, the best known and strongest currents are found on the western side of the oceans and these include the Gulf

Box 1.9
The Coriolis effect

An object at the equator where the circumference of the earth is the greatest and an object in higher latitudes where the circumference is less, will both make one full revolution in one day. But as the object at the equator has further to travel it will have a faster easterly motion. The earth's surface has a velocity of 1600 km/hr at the equator and 1400 km/hr at 30 degrees of latitude.

In reference to Fig. B1.9.1, an object travelling from point A on the equator directly north to point B, would have an initial sideways momentum to the east of 1600 km/hr. As the object moves further north, it will be travelling sideways to the east faster than the earth's surface. Therefore, instead of travelling due north in a straight line, the object will curve to the right. Similarly, an object travelling from A due south to point C will be deflected to the left. The Coriolis effect is a result of the earth's rotation towards the east and the difference in the speed of the earth's surface at

different latitudes; its effect is to deflect all moving entities, including winds and currents, on the earth's surface towards the east.

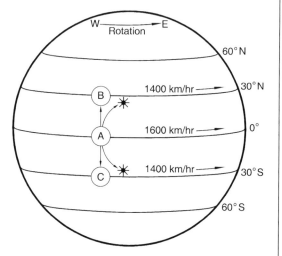

Fig. B1.9.1 The Coriolis effect on objects travelling from point A due north towards point B and due south towards point C. In each case, the objects are deflected to the points marked by asterisks.

Stream in the Atlantic and the Kuroshio in the Pacific. In the Indian Ocean, the Somali Current is dependent on the monsoons which change direction seasonally. In the southern hemisphere, strong currents slide up the eastern side of the oceans and the best known of these are the Humboldt or Peru Current off South America and the Benguela Current off Africa. Westward-flowing equatorial currents occur in the Atlantic, Indian and Pacific oceans and are separated by an eastward-flowing equatorial countercurrent. In the Southern Ocean, unhindered by continental barriers, the Antarctic Circumpolar Current (called the West Wind Drift by sailors) flows from west to east around the globe.

Both warm and cold currents have effects on adjacent land masses. Cold currents flowing from higher latitudes include the Humboldt Current, which bathes the tropical regions of the Eastern Pacific Ocean in cold water. Here, unlikely animals such as seals and penguins are found in the

Galapagos Islands that lie on the equator. On the other hand, warm currents flowing from low latitudes make the climate of some high-latitude countries more moderate. In the Atlantic, for example, the harbours of Norway are usually free from ice because of the flow of relatively warm Atlantic water into the Norwegian Sea. On the other side of the Atlantic, the harbours in Labrador, located further south than Norway, are frozen for several months of the year (see also the conveyor-belt circulation, involving deeper currents, in Section 1.5.5).

Upwellings and downwellings refer to the vertical movement of water between deeper and shallower water and both are important oceanic mixing mechanisms. However, upwellings are of more interest from a productivity viewpoint as they are associated with bringing cold, nutrient-rich water to the surface. An upwelling in a particular area is created by surface waters moving away and being replaced by deeper water

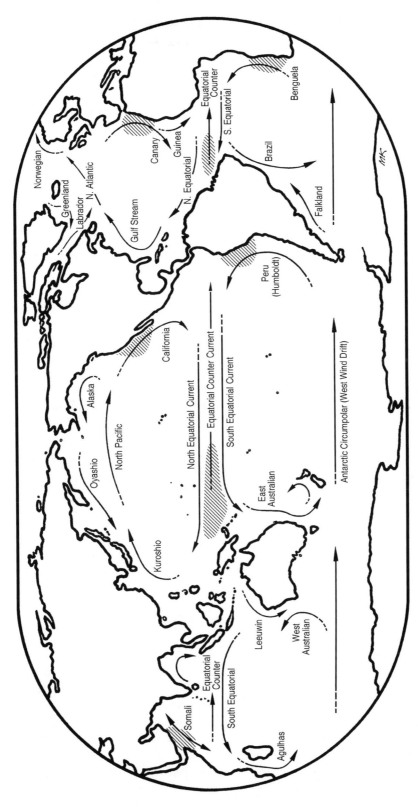

Fig. 1.18 Oceans of the world showing major surface currents (arrows), and main upwelling areas (shaded).

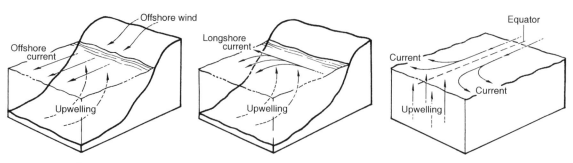

Fig. 1.19 Conditions causing upwellings, including (from the left) offshore winds, longshore currents and diverging currents.

(Fig. 1.19). Upwellings in coastal areas may be caused simply by offshore winds. The friction of the wind moving across the sea drags surface waters away from the shore and these are replaced by deeper nutrient-rich water moving up from the sea floor. For this reason, fishers on islands often take advantage of better fishing (and, of course, more sheltered waters) on the leeward side of the islands.

Upwellings may also be created by currents flowing along a coast, either north to south in the northern hemisphere, or south to north in the southern hemisphere. Off the Pacific coast of South America, for example, the Coriolis effect causes the strong north-flowing Humboldt Current to veer to the left, away from the coast, where it is replaced by an upwelling of deeper water. This upwelling results in one of the world's most productive fishing areas off Peru (see Section 1.8). In general, currents on the western sides of oceans (e.g. the Gulf Stream in the Atlantic and the Kuroshio in the Pacific) contain warm, nutrient-poor, water from the tropics and have low productivities. However, their movement along coasts may result in upwellings at their boundaries by a similar mechanism to the one that causes the Peru upwelling.

Divergences occur when surface currents move away from each other and are common on the equator around the western side of oceans where parallel currents diverge. An equatorial upwelling in the Western Pacific results from two parallel currents (the North and South Equatorial currents) diverging: one north towards Japan, and the other south down the east coast of Australia. As these two currents move apart, deeper water moves up to replace surface waters.

Major upwellings, and not coincidentally some of the world's largest fisheries, are found in coastal upwellings associated with the Canary, Benguela, California and Peru currents. Equatorial upwellings, particularly in the Western Pacific, result in large catches of tuna. Seasonal upwellings also occur, such as off the coast of northeast Africa, where strong monsoons blow northwards along the coast of Somalia in summer months. Away from coasts and the equator, the central gyres of the oceans are among the world's least productive ecosystems.

1.5 Human impacts on marine ecosystems

Inshore marine ecosystems are used and affected by many activities other than fishing, including coastal development, forestry, agriculture, shipping and industry. Many of these activities result in coastal waters being the depositories of runoff, waste and pollutants. Because of currents and the interconnectivity of ecosystems, even the most remote areas, including the deep ocean and polar seas, are now affected. This section briefly summarizes the impacts and pollutants that affect marine environments – effects that must be taken into account by those managing fisheries and ecosystems. More extensive information is given in Clark (2001) and other references in this text.

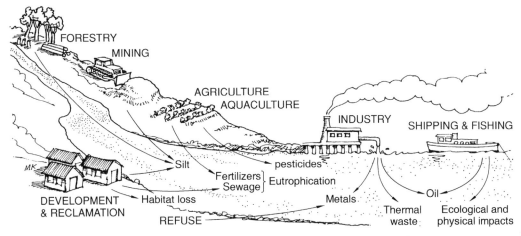

Fig. 1.20 Some impacts and pollutants (lower case) affecting the marine environment with common sources (upper case).

Pollution can be defined as the introduction of any substance or energy into the marine environment which results, or is likely to result, in harm to marine life and hazards to human health. This definition, précised from the World Health Organization, presents some difficulties. Problems include determining which substances, and what concentrations, are harmful. Some substances such as sewage and fertilizers entering seawater are beneficial to primary productivity but large quantities cause excessive plant growth, the decay of which results in oxygen depletion (eutrophication). Under the definition, litter floating in the sea or stranded on a beach, although aesthetically displeasing, could not be described as pollution unless it causes harm to marine life. Only relatively recently has it been recognized that discarded packaging, typically plastic and polystyrene, is harmful to marine species. Large fish and turtles mistaking litter for food, choke on plastic bags (which resemble jellyfishes) and disposable cups. Entanglement in plastic packaging and discarded fishing gear has also been responsible for deaths in species such as turtles, seals and dolphins.

The urgent requirement for environmental protection has led to the need for fisheries managers to adopt an ecosystem-based management approach. This approach extends the conventional principles for sustainable fisheries development to cover the ecosystem as a whole. It aims to ensure that aquatic ecosystems not only maintain their capacity to produce seafood, but are sustained indefinitely for the benefit of present and future generations. Fisheries management, including the management of ecosystems, is discussed in Chapter 6.

Major impacts on marine ecosystems result from fishing and coastal development, both of which affect fish stocks, fish habitats and ecosystems. All fishing affects the environment to a greater or lesser degree and the effects are discussed in Chapter 3 'Fishing and fishers'. Pollutants reaching the sea include organic material and sewage that creates eutrophic (oxygen-depleted) conditions as well as silt and heat. More serious pollutants include toxic organic compounds, pesticides, petroleum products and heavy metals. Impacts and pollutants are caused or derived from municipal, industrial, and agricultural sources, some of which are summarized in Fig. 1.20. Climate change and ozone depletion are global factors that also have the potential to affect marine ecosystems. These are discussed in the following subsections.

1.5.1 Habitat modification and loss

Coastal development, harbour and marina construction, reclamation, coastal aquaculture and the demand for coastal living space have all taken their toll on aquatic ecosystems and affected key habitats of marine species. Inshore areas, particularly estuaries, salt marshes, wetlands and mangrove forests, are known to provide food and shelter for many aquatic species, including many of commercial importance. Yet the view that many of these coastal areas are wastelands still persists and is often actively promoted by property developers keen to gain access to valuable waterfront land.

In developed countries, estuaries and wetlands are being used for the construction of canal-based housing projects and marinas as well as being 'reclaimed' or landfilled for industrial areas and housing estates. About one-third of all estuaries in the United States have disappeared. In parts of Southeast Asia, large areas of mangroves have been replaced by aquaculture ponds to farm penaeid shrimps. In small island countries, where populations are increasing at high rates and the demand for living space is high, coastal areas are under increasing threat. On the island of Tutuila in American Samoa, for example, over 25% of coastal wetlands and mangrove areas have been converted to dry land for buildings.

In some cases, marine habitats are lost through poor planning. In many subtropical and tropical countries, mangrove areas have been destroyed indirectly by the construction of coastal roads that interrupt the mixing of fresh and salt water. Unless large pipes are placed beneath the road, mangroves on the seaward side of the road die in water that is too saline and those on the landward side die in water that is too fresh. Wharves, breakwaters and causeways as well as landfill areas are all likely to interfere with water currents, the replenishment of beaches and the natural movements of fish. Even the best intentioned of development projects can affect marine life; the construction of causeways between islets in the atoll of Kiribati, for example, is believed to have blocked the spawning migration of some fish species.

In addition to the direct loss of habitats, development, increasing populations and industry often result in high nutrient loads, silt and heated water entering coastal waters and affecting marine life; these are discussed in the following section.

1.5.2 Eutrophication, siltation and heat

Small quantities of nutrients entering bodies of water are often beneficial. Nutrients are required by all marine plants including phytoplankton, macroalgae (seaweeds) and seagrasses. However, larger quantities of nutrients and organic material result in increases in aerobic bacteria which use up oxygen dissolved in the water. The problem is exacerbated by the excessive growth of plants, the decomposition of which uses up even more oxygen.

Waters that are very rich in nutrients are called eutrophic and the associated depletion of oxygen is a common cause of mass deaths in shellfish and fish. The concentration of organic matter in water is often measured by its biological oxygen demand or BOD_5 which measures the oxygen used over a five-day period by micro-organisms as they decompose the organic matter (usually in sewage) at a temperature of 20°C. The greater the demand for oxygen, the more organic material is present in the water, and waters with a BOD_5 greater than a particular prescribed limit are regarded as eutrophic. The most common anthropogenic sources of high nutrient loads in coastal waters are livestock and human sewage and runoff containing agricultural fertilizers.

With rapidly growing populations, the disposal of sewage is of increasing concern. Sewage may be treated to varying degrees, but this is expensive and requires areas of land for the construction of ponds. Primary sewage treatment involves grit removal, screening, grinding, flocculation, and sedimentation. Secondary treatment involves the oxidation of dissolved organic matter to form a biologically-active sludge and tertiary treatment involves nitrogen removal. Up to 50% of the costs of operating a treatment plant are involved in the handling and disposal of solid wastes. The use of septic tanks, underground tanks in which the organic matter in sewage is decomposed through bacterial activity, is common in some coastal areas away from centres of high population density. However, regardless of the degree of treatment, sewage disposal often results in quantities of nutrients, particularly nitrates and phosphates, entering coastal waters.

In areas of intensive farming, runoff from the land may contain large quantities of nitrates and phosphates from fertilizers. Intensive aquaculture operations can also produce as much organic waste as a moderately large town. Large quantities of faeces and unused food settle beneath salmon and tuna farming pens, under rafts used to culture mussels and oysters, and near the outlets of ponds used to farm fish and prawns.

One of the first indications of eutrophic conditions is the excessive growth of marine algae such as green sea lettuce, *Ulva*, and green thread alga, *Enteromorpha*, in cooler waters and Sargasso weed, *Sargassum*, in warmer waters (see Fig. 1.22 in Section 1.6.1). Excessive algal growth may result in changes in the structure of marine communities, the most obvious being an increase in the abundance and diversity of herbivorous fish attracted by the plant growth. In tropical waters, overgrowths of *Sargassum* with their hard, swaying fronds are also responsible for damaging corals through abrasion and shading. Corals in eutrophic lagoons have been smothered by algal growth which prevents light reaching the coral's symbiotic zooxanthellae.

Phytoplankton growth is usually limited by the availability of nitrogen, phosphorus and silicon. The addition of these limiting nutrients to coastal waters may result in rapid increases (or blooms) of phytoplankton, some species of which are harmful to marine life and humans. Red tides and several types of fish and shellfish poisoning (discussed in Section 1.6.3) are often attributed to high nutrient loads in coastal waters.

Fish kills are commonly reported near sources of high nutrient loads where eutrophication results in the suffocation of invertebrates and fish through severe oxygen depletion. Even at lower levels of eutrophication, the species diversity in the affected area is often greatly reduced. However, a small number of more tolerant or opportunistic species can thrive in nutrient-rich conditions and their

numbers may increase dramatically. For example, the abundance of the polychaete worm, *Capitella*, which increases to take advantage of large quantities of organic material, is used as an indicator of organic pollution.

Human and livestock sewage contains bacteria, viruses and parasite eggs that, contrary to common belief, remain viable in seawater; this is particularly so in tropical areas where seawater temperatures approach human internal temperatures. Pathogens, such as faecal coliforms and enterococci, threaten human health causing gastroenteritis, hepatitis and other diseases. The presence of the bacterium *Escherichia coli*, which is found in human intestines, is used as an indicator of water contamination by sewage. A faecal coliform count of less than 10 per 100 ml of water in 80% of collected samples is the World Health Organization's standard for areas in which shellfish are collected. Unacceptably high faecal coliform counts may result from inadequate wastewater treatment, aging sewage infrastructure, and runoff after heavy rains. Actions to reduce contamination may include improving sewage treatment and moving the outfalls of discharge pipes further offshore.

Sewage-derived nutrients may also enter marine ecosystems indirectly from groundwater. In some coastal areas, and particularly in small islands, drinking water is obtained from a pocket or lens of freshwater that floats on top of the seawater beneath the shoreline. Inadequate sewage disposal can result in groundwater becoming contaminated. Besides the obvious human health hazards associated with using contaminated water, groundwater leaches out into coastal waters causing eutrophication. Island countries with adjacent shallow lagoons are particularly at risk and some attempts are being made to counteract this. In Fiji, for example, community groups have planted thousands of coconut seedlings along some parts of the coast to absorb nutrients before they enter lagoons.

Animal sewage from farming operations are also of concern in some areas. In some developing countries, where pig rearing is commonly carried out on the banks of rivers, large quantities of

---

Box 1.10
Shellfish contamination

All bivalve molluscs including oysters, clams, cockles and mussels, pump large quantities of seawater through their systems to obtain oxygen and food. They feed by filtering microscopic plants from the water or organic material from the sea floor, and are likely to pick up substances in the water or in the substrate, including bacteria, toxic plant material and heavy metals.

New species of bacteria belonging to the genus *Vibrio* are being discovered in many different marine environments and some have been implicated in causing diseases in fish, crustaceans and corals as well as humans. *Vibrio vulnificus* occurs naturally in low-salinity marine environments, such as estuaries, and reproduces rapidly when temperatures are high. In the Gulf of Mexico, levels of bacteria can reach 1000 cells/ml

in seawater and up to 100 000 cells/g in oysters. Humans are infected by the bacteria through open wounds exposed to seawater and by eating raw shellfish including oysters, clams and mussels. The infection, which causes diarrhoea and vomiting (and may be fatal in those with liver conditions) is common in countries such as Japan where raw shellfish is consumed regularly. Numbers of bacteria, in general, can be high in coastal waters near sewage outfalls, dairies, pig farms, chicken batteries and intensive aquaculture ponds. Bacteria in shellfish can be destroyed by cooking, or by a process called depuration (in which the shellfish are placed in clean or sterilized running water for a period). High nutrient loads in coastal waters can also result in blooms of plankton species, some of which produce potent toxins that accumulate in the flesh of filter-feeding molluscs (see Section 1.6.3).

---

pig faeces are washed into inshore waters. In addition to the creation of eutrophic conditions, a particular strain of *Escherichia coli* from pigs can transmit gastroenteritis to humans and result in infant deaths. The consumption of shellfish from polluted waters is also a serious health risk. Filter-feeding shellfish, particularly bivalve molluscs such as oysters and mussels, accumulate human pathogens from sewage as well as other toxins (see Box 1.10 'Shellfish contamination').

Silt is formed by natural processes including the erosion of land and coasts by rain, rivers, wind and the sea. Anthropogenic sources of silt include coastal development, harbour construction, dredging, mining, agriculture and forestry. Waterborne silt in large quantities damages the delicate gills of fish and reduces the amount of sunlight reaching light-dependent animals (such as corals) and plants. Siltation has changed the composition of benthic species in many estuaries and lagoons. Because of sedimentation, some tropical lagoons are becoming measurably shallower each year and coral formations are being smothered and replaced by seagrasses.

In many parts of the world, dredging in shallow water for the extraction of materials including sand and gravel produces a plume of silt that affects many shallow-water communities. Such mining activities have resulted in the loss of spawning and nursery grounds for herring (*Clupea harengus*) in the Baltic. The construction of shore-based facilities and the continual dredging of shipping channels produce large quantities of sediment that is dumped at distant sites. About 80% of the more than 45 million tonnes of waste discharged into United States marine waters consists of waste produced by dredging. In the United Kingdom, a similar percentage of discharged wastes has resulted mainly from dredging ports and estuaries. Besides disrupting marine communities, the dumping of dredging spoils may spread unwanted species, including the cysts of toxic dinoflagellates, to new areas. In riverside and coastal mining operations, mine tailings are often released as slurry, often contaminated with metals, into nearby waters. In Papua New Guinea, mining waste has devastated 200 km of the Fly River, forming shallow banks in coastal deltas, and causing the collapse of some valuable fisheries, such as the one for barramundi.

Mountainous coastlines, particularly those subject to torrential rain in the tropics, are susceptible to erosion that causes not only coastal siltation but landslips capable of washing away entire villages. These effects are often due to poor land management practices including deforestation, agriculture and clearing land near river banks. Forest trees cut down for local use and for export have been disappearing at such a rate that the resource is unsustainable without replanting. The Philippines, for example, has one of the highest rates of deforestation in the world with less than 5% of the country protected in parks and other reserves.

Without the root systems of trees to stabilize soils, erosion often results in the siltation of nearby rivers and coasts. Clearing land on slopes for agriculture creates erosion problems similar to those created by mining and forestry operations. Slash-and-burn methods, often used on newly cleared land, produces soil that becomes depleted of nutrients within a few years and the abandoned land is subject to severe erosion. In the case of agriculture, erosion can be avoided by sensible management practices, including planting on terraces along contours, and by leaving belts of trees across the slope of the land (see Box 1.11 'Controlling erosion').

Heat is considered a pollutant when increased water temperatures result in harm to aquatic species and ecosystems. A major source of heat results from the discharge of cooling water by factories and power plants. Many plants and animals are sensitive to temperature changes as small as 5°C, and the release of warm water causes dramatic changes in marine populations. Often, affected areas become dominated by a smaller number of more tolerant species.

Coastal coal and oil powered electricity-generating installations, in particular, use large volumes of seawater to cool generators. Water is discharged at temperatures that are often 10°C above those of adjacent waters and increased temperatures often extend several kilometres from the point of discharge. In the interests of heat exchange efficiency, the water intake is usually placed some distance from the discharge point to avoid the recirculation of heated water. In warmer countries, air-conditioning places peak demands on power just at the time when sea temperatures are at their summer maximum and the effects of hot water discharges are more damaging to organisms living close to their maximum levels of thermal tolerance. In colder countries,

---

Box 1.11
Controlling erosion

Erosion on steep coastlines can be reduced by contour farming involving cultivation along the contours of hillsides and the construction of terraces to reduce runoff (Fig. B1.11.1). A bench terrace consists of a series of step-like, flat areas cut into a hillside and a broad base terrace consists of a ridge of soil or a band of leguminous shrubs placed around the contour of the hill. Besides increasing soil fertility and crop yields, the reduction in water runoff and soil erosion helps to reduce siltation in nearby waterways.

Fig. B1.11.1 Contour farming, in which cultivation is on terraces edged by trees, reduces the risk of erosion, landslip and water siltation.

peak power demands are due to heating require-ments and, as this occurs during winter months, the effects of discharging hot water are less.

Cooling water from shore installations may be used with benefit in aquaculture in cooler clim-ates. Heated water is released into ponds used to farm various aquatic species, which conse-quently grow more quickly and reproduce earl-ier. Releasing warm water into ponds, whether used for aquaculture or not, has the benefit of allowing the water to cool before reaching the sea. An additional problem is that cooling water is usually treated with chlorine to inhibit the set-tlement of organisms which would otherwise restrict water flow through the heat exchange system. Containing warm chlorinated water in ponds will allow some of the chlorine to be broken down by sunlight before the water is discharged back into coastal waters.

1.5.3 Petroleum, metals, toxic chemicals and solid waste

Oil and petroleum products can reach coastal waters from oil refineries, offshore wells, tanker operations and accidents, as well as from indus-trial and urban wastes. Although tanker accid-ents are distressingly spectacular and result in local devastation, they do not account for more than a small proportion of the global input of petroleum hydrocarbons. In sea transportation, the loss of the cargo due to evaporation and its eventual return to the sea accounts for a much higher input of petroleum. Often, discharges from industry and urban stormwater result in most of the oil and hydrocarbon pollution of coastal waterways.

Many of the compounds in crude oil and other petroleum products affect organisms by suffoca-tion, lowering fertility and causing diseases. Oil forms a thin film on the surface of the water and may cover and kill intertidal animals attached to rocks. Even minute quantities of hydrocarbons in the sea can cause seafood to become contamin-ated with what has been called 'kerosene taint'.

Oil spills from tanker accidents may be treated by dispersants sprayed on the slick and contained with floating booms. These methods are often ineffective when dealing with large quantities of oil but are often used in spills close to shore where the slick has the potential to pollute beaches or affect aquaculture operations. The rate of disper-sal of oil from spills is highly dependent on cur-rents and wave action. The 37 000 tonne oil spill from the tanker Exxon Valdez in the Gulf of Alaska in 1989 caused a huge slick that affected marine species in the entire gulf area. However, the 85 000 tonne spill from the Braer on the coast of the Shetland Islands in 1993 was dispersed within a few days by offshore currents and high wave action. Many of the smaller molecules in oil are degraded by bacteria present in the sea but the larger molecules of resins are degraded more slowly, and may reach beaches as tar balls.

Heavy metals enter marine and estuarine ecosystems through the discharge of industrial waste, treated sewage, storm water runoff, min-ing operations, fuel use, and the corrosion of pipes and roofs. The marine environment also receives natural inputs of metals from the weath-ering of ore-bearing rocks and volcanic activity. Minute quantities of some metals are required by organisms for metabolic activity. Vertebrates and some invertebrates use the iron-containing pigment haemoglobin to transport oxygen and some molluscs and crustaceans similarly use the copper-containing pigment haemocyanin, which is blue when oxygenated. Mercury, on the other hand, has no place in metabolic activity and is toxic at even very low levels. Even metals re-quired in small quantities may be toxic in high concentrations. Copper, for example, in spite of being a component of haemocyanin, is used in antifouling paint to discourage the settlement and growth of marine organisms on the hulls of ships (see Box 1.12 'Foul play').

Heavy metals are particularly dangerous because they are cumulative – that is, even small amounts ingested with food material will, over a long period, build up to high concentrations in the organs and flesh of marine animals. This process, called bioaccumulation, results from the limited ability of most animals to excrete excess ingested metals. Filter-feeding organisms

Box 1.12
Foul play

Many sessile organisms settle on underwater structures including wharves, floating docks and ship's hulls causing some damage and much expense. In spite of the plant-like appearance of some, all of the fouling organisms shown in Fig. B1.12.1 are animals and, as they do not require light, can settle and prosper in dark places such as in the water intake pipes of power stations and ships. The settlement and growth of fouling organisms is reduced by the use of antifouling paint on the underwater parts of vessels; a ship with a hull laden with marine growth will be slower and use more fuel than one with a clean hull. The effectiveness of antifouling paint relies on its ability to release poisonous material at a low rate and form a barrier layer between a ship's hull and the creatures trying to get onto it or into it.

Antifouling paints usually contain metals in forms that are toxic to marine creatures, and one of the most effective ones is tributyl tin, or TBT. But TBT is too effective and poisons organisms well away from the antifouled ship. Concentrations as low as one part in a thousand million of water has been found to kill molluscs or interfere with their reproduction. The use of TBT is banned in all enlightened countries but unfortunately not worldwide: tankers and freighters, particularly those sailing under flags of convenience, still use the dangerous paint. Ports with a high density of shipping are particularly at risk. Suva Harbour in Fiji, for example, has had the unfortunate distinction of having the highest ever recorded concentration of TBT in its waters. This leaves copper-based antifouling paints for general use. Copper is not as effective as tin but in the layer closest to the boat's hull, high concentrations will kill or prevent organisms settling. The wide use of copper-based antifouling paints does not help the owners of aluminium boats as the two dissimilar metals produce a strong and destructive electrochemical reaction.

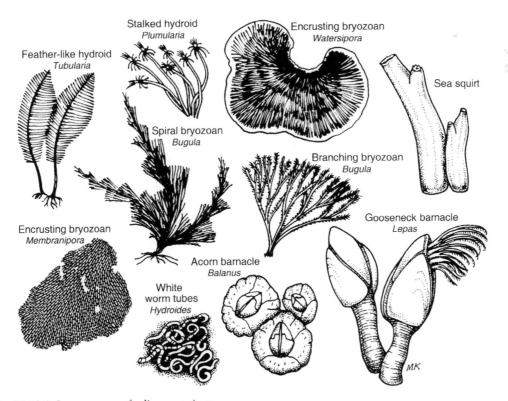

Fig. B1.12.1 Some common fouling organisms.

Box 1.13
Mercury rising

Fish of the open ocean live in one of the most pristine of environments, and it is surprising to find that many of them contain mercury. The dangers of mercury to human health have been known for some time, at least since the discovery in the nineteenth century that its use as sizing (a stiffening agent) in the manufacture of felt hats resulted in workers in the industry suffering from brain disease. Hat makers continually licked threads used in sewing thereby accumulating large quantities of mercury, and the unfortunate results gave rise to the expression 'as mad as a hatter'.

Methyl mercury is accumulated and stored within the fat of predatory marine fish. Because of the concentrating effect within food chains (see Section 1.7) mercury levels are higher in top carnivores including shark, tuna, marlin and swordfish. Marlins have been recorded with levels of mercury in their flesh of over 16 parts per million (ppm) whereas the internationally approved limit is 0.5 ppm (Fig. B1.13.1). Because mercury accumulates, larger fish within a given species are likely to contain more of the substance. Hence it

has been made illegal in some countries to market sharks larger than a certain maximum length. Canneries keep mercury levels within legal limits by mixing smaller tunas (skipjack tuna and small yellowfin) into cans of larger tunas.

Interestingly, high mercury levels have been found in old museum specimens of fish, suggesting that its presence is not a result of modern industrial pollution. Most mercury in seafood is derived from natural sources including habitats such as volcanic formations and islands. To put the danger of mercury content into perspective, people would have to regularly eat quite a lot of large fish (greater than 12 kg) to be in danger. In addition, the risks must be weighed against the health benefits of eating fish.

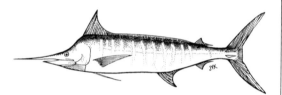

Fig. B1.13.1 The marlin has been recorded with up to 16 ppm of mercury in its flesh. Drawing by Jennifer Kallie.

accumulate more than any other group and bivalve molluscs may contain metals in concentrations of up to a 1000 times higher than that in fishes living in the same area. Carnivorous species feeding on contaminated prey will gain even higher concentrations by biomagnification through the food chains and become extremely toxic to human beings (see biological accumulation and magnification in Section 1.7.3). Many large pelagic fishes, in particular, contain considerable quantities of mercury (see Box 1.13 'Mercury rising'). Although humans gain most cadmium and lead from the consumption of terrestrial food, particularly plants, mercury is obtained directly from eating fish or indirectly by eating meat such as pork and poultry fed on stock food containing fishmeal. Unlike contamination by bacteria, heavy metals in food cannot be destroyed by cooking.

Mercury is the most publicized contaminant in the marine environment as it has been responsible for serious health problems. In Japan in the 1950s, people eating shellfish caught in Minimata Bay suffered from severe mercury poisoning, the effects of which included brain and genetic damage. In this case, mercury used as a catalyst in the manufacture of formaldehyde had been released into the bay, and the poisoning resulted in mercury contamination coming under public scrutiny in the early 1960s. In the USA, 48 tonnes or about 40% of the total mercury output comes from coal-fired power plants. Medical waste incineration is believed to be responsible for about 10% of the total amount of mercury released into the air each year, and some hospitals have been phasing out mercury thermometers, blood pressure cuffs and laboratory materials using the element.

Much of the anthropogenic input of metals into the marine environment is from gases released into the atmosphere, and discharges or dumping from industries such as manufacturing plants, pulp mills and mines. Cadmium inputs are from zinc and lead mining and from industries involved in electroplating and the manufacture of nickel cadmium (NiCad) batteries. Copper, used extensively in manufacturing electrical equipment, enters the marine environment in waste-water runoff and sewage. Lead, in spite of causing serious problems to human health on shore where it is derived from vehicle exhausts, old paint and atmospheric fallout, is not particularly toxic to marine organisms. However, it is now so widespread that lead has been found deposited in Arctic ice. Most countries have set maximum allowable levels of heavy metals in food and recommended maximum levels of human consumption. Those set by the Canadian government are given in Table 1.4 as an example.

Chlorinated hydrocarbons, or organochlorines, are synthetic compounds developed for agriculture and industry and, unlike metals, do not occur naturally in the marine environment. Low molecular weight organochlorines include many volatile industrial solvents as well as chlorofluorocarbons (CFCs) used as refrigerants and aerosol propellants. Although volatile organochlorines are lost to the atmosphere with little direct harm to marine ecosystems, CFCs are the main cause of the depletion of the ozone layer and the resulting increased ultraviolet radiation may be affecting marine life (see 'Ozone depletion' in Section 1.5.6).

Table 1.4 Indicative recommended maximum consumption of metals expressed in mg of metal consumed per kg of body weight per day. Health Canada (1990).

Cobalt	0.01
Copper	0.02
Nickel	0.02
Cadmium	0.0002
Mercury	0.0001
Lead	0.0000785

The larger molecular weight organochlorines, including the pesticides dichlorodiphenyl-trichloroethane (DDT) and dieldrin as well as polychlorinated biphenyls (PCBs), used in the manufacture of paint and plastic, are serious pollutants. The pollution caused by these organochlorines is regarded as so dangerous that their use in most developed countries is either restricted or discontinued. However, because of their persistence, they continue to appear in the marine environment and some are still being used in developing countries. Other than from runoff from urban and agricultural areas, organochlorines reach the sea through the evaporation of pesticides and by subsequent fallout from the atmosphere.

Like heavy metals, organochlorines are not broken down by bacterial action and they accumulate with each intake of contaminated food over the organisms' lifespan. Even though present in only extremely low concentrations in seawater, they can accumulate to toxic levels in marine organisms. The widespread use of pesticides came under public scrutiny in the early 1960s with the publication of the book *Silent Spring* by Rachel Carson in 1963. Both DDT and PCBs interfere with calcium metabolism in birds, causing soft-shelled eggs and malformed young and have caused the deaths of fish, sea birds and sea lions during the 1960s. Even though the use of dangerous organochlorines has been discontinued in many countries, residues remain in sediments and, through the atmosphere and food chains, they are distributed over the entire planet. Pesticides including dieldrin are present in marine sediments of the Great Barrier Reef, and significant levels of residues have been found in dugongs and dolphins. Organochlorines have also been implicated in causing kidney damage in Arctic birds and mammals.

Other persistent and dangerous pollutants include radioactive substances, some of which have a half-life of many thousands of years. There were considerable inputs of radioactivity to the atmosphere and the sea from World War II until the 1970s from the use and testing of above-ground nuclear weapons. Fallout reached a maximum in the 1960s, when levels of contamination were found in many marine species, and has been decreasing ever since.

Seawater is naturally radioactive mainly due to the presence of radioactive isotopes of potassium and hydrogen; the latter, named tritium (^{3}H) is formed through the action of cosmic radiation. Although cooling water and liquid effluents from nuclear power stations, reprocessing plants, and other nuclear facilities are treated to remove radioactivity, and strict limits are applied, some low residual radioactivity enters the sea. Accidents, including the explosion at Chernobyl in 1986 and the loss of nuclear powered submarines, have resulted in short-term effects. The disposal of waste presents a continual challenge and ever more strict regulations are being applied. One of the most notorious cases involving hazardous wastes occurred in the Love Canal area of Niagara Falls in the late 1970s when residents had to evacuate their homes due to wastes leaking from a former disposal site.

The disposal of waste from coastal cities and communities affects the marine environment both through the placement of disposal areas and because of harmful residues leaching from increasing volumes of waste. Coastal disposal sites have often included wetlands and mangrove areas and pollutants leaching from the sites include oil, heavy metals and paint residues. Discarded packaging including cans, bottles and plastic often end up on coastlines and in the sea. Besides being visual pollution, even relatively inert material can choke or trap marine life.

Antibiotics are used in terrestrial animal production and in aquaculture, mainly to prevent and treat bacterial diseases. They reach the sea in runoff from agricultural areas and more directly from their use on species used in aquaculture. There are some public health concerns regarding the unintentional ingestion of antibiotic residues, which may cause aplastic anaemia in humans as well as antibiotic resistance in bacteria that are pathogenic to humans (FAO, 2002). Many governments are placing limits on the type and doses of antibiotics used in aquaculture and are monitoring the residues in products and nearby

substrates. The antibiotics flumequine and oxytetracycline used to treat diseases in farmed fish, for example, have been found in sediments near trout and sea bass farms (Lalumera et al., 2004).

1.5.4 Species invasions, introductions and translocations

Many marine species have been either deliberately or accidentally introduced into different parts of the globe. Alien or exotic species can affect fisheries either directly by preying on and competing with native target species or indirectly by disrupting existing aquatic food webs and ecosystems.

In a relatively few cases, larger introduced species, including sea stars and crabs, have predated directly on native species that are targeted in local fisheries. Invading sea stars, for example, have devastated local beds of clams, cockles and mussels. More often, larger invaders affect fisheries indirectly by competing with the target species and by disrupting marine ecosystems. On the assumption that all available niches in an ecosystem are occupied, the introduction of an exotic species will result in intense competition. A basic principle in biology (the competitive exclusion principle) is that two species cannot simultaneously occupy the same niche in the same place over an extended period. Either the invader or the native species with which it competes is likely to be displaced. If the invader has some advantage, say it is less susceptible to existing predators, it will survive at the expense of the native species. Competition appears to be greatest between sessile species where the availability of food is a limiting factor. Some invasive species also appear to be more tolerant to pollutants in the marine environment than native species. In some cases both the invader and the native species can survive, particularly in complex ecosystems where, perhaps, a greater diversity of niches provides more opportunities. An existing predator, which preys on both the invader and the native species, may prevent the invader becoming dominant. However, in many cases the invading species has an advantage, particularly if it was introduced without its natural predators and competitors.

Human activities that provide 'transport vectors' for the movement of species across natural barriers into new environments and habitats include the international shipping trade, the aquarium trade, aquaculture activities and the construction of canals joining previously separated bodies of water.

Many of the unintentional introductions have resulted from exotic species being transported either on ships' hulls (as fouling organisms) or, perhaps more commonly, in ships' ballast water. Ballast water is carried to maintain an unloaded vessel's stability and is discharged in ports before taking on cargo. Most of the plankton, including the larvae of larger organisms, must surely die in ballast water during a long voyage through waters of varying temperatures. But some hardy organisms can survive the journey and have the ability to thrive in a new area. As international sea trade is increasing, the potential for exotic species arriving in ports around the world is also increasing. The zebra mussel, *Dreissena polymorpha*, which was introduced into the Great Lakes from Europe in 1986, was probably carried by a ship. The species grew rapidly in its new location and now covers pilings and clogs power station water intakes. The marine mussel, *Perna viridis*, is now established in the Gulf of Mexico and northern Australia where it is presently a fouling organism and competes with local species. The Northern Pacific sea star, *Asterias amurensis*, which occurs naturally from China across the Bering Sea to northern Canada, was introduced into Tasmanian waters, most likely via ballast water. It has now spread across the southern part of Australia from Western Australia to northern New South Wales and is threatening biodiversity, aquaculture farms and scallop and abalone fisheries.

Some invasions have occurred because of the construction of canals connecting lakes and seas. At least 45 species of Red Sea fishes have migrated into the Mediterranean Sea through the Suez Canal (Boudouresque, 1999). The Spanish mackerel, *Scomberomorus commerson*, a fast predatory fish with a natural distribution in the Indo-Pacific, appears to have entered the Mediterranean Sea in this way (Frimodt, 1995). The parasitic sea lamprey entered the Great Lakes via man-made canals in the 1930s and virtually eliminated lake trout from Lakes Huron and Michigan. In the absence of trout, populations of another invader, the alewife, increased until they were bought under control by the introduction of coho salmon as a predator in the 1970s.

Most public and scientific attention has been on large conspicuous invaders, but protists, bacteria, and viruses, have several attributes that provide them with an increased ability to invade new areas (Ruiz & Dobbs, 2003). First, the magnitude of organism transfer (or propagule supply) increases with decreasing size. Second, many micro-organisms reproduce asexually, minimizing the numbers required for establishment. Third, many micro-organisms have broad environmental tolerances, increasing the geographic range and seasonal window available for colonization. Fourth, the life histories of micro-organisms often include resting stages or cysts that can further extend environmental tolerance by surviving through conditions that are occasionally unsuitable. These attributes suggest that invasions by micro-organisms are widespread, occurring largely without detection, and may contribute to the observed increase in emerging diseases in marine ecosystems.

Introductions of micro-organisms have included toxic species which have the potential to devastate fisheries and affect human health. The dinoflagellates, *Gymnodinium* and *Alexandrium*, for example, produce toxins that cause paralytic shellfish poisoning in people eating affected shellfish (see Section 1.6.3 on harmful algal blooms). In Australia, for example, cysts of these toxic dinoflagellates have been found in sediments in ballast tanks and this is believed to have resulted in their introduction and establishment in local waters.

Many aquatic species have been deliberately introduced for aquaculture, sometimes to the considerable benefit of the receiving country. The trochus or top shell, *Trochus niloticus*, for example, has been introduced to small island countries to the east of its natural distribution in the Western Pacific Ocean. Trochus has been introduced into the Cook Islands where it is now the basis of a well-managed and valuable fishery for its meat as well as the shells, which are exported for the manufacture of pearl buttons; no negative effects have been observed although the species may compete for food with green sea snails (*Turbo* spp.). Oyster species, in particular, have been moved around the world for farming. The European oyster, *Ostrea edulis*, was introduced into the Gulf of Maine, USA, in the 1950s and forms the basis of fisheries there. The Pacific oyster, *Crassostrea gigas*, has been introduced into many countries and is now one of the most widely distributed of all marine species.

However, many aquatic species introduced for aquaculture have subsequently become established in the wild. The Pacific oyster, which competes with native oysters for settlement space, is regarded as a nuisance and even a noxious species in some areas. Atlantic salmon that are farmed in net cages in Canada have escaped and, as they are now found established and spawning in every river system on the British Columbian coast, may displace native salmon and the steelhead trout.

Species of *Tilapia*, a freshwater fish, are now farmed in many parts of the world. Genetically improved farmed tilapia (GIFT) have been developed by M.V. Gupta of the WorldFish Centre (formerly ICLARM) by the selective breeding of several strains of Nile tilapia. The fast growing GIFT or super tilapia has been distributed around the world where it contributes to food security in many under-developed countries (Anon, 2005). However, there are concerns that tilapia farmed near the coast have become adapted to brackish water and may become established in coastal waters.

All introductions and translocations of species, usually for aquaculture or the aquarium trade, are potentially dangerous. The introduced species may bring with it diseases and parasites that threaten native species. In the absence of its usual predators and competitors, the introduced species may spread rapidly in its new location, displace native species and become regarded as a

pest. Once introduced, the eradication of noxious species is all but impossible and many countries have introduced strict protocols for the import of live species. Such is the concern regarding introduced species that many countries have established centres to conduct research, monitor and publicize the threats of introduced exotic species; these include the National Introduced Marine Pest Information System in the USA, and the CSIRO Centre for Research on Introduced Marine Pests (CRIMP) in Australia.

1.5.5 Climate change – the greenhouse effect and global warming

The greenhouse effect relates to the trapping of the sun's warmth in the lower atmosphere by carbon dioxide, methane and other gases. These greenhouse gases allow the passage of incoming solar radiation but restrict the outgoing infrared radiation reflected from the earth.

The trapping of heat on the planet is a natural effect, allowing the earth's surface to stay at habitable temperatures. Without this effect, life would not exist. The problem is that the activities of humans have increased the levels of carbon dioxide and other greenhouse gases such as methane and nitrous oxide in the atmosphere. Global warming is a term used to denote the accelerated warming of the Earth's surface due to anthropogenic releases of greenhouse gases that are often derived from industrial activity and deforestation.

Levels of carbon dioxide in the atmosphere, from burning forests, grasslands and particularly fossil fuels such as coal and oil, have been increasing steadily. The burning of coal, oil and gas has poured 150 billion tonnes of carbon dioxide into the atmosphere since the industrial revolution. At present about 24 million tonnes are being produced each year and the USA accounts for about 25% of this. Early in 2001, data from satellites were used to confirm that greenhouse gases are increasing and that they are producing long-term changes in the Earth's atmosphere.

Since the industrial revolution, the earth's average temperature has risen by 0.6°C and some of the more extreme predictions are that global temperatures could rise by over 5 degrees during the next century. Various factors may either increase or decrease this figure including changes in vegetation, cloud formation and ocean circulation so that an average temperature change of about 2°C by 2100 might be expected. There are already signs of changes in disease patterns that are believed to be due to global warming. Malaria appears to be moving into areas that were previously too cool for the carrier mosquitoes of the genus *Anopheles*. Tick-borne encephalitis has spread northwards into Sweden, and European summer seasons of food poisoning are extending. Climate warming can increase pathogen development and survival rates, disease transmission, and host susceptibility (Harvell et al., 2002).

It is unlikely that fish stocks will remain unaffected by shifts in climate and this suggests that long-term fisheries management plans need to take into account the possibility of global warming. An ostensibly positive result of higher water temperatures is that warm water species may extend their distribution into waters that were previously too cool. Mangroves, for example, have expanded their range into more temperate habitats (Santilan & Williams, 1999) and presumably this has increased the available habitat areas for many marine species of economic importance. Although the latitude ranges of some organisms are increasing under the influence of global warming, those of other species are reducing. The stock of the North Sea cod, *Gadus morhua*, for example, is believed to be shrinking back from the southern limit of its distribution (O'Brien et al., 2000).

The increasing temperature of tropical seas, about one degree over the last 100 years, has had deleterious effects on those marine species that were already in temperatures close to the upper limit of their tolerance. Many coral polyps, for example, are sensitive to elevated temperatures and warmer water is believed to be the cause of the current extensive bleaching and damage to coral reefs around the world. Reports of mass coral bleaching events have increased since the mid-1970s (e.g. Glynn, 1996) coinciding with an increased rate of global warming (Hoegh-Guldberg, 1999). These large-scale coral bleaching events appear to be primarily due to unusually warm sea surface temperatures during the warm-water season at affected coral reef sites.

From a fisheries viewpoint, more serious effects of rising sea temperatures may include changes in surface currents and therefore upwellings. Many upwellings that are important in enhancing nutrient availability and therefore supporting exploited fish stocks rely on surface currents either diverging or being deflected away from continents (see Section 1.4.3). The failure of these currents would reduce primary productivity in the surface layers of the sea and cause the stocks that rely on this productivity to collapse. A more global effect could involve ice in the North Atlantic melting and the low density water stalling the undersea conveyor belt current that not only affects terrestrial climates but is responsible for moving nutrients from the productive waters of Antarctica into the other oceans (see Box 1.14 'The global conveyor belt').

In a FAO review of the possible effects of climate changes on fisheries, Csirke and Vasconcellos (2005) suggest that management strategies must be developed that allow for the uncertainties associated with temperature change. This may involve adopting a precautionary approach under which the conservation of fish stocks is given priority over the maximization of fisheries yield. The authors identified the lack of long-term data as one of the major constraints in understanding the causes of regional variations in fisheries production and forecasting future regimes.

If the world's temperature is raised by the greenhouse effect, the volume of seawater in the seas would expand and melting ice from the poles would add to the volume. Over the past 40 years there has been a 40% reduction in the thickness of ice in the western Arctic, an area that is warming three times faster than the global average. An average increase in world temperatures by 1.5–4.5°C over the next 50 years could cause a rise in sea levels of 20–140 cm. As a result, many low-lying coastal areas would be inundated and

Box 1.14
The global conveyor belt

Subsurface currents are generated by differences in density which are related to the temperature and salinity of bodies of water. These differences cause masses of water to rise and sink and drive a belt of water that circulates around the globe. As water in the northern Atlantic cools, it becomes denser and sinks to form a south-flowing cold water current. This slow-moving belt of deep cold water joins the deep water encircling Antarctica and twists through the Indian and the Pacific oceans. In tropical waters it gradually warms and rises before circulating back to the North Atlantic. A parcel of water may take over a thousand years to complete this odyssey (Fig. B1.14.1).

Cold water flowing south, and warm water flowing north is an essential part of the global heat exchange between high and low latitudes, and between hemispheres. Via exchange with the atmosphere, the heat from northerly flowing currents, for example, allows northern European countries to have relatively mild climates. This process also moves nutrients from the productive waters of Antarctica into the other oceans, allowing a higher abundance of marine life.

If global warming changes the rate at which glaciers and icebergs in the northern Atlantic melt, the freshwater will decrease the density of surface waters. This low-density water will remain on the surface instead of sinking and driving the conveyor belt system of ocean circulation. The stalling of the system would result in a colder climate and severe winters in Europe and North America, and a corresponding heating in the southern hemisphere. The abundance and diversity of marine life in many areas may be reduced.

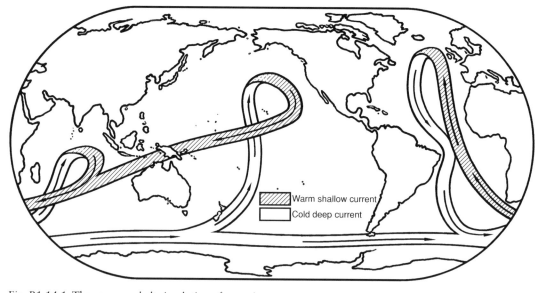

Fig. B1.14.1 The conveyor belt circulation of oceanic water.

the shapes of some continents would change markedly. Many low coral islands and atolls would be covered by the seas.

Although there is general agreement that the presence of greenhouse gases will affect the climate in some way, there is some uncertainty whether this will lead to a general global warming or affect areas of the globe in different ways. This is because heat is distributed around the world in quite a complex way, and the sea is largely responsible for this (Box 1.14). Whether the greenhouse effect will result in a warmer climate with rising seas or a new northern ice age is not known. What is known is that human activities are likely to affect the planet in some undesirable way. However, countries have not been able to agree on how to implement the 1997 Kyoto Protocol, which calls for reductions in greenhouse-gas emissions. Concerns expressed relate to the costs of meeting the lower emission targets; reducing carbon dioxide emissions from coal-burning power plants, for example, would lead to higher electricity prices. On the positive side, concerns over greenhouse gases have resulted in a renewed focus on renewable energy, fuel-efficient transport and economies based on hydrogen rather than on carbon fuels.

1.5.6 Ozone depletion

Ultraviolet radiation from the sun splits oxygen molecules into individual atoms which combine with molecular oxygen (O_2) to form ozone (O_3). Although ozone close to ground level is a major constituent of photochemical smog, its presence in the stratosphere, at altitudes from 20–50 km above the earth, results in the absorption of harmful ultraviolet radiation that would otherwise damage plant and animal life.

In the 1980s, scientists discovered that a hole in the ozone layer formed over the Antarctic every spring. Satellite images were showing that the hole in the ozone layer was growing to eventually cover an area equivalent to the entire North American continent and occasionally wobbling over New Zealand, Australia, southern Chile and Argentina. More recently, ozone depletion has been observed over the Arctic. The danger is that these holes in the ozone layer allow increased levels of ultraviolet radiation to bombard the earth's surface. Besides increasing the incidence of skin cancer in humans, ultraviolet radiation changes some substances found dissolved in seawater into hydrogen peroxide and other compounds hazardous to living tissues. It is also known to inhibit the growth of bacteria (and is often used to sterilize the bottled water now commonly sold around the world) and some types of bacteria are beneficial to life on earth.

From a fisheries production point of view, there is concern that increased levels of radiation due to ozone depletion might affect marine food chains by inhibiting photosynthesis in phytoplankton (see Section 1.6.2).

Ozone is destroyed by reactions involving chemicals such as chlorine and chlorofluorocarbons (CFCs). These chemicals are used in air conditioners, refrigerators, and aerosol cans, and are blamed for the destruction of the protective ozone layer. This destruction is worst at the poles, where cold stratospheric temperatures promote ozone-destroying chemical reactions. Virtually all of the chlorine in the atmosphere is due to human activity and there are now international agreements to discontinue the production of CFCs. Since the late 1990s, the CFCs have been replaced by alternatives such as hydro-fluorocarbons (HFCs) which, as they do not contain chlorine atoms, do not destroy ozone. However, due to the extended life of CFCs in the atmosphere (up to 100 years), ozone in the stratosphere will recover to former levels very slowly. Although chlorine in the stratosphere is decreasing, recovery will be affected by the use of other ozone-depleting chemicals, particularly those containing bromine, used commonly in fire retardants.

1.5.7 Assessing and minimizing environmental impacts

From a global viewpoint, excessive fishing is the major cause of depleted fish stocks. However, many coastal, estuarine and freshwater stocks in particular are being additionally threatened by damage to the marine environment. In addition to fishing, recreation, development and industry all make demands on the marine environment. Coastal marine ecosystems are suffering from habitat destruction, coastal development, pollution and eutrophication in addition to the damage caused by fishing operations.

Depleted fish stocks will not recover under conventional fisheries management measures, such as reducing fishing effort and imposing lower catch quotas, if the major constraints to recovery are damaged ecosystems. This realization has led many fisheries managers to advocate the setting of broader management objectives to protect not only fish stocks but the ecosystems that support them. Fisheries management that has a broader ecosystem focus has been referred to as an ecosystems approach to fisheries (EAF) or ecosystems-based management (EBM) and is discussed in Section 6.5.3, 'Controls to protect ecosystems'. Implementing ecosystem-based management involves identifying environmental threats and providing protection for marine ecosystems.

Many human activities threaten coastal ecosystems and these may be viewed at different ethical levels. At an anthropocentric level, there is concern only with environmental impacts that affect food safety and human well-being. A broader-level view is that the environment should be protected, regardless of its use to humanity, because it has an independent value. Actions taken to mitigate environmental effects, of course, may satisfy aims at both levels. The protection of a mangrove area for its intrinsic value, for example, may also serve to protect fishes that use the mangroves as nursery areas.

Implementing ecosystem-based management involves identifying environmentally damaging actions and mitigating their impacts, which may be assessed either at the individual organism level or at the community level. At the individual organism level, laboratory studies may be used to test the effects or toxicity of pollutants to organisms. At the community level, field surveys may be used to examine the response of marine communities to various environmental threats.

Laboratory studies on a sample of individuals are often used to test the effects of a particular pollutant. As the toxicity of a substance depends on its concentration and individuals of the same species have differing degrees of tolerance to pollutants, the concentration of toxin that kills 50% of the sample within a specified time, often either 48 or 96 hours, is used as a measure. If samples of organism are exposed to various concentrations of the toxin for 96 hours, the concentration that results in the death of 50% of the sample within the specified time is known as the median lethal concentration (96 h LC_{50}). Problems with measuring the toxicity of substances include the possible combined effect of two or more toxins in the environment; one toxin may either increase or decrease the toxic effect of the others. Another problem is that some toxins produce sublethal effects, including reproductive impairment, that are not detected in such toxicity studies.

A more common method of assessing environmental impacts involves conducting surveys in both impacted (experimental) and unimpacted (control) sites, both with ecological characteristics that are otherwise similar. Alternatively, surveys can be conducted at the same site before and after an environmental impact. Such surveys involve determining the relative abundance (say, numbers per square metre) of a range of species before and after the impact. In the case of naturally fluctuating populations, the results from single surveys before and after the impact, (or single surveys in experimental and control sites) can be misleading, and a time series of surveys is necessary. Fig. 1.21 shows the relative abundance of a population over time and the numbered black squares represent the results (the mean number of organisms per square metre) from 8 surveys. Comparing the results from the surveys before and after the impact, correctly suggests that the mean number per square metre has been reduced from 15 to 10 individuals per square metre by the environmental impact. However, if just two surveys had been conducted, one before and one after the impact (surveys 4 and 5), results would indicate (incorrectly) that the average population size had increased.

The precautionary principle suggests that no material should be released into the sea and no development should be undertaken unless it can be shown to cause no harm. However, all proposed developments and actions are likely to affect the environment in some way and it is often difficult to prove that some of the effects will not be negative. A reasonable approach is to require an environmental impact assessment (EIA) for each proposed development and activity that has the potential to affect the environments. An EIA is a formal process for evaluating the likely and

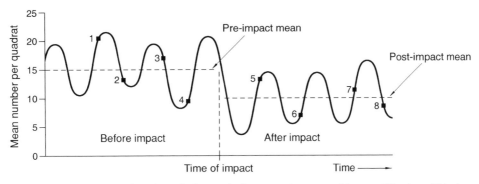

Fig. 1.21 Measuring differences in abundance before and after an environmental impact. Numbered black squares represent the results from 8 surveys to estimate abundance of a population which is fluctuating around a mean of 15 individuals per square metre before the impact and 10 after the impact.

possible risks or impacts on the environment of a proposed activity or development. The requirement for EIAs, first introduced in the United States in 1969, has been adopted by many countries. In most cases, government agencies with responsibility for the environment are the implementing bodies.

In many countries, regulating authorities require EIAs to identify and mitigate any adverse ecological effects of all development proposals. In coastal and marine activities, an EIA process is undertaken for all types of development including the construction of marinas, hotels, harbours, breakwaters, wharves, coastal roads, bridges, and pipelines as well as activities including fishing, aquaculture, boat charters, dredging, sewage release and dumping. Besides assessing the adverse ecological effects, some authorities require an EIA to assess social and economic impacts of the proposed development. The benefits of requiring EIAs is that all development plans have to conform to particular standards, consider less hazardous alternative ways of achieving the same objective, and ensure that the ecological, social and economic interests are addressed. Under EIA requirements, development plans are proactive rather than reactive with respect to any environmental effects. Plans must address the mitigation of negative environmental and social impacts.

Because of the high cost and time involved in carrying out an EIA, a preliminary environmental impact report is sometimes required to determine if a full assessment is necessary. If impacts are considered to be minimal or even enhance existing conditions, a full EIA may not be necessary. If an EIA is required this involves a sequence of steps which culminates in a full EIA report (see Box 1.15 'Environmental impact assessments').

1.6 Photosynthetic marine organisms

Virtually all forms of life gain their nutrients either directly or indirectly from photosynthesis. The exceptions are the few chemosynthetic (or chemoautotrophic) organisms that can gain energy from particular chemical compounds. In abyssal, hot-water vents for example, certain bacteria use sulphur expelled from beneath the earth's crust and in anaerobic sediments on muddy shores chemoautotrophic bacteria use sulphides such as hydrogen sulphide formed beneath the substrate.

Plants are autotrophic organisms, capable of producing organic compounds from inorganic raw materials. Energy from the sun and food material in the sea (mainly nitrates and phosphates) are used by plants to form tissues through photosynthesis. The rate of formation of organic compounds from inorganic material is known as primary productivity. As plants produce organic matter using energy obtained from sunlight and chemosynthetic bacteria produce organic matter

---

Box 1.15
Environmental impact assessments

Requirements differ from country to country, but the actions involved in an EIA are suggested in the following generalized summary of the contents of an EIA report:

- Description of the proposed project, including phases from construction to operation and maintenance of the facility.
- Summary of any existing conventions, regulations, standards and guidelines that are applicable.
- Description of the environment affected (which also acts as a benchmark against which future environmental impacts can be assessed).
- Survey of possible environmental impacts, with an assessment of the risks and effects associated with each and how these may be mitigated.
- Survey of possible socio-economic impacts, with an assessment of the effects on the public, land-use and economic activities.
- Analysis of alternative methods of achieving the desired outcome with each alternative

evaluated in terms of potential environmental impacts and costs.
- Management plan including statements of how the project will be monitored, how any adverse impacts will be mitigated and how actions will be funded.
- The identification of training required by those involved.
- Stakeholder input to the proposal obtained from publicity, advertisements, household surveys and public meetings, with the inclusion of opinions of non-government organizations, the local community and visitors to the area.

After considering the EIA, the regulating authority may advise that the project or activity either 1) proceeds as planned as it represents no threat to the environment, 2) does not proceed as it presents insurmountable threats to the environment, or 3) proceeds with modifications. Alternative 3 means that certain actions or features must be included in the project design to safeguard the environment and to address community concerns. After a project is completed, a post-audit is sometimes required to determine how closely the EIA's predictions matched the actual impacts and to ensure that mitigation measures are in place.

---

from energy obtained from the oxidation of sulphides, both are regarded as primary producers. All other non-autotrophic organisms, including animals, fungi and most bacteria are heterotrophic. These organisms are incapable of directly using inorganic food materials, and must obtain their food from either plants or animals (or their waste products and remains). This section discusses the plants in the sea that are responsible for primary productivity and Section 1.7, 'The flow of energy and material', describes how plant material is incorporated into marine food chains.

Many organisms in the sea can photosynthesize, including flowering plants, larger forms of algae (macroalgae) and, most importantly, a large range of microscopic plants collectively

referred to as phytoplankton. Additional productivity in coastal areas is provided by shoreline plants including cordgrasses and mangroves. Even though plants do not form fossils easily, branching masses of silica from primitive plants have been found in Precambrian rocks over 2000 million years old. The first plants originated in the primeval ocean and the earliest of these are known from the rock structures that they produced. Structures called stromatolites were created by the fossilization of concentric layers of blue-green algae (cyanobacteria) and trapped sediment, and are found exposed in lagoons in Australia as well as in Precambrian rocks elsewhere. Some ancient marine plants became adapted for life in freshwater, and it is from these that terrestrial plants are thought to have eventually evolved. Others remained in the sea and evolved into the variety of marine algae found in the world today and the simplest of these, the floating phytoplankton, are some of the most important plants on the planet.

1.6.1 Marine macroalgae – seaweeds

Seaweeds differ from land plants in that they absorb nutrients directly from the surrounding water. Although they have no need for roots, many species attach themselves to the substrate by means of root-like holdfasts. There are no flowers, and the tips of certain fronds are specialized to produce male and female reproductive cells. As in terrestrial plants, algae photosynthesize by absorbing light energy from the sun in chloroplasts, and these contain pigments that give plants their colour. The green of chlorophyll

is the most familiar, but this may be masked by the presence of additional red or brown pigments. Larger seaweeds are grouped, according to their dominant pigment, into either brown, red or green algae (Box 1.16).

Overall, brown algae (phylum Phaeophyta) are usually the most abundant and conspicuous of the seaweeds. Brown algae contain fucoxanthin, which imparts their characteristic brown colour, and store their carbohydrates, not as starch, but as mannitol and laminarin. In cooler waters, the familiar brown seaweeds include *Fucus* (or related fucoid species) which can be seen attached to rocks in the lower intertidal zone (Fig. 1.22). Below the low tide mark, the larger flat-bladed brown seaweeds are common. These include the leathery oarweed, *Laminaria*, in the northern hemisphere and the strap-weed, *Eklonia*, in the southern hemisphere.

Box 1.16

Light penetration in the sea

Sunlight penetrates the sea, but the colours making up white light are selectively filtered out with depth. The first to disappear are the longer wavelengths of red, orange and yellow – deep-water species are often red in colour and appear black in deep water due to the lack of red light. The

last of the components of white light to disappear are the shorter wavelengths of green, blue and violet. As algae use complementary colours for photosynthesis, green algae are more common in shallow water and red algae are more abundant in shady caves and in deeper water. A coralline red alga has been found off the Bahamas at a depth of 280 m, the deepest of any photosynthesizing plant yet discovered.

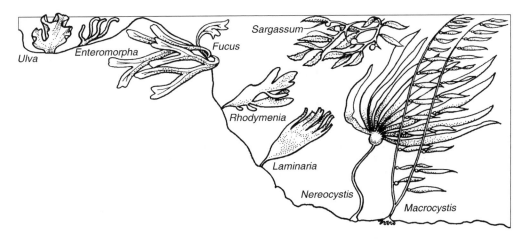

Fig. 1.22 Seaweeds of cooler waters. The drawings are not to scale and the species shown do not necessarily all occur in the same area.

The large brown seaweeds that make up kelp forests grow in temperate and cold waters of coasts around the world. The giant kelp, *Macrocystis pyrifera*, is the most widely distributed of all the kelp species and is indeed huge. These plants may be attached to the sea floor in depths of over 30 m with the fronds reaching up to form extensive canopies on the surface of the sea. The fronds can grow at the remarkable rate of over 30 cm per day making it one of the fastest growing of all plants. Because of their remarkable growth, some giant kelps are harvested commercially, and can be heavily cropped (see Box 1.17 'Human use of marine algae').

In tropical areas, there are fewer and less striking species of marine algae than in cooler waters, and some are shown in Fig. 1.23. A few smaller brown seaweeds attach to coral rubble, including the funnel weed, *Padina*, and the spiny top, *Turbinaria*. Some of the largest of the warm-water brown seaweeds belong to the genus *Sargassum* and these have evolved float bladders that, in the case of attached species, keep the plant off the sea floor. *Sargassum* can also be found free-living, using its bladders to float and drift on the surface of the sea. Great rafts of *Sargassum* float in the Sargasso Sea, the calm central gyre formed by circulating currents including the Gulf Stream (see Fig. 1.18 in Section 1.4.3). These floating rafts support an ecosystem of distinctive animals that use the seaweed for food and shelter.

Although there are more species of red algae (Rhodophyta) than all the browns and greens put

Box 1.17
Human use of algae

Several algal species are used by humans as foods and food additives. The main substances extracted from brown algae are the jelly-like alginates that are derived from the cell walls. Alginates are colourless, tasteless, and odourless and used to make non-fat aspic for canned meats, thickeners for ice cream and cosmetics, and emulsifiers in medicines and paints. Mannitol is also extracted from giant kelps for use as an artificial sweetener.

Species of red algae used as human food include *Porphyra*, which is farmed for its sheet-like fronds called nori in Japan and used wrapped around rice in sushi. Other species of *Porphyra* have been used as food – in Britain, for example, the seaweed is called laver, and Welsh laver bread is made from the seaweed's fronds that have been boiled, dipped in oatmeal, and fried. Irish moss or carrageen, *Chondrus crispus*, which grows on rocky coasts on both sides of the Atlantic in the northern hemisphere, is used to extract the phycocolloid carrageenen used as a thickening or emulsifying agent in food products. In tropical regions, including Southeast Asia and some Pacific islands, the red seaweed *Eucheuma* (Fig. 1.23) is farmed for the same purpose and is grown on floating ropes before being harvested, dried and processed; production in some countries is considerable with the Philippines producing between 90 000 and 115 000 tonnes per year (Garcia, 2005). The green alga known as sea grapes (*Caulerpa*) are collected from fringing coral reefs in many tropical countries. In Fiji, women who collect the crisp and tangy sea grapes place small pieces of the seaweed in new reef crevices to be assured of future crops.

Seaweeds contain many minerals, including traces of elements that are required by humans in very small quantities. Iodine is one of these, and the lack of it causes the condition known as goitre. The thyroid gland, a large ductless gland in the neck of all vertebrates, secretes hormones that regulate the rate of metabolism and hence growth and development. In goitre, there is a thyroid hormone deficiency, and the typical swelling of the neck results from a futile attempt by the gland to produce more hormone by enlarging itself. It appears that iodine is necessary for the thyroid gland to produce the hormone. Goitre is almost unknown in Japan where seaweeds are regularly eaten but is common in many countries where seaweeds are not part of the diet.

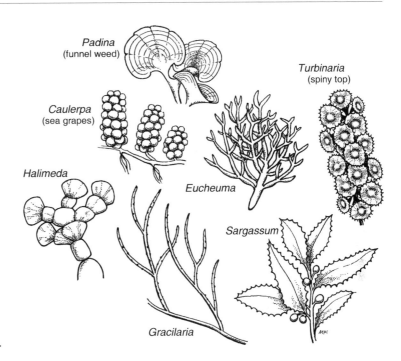

Fig. 1.23 Seaweeds of tropical waters. The drawings are not to scale and the species shown do not necessarily all occur in the same area.

together, many are inconspicuous or grow in deeper water. Red seaweeds are either filament-ous or flattened and may appear either red, vio-let or brown even though they all contain the pigments phycoerythrin and phycocyanin. Some red algae have a remarkable capacity to extract calcium from seawater. The so-called coralline red algae become stiff and even stony with the calcium in their tissues, and some, hardly recog-nizable as plants, grow as pink incrustations across the surface of rocks. The more plant-like species include the small branching *Corallina*.

Green seaweeds (Chlorophyta) generally pre-fer shallow sunlit water. Small ribbon-like species such as *Enteromorpha* are found high in the intertidal zone on reef flats and in tide pools. The bright green, sheet-like sea lettuce or green laver, *Ulva lactuca*, is found in cool and temperate waters around the world – its translucent and sheet-like fronds are only two cells thick. On tropical reefs, bunches of green sea grapes, *Caulerpa*, can often be found growing in pools and channels among the coral. Dense bunches of the small grapes grow on stalks attached to a hor-izontal stem that winds along crevices in rocks

(Fig. 1.23). Several species of tropical green sea-weeds extract calcium from seawater including the common coral reef species *Halimeda*, which has branched chains of flat segments that are often calcified and therefore gray in appearance. As they grow and die, these calcified and stony species contribute considerably to the mass of coral reefs.

1.6.2 Microalgae – phytoplankton

Most primary productivity in the sea is due, not to large plants such as seaweeds and seagrasses, but to the numerous microscopic plants that are part of the plankton. The word 'plankton', like 'planet', comes from the Greek root, meaning wanderer. And, just as the planets in their orbits around the stars appear to wander across the sky, the plankton drift across the waters of the world. The term plankton is used to refer collectively to the huge variety of microscopic algae and (mostly) small animals that are unable to swim to any great extent.

The plankton are separated into phytoplank-ton, which can photosynthesize and are therefore

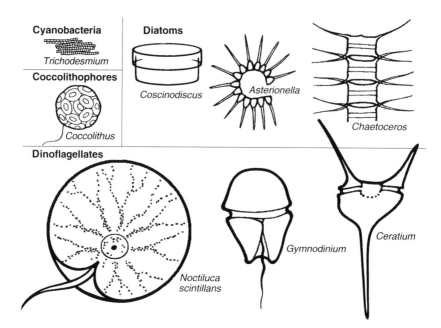

Fig. 1.24 Phytoplankton including cyanobacteria, coccolithophores, diatoms and dinoflagellates. Not to scale: sizes range from about 10 microns in coccolithophores to over 1000 microns (1 mm) in *Noctiluca*.

regarded as plants, and zooplankton that cannot photosynthesize and are regarded as animals. However, the division between the two is often blurred as some small phytoplankton can actively capture food as well as photosynthesize. The phytoplankton are reviewed in this section and zooplankton are described in Section 1.7.1.

Phytoplankton are generally less than a quarter of a millimetre in length, with a few reaching about 4 mm. But in spite of their small size, the total living weight, or biomass, of these small plants in the seas is perhaps five times larger than that of all vegetation on land. The different classes of phytoplankton (Fig. 1.24) include photosynthetic bacteria, diatoms, coccolithophores, and dinoflagellates.

Photosynthetic marine bacteria, including the cyanobacteria (formerly called blue-green algae) and prochlorophytes, are increasingly being found to be important primary producers in oceanic food webs. In warmer waters, cyanobacteria occasionally occur in such abundance that they form dense mats of filaments that colour the sea (see Box 1.19 'Red tides'). Coccoliths make up another part of the phytoplankton and some

of them form beautiful and complex capsules of calcium carbonate that resemble cut glass (Fig. 1.24). The remains of these microscopic plants form limestone deposits which are sometimes immense, such as the picturesque white cliffs of Dover in Britain.

All diatoms are similar in that the walls, or tests, of their single cells are made of silica, and three different types are shown in Fig. 1.24. Sea floor sediments formed by a rain of the remains of these small but abundant plants are often rich in silicates. Indeed, compounds of silicon, including silica (silicon dioxide) occur as sand, quartz, and feldspar, and are so common that, second to oxygen, silicon is the most abundant element on earth. Because of the hardness and durability of their cell walls, diatoms have formed extensive fossil deposits – the fact that some of these deposits are over 900 m thick provides an idea of their abundance. Diatomaceous earth (or kieselguhr) mined from these fossil deposits is still used as a very fine grinding powder and as a mild abrasive in toothpaste.

Dinoflagellates, mostly less than 1 mm in size, have tail-like flagella that allow them to move vertically up and down in the water column. Many dinoflagellates are bioluminescent, and *Noctiluca scintillans* is one of the main phytoplanktonic species responsible for bioluminescence in temperate waters. With a gelatinous sphere over 1 mm in diameter, *Noctiluca* is just visible to the naked eye and is often responsible for the wake of bioluminescence trailing behind vessels on night passages (see Box 1.18 'Cold light – bioluminescence'). Although most dinoflagellates can photosynthesize, many can ingest food, and sometimes the same species can feed either as an animal or as a plant. Several species of dinoflagellates are responsible for the harmful algal blooms discussed at the end of this section.

Because of the huge biomass of the phytoplankton, it has been suggested that the planet could survive even if all terrestrial vegetation was removed, as long as the microalgae remained intact. This most unattractive scenario underscores the importance of the oceans in ensuring the health of our planet. Of the plants and animals that are under threat from human activities, one could be led to believe that at least the phytoplankton, out in open ocean and away from human influence, are safe. But this may not be so.

Box 1.18

Cold light – bioluminescence

The term phosphorescence has been commonly used for the light produced by living creatures, and this is probably a hangover from the days when people were familiar with the phosphorus used in matches. Phosphorus produces light and heat, and when subjected to friction, bursts into flame. But the cold light produced by luminescent creatures is produced by different chemical reactions altogether. Several substances, collectively called luciferins, are oxidized by the enzyme luciferase and this reaction produces light. The correct term for the production of light by living organisms, therefore, is bioluminescence. Bioluminescence should not be confused with fluorescence, which is the emission of light by some organisms (such as many corals) when subjected to ultraviolet light.

Bioluminescence is rare in freshwater, and on land is most common in a few families of insects. But in the sea, a range of bacteria, plankton, ctenophores, annelid worms, molluscs and fishes are bioluminescent. Most bioluminescent organisms produce light in short pulses, but some glow for longer periods. Some creatures emit clouds of luminescent material. Unlike bioluminescent terrestrial organisms, which emit a variety of colours, most marine organisms emit light at the blue and green end of the spectrum, the wavelengths that have the greatest penetrating power in water. Some fish, however, emit red and infrared light. Bioluminescence appears to have evolved many times in the sea because of the large number of distantly-related taxonomic groups with bioluminescent species and the several distinct chemical mechanisms of light production involved.

Bioluminescence may be a means of attracting prey or it may be just to illuminate the surrounding water to find food. The dorsal fin of the deep-sea anglerfish has a rod-like extension ending in a light-emitting lure that dangles enticingly in front of its mouth. In schooling species, bioluminescence may be a means of the school keeping together. In some species, such as the small crustaceans called ostracods, light may be used to attract or recognize mates. It is possible that light is used by some species to frighten predators. Some squids and small crustaceans emit a cloud of luminescence to confuse or repel a potential predator. Even in small bioluminescent creatures, a combined burst of light from several individuals close together may create the illusion of a much larger organism, one too large for a would-be predator to tackle. Still another possibility is that the production of light actually provides camouflage. This works in a similar way to the counter-shading of fish described in Chapter 2. A small animal close to the surface can be seen from below as a silhouette against the sky. If the animal produces light to match this source, an effect called counter-illumination, it becomes less visible to a predator swimming below.

The hole in the ozone layer (discussed in Section 1.5.6) is allowing increased levels of ultraviolet radiation to bombard the earth's surface. Estimating the extent of the effects of increased ultraviolet levels is difficult. The plants and animals that it is thought or known to affect fluctuate naturally in response to physical factors such as ocean currents, temperatures and the extent of annual sea ice. Directly under the southern hole in the ozone layer, the phytoplankton provide food for zooplankton such as krill that sustain many of the Antarctic's fish and baleen whales. The Southern Ocean is one of the world's most productive marine ecosystems, home to huge numbers of penguins, seals, and bottom plants, and a major supplier of nutrients carried to other parts of the world by undersea currents. Thus the ozone hole problem becomes a global one as changes in the Southern Ocean may affect the entire planet.

Oxygen is depleted by the metabolism of animals and is replenished by the activity of plants. This is basic biology. As the drifting phytoplankton are the most abundant of all the earth's plants, anything that is likely to affect their well-being is indeed cause for concern. It is inarguable that all life in the sea is ultimately dependent on marine plants, and phytoplankton in particular. It may not be stretching things too far to suggest that all life on earth is dependent on those tiny plants floating in the surface of the seas.

1.6.3 Harmful algal blooms

Populations of phytoplankton periodically go through massive increases in abundance, referred to as plankton blooms, and a few species, dinoflagellates in particular, produce potent toxins. These blooms of toxic species (called harmful algal blooms or HABs) are responsible for fish and shellfish poisoning in humans in many parts of the world.

Ciguatera fish poisoning (CFP) is responsible for major public health problems in inshore tropical regions. CFP results from the consumption of fishes that have accumulated ciguatoxin and maitotoxin, which are produced by several

organisms including the benthic dinoflagellate, *Gambierdiscus toxicus*. This dinoflagellate occurs as a film on corals and seagrasses and contains the precursor to ciguatoxin, to which is it is converted in the livers of herbivorous fish. Usually the abundance of *Gambierdiscus* is low but, like other phytoplankton species, their numbers increase dramatically when there are high levels of nutrients available. This happens during the wet season when nutrients are washed from the land by rain and during cyclones when nutrients are released from damaged shorelines and coral reefs. In many cases, outbreaks of ciguatera have been associated with human activities such as harbour dredging and the illegal use of explosives for fishing.

By magnification up the food chains, ciguatoxin reaches dangerous levels in top carnivores, such as some emperors, red snappers, barracudas, moray eels and large mackerels (see Box 1.21 'Biological accumulation and magnification' in Section 1.7.3). People eating these usually edible fishes become affected by ciguatera, and suffer from tingling, numbness, muscle pains, and a curious reversal of temperature sensations (cold objects feel hot to touch). In extreme cases, death occurs through respiratory failure. Fig. 1.25 shows a cartoon designed to raise community awareness of ciguatera in Pacific island countries.

A poorly documented but possibly large number of people in tropical areas die through the condition. Just the probability of fish being ciguatoxic is sufficient to cause severe marketing problems in many tropical coastal fisheries. After an outbreak of ciguatera, there is a reluctance to buy even fishes that never get affected by ciguatera, such as tuna and other oceanic fish that live in areas that are sufficiently distant from coral reef ecosystems where *Gambierdiscus* grows. Unfortunately, and in spite of widespread folklore on the subject, there is no reliable, cheap test to determine whether or not a particular fish is ciguatoxic before consumption. One common belief is that toxic fish can be recognized by exposing a fillet of the fish to flies – the flesh is regarded as poisonous if the flies avoid it. Another belief is that a toxic fish can be

Fig. 1.25 A cartoon used to raise community awareness of ciguatera in Pacific island countries. The sequence of events shown are: (a) the dinoflagellate, *Gambierdiscus*, is associated with seagrass beds; (b) increased nutrients (e.g. released from damaged corals) allow *Gambierdiscus* to increase in abundance and toxins are concentrated in small grazing fish; (c) carnivorous fish feeding on the small fish further concentrate ciguatoxin in their own flesh; (d) people eating affected fish suffer from ciguatera.

recognized by placing a silver coin on the flesh – if the coin turns black, the flesh is not safe to eat. Unfortunately these tests, and many other widely trusted ones, do not work.

Shellfish poisoning is classified by the symptoms caused as paralytic shellfish poisoning, diarrhetic shellfish poisoning, amnesic shellfish poisoning and neurotoxic shellfish poisoning. Each of these syndromes is caused by different species of toxic algae that occur in coastal waters.

Paralytic shellfish poisoning (PSP) results from people consuming shellfish affected by toxins produced by the dinoflagellates *Gymnodinium catenatum*, *Pyrodinium bahamense* and several species belonging to the genus *Alexandrium*. The toxin produced, saxitoxin, causes neurological symptoms that include tingling, numbness and nausea. Severe cases result in respiratory arrest within 24 hours of consumption of the toxic shellfish. There is no antidote and effects in non-lethal cases persist for a few days.

Diarrhetic shellfish poisoning (DSP) results from people consuming shellfish (often bivalves such as mussels) containing a toxin produced by dinoflagellates belonging to the genera *Prorocentrum* and *Dinophysis*. The toxin, okadaic acid, inhibits enzymes and produces gastro-intestinal symptoms soon after consumption. Poisoning is characterized by incapacitating diarrhoea as well as nausea, vomiting, abdominal cramps, and chills. Although severe diarrhoea may lead to death from dehydration, recovery usually occurs within three days.

Amnesic shellfish poisoning (ASP) results from people eating shellfish containing a toxin, domoic acid, derived from some *Pseudo-nitzschia* diatoms. Domoic acid competes with glutamic acid and binds to chemical receptors in brain cells and causes their dysfunction. Symptoms begin with nausea, vomiting, and abdominal cramps. Neurological symptoms may include seizures, disorientation and respiratory difficulty. In severe cases, brain damage may result in coma and death.

Neurotoxic shellfish poisoning (NSP) results from consumption of shellfish contaminated with brevetoxin produced by the dinoflagellates *Karenia brevis* and *Gymnodinium breve*. Brevetoxin affects the human nervous system and symptoms are similar to (but not as life-threatening and long-lasting as) those produced by ciguatera. Effects include tingling, dizziness, fever, chills, muscle pains, nausea, diarrhoea, vomiting, headache, reduced heart rate and pupil dilation. Interestingly, the toxins can become airborne (as toxic aerosols) because of wave action and cause people swimming and walking on the shoreline to suffer respiratory asthma-like symptoms from inhaling the airborne droplets.

The reasons that dinoflagellates produce toxins are unclear. The toxins may serve to discourage predators but the possibility exists that the toxins are essential in dinoflagellate metabolism and are only coincidentally harmful to other species.

In the United States about half of all seafood intoxications results from ciguatera and incidents of shellfish poisoning are increasing. Such is the concern over shellfish poisoning and its increasing occurrence that many countries conduct monitoring of shellfish toxicity. Canada, for example, has been carrying out extensive laboratory monitoring since 1943 and closes beaches and estuaries to shellfish harvesting when high concentrations of either the causative organism or the toxins in shellfish are detected. Algal blooms can also cause other problems besides toxicity. Large concentrations of phytoplankton can reduce the amount of dissolved oxygen in the water or simply clog the gills of marine species. Some dense blooms of non-toxic species such as the dinoflagellate, *Noctiluca scintillans*, may also threaten fish by grazing on larval fish and zooplankton (Hallengraeff et al., 2005). Algal blooms are responsible for large economic losses

in aquaculture operations, causing asphyxiation in fish kept in coastal pens and losses in farmed shellfish.

1.7 The flow of energy and material

A brief examination of the production and structure of life in the sea is a necessary precursor to an understanding of the distribution and characteristics of fisheries resources. The details in this section are intended as an introduction and more details can be gained from review papers (e.g. Polunin & Pinnegar, 2002).

Within and between the various marine ecosystems described previously, there is a flow of matter and energy from one living organism to another. All animals are consumers of some organisms, and are prey to others. As in terrestrial ecosystems, nearly all life in the sea is dependent on sunlight and the plants described in the previous section. Plant material is eaten by herbivorous animals which themselves are eaten by other, usually larger, animals – this flow of

---

Box 1.19
Red tides

In certain conditions, phytoplankton species increase in numbers, or bloom, to such an extent that their density colours the sea. The sea may be coloured pink, violet, orange, yellow, blue, green, brown, or red depending on the pigment used to trap sunlight by the various species. As red is the most common pigment, the phenomenon has come to be called a red tide, even though it has nothing to do with tides. Most species causing discolouration of the sea are not harmful and those species that are harmful may never reach the densities required to discolour the water. However, as with other phytoplankton blooms, the decomposition of the plants can result in oxygen depletion.

Red tides appear to be a natural phenomenon occurring in response to seasonal increases in nutrient levels in the sea. Coastal nutrients,

particularly phosphates, are known to stimulate or prolong blooms. It is believed that the Red Sea was so-named because of frequent dense blooms of a red cyanobacterium, *Trichodesmium*, and red tides due to a related species were observed in Australia as early as 1770 during Captain Cook's voyage through the Coral Sea. However, the incidence of red tides may be increasing for two reasons. First there are increasing nutrient levels in many coastal areas related to fertilizer use and sewage treatment, and of course these are related to increasing human populations. Second, plankton may be introduced to distant ports after being carried in the ballast water of freighters: when the freighters discharge their ballast water prior to loading cargo, plankton may become established in an area outside their natural distribution. The dinoflagellate *Gymnodium catenatum*, which was unknown in the Australian region until the late 1980s, is believed to have been introduced in the ballast water of ships from Korea and Japan.

material from plants to herbivores to carnivores is discussed in the following subsection on food webs.

Larger plants in the sea are consumed by a range of herbivores from molluscs to fishes but, because of the huge biomass of phytoplankton, the consumers that can use this resource are often the most important in marine food chains. Several larger animals have evolved to take advantage of drifting phytoplankton. The most notable of these are the bivalve molluscs, the cockles and clams, which pump water over gills that filter out the phytoplankton. But the most important consumers of phytoplankton are the small animals, collectively called zooplankton.

1.7.1 Zooplankton

Many different types of animals are represented in the zooplankton and it is convenient to divide them into two large groups – the meroplankton and the holoplankton. The meroplankton are species that are planktonic for only part of their lives and the holoplankton are species that remain drifting in the plankton for their entire lives.

The meroplankton includes the larvae of many marine animals (Fig. 1.26). Oysters, mussels, crabs, lobsters, coral and most fishes produce larvae that drift with the currents for periods varying between a few days and several months. At the end of the planktonic phase the larvae either settle on the sea floor or swim as adults, often to places some distance from the spawning grounds. The larval phase thus allows for the colonization of new areas, and prevents competition between parents and offspring for resources such as space and food.

The holoplankton (Fig. 1.27) includes the protozoa known as foraminiferans, or forams, and radiolarians that are about the same size as the largest of the phytoplankton. Since improved sampling methods have allowed organisms less than 1 micron (one thousandth of a millimetre) in diameter to be caught in sampling gear, the abundance of heterotrophic marine bacteria (about 0.4 microns in diameter) has been appreciated and their importance in recycling carbon in food chains is a relatively recent discovery. Foraminiferans consume bacteria and diatoms and move with amoeba-like appendages that emerge through their perforated shells of calcium carbonate. As they die, their shells or tests collect on the sea floor and become incorporated into rock as limestone – about one-third of all sea floor sediment is composed of their tests. Indeed, the pyramids in Egypt are faced with foraminiferan limestone. Forams tests contribute to marine sediments (as a major constituent of globerina ooze) and the beaches in some parts of the world.

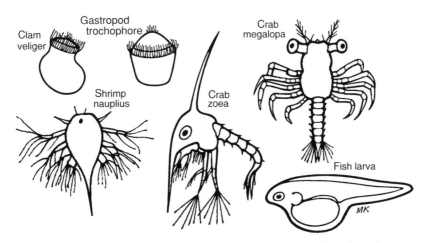

Fig. 1.26 Zooplankton. The meroplankton include species that are planktonic only in the early stages of their lifespan: the larval stages of molluscs, crustaceans and fishes. Sizes range from about 0.15 mm in the trochophore to over 4 mm in the fish larva.

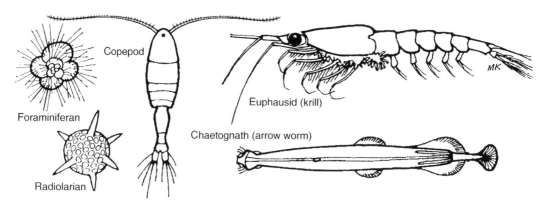

Fig. 1.27 Zooplankton. The holoplankton include species that remain planktonic for their entire lives. Sizes range from about 1 mm in the foraminiferan to over 40 mm in the euphausid.

Radiolarians have silicon-containing tests with radiating thread-like appendages and, unlike foraminiferans, are found in the upper sunlit layers of the oceans.

Larger species in the zooplankton include the arrow worms, or chaetognaths – free-swimming predators shaped like finned darts with a length of about 10 mm. Not only are these animals quite common in the plankton, but species of arrow worms in a particular area are often specific to that region; thus arrow worms are often used as indicator organisms, and the movement of bodies of water can be traced by the species of arrow worms contained therein. Crustaceans, which are common in the zooplankton, include marine copepods that are virtually the sole food of herrings in the North Atlantic, and the small shrimp-like animals called krill are present in such large numbers that they provide food for large populations of birds, fish and the large blue and finback whales.

Highly-modified planktonic gastropods (sea snails) include the small pteropods and heteropods, as well as some odd larger species. One of the oddities is the violet sea snail, *Janthina janthina*, that secretes a foamy mucus that traps bubbles of air before hardening, and floats on the sea's surface on a raft of bubbles (Fig. 1.28).

The giants of the plankton are the animals commonly called jellyfish. The true jellyfish (the Scyphozoa) are free-swimming individual animals with a bell or umbrella beneath which is a set of tentacles around the bell margin and a separate set around the mouth. The largest of all jellyfish belong to the genus *Cyanea* and some of these have bells that are huge masses of jelly measuring up to 4 m across with tentacles more than 30 m long. The common moon jellyfish, one of the world's most widespread animals, has a transparent bell of less than 20 cm in diameter, and is recognizable by the four horseshoe-shaped reproductive organs visible through the top of the bell. Jellyfish are capable of limited movements – perhaps insufficient to move against currents but certainly sufficient to move up and down in the water body and perhaps across currents. In fact, as the density of the jelly is slightly greater than water, the jellyfish would sink if it was not for this ability. Movement is achieved by successively relaxing then contracting the bell which forces water down and the jellyfish up; if the body is orientated at an angle, the jellyfish will move sideways as well. This sequence is shown in the two drawings of the compass jelly, so-called because the bell has markings resembling the points of a compass, in Fig. 1.28.

A second group of planktonic animals, also commonly called jellyfish, is the siphonophores, and the animals in this group are floating colonies of polyps. Perhaps the best known of the colonial jellyfish is the Portuguese man-of-war, which is found in all warm seas and is occasionally blown into cooler waters. These striking creatures are driven by the wind with their blue gas-filled floats

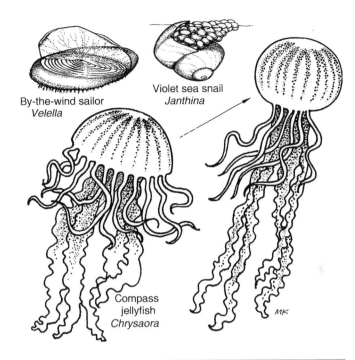

Fig. 1.28 Larger representatives of the meroplankton. From top left and clockwise, the by-the-wind sailor, *Velella*, the violet sea snail, *Janthina*, hanging beneath its raft of bubbles and the compass jellyfish, *Chrysaora*. The two drawings of the jellyfish show it 'swimming' by successively relaxing and contracting its bell.

Box 1.20
Human use of plankton

Several larger species of plankton are used as human food. These range from krill (euphausids) that are commercially fished in Antarctic waters and some small jellyfishes that are dried and used as a flavouring in soup. Plankton have also been accredited with the ability to directly support human life. In 1888, Prince Albert I of Monaco, the founder of the famous *Musée Océanographique de Monaco*, suggested that shipwrecked survivors in a lifeboat could escape death by starvation if they had several items of equipment with them including 'one of several nets of straining cloth . . . on about sixty feet of line in order to pick up the smaller sea fauna'.

A French doctor, Alain Bombard was concerned that people in a lifeboat generally died within a very short time – sometimes within a day or two of a

shipwreck. So he crossed the Atlantic in a rubber raft to show that that people died, not from thirst or hunger, but by giving up hope. In his voyage across the Atlantic, Bombard survived by catching fish to squeeze for drinking water and by catching plankton in a towed net as a source of food and vitamins. Humans and whales are two of those unfortunate animals that have lost the ability to manufacture their own ascorbic acid (vitamin C) and must take it in regularly as food. Vitamin C deficiency, or scurvy, was the scourge of ancient sailors, who wandered about the world dying of anaemia and internal bleeding until the British Navy enforced the consumption of lime juice. Of the four vitamins A, B, C and D that our bodies cannot do without, even for a short time, Bombard obtained vitamins A and D from fish livers, some parts of the vitamin B complex from fish flesh and vitamin C from plankton.

---

acting as sails. The man-of-war's tentacles may be over 18 m long and bear nematocysts that are capable of paralysing quite a large fish. But in spite of their powerful and painful stings there

appears to be no record of any deaths in human victims. A smaller colonial jellyfish is the by-the-wind sailor, *Velella*, which has a transparent sail mounted on top of a blue oval disc that lies flat on the surface of the water (Fig. 1.28). The vertical sail appears to be set at angle to the fore and aft axis of the oval disc, which is generally less than 6 cm in length. According to popular belief the angle of the sail is supposed to allow the by-the-wind sailor to tack to windward like a sailing boat.

1.7.2 Daily migrations and the seasonal distribution of zooplankton

In the open sea, there is an array of zooplankton, both permanent and temporary, feeding on a huge living mass of phytoplankton. But zooplankton feeding is not continuous. Zooplankton shun the light and retreat to deeper waters during the day and rise to graze on the phytoplankton only during the night (Fig. 1.29). The density of zooplankton in deeper water is such that it often forms a layer (the so-called deep scattering layer) that gives the impression of a false sea floor on echo sounders and sonar.

The purpose of this daylight migration into the depths may be either to avoid predation by surface-feeding carnivores including seabirds, to avoid damage from ultraviolet radiation or to allow plankton to take advantage of different subsurface currents. As deeper currents are slower and flow at an angle to surface currents, this last possibility results in zooplankton being in a different location when they rise to the

surface on the following day, and thus being in new pastures of phytoplankton. Whatever the purpose, the vertical migration of zooplankton allows phytoplankton to regenerate, under reduced predator pressure, during the day before being exposed to each nocturnal onslaught.

The abundance of phytoplankton depends on the presence of plant nutrients, mainly nitrates and phosphates, as well as sunlight for photosynthesis. In surface waters of the sea, nutrients are generally used up quite quickly by the phytoplankton and, in the absence of mechanisms to replace these nutrients, many waters of the world are barren places. Light and nutrients are limiting factors, and the absence of either one or both will affect the abundance of phytoplankton and therefore their predators, the zooplankton.

In tropical regions, sunlight is not limited but surface nutrients are; these are locked below a permanent thermocline (see Fig. 1.16 in Section 1.4.3). In most tropical areas, phytoplankton abundance is low and variable over the year. In polar regions, on the other hand, light is more of a limiting factor than nutrients, and there is a dramatic increase in phytoplankton abundance in summer, when light is sufficient for a net increase in primary productivity.

In temperate regions, both light and nutrients may be limiting factors. During the winter, the effect of rough seas on the water column provides mixing and brings nutrients up from deeper water. However, short days and low light may be insufficient to allow high primary production. In spring, as the days get longer and seas get warmer, there is a large increase in the abundance of phytoplankton taking advantage of the available nutrients. As the nutrients are used up, phytoplankton abundance drops to low levels during the calmer summer months. A second lower peak in abundance may occur in autumn when there is still sufficient light and nutrients again become available through rough seas and water mixing (Fig. 1.30). Zooplankton populations, with some delay, follow the phytoplankton in abundance over the year. Many temperate water marine animals that produce plankton-eating (planktotrophic) larvae spawn in the spring or autumn

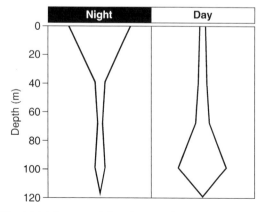

Fig. 1.29 The vertical distribution of copepods over a 24 hour period.

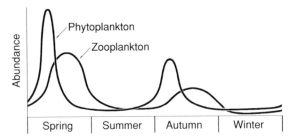

Fig. 1.30 The seasonal abundance of phytoplankton and zooplankton in temperate waters.

(or sometimes both) when larvae can take advantage of the seasonal abundance of plankton.

The generalized seasonal patterns in plankton abundance are modified by mechanisms that affect nutrient levels in surface waters. These include eddies created by land masses and seamounts, gyres created by major currents and upwellings of nutrient-rich water resulting from several factors.

1.7.3 Food relationships, trophic levels and food webs

Previous sections of this chapter have described the plants responsible for primary productivity. The autotrophic organisms include flowering plants, algae and phytoplankton. All other marine organisms, including bacteria, fungi, herbivores and carnivores, are heterotrophic and must obtain their food from plants or other animals and their remains.

Many smaller animals, from sea snails to fishes feed on the larger algae, but the most important herbivores in the sea are the ones that can take advantage of the phytoplankton. In the marine environment, the niche of the large terrestrial herbivores, such as rabbits and antelopes, is taken over by large numbers of usually quite small herbivores, the zooplankton. It is the zooplankton grazing on the phytoplankton that provides the crucial link between plants and carnivores in the sea. A few larger marine invertebrates can feed directly on the phytoplankton. The best known of these are the bivalve molluscs, including mussels, oysters, cockles, and clams, that pump

Species at lower trophic levels, such as molluscs, prawns and small crabs, are often abundant in marine ecosystems and are here being sold by a fisher's family in Malaysia.

seawater over gills that trap phytoplankton. This feeding on material suspended in the water, suspension feeding, is not limited to herbivores.

Many animals, from barnacles and corals to whales, have specialized structures to filter zooplankton from the sea. Barnacles use feathery appendages to filter material from the water and coral polyps trap plankton in sheets of mucus or with their tentacles. Baleen whales have fringed plates, the baleen, that forms a plankton sieve. The possession of specialized filtering apparatus accounts for the large numbers of sessile species, ones that have to wait for food particles and plankton to be transported to them by water currents.

Detritus is digested by bacteria and other micro-organisms and some larger species feed directly on deposits on the sea floor. Deposit feeders include polychaete worms that burrow in and feed on the surrounding detritus and sea

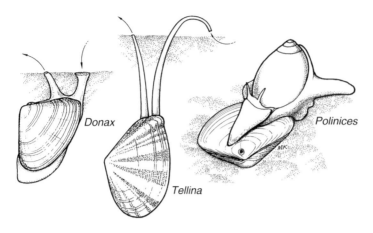

Fig. 1.31 Feeding strategies in molluscs. From the left, a suspension feeder (the surf clam, *Donax*), a deposit feeder (the tellinid clam, *Tellina*) and a carnivore (the moon snail, *Polinices*). Water flow is shown by arrows. The moon snail uses its radula teeth and acid secretions to drill the bevelled hole shown in the shell of its prey, a bivalve mollusc.

cucumbers that move slowly across the sea floor vacuuming up material. Suspension and deposit feeders are often the most numerous of all species in shallow-water habitats, and these provide food for vertebrate and invertebrate predators.

Many species groups have evolved an array of feeding methods. In molluscs, for example, many species of bivalves are suspension feeders with siphons through which seawater is pumped to filter out food material. Other bivalve molluscs are deposit feeders, and have an extended inhalant siphon that vacuums up surrounding deposits from the sea floor. A large number of gastropod molluscs (sea snails) are active predators and feed on other invertebrates (Fig. 1.31). The moon snail (*Polinices*, family Naticicae) ploughs through sandy beaches hunting clams on which it uses its radula teeth and acid secretions to drill a bevelled hole to suck out the clam's flesh.

As one would expect in such a diverse group, fishes include suspension feeders, deposit feeders, grazers, predators, and parasites. Filter-feeding fishes range from clupeids (sardines and anchovies) to large plankton-feeding sharks and rays. Typically, these species swim through the water with their large mouths open to strain plankton through specialized sieves such as fine gill rakers. Plankton-feeding fishes are some of the most abundant of all fishes in the sea.

Fish feeding on deposits and small organisms in the deposits often have downward pointing mouths adapted to vacuum food from the sea floor. Many species have other adaptations for deposit feeding. These include goatfish or red mullet (family Mullidae) with barbels to 'feel' through deposits and species such as ponyfish (family Leiognathidae) with extendable mouths. Fish that prey on other fish secure their prey by a range of tactics from camouflage in a stonefish waiting for its prey to get within striking distance, to enticement in an anglerfish using a lure to attract its prey. And many fish of the open sea rely on speed to both capture prey and escape predators.

Feeding relationships are commonly influenced by symbiotic relationships, close associations between two different species, and these are common in the marine environment. Symbiosis is divided into mutualism in which both species gain advantages from the association and commensalism in which one gains advantages while not harming the other. Reef-building corals and giant clams (*Tridacna* and *Hippopus*), for example, have a symbiotic relationship with photosynthetic zooxanthellae (dinoflagellates). In corals, zooxanthellae are associated with the coral polyps and in giant clams they are embedded in the tissues of their brightly coloured mantles. During photosynthesis, zooxanthellae pass some resulting organic material to their hosts.

Most marine organisms have one or more parasites. Parasitism is an association between two species in which one (the parasite) draws nourishment from the other species (the host) to the latter's detriment. Parasites are heterotrophs, usually smaller ones, which feed at the expense of a larger organism. In general, parasites do not kill their hosts (to do so would destroy their food source) but may debilitate them and affect growth and reproduction. Parasitism affects food webs by reducing the matter and energy that the host can pass on to predators. However, the parasites themselves divert material to other organisms that prey on the parasites.

Parasitism is sufficiently common for a number of smaller fishes to have evolved to take advantage of the parasites as a food source. Species of cleaner fishes, including species of gobies and wrasses, clean other fishes of external parasites (ectoparasites) including copepods and isopods. In coral reef ecosystems, cleaner fishes are particularly common and have evolved distinctive colours and displays to advertise their services to other fishes including many large species that would generally be regarded as voracious predators. Species from moray eels to manta rays visit 'cleaning stations' on the reef where they open their mouths and expose their gills for the small fish to remove debris and external parasites. Convergent evolution has resulted in quite different families of fish evolving the form, colour and structures necessary for cleaning larger fish. The neon goby cleaner fish of the Caribbean and the cleaner wrasse of the Indo-Pacific have similar cleaner uniforms – blue with a longitudinal black stripe and a seesaw dancing routine in front of potential clients. In the Indo-Pacific, a particular blenny mimics the colouration of the cleaner fish, and this similarity allows it to approach large fish expecting to be cleaned. However, the sabre-toothed blenny is a parasite, and it rushes in to nip pieces from the skin and fins of the large fish (Fig. 1.32).

Organisms can be thought of as gaining nourishment at different trophic levels and these may be depicted as an energy pyramid (Fig. 1.33). The first or lowest trophic level in the energy pyramid, the primary producer level, consists of marine plant material. The three classes of primary producers include macroalgae, marine flowering plants and phytoplankton. In inshore areas, microscopic photosynthesizing algae and cyanobacteria (microphytobenthos) will contribute

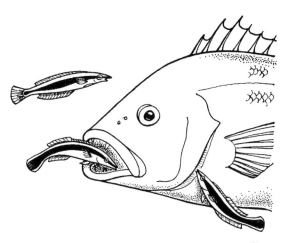

Fig. 1.32 Cleaner wrasses, *Labroides dimidiatus*, cleaning the mouth and gills of a large snapper. Above left is the mimic, a parasitic sabre-toothed blenny, *Aspidonotus taeniatus*. Both fishes are about 8 cm in length.

considerably to food webs. In oxygen-deprived substrates, anaerobic and chemosynthetic bacteria may also contribute to production.

Plant material is fed upon by animals at the next trophic level (the herbivore level) and these herbivores become prey species for carnivores (the carnivore level). As some fish feed on other carnivores, there may be several trophic levels of carnivores.

The species that occupy each trophic level are different in different ecosystems. Some animals feed at different trophic levels either continually or at different stages of their life cycle. Many mullets, for example, are omnivores and will eat detritus, algae and small crustaceans. Some animals with symbiotic relationships with algal cells, such as the reef-building corals discussed previously, can feed either indirectly from sunlight or, perhaps preferentially, on drifting zooplankton. To make matters even more complicated, corals are among the relatively few animals that can absorb dissolved organic matter directly from the water.

In spite of the difficulties of categorizing some species, the concept of trophic levels is useful, particularly in examining the abundance and biomass (the total weight of all the organisms)

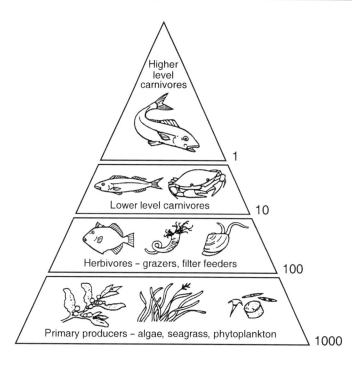

Fig. 1.33 An energy pyramid. Numbers at the right of the pyramid represent the relative biomass at each trophic level assuming an ecological efficiency of 10%.

able to be supported at different levels. At each level, most of the total weight of material (the biomass) or energy is lost. The loss is due partly to the organisms' failure to digest all that is consumed and partly due to the expenditure of energy on respiration, movement and reproduction. As a result, only a small proportion of the food consumed is devoted to somatic growth that may be passed on through the food chain to the next trophic level. There is, therefore, a large decrease in total biomass of organisms at each succeeding trophic level.

The efficiency of this process, termed the ecological efficiency, is defined as the ratio of the weight in one trophic level to that in the one below. This efficiency is often assumed to be about 10%, but may be as low as 5% or as high as 20% depending on the trophic level and the ecosystem under consideration. The biomass values shown to the right of the energy pyramid in Fig. 1.33 arbitrarily assume a 10% level of ecological efficiency – that is, the energy passed from one trophic level to the next. It therefore takes 1000 kg of plant material to produce 1 kg of a higher-level carnivore such as a snapper. As

overall biomass decreases, the concentration of non-excretable substances can increase from lower to higher trophic levels (see Box 1.21 'Biological accumulation and magnification').

Because of this loss of gross productivity at each succeeding trophic level, animals at high trophic levels are unable to maintain very large populations. A top carnivore such as a large tiger shark is, perhaps thankfully, not common at all and most sharks needs to swim over a huge territory in order to have access to all the food that they require. The inefficiency of biomass transfer means that there are seldom more than five trophic levels in an ecosystem. This reduction in biomass at higher trophic levels has implications in species that can support fisheries discussed later in this section.

The connections between organisms through which energy and material move is called a food chain, which in its simplest form can be depicted as plant > herbivore > carnivore. However, connections between organisms and their food are so complex that they form a food web, the depiction of which is always a gross simplification of reality. Nevertheless, all food webs begin with the

Box 1.21
Biological accumulation and magnification

Biological accumulation, or bioaccumulation, refers to the concentration of stable substances found in minute quantities in seawater in the tissues of organisms. In the marine environment, bivalve molluscs and other filter-feeders are extremely effective at accumulating some stable and non-excretable substances. Example of persistent contaminants include many metals (including mercury, lead, cadmium) and chlorinated hydrocarbons (including dichloro-diphenyl-trichloroethane (DDT), chlordane and dieldrin). These non-biodegradable substances are not readily excreted by organisms and hence accumulate in their tissues. Some toxins, or the precursors to them, produced by some species of phytoplankton also accumulate in flesh of fishes and shellfishes and are responsible for fish and shellfish poisoning in human consumers.

In a similar way that biomass decreases up the food chains, relatively stable compounds increase in concentration. This effect, termed biological magnification, results when small a amount of a non-excretable substance accumulates in algae or initial consumers, and is passed on to, and

incorporated into the tissues of, higher-order consumers. As a carnivore may eat many contaminated prey species, the contaminant can reach very high concentrations in top carnivores; in one meal a large carnivorous fish can gain the toxins accumulated by smaller fishes over their entire lifespan.

Pesticides such as DDT have entered the marine environment from runoff from land and dumping. In the environment, DDT is very persistent and it, and its breakdown products, remain in the environment for many years. Even though its production has been discontinued, DDT remains in the environment and has been detected in organisms in places as remote as Antarctica. Chlorinated hydrocarbons are not metabolized, and are therefore concentrated by biological magnification through the food webs. Pathways from phytoplankton, to zooplankton, to planktivores, to fishes and to seabirds have resulted in increased concentrations from 0.1 parts per million (ppm) in phytoplankton, up to 25 ppm in seabirds. The effects on animals at higher levels of the food chain are most serious. The reproduction success of birds, for example, is dramatically reduced by DDT, which interferes with the deposition of calcium in the shells of eggs.

primary producers in the community and end with the decomposers, including fungi and bacteria that break down the wastes, faeces and remains of plants and animals into simple organic substances. These substances include dissolved organic matter (DOM) and particulate matter (detritus) that can be used by primary producers and deposit feeders. The connections between the producers and the decomposers are more variable and include herbivores and often several levels of carnivores. A simplified tropical food web is shown in Fig. 1.34. In this food web there are three pathways to top carnivores involving large plants, phytoplankton and organic material (dissolved organic matter and particulate matter or detritus). These pathways can be depicted as food chains as follows:

1 large plants > herbivores (browsers and grazers) > medium-sized fishes > large piscivores
2 phytoplankton > zooplankton > planktivores (clupeids, corals) > medium-sized fishes > large piscivores
3 detritus > detrital feeders (invertebrates, vertebrates) > medium-sized fishes > large piscivores
Both autotrophic and heterotrophic bacteria are ubiquitous in seawater and play an important part in primary production and detrital pathways respectively. Detrital pathways lead to the recycling of excreted wastes and decaying material in food chains.

The classical food web of the open sea is one that contains three major groups. Large phytoplankton including diatoms and dinoflagellates are grazed on by zooplankton (notably

At the top trophic level of the food chain, large seabirds such as these Patagonian pelicans feed on carnivorous fish.

copepods) which are, in turn, consumed by fishes. More recently, the importance of much smaller organisms, including photosynthetic and heterotrophic bacteria, in food chains has been recognized (Azam et al., 1983; Fenchel, 1988). In a food chain known as the microbial loop, dissolved and particulate organic matter is used by heterotrophic bacteria that are consumed by protozoans (including foraminiferans and radiolarians) that are consumed by metazoans. Dissolved organic carbon that drives the microbial loop may be derived from cell disintegration and leaching from faeces and wastes.

A combination of the classical and microbial food chains is shown in Fig. 1.35. The classical and shorter herbivorous food chain, shown to the right of Fig. 1.35, is likely to prevail in nutrient-rich situations such as those that occur in the spring in temperate waters and in upwellings. The longer microbial food chain, shown to the left of Fig. 1.35, is likely to prevail in nutrient-poor, or oligotrophic, conditions that occur in the summer of temperate regions and year round in many tropical regions of the ocean. Heterotrophic bacteria are now believed to play an important role in the regeneration of nutrients in the photic zone and allow the continuing productivity of phytoplankton. This may be particularly important in tropical waters where strong thermoclines prevent convection currents from bringing nutrients up from deeper water.

Zooplankton such as copepods, which in the classical model graze on diatoms and dinoflagellates, also consume other heterotrophs such as large ciliates. These in turn feed on small flagellates that feed on heterotrophic bacteria. Because the microbial loop (shown on the left of the figure) involves a longer food chain, the food available for planktivorous fish is less than that provided by the classical one. However, trophic levels are often blurred with some species feeding at more than one level.

Fig. 1.34 A simplified, tropical, marine food web. Plants include mangroves (1), algae, seagrasses (2) and phytoplankton (greatly magnified in 3). Corals (4) and giant clams (5) with symbiotic zooxanthellae also use sunlight indirectly. Large plants are eaten by herbivorous animals such as rabbitfishes (6) and sea urchins (7). Phytoplankton are consumed by zooplankton (magnified in 8) and filtered from the water by animals including giant clams. Zooplankton are consumed by carnivores such as corals and small fishes, including sardines (9). Coral grazers, such as parrotfishes (10) feed on coral polyps. Invertebrates and smaller fishes, are preyed upon by medium-sized fishes including carangids and serranids (11) which are preyed upon by top carnivores such as groupers, barracudas and sharks (12). Decomposers, including bacteria and fungi, break down wastes to form dissolved organic carbon as well as detritus (13) that is consumed by a wide range of invertebrate and vertebrate deposit feeders such as the sea cucumber (14) and mullet (15).

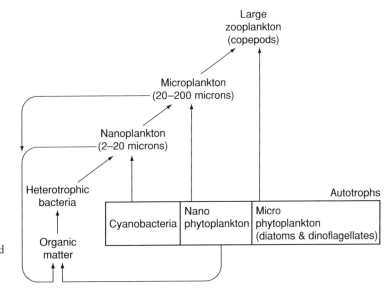

Fig. 1.35 Pathways in oceanic food webs. The classical model from diatoms and dinoflagellates to zooplankton is shown on the right. The microbial loop involving heterotrophic bacteria using dissolved and particulate organic matter is shown on the left.

1.8 Productivity and fisheries

1.8.1 Primary productivity and yield

Almost all marine life relies directly or indirectly on plants and therefore depends on the availability of plant nutrients. The presence of mechanisms to provide plant nutrients explains why upwellings and some inshore areas are some of the most productive of all marine environments. The absence of such mechanisms is the reason why vast regions of the open sea are the equivalent of terrestrial deserts, and extremely poor in marine life.

Terrestrial grasses and plants are eaten by herbivores that are often quite large – animals that include rabbits, sheep, cattle and antelopes – and these herbivores are eaten by terrestrial carnivores including humans. But in the sea, there are very few large herbivores and most, such as the dugong and the manatee, are not common and have a limited distribution. In the marine environment, the role of the large terrestrial herbivores feeding on grasses is taken over by a very large number of small herbivores, the zooplankton, that graze on phytoplankton.

Although most of the terrestrial animal protein consumed by humans is from herbivorous animals such as sheep and cattle, most of the preferred protein from the sea is from carnivorous animals such as mackerels, snappers, tunas and even sharks. In terms of food chains and ecology, eating such carnivores is the marine equivalent of eating dogs and lions! As discussed in Section 1.7.3, for a given amount of primary productivity, the environment is able to support a much greater biomass of animals at a lower trophic level than it can those at a higher trophic level. And, a given amount of primary productivity will support a much larger number of humans living on a diet of clams and cockles than those eating carnivorous fish. Carnivorous fish are several energy-using steps away from the primary productivity that is used directly by filter-feeding bivalves. World catches are now being dominated by species at lower trophic levels as larger and more valuable predatory species are being removed from ecosystems. This process, referred to as 'fishing down marine food webs' (Pauly

Table 1.5 Indicative mean yields (t/km^2/year) from various marine ecosystems. Note that individual yields within each category are highly variable. From various sources including Gulland (1977) and Marten and Polovina (1982).

Open ocean (temperate)	0.5
Open ocean (tropical)	0.02
Shallow banks (temperate)	3
Reefs (tropical)	4
Continental shelf (temperate)	2
Continental shelf (tropical)	6
Estuaries (temperate)	10
Estuaries (tropical)	15
Upwellings	18

et al., 1998; Pauly & Palomares, 2005) is discussed in Section 1.8.2.

The potential amount of seafood available for harvesting on a sustainable basis depends on the productivity of the fishing area and on the trophic levels of the targeted species. Although the productivity of the open sea beyond the continental shelves is generally low, pelagic species such as horse mackerels are hunted using pelagic trawls and tunas are exploited using seine nets and longlines. A small number of seamounts are targeted for some long-lived and fragile deep-sea species including orange roughies and deep-water snappers. Upwellings at the edge of continental shelves, usually at the eastern boundaries of the oceans (see Fig. 1.18 in Section 1.4.3), bring deep, nutrient-rich water to the surface and result in much higher production. Upwellings support high concentrations of small pelagic species that are exploited using purse seines and mid-water trawls. Indicative values of mean yields from various marine ecosystems are given in Table 1.5 but in the light of the current depleted condition of many fish stocks these may not be sustainable in the long term (see Section 1.8.2).

1.8.2 Productivity from fisheries and aquaculture

On the basis of marine food chains and the number of trophic levels, Ryther (1969) estimated

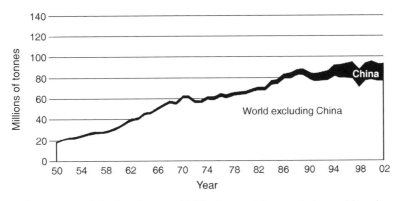

Fig. 1.36 Production from capture fisheries of the world. The lower white area is the world catch excluding China. The black band shows the reported catch from China. FAO (2002).

production from the world's seas to be about 240 million tonnes, of which about 100 million could be caught as a sustainable catch. On the basis of primary productivity and transfer efficiency between trophic levels, Gulland (1970) estimated a sustainable catch to be between 102 and 142 million tonnes.

The reported landings from marine capture fisheries are shown in Fig. 1.36 and Table 1.6.

Since the early 1990s, catches of marine species have fluctuated around an average of 84 million tonnes with an additional 9 million tonnes from inland waters (FAO, 2002). Catches reported by China, the world's largest fishing nation, are believed to have been exaggerated (Watson & Pauly, 2001) possibly by managers under pressure to meet state production goals. If this is so, the total world catch has been declining since the

The production of farmed salmon is very high in many cooler parts of the world. This circular support float for a salmon enclosure is being constructed on a beach near Puerto Montt, Chile. After launching, the circular float will support a net enclosure containing salmon.

Table 1.6 Capture production (in tonnes) by groups of species. FAO data.

Species group	1998	1999	2000	2001	2002	2003	2004
Carps, barbels and other cyprinids	605 430	616 768	574 292	569 838	616 874	631 455	634 688
Tilapias and other cichlids	597 988	636 758	680 004	678 157	662 276	687 683	720 270
Miscellaneous freshwater fishes	5 049 323	5 594 704	5 631 071	5 796 189	5 480 575	6 329 264	6 519 307
Sturgeons, paddlefishes	3 858	2 851	2 603	2 313	1 908	1 788	1 479
River eels	12 047	11 939	16 138	12 374	11 443	10 517	10 118
Salmons, trouts, smelts	889 983	913 309	805 138	891 026	809 854	964 016	878 609
Shads	707 411	788 770	860 346	665 284	589 692	524 875	581 003
Miscellaneous diadromous fishes	79 373	75 921	83 328	75 650	79 538	79 467	83 944
Flounders, halibuts, soles	936 868	956 926	1 009 392	948 316	915 298	917 591	874 929
Cods, hakes, haddocks	10 426 230	9 431 145	8 696 159	9 304 921	8 474 299	9 385 631	9 431 550
Miscellaneous coastal fishes	6 215 346	6 113 345	6 099 838	6 298 508	6 294 529	6 813 211	6 992 491
Miscellaneous demersal fishes	2 980 574	2 955 562	3 032 799	3 007 377	3 061 811	3 070 590	3 191 812
Herrings, sardines, anchovies	16 691 528	22 671 343	24 914 890	20 638 261	22 284 707	18 822 490	23 262 783
Tunas, bonitos, billfishes	5 728 630	5 901 440	5 784 336	5 737 423	6 108 814	6 248 016	6 024 620
Miscellaneous pelagic fishes	11 232 039	10 715 033	10 652 971	12 332 079	11 772 515	11 533 859	11 175 995
Sharks, rays, chimaeras	836 072	857 768	870 329	845 686	843 413	881 161	810 322
Marine fishes not identified	10 738 633	10 767 224	10 784 771	10 635 314	10 560 022	9 726 050	9 790 738
Freshwater crustaceans	642 417	494 111	563 641	626 407	816 405	366 450	355 356
Crabs, sea spiders	1 077 051	1 061 010	1 102 110	1 093 550	1 121 493	1 338 060	1 360 953
Lobsters, spiny-rock lobsters	217 083	229 020	227 676	222 025	225 549	225 132	232 922
King crabs, squat-lobsters	79 244	77 644	67 933	46 385	41 851	43 993	36 449
Shrimps, prawns	2 738 763	3 031 616	3 099 462	2 957 794	2 972 929	3 529 992	3 602 942
Krill, planktonic crustaceans	90 441	101 957	114 426	104 218	125 987	117 982	118 165
Miscellaneous marine crustaceans	1 402 859	1 290 805	1 372 079	1 427 312	1 359 158	450 355	489 075
Freshwater molluscs	580 047	552 452	595 286	628 205	631 444	436 274	428 401
Abalones, winkles, conches	115 713	120 267	119 157	130 990	112 281	121 216	138 822
Oysters	160 285	158 187	249 647	198 132	185 072	197 754	151 941
Mussels	227 197	207 470	261 635	240 718	224 741	187 655	190 202
Scallops, pectens	554 310	609 418	665 569	706 295	744 744	804 291	800 542
Clams, cockles, arkshells	838 118	841 658	798 069	822 529	799 377	898 846	846 914
Squids, cuttlefishes, octopuses	2 864 954	3 602 673	3 679 289	3 348 274	3 261 396	3 611 940	3 775 173
Miscellaneous marine molluscs	1 594 064	1 564 499	1 510 007	1 505 637	1 491 835	936 113	987 296
Frogs and other amphibians	3 009	1 807	2 328	2 486	2 463	2 988	3 014
Turtles	1 182	1 243	1 010	818	1 444	1 498	405
Sea squirts and other tunicates	3 443	3 905	3 858	2 427	2 320	2 951	2 497
Horseshoe crabs and other arachnoids	3 252	2 397	1 696	1 299	1 387	1 190	519
Sea urchins and other echinoderms	110 298	121 567	122 367	107 135	123 831	106 671	111 183
Miscellaneous aquatic invertebrates	527 037	542 959	556 887	474 945	454 272	521 469	389 379
World total (tonnes)	87 562 100	93 627 471	95 612 537	93 086 297	93 267 547	90 530 484	95 006 808

late 1980s. By 2002, according to FAO statistics, aquaculture was adding over 50 million tonnes (including aquatic plants) to total production of which 70% was produced by China.

Excessive fishing has reduced stocks of exploited marine species to dangerously low levels. Not only have catch weights been falling but the mean sizes of fishes in the catches have been decreasing. Individuals of the same species in catches are getting smaller in size as larger specimens are removed from the populations. This has important implications in terms of reproductive potential as larger females produce a disproportionately large number of eggs (egg-carrying capacity is related to volume which has a cubic relationship to fish length – see Section 4.3.1 'Size and growth'). Of even more concern is the removal of larger and usually more marketable predatory

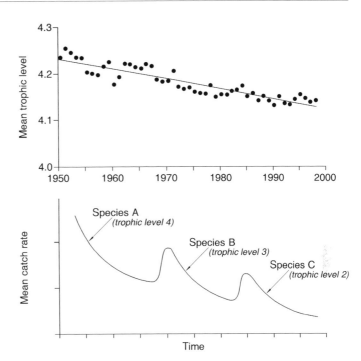

Fig. 1.37 Top: trend of mean trophic level of fishes in catches from an open-water ecosystem indicative of a transition to a greater proportion of smaller pelagic fish in the total catch. From an analysis of FAO data by Pauly and Palomares (2005). Bottom: a theoretical curve illustrating the sequential depletion of stocks at successively lower trophic levels. The depletion of the higher trophic level stock (at the left) is followed by fishers targeting species at lower trophic levels. For each successive species targeted, catch rates are initially high but decrease as the stock is depleted.

species from ecosystems: the decline of these populations is followed by fishing on smaller and usually less desirable species at lower trophic levels. This effect has been referred to as 'fishing down marine food webs' (Pauly et al., 1998).

Disturbingly, the trend of fishing at successively lower trophic levels has been observed in open-water ecosystems where large pelagic predators including billfishes and sharks are being depleted and replaced by catches of successively smaller tuna and mackerel species (Pauly & Palomares, 2005). The graph at the top of Fig. 1.37 shows the trend in mean trophic levels of open-water catch species and this is indicative of an increasing proportion of smaller pelagic fish in the total catch. The numerical value of mean trophic level (TL) is indicative of how many steps a species is away from the marine food web's primary producers including benthic algae and phytoplankton. Primary producers and detritus are assigned TL values of 1 and larger order carnivores may have a TL value from 4–4.5.

The graph at the bottom of Fig. 1.37 illustrates a hypothetical fishery in which there is a sequential depletion of stocks, reflected in a reduction in catch rates, at successively lower trophic levels.

The depletion of the higher trophic level stock (at the left) is followed by fishers targeting species at lower trophic levels. For each successive species targeted, catch rates are initially high but these decrease as the stock is depleted. Fishing down marine food webs has resulted in world catches being increasingly dominated by species at lower trophic levels.

About 16% of the world's total animal protein consumption is from fish. In 1997 the average consumption was about 16 kg per person per year (FAO, 2000) although this varied greatly between countries. About 75% of the world fisheries production is used for direct human consumption with the remainder used mainly for the manufacture of fishmeal and oil. Catches from just ten species collectively account for about 30% of the world catch in terms of quantity (Fig. 1.38). Some of the world's largest single-species fisheries are based on clupeoids including herrings, pilchards, anchovies, sardines and most famously the Peruvian anchoveta, *Engraulis ringens*.

The Peruvian anchoveta exists in a small area of upwelling associated with the strong north-flowing Humboldt (Peru) Current (see Fig. 1.18).

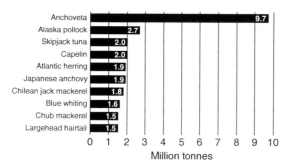

Fig. 1.38 The top ten individual species in world fish catches. FAO (2004).

The annual catch had been increasing and, in 1972, reached a peak of over 13 million tonnes, equivalent to about 15% of the total world catch and a remarkable 40 tonnes of fish per km^2. A strong El Niño effect (see Box 1.22 'El Niño and La Niña') in 1972 caused the Humboldt Current to reverse, the upwelling to fail, and stocks of anchoveta to suffer food shortages and high mortality rates. Pressed in against the coast by a tongue of warm water, the fish, already reduced by heavy fishing, concentrated in schools and became even more vulnerable. The decimated population and the lack of large mature individuals caused recruitment to fail and the world's largest fishery to collapse.

The collapse caused the death of millions of guano-producing seabirds which use the anchoveta for food and there was some concern that the existing ecology of the area had been irreversibly affected, possibly allowing sardines to become the dominant organisms. Annual catches of the anchoveta remained below 3 million tonnes for many years, and in 1990, almost 20 years later, the catch was only 3.8 million tonnes. The annual catch recovered to a secondary peak of over 11 million tonnes in 2000.

Catches of many upwelling and other pelagic species including the Chilean jack mackerel and the South American pilchard (or sardine) fluctuate greatly with periodic climatic events associated with the El Niño Southern Oscillation. The dip in the total world catch for 1998 (Fig. 1.36), for example, was due to very low catches of the Peruvian anchoveta during the severe El Niño

event of 1997/98. Other small pelagic species of commercial importance include capelin, chub mackerel and the California pilchard (or sardine).

The Northwest Pacific is the most productive fishing area of the world, with catches oscillating between 20 and 24 million tonnes (including China) since the late 1980s. But even here, catches of Alaska pollock have been decreasing since the late 1980s. The Japanese sardine (or pilchard) fishery collapsed in the mid-1990s and was followed by a strong recovery of the anchovy population, which has been supporting catches of close to 2 million tonnes since 1998. This alternation between sardine and anchovy stocks follows a pattern observed in many other regions of the world (FAO, 2004).

Total catches of tunas and tuna-like species account for over 10% of the total value of landings for human consumption. The world catches of tunas increased from less than 0.5 million tonnes in the early 1950s to a peak of 4 million tonnes in 2002. Skipjack tuna accounts for about 50% of this total, and is one of the top species in world fisheries production. In some areas of the Atlantic and Pacific, straddling stocks of jack mackerel, squids and demersal fish on seamounts also contribute significantly to production.

The contribution of sharks and rays to the total reported catch is minor compared with that of other oceanic resources although there may be considerable under-reporting of sharks caught as bycatch or for the shark-fin soup market.

In many areas, according to FAO data, less valuable and less exploited species are being targeted following the depletion of more traditional stocks. In the Northeast Atlantic, for example, the continuous decline in cod catches has been counterbalanced by increasing catches of formerly low-valued fish species such as the blue whiting and sand eels. In the Southwest Atlantic, the decline of the Argentine hake has been followed by an increasing trend in catches of shortfin squid. The decline in catches of pilchards (or sardines) and pollock in the Northwest Pacific has been partially compensated for by increases in catches of the Japanese anchovy, largehead hairtail and squids. The Alaska pollock and blue

Box 1.22
El Niño and La Niña

Southeast trade winds normally blow across the South Pacific from a high-pressure area over the Southeastern Pacific to a relatively stable low-pressure area to the north of Australia (Fig. B1.22.1 (top)). For reasons that are not known, these two areas alternate approximately every two to eight years; high-pressure builds up in the west and the trade winds reverse and blow from west to east (Fig. B1.22.1 (bottom)). This alternating in atmospheric pressure areas is called the Southern Oscillation.

In normal years the trade winds drive a strong surface current from east to west. When the trade winds falter or reverse, the warm-water pool that has built up on the western side of the Pacific flows back towards the east. Sea levels in the Eastern Pacific can rise up to 20 cm as a tongue of warm water arrives off the South American coast pushing back the northern limit of the Humboldt Current. Near the coast of Peru, fishers and sailors call the unusual current El Niño, after the Christ Child because it appears soon after Christmas (although this timing does vary). As El Niño and the Southern Oscillation are linked, the acronym ENSO is used to describe the event. Sometimes when normal

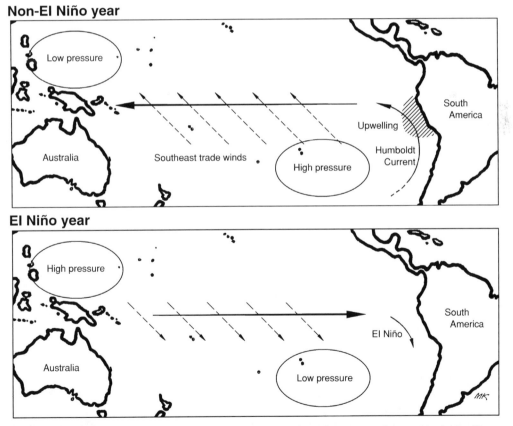

Fig. B1.22.1 Conditions associated with a normal year (top) and an El Niño year (bottom) in the Pacific Ocean. Normally the southeast trade winds (broken arrows) drive a current from east to west (heavy arrow) and the Humboldt Current drives an upwelling off Peru. In an El Niño event, the winds and currents are reversed and the upwelling disappears.

atmospheric pressures return after an El Niño event, currents reappear with unusual strength and with colder and stormier conditions. These opposite conditions to El Nino are given the contrasting name of La Niña.

El Niño events typically last about a year, although some can be longer, and most recent events occurred in 1986/87, 1991/92, 1993, 1994, 1997/98, and 2002/03. In Australia, trends in the Southern Oscillation Index (SOI) which is calculated from air pressure differences between Tahiti and Darwin are provided with weather forecasts by the Bureau of Meteorology (www.bom.gov.au). Sustained negative values of the SOI often indicate El Niño episodes with weakening trade winds and an increased probability of droughts in eastern and northern Australia. Positive values of the SOI are associated with the stronger trade winds and warmer seas of a La Niña episode and an increased

probability that eastern and northern Australia will be wetter than normal.

Every several years the pressure systems in the Pacific reverse rather than just weaken, causing dramatic El Niño events such as the one in 1997/98. The warm water off Peru fuels evaporation, cloud formation and storms, and then proceeds to affect weather in different parts of the globe. Devastating regional effects have included rainstorms in Chile and Peru (usually dry places), droughts in Indonesia, dust storms and bush fires in Australia, floods in East Africa, and a rise in bubonic plague cases in New Mexico. Reversals in sea temperatures in the Pacific are believed to alter wind directions globally, bringing freezing conditions and floods to Central Europe and droughts to South Africa. During El Niño events, corals in the Western Pacific are badly affected by exposure at the unusually low sea levels.

whiting are now two major species in terms of catch weight (although not in value) and overall catches of the more valuable gadiformes (the cods, hakes and haddocks) have been decreasing since the late 1960s. Northwest Atlantic fisheries production reached its lowest level in 1994 and again in 1998 with the collapse of groundfish stocks off eastern Canada.

Marine crustaceans and molluscs make up a small but commercially valuable part of the world catch. Catches of oceanic squids, particularly the Eastern Pacific jumbo flying squid, the Western Pacific Japanese flying squid and the Argentine shortfin squid contribute about 3 million tonnes to world catches and catches have been steady since a low catch in the El Niño-affected year of 1998. Catches of the Argentine shortfin squid in the Southwest Atlantic, however, have been falling since 1999, when catches of around 1.1 million tonnes were reported.

Of the top ten species that collectively account for about 30% of the world catch in terms of quantity (Fig. 1.38), seven are from stocks that are considered to be fully exploited or over-exploited. There is believed to be no potential for increasing catches of anchoveta, Chilean jack

mackerel, Alaska pollock, Japanese anchovy, blue whiting, capelin and Atlantic herring. Catches of the Chilean jack mackerel have decreased by about 50% from the fishery's historical peak production reached in 1994. Only two of the other top ten species, the skipjack tuna and the chub mackerel, may be able to support higher fishing pressure in some areas. The status of the remaining species (the largehead hairtail) is unknown (FAO, 2004). A recent study of global fisheries data (Worm et al., 2006) suggests that the loss of marine biodiversity is increasing the rate of ecosystem collapse (see Section 6.5.3 'Ecosystems-based fisheries management'); their somewhat pessimistic projections are that all major fisheries will be based on stock sizes of less than 10% of their original size by the middle of the present century. The discovery of new resource species to exploit is becoming increasingly unlikely (although see Box 1.23 'New fisheries') and the status of existing stocks monitored by FAO is summarized in Table 1.7.

Nearly all the world's most important commercial fisheries, except those based on tunas and some species of carangids (jacks and trevallies) are based in temperate waters. However, the

Table 1.7 Status categories of stocks monitored by
FAO (2003).

Underexploited	3%
Moderately exploited	21%
Fully exploited	52%
Overexploited	16%
Depleted	7%
Recovering from depletion	1%

catches made by subsistence fishers, who catch
fish to eat rather than to sell, is largely
unquantified, and unlikely to be included in
official catch statistics. In many tropical countries
in particular, where fish is the most important
source of protein in the absence of intensive agriculture, subsistence fishing results in a total catch
that is often several times larger than that from
commercial fishing.

In 2002, the total world aquaculture production (including aquatic plants) was 51.4 million
tonnes, of which China produced 71.2% (FAO,
2004). The largest production for an individual
species was from the Pacific cupped oyster,
Crassostrea gigas, for which catches totalled 4.2
million tonnes in 2002. However, over 40% of
total aquaculture production is from carps and
other cyprinid fishes. The production of aquatic
plants, particularly Japanese kelp, *Laminaria
japonica*, followed by nori, *Porphyra tenera*,
totals about 12 million tonnes, with China, the
Philippines and Japan being the major producers.
Over 90% of aquaculture production is from
developing countries where there have been
environmental concerns, particularly with shrimp
or prawn farms. Mangrove areas have been
destroyed to construct ponds and nearby waters
have become polluted with food and faeces.
Developments include the farming of Atlantic cod,
Gadus morhua, in Norway and Iceland as well as
farming operations that fatten fish usually harvested directly from the wild. The aquaculture of
wild-caught tunas by fattening them in sea cages,
for example, is an increasingly important activity
in Australia, Mexico and the Mediterranean region.

Up to 80% of aquaculture relies on the farming
of hatchery-reared larvae and juveniles. The
balance consists of capture-based aquaculture
operations in which larvae, juveniles or adults
are collected from the wild and grown to marketable size in captivity, using aquaculture
techniques. Species grown this way include many
shellfishes and carnivorous finfishes including
milkfish, groupers, tunas, yellowtails and eels
(FAO, 2004).

It is doubtful that aquaculture production can
make up for the decreasing catches from wild
fisheries. Much aquaculture depends on expensive foods that contain fishmeal produced from
wild fisheries: in these cases aquaculture exacerbates the problems of overexploited wild fisheries.
Exceptions are farming operations based on filter-
feeding invertebrates such as mussels and oysters,
as well as on herbivorous fishes such as catfish
and tilapia.

All the information available tends to confirm
the estimates made by FAO in the early 1970s
that the global potential for marine capture

Box 1.23
New fisheries

New fisheries are a rare thing but in 1997 the
Patagonian toothfish, *Dissostichus eleginoides*,
and the mackerel icefish, *Champsocephalus
gunnari*, began to be exploited in sub-Antarctic
waters 4000 km south-west of Albany in Western
Australia. The fishery, worth about US$30 million,
operates under one of the world's most stringent
management regimes for a commercial fishing
operation. Measures include strict catch quotas
for all species, controls on bycatch, a maximum
of three fishing boats and the inclusion of two
government observers on each expedition.
However, illegal fishing in the area represents a
major threat to the sustainability of the fishery.
Numerous fishing vessels have been apprehended,
and the crews arrested and prosecuted.

fisheries is about 100 million tonnes, of which only 80 million tonnes are probably achievable. It also confirms that, despite local differences, overall, this limit has been reached. These conclusions lend support to the call for more rigorous recovery plans to rebuild stocks that have been depleted by overfishing and to prevent the decline of those being exploited at or close to their maximum potential.

Excessive fishing is undoubtedly the cause of the crisis in world fisheries. Although allowing depleted stocks to recover is most directly achieved by reducing the amount of fishing, there is also a need to protect threatened marine ecosystems. The catches of many species may be declining, not only due to excessive fishing pressure, but because the environment is deteriorating, or because the tenuous links between species is unbalanced. The marine environment is in demand, and under pressure, from many different sources, including commercial fishing, subsistence fishing, recreational fishing, aquaculture, tourism, water sports, shipping, coastal development and industry. Many of these uses threaten the marine environment in various ways. In the face of decreasing catches, even in strictly managed fisheries, many scientists are advocating that the narrowly-based management of a single resource species be replaced by the more broadly-based management of the ecosystems that support all marine species. These aspects of fisheries management are discussed in Chapter 6.

The degree of our ignorance regarding fish stocks and ecosystems is high, particularly in relation to natural variability, and there is always good reason to do more research. But as pointed out by Ludwig et al. (1993), the need for more research should not be used as an excuse to avoid addressing the more difficult problems of population growth and the excessive use of resources.

Exercises

Exercise 1.1

Map the distribution and areas of either wetlands, salt marshes, mangroves, seagrass beds or coral reefs adjacent to your town or community. Consult records and interview older local inhabitants to discover if these areas are increasing or decreasing in size. What changes have occurred and why? Are the areas adequately protected at present?

Exercise 1.2

Obtain recent fish catch statistics for your country and review the status of the five most important species or fisheries. Consider the status of the resources (underexploited, fully exploited, or overexploited) and whether rules and regulations applied to the fishery are effective.

Exercise 1.3

Review the most serious threats to the coastal environment in an area with which you are familiar. What regulations apply to the area – and are these being enforced? What actions are being taken, or should be taken, to counteract these threats?

2 Exploited species

2.1 Introduction

The proper assessment and management of a fishery requires an understanding of the biology, life cycle and distribution of the species on which it is based. In spite of the large diversity of marine life, most important fisheries are based on members of just three large scientific groups.

Species are included in a particular taxonomic group on the basis of having similar characteristics and larval stages, as well as having what is believed to be a common ancestor, perhaps many millions of years ago. The subsections of this chapter describe some members of the phylum Mollusca (including clams, sea snails and squid), the subphylum Crustacea (shrimps, prawns, lobsters, and crabs) and the vertebrate classes that include fishes. Each subsection provides a brief description of the taxonomy and general biological characteristics of the group, with examples of some commercially important species, representative life cycles, and examples of management controls that have been applied to fisheries based on the species.

2.2 Invertebrates

Invertebrates are often the bases of very valuable fisheries with some species, such as abalone and lobsters, being of higher value per unit weight than most species of fish. In some countries where seafood is the main source of daily protein, intertidal invertebrates, particularly molluscs, are particularly important when bad weather prevents fishing from boats at sea. The two most important groups of marine species, from a fisheries viewpoint, are molluscs and crustaceans and

these are discussed in separate subsections. A third subsection includes other invertebrates that are locally important in some parts of the world.

2.2.1 Molluscs

Many different species of intertidal and subtidal molluscs, cockles, clams and sea snails are harvested from a range of inshore habitats including estuaries, mud flats, sandy beaches and reef flats. Mechanical harvesting methods for burrowing molluscs include the use of towed rakes or suction pumps in conjunction with water jets to liquefy the sand or mud containing clams or cockles. Coastal communities have used clams and sea snails as food since early times, and discarded shells have been found in the middens and wasteheaps of prehistoric settlements. In countries where seafood is still collected as an important source of protein, the catch of molluscs is often greater in weight than that of any single family of fish species.

With over 50 000 different living species, the phylum Mollusca is second in size to the largest of all animal phyla, Arthropoda, which contains the crustaceans discussed in the next subsection. The phylum Mollusca includes creatures as different as the periwinkle, giant clam and octopus. These diverse animals are related by being soft bodied with organs covered by a sheet of tissue called the mantle, and in having similar larval stages. In most molluscs, the eggs develop into trochophore larvae which pass through veliger larval stages to adult forms. With the exception of the bivalves, all molluscs have teeth arranged in rows along a file-like flexible strip called the radula, which is used in feeding. Although the

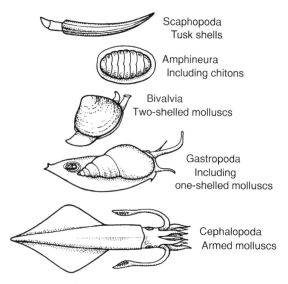

Scaphopoda
Tusk shells

Amphineura
Including chitons

Bivalvia
Two-shelled molluscs

Gastropoda
Including
one-shelled molluscs

Cephalopoda
Armed molluscs

Fig. 2.1 The five ecologically significant classes of the phylum Mollusca. Of these, the bivalves, gastropods and cephalopods are of great importance in fisheries.

most obvious molluscan characteristic is often one or two shells secreted by specialized cells in the mantle, not all molluscs have shells. Of the seven classes of the phylum Mollusca, the five

ecologically significant ones are shown in Fig. 2.1: of these, the bivalves, gastropods and cephalopods are important in fisheries and are discussed in the present section.

Bivalves – clams and cockles

The class Bivalvia is familiar because it contains the highly-regarded and ubiquitous food species such as cockles, clams, oysters, mussels, and scallops. Bivalve molluscs have two shells, or valves, joined together dorsally by a horny ligament called the hinge. The two shells are usually prevented from sideways movement by sockets and grooves ('teeth') located on the hinge line of the shells, and are held together by one or two adductor muscles. Most bivalves, including mussels and giant clams, have an anterior and a posterior adductor muscle, although these may be unequal in size. Species such as scallops and oysters have only a posterior adductor muscle (Fig. 2.2). When a bivalve's adductor muscles are relaxed, the interior ligament forces the shells to gape open.

The soft body of a bivalve is enclosed by two lobes of sheet-like tissue (the mantle), which lie

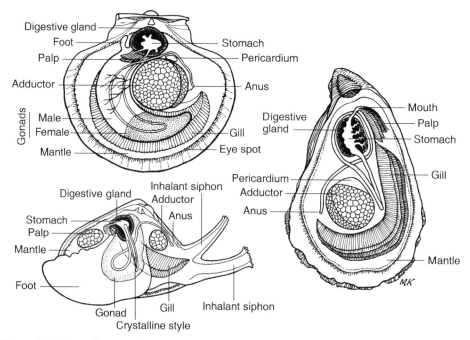

Fig. 2.2 Views of bivalve molluscs with one shell, mantle, and gill lobes removed. Clockwise from bottom left: a clam, a scallop and an oyster.

between the body and the shells. At its margin, each mantle lobe has three folds, the outer of which secretes material to increase the shell's size and thickness; the shell therefore grows from the hinge area, or umbo, which is the oldest part of the shell. The inner fold of the mantle is largely muscular, and is responsible for forming a seal at the joint of the left and right lobes of the mantle. The middle fold may contain sensory organs, including the rows of eyes in some scallops (*Pecten* spp.). Each mantle lobe is attached to the shell a short distance in from the shell margin, along the pallial line. In many species, the mantle lobes are joined together at their outer perimeter except in two areas that form the inhalant and exhalant openings. In some burrowing species, these openings are extended into two tube-like processes, the inhalant and exhalant siphons (Fig. 2.2).

Bivalves are filter feeders, and most species obtain food by sifting either microscopic plants from the surrounding water (suspension feeders), or organic material from the substrate (deposit feeders). Most bivalves have enlarged gills which have the dual function of food collection and gaseous exchange. Water is drawn in through the inhalant aperture into the mantle cavity by the action of hair-like cilia which cover the gills, and is pumped out through the exhalant aperture. Food material is passed along the gill surface to ciliated labial palps, one pair on each side of the mouth. The food in mucous strings passes into the stomach, which is surrounded by a digestive gland producing enzymes. In most species, the food is wound around a rotating rod-like apparatus called the crystalline style, and ground against a hard gastric shield. The style, which releases enzymes as it wears away, is produced by a style sac, and is rotated by cilia within the sac. The long intestine may contain one or more loops before passing above the posterior adductor muscle as the rectum. The anus is positioned so that waste material and faeces are expelled into the water which passes out through the exhalant aperture (Fig. 2.2).

Although most bivalve species are dioecious (have separate sexes), several commercially important species are hermaphroditic (have both male and female gonads). Hermaphroditic species may be either sequential hermaphrodites, such as some oysters (*Crassostrea* spp.), which have gonads that alternate between being male and female, or simultaneous hermaphrodites, such as some scallops (*Pecten* spp.) and giant clams (*Tridacna* spp.), which have gonads that function as ovaries and testes at the same time. Note the male and female part of the curved finger-like gonad of the scallop shown in Fig. 2.2.

In most bivalves, eggs and sperm are released through the exhalant aperture directly into the sea where fertilization takes place. After fertilization, eggs hatch to trochophore larvae, which are succeeded by veliger larvae, and often remain in the plankton for several weeks before settling as adult forms. Of the large numbers of eggs released, many millions in some species, a much smaller number become fertilized, and the resulting larvae have to survive high predation rates while in the plankton. In addition, the floating larvae are subject to variable winds and currents which affect the probability of reaching suitable substrata on which to settle. An example of a bivalve life cycle is provided in Box 2.1 'The life cycle of a giant clam'.

In the few species of bivalves which brood their eggs, fertilization occurs when sperm is carried into the mantle cavity with water entering through the inhalant aperture. The eggs are incubated internally until they reach a more advanced stage of development. In some commercial oysters, those of the genus *Ostrea*, fertilization takes place within the mantle cavity of the female, and larvae are incubated for one week or more before being released at the shelled veliger stage. Almost all freshwater bivalves also incubate their eggs. In freshwater mussels, such as *Unio* and *Anodonta*, the eggs are incubated within the gills, where they develop and are released at a parasitic larval stage, the glochidium. The glochidea attach themselves with spines, byssus threads, or hooks to the fins and gills of fish, and become encysted by the host tissue on which they feed. After a parasitic period of up to a month, the immature bivalves break free of the cysts to lead an independent existence.

Box 2.1
The life cycle of a giant clam

Giant clams, *Tridacna* spp., have been the subject of intensive research (Copeland & Lucas, 1988), and a typical life cycle is shown in Fig. B2.1.1. Giant clams begin life as males, and mature at about two years of age, after which they function as both males and females. At low latitudes, spawning occurs during the summer months when clams, detecting the presence of eggs in the water, forcibly eject sperm through the exhalant aperture. About 30 minutes after releasing sperm, an individual clam releases its own eggs, thereby avoiding self-fertilization. This epidemic spawning, where individuals are stimulated to spawn by the

presence of either eggs or sperm in the water, has been recorded for many other species. The number of eggs released by each individual varies between species, and hundreds of millions are produced by large specimens of *T. gigas*. The eggs hatch into ciliated trochophore larvae after about 12 hours, and these develop into a shelled veliger stage about two days after fertilization. The veliger develops a foot (becomes a pediveliger) and alternates between being in the water column and on the substrate. Advanced veliger larvae settle permanently on the substrate within ten days of being released.

All giant clams are initially attached to the substrate by byssal threads, but the larger species, including *Tridacna gigas* and *T. derasa*, lose this attachment and remain in place by virtue of the weight of the massive shells. Both the great clam, *T. maxima* and the boring clam, *T. crocea* bore into the coral reef, the latter to the upper margins of its shells. Giant clams are unique among bivalves in that they are able to obtain nutrients from a relationship with microscopic plant cells, zooxanthellae. Free-living zooxanthellae become established in the mantle of the larval clam after settlement. After this, the now symbiotic zooxanthellae photosynthesize and produce nutrients which are used as food by the clam. This symbiotic relationship is evident in a few other tropical invertebrates, including most shallow-water corals.

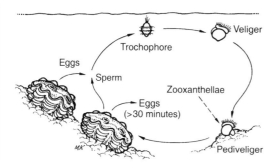

Fig. B.2.1.1 The life cycle of a giant clam, *Tridacna*. Giant clams are hermaphroditic: each clam releases sperm and, after about a 30 minutes delay, its own eggs. Zooxanthellae become established in newly-settled veliger (or pediveliger) larvae.

Most adult bivalves remain free-living on, or burrowed in, sandy or muddy substrata, and other species become attached to hard objects. The foot of burrowing bivalves is large and blade-like, and adapted for digging into the substrate; the length of the siphons of these species reflects the depth to which they burrow. Some bottom-dwelling species of file shells (*Lima*) and scallops (*Pecten*) can swim over short distances. In the presence of predators, scallops swim by rapidly opening and closing their shells, forcing a jet-stream of water through apertures on either

side of the hinge, and move jerkily (shell margin first) through the water.

In attached bivalves, the foot and the anterior adductor muscle are greatly reduced or absent. Species such as mussels (*Mytilus*) use a special byssus gland to secrete a cluster of flexible byssal threads which form an attachment to solid objects. In some species, including the oyster, a small quantity of adhesive fluid is released by the byssus gland and one valve (usually the right-hand one) becomes cemented to the substrate. A few species are capable of boring into stone and

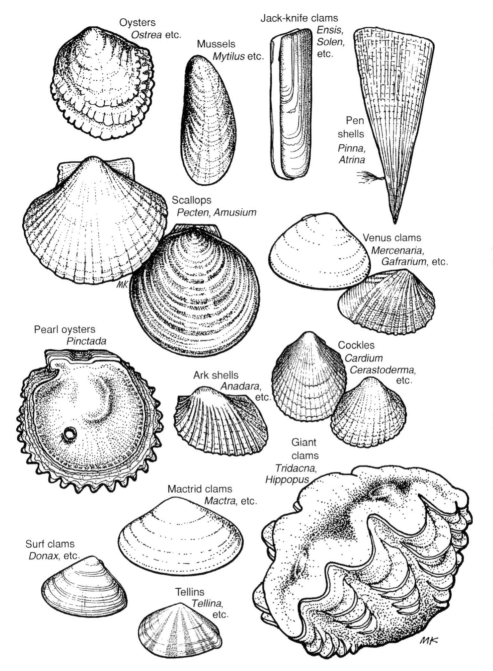

Fig. 2.3 Representative groups of bivalve molluscs.

wood. Piddocks of the family Pholadidae bore into sandstone and the shipworm, *Teredo*, makes long burrows in timber (see Box 2.2 'The teredo or shipworm').

The recorded world catch of bivalve molluscs,

mainly clams, cockles, oysters, mussels and scallops is about one million tonnes per year, and a large but unquantified catch is taken by subsistence fishers. Some representative species are shown in Fig. 2.3.

Box 2.2
The teredo or shipworm

The long and worm-like *Teredo navalis*, known to mariners as the shipworm, is a bivalve mollusc and bores into submerged timber using its two shells. Under the influence of the adductor muscles, the sharp edges of the two shells move through an arc of 180 degrees and then reverse, chiselling away wood to create a burrow that may be up to a metre in length. Most of the animal consists of the greatly extended siphons that secrete a calcareous layer on the burrow wall. Water is drawn in over gills which extend into the inhalant siphon and is expelled, along with wood debris, though the exhalent siphon. At the burrow entrance, both siphons terminate in shelly plates called pallets. The teredo causes considerable damage to timber structures and wooden vessels (Fig. B2.2.1).

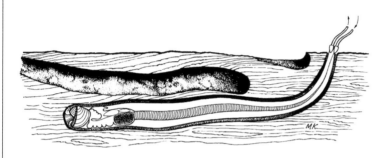

Fig. B2.2.1 The teredo or shipworm, *Teredo navalis*, boring into timber. The arrows show water movement in and out of the siphons.

Wild fisheries for mussels and oysters still exist but stocks have been reduced by excessive harvesting, reduced habitat areas and polluted waters. The fishery for the American oyster, *Crassostrea virginica*, which was once the main oyster marketed in North America, for example, has suffered from disease and pollution, especially in Chesapeake Bay. However, because of their sessile nature and ability to filter food from passing water currents, many bivalves are ideal candidates for aquaculture and oysters and mussels are two of the most widely farmed marine organisms in the world. Bivalves are farmed in highly productive inshore waters, either by allowing the natural settlement of larvae (commercially called spat or seed) on artificial substrates (called cultch) such as wooden slats or ceramic tiles or by the production of spat in shore-based hatcheries. The blue mussel, *Mytilus edulis*, has a wide distribution from the North Atlantic to the North Pacific oceans and is marketed in North America, Spain, the Netherlands, Denmark and France. Wild mussels are harvested with dredges, rakes and tongs and farmed mussels are grown either suspended from rafts and poles or on the sea floor. Mussel culture has a long history in France where they are grown on rope wrapped around oak poles set in intertidal areas.

Because of their value in aquaculture, bivalves such as oysters and mussels have been transplanted to areas beyond their original distribution. The Pacific or Japanese oyster, *Crassostrea gigas*, for example, has been introduced into many parts of the world, including Europe and North America, and is now one of the most widespread of all cultured marine organisms. Although many of these transplantations have provided considerable economic benefits, the introduction of exotic species to an area is not without risk. Beside the possibility of introducing parasites and disease, there is an even greater likelihood that the introduced species will outcompete endemic species for settling space, and eventually displace them.

The family Cardiidae includes many species, commonly called cockles, which are distributed in all seas and commonly used as food. The common European cockle, *Cerastoderma edule*, is an important food item in northwestern Europe, and related *Cardium* species are found throughout tropical areas. Venus clams are the most successful of all bivalve molluscs with over 400 different species found around the world. In North America, the ocean quahog, *Arctica islandica*, and the northern quahog, or hard clams, *Mercenaria mercenaria*, have been used as food ever since the continent was populated; indeed, the scientific name of the latter, *Mercenaria*, is based on the use of this species shells as money, or 'wapum', by native tribes. Quahogs now come from wild fisheries and farms and are used to make the thick soup known as clam chowder. Tropical venus shells include species such as *Gafrarium* that are collected as food.

Several species of surf or wedge clams, particularly *Donax* species, are found only on high energy beaches swept by heavy surf, where mechanical harvesting is difficult. The plump rounded ark shells, *Anadara* species, are mostly tropical and are gathered from the wild as well as being farmed in Southeast Asia. Jack-knife clams, sometimes called razor shells or Chinaman's fingernails and large pen shells (Fig. 2.3) are harvested in shallow silty areas. Byssal threads produced by pen shells bind them to shells and grit under the sand: the 'golden fleece' of ancient times is claimed to have been made from the silky byssal threads of the giant Mediterranean pen shell, *Pinna nobilis*. Tellins or sunset shells are mainly tropical sand-dwelling clams that have elongate fragile shells with rays of colour radiating out from the hinge area.

Scallops of the genus *Pecten* and related genera are the bases of valuable temperate-water fisheries in many parts of the world. The Atlantic sea scallop, *Placopecten magellanicus*, is the main scallop harvested in the coastal areas of the United States and is mainly harvested with dredges and otter trawls. Scallops are hermaphroditic, with the curved reproductive organ consisting of both male and female parts. This organ (commercially called roe) and the single large adductor muscle are the only parts marketed, although in scallops intended for some markets, the roe is removed. Saucer scallops of the genus *Amusium*, are

trawled in some tropical and subtropical areas of the Indo-Pacific.

Giant clams (*Tridacna* and *Hippopus* species) include the world's largest bivalves, with *T. gigas* reaching weights of up to 250 kg in a lifespan of over 100 years. Only eight species of giant clams are known and all are confined to the Indo-Pacific region, where they are traditionally used as food by many coastal communities. In Southeast Asia, the dried muscles of giant clams are highly valued, and boats from this region have historically fished in the tropical Indo-Pacific. They still fish illegally for clams in several areas, including the Pacific islands, and must have contributed to the decline of clam stocks in some localities. As clams require clear, shallow water, they can be easily collected and hence overexploited. Some species are now locally extinct in Micronesia, and *T. gigas* appears to have disappeared from the islands of Vanuatu and Fiji since the 1970s. The farming of giant clams in many areas has been aimed at restocking reefs where they have been overfished (Copeland & Lucas, 1988). Through their symbiotic relationship with zooxanthellae, giant clams are unique in being the only farmed marine animals that can feed indirectly from sunlight.

Some bivalves are exploited or farmed because of their lustrous shells or because they produce valuable pearls. A cross-section of a typical mollusc shell is shown in Fig. 2.4 and consists of three separate layers, an outer periostracum, a middle prismatic layer, and an inner nacreous layer. The periostracum is a thin protective layer, composed of a flexible horn-like material called conchiolin. The two other layers consist of calcium carbonate, arranged as either vertical crystals in the middle prismatic layer, or as flat crystals of aragonite in the nacreous inner layer. The inner layer, known as mother-of-pearl, displays iridescent colours that are created by the interference of light waves reflected from the thin overlapping layers. In some cases, sand grains or parasites which manage to get past the attachment between the shell and the mantle form an irritant. The mantle may then produce concentric layers of nacreous shell around the object. These deposits

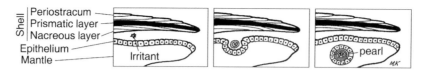

Fig. 2.4 Pearl formation in bivalves. An irritant between the shell and the mantle is enveloped by the mantle which produces concentric nacreous layers, a pearl, around the object.

eventually become a blister, or half-pearl, on the inside of the shell, or less commonly, a free spherical pearl (Fig. 2.4). Although pearls may be formed by many bivalve species, commercially valuable pearls are only produced by those whose shells have nacreous layers with desirable reflective properties, particularly pearl oysters of the genus *Pinctada*.

In the early 1900s pearl oysters were collected by divers operating from sailing vessels in many tropical areas. Broome in Western Australia was the base of the world's largest pearl fishery with up to 400 pearling luggers working with crews of divers. As a result of the demand for pearl shell, wild stocks of pearl oysters have been overexploited in many areas. The technique of producing spherical cultured pearls was developed in Japan in the 1920s. Pearl oysters are now produced in hatcheries and cultured pearls are grown by deliberately 'seeding' the oyster with an irritant around which a pearl may form (Fig. 2.4). The shells of the pearl oyster are wedged apart to allow the implantation of a small piece of mantle tissue and a shell fragment nucleus into the gonad. The mantle tissue grows around the nucleus, and deposits an increasing number of layers of nacreous shell until a pearl is formed. The goldlip pearl oyster, *Pinctada maxima*, which produces gold, silver or pink coloured pearls, is cultured in northern Australia and Southeast Asia, and the blacklip pearl oyster, *P. margaritifera*, which produces grey to black coloured pearls, is grown in Japan, the Philippines, French Polynesia and the northern Cook Islands. The winged pearl oyster, *Pteria penguin*, which cannot be used for the production of round pearls, as its gonad is too small for the successful insertion of a pearl nucleus, is used in the production of 'blister' or half-pearls. In this case the nucleus is inserted

between the mantle and the shell, and a half-pearl grows attached to the shell.

Gastropods – sea snails

The class Gastropoda is the largest class of molluscs. Gastropods have evolved for feeding on a much wider range of food than bivalves, and are found in the open ocean, in fresh water and on land. Most species used as food, however, inhabit shallow coastal areas, particularly rocky or coral reefs. Although some gastropods, such as sea slugs (nudibranchs), have no shell, most species of commercial interest have a single shell. The typical shell is conical with the oldest part of the shell at the apex. The shell grows in a series of successively larger whorls around a central axis, and terminates at an opening. The mantle covers the visceral mass which is contained within the shell, and the head and foot are often exposed ventrally at the shell opening. The head often has tentacles and eyes, and the foot, in many species, bears a disk, called the operculum, on its posterior dorsal surface; as the foot is retracted into the shell, the operculum forms a protective door to the shell opening. Most gastropods crawl or glide by means of their muscular foot which is lubricated by special mucus glands. Some, such as the stromb, move by using their opercula; the spider shell, *Lambis*, for example, moves by jabbing its pointed operculum into the substrate and pulling itself forwards (Fig. 2.5).

The radular teeth of gastropods are adapted for a variety of feeding methods in different species. The re-curved radular teeth are arranged on a membranous belt supported by the cartilaginous odontophore (Fig. 2.6). The number of teeth on the radula varies between species from several thousand teeth in some herbivorous gastropods which rasp algae from the substrate, to

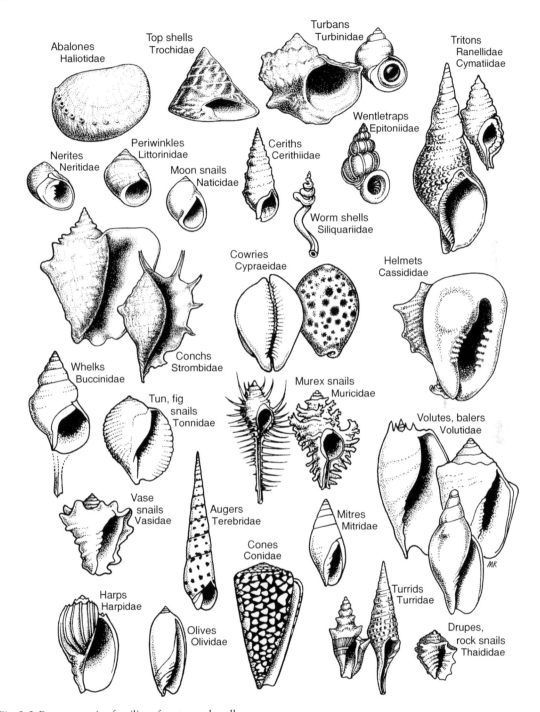

Fig. 2.5 Representative families of gastropod molluscs.

only a few specialized teeth in some carnivorous forms. In herbivorous gastropods, the odontophore is pressed against the substrate and retracted by muscular action. During retraction, the radular teeth scrape particles and algae from the substrate. As teeth are worn away, more are produced in the radular sac.

The radula of some carnivorous gastropods

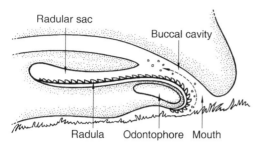

Fig. 2.6 The mouth cavity and radula of a grazing gastropod.

such as *Murex*, and the moon snail, *Polinices*, are adapted for drilling holes in the shells of bivalves, and these species are responsible for devastating commercial oyster and cockle beds. In these species, the drill-like action of the radula cuts a bevelled hole into the area near the umbo, and penetration is aided by the secretion of acid from a gland on the gastropod's foot. The proboscis is inserted into the hole, and the soft tissues of the bivalve are slashed by the radula and ingested (see Fig. 1.31 in Chapter 1). Tropical cone shells (*Conus* spp.) feed on worms, other gastropods and fish which they harpoon and poison by firing a single barbed radula tooth into their prey. A nerve poison is introduced via a hollow cavity in the tooth, and some species, if handled carelessly, are dangerous to humans (see Box 2.3 'Beautiful but dangerous').

Most gastropods are dioecious, although a few species, such as the slipper limpet, *Crepidula fornicata*, are hermaphroditic. In dioecious species, the single gonad, either an ovary or a testis, is located in the spiral of the visceral mass with the digestive gland. The eggs of primitive gastropods, such as abalone (*Haliotis* spp.), are fertilized externally in the surrounding water, and grow through a free-swimming trochophore larval stage (see Box 2.4 'The life-cycle of an abalone'). Internal fertilization, however, is most common. In this case, fertilized eggs are released either in

---

Box 2.3
Beautiful but dangerous

Cone shells are highly prized by shell collectors and beachcombers but handling live individuals of some species can be dangerous. There are over 400 different species of cones, and the particularly striking species known as the glory-of-the-seas, *Conus gloriamaris*, is legendary. Although superficially similar to the textile cone shown in Fig. B2.3.1, it has a much finer network of colouration. Only a few specimens were found during the hundred years following its original description and in the 1950s one shell sold for US$2000. But the activities of sports divers and the discovery of many more specimens has resulted in present prices decreasing to less than US$100. A cone secures its prey using a harpoon fired from a tube-like proboscis. In some species, the harpoon contains extremely potent venom and can penetrate clothing and human skin. Holding the cone by the wide end is less dangerous but not advisable – although the proboscis emerges from the narrow tapered end, it is quite extendable and can reach some distance from the shell.

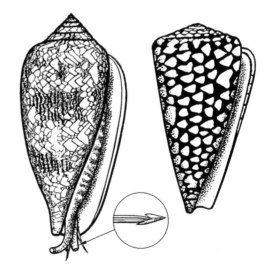

Fig. B2.3.1 Cone shells known to be dangerous to humans. The textile cone, *Conus textile*, and the marbled cone, *Conus marmoreus*. Sizes range up to 13 cm. The inset shows the poisonous dart contained within the proboscis (arrowed).

gelatinous strings, cases, or capsules, and the trochophore larval stage is not evident. About 25% of the global production of abalone comes from Tasmania, where the 125 divers in the fishery (in 2006) are required to hold a commercial dive (abalone) licence and authorization by the holder of the quota unit. The two main species of abalone commonly found in Tasmania are the blacklip abalone and the greenlip abalone. Populations of at least five abalone species on the US Pacific coast have been depleted and because of their susceptibility and value, the farming of various species is becoming more common.

Many other gastropods are used as food in coastal areas. Periwinkles (littorinids) and turban shells (*Turbo* spp.) are collected for food on temperate and tropical rocky shorelines in many parts of the world. In tropical regions, species such as conchs (*Strombus* spp.), and spider shells (*Lambis* spp.), are collected for food by subsistence fishers on rocky shores and coral reefs.

In the Mediterranean, the dye murex has been collected for making the crimson/purple dye known as Royal Tyrian purple since the days of the Phoenicians (see Box 2.5 'The first recorded case of overfishing?'). A wide range of gastropod molluscs are exploited just for their shells. The fascination with, and value placed on shells dates from the earliest of times. Shells have been used as currency by many civilizations and, in 600 BC, the first Chinese metallic coins were cast in the shape of small cowries. The scientific name of the money cowry, *Cypraea moneta*, reflects its use as currency in parts of Asia and in islands of the Indian Ocean.

The exploitation of gastropods for their shells is unquantified, but may account for a much larger live catch weight than does their exploitation for food. Because of the diversity in form and colour of their shells, many gastropod species are collected for the ornamental shell trade and for sale to tourists. The gastropod groups most in demand are the large, colourful species of cowries, cones, helmets, volutes, conchs and spider shells (Fig. 2.5). Gathering desirable species of gastropods provides an income for many tropical, coastal communities, and many development programmes have encouraged this activity (e.g. in Fiji, Parkinson, 1982). However, many of the shells sold to tourists in tropical countries are

---

Box 2.4
The life cycle of an abalone

Abalone or ear shells, *Haliotis* spp., are the bases of valuable dive fisheries in temperate waters, and have therefore been the subject of intensive research (Shepherd et al., 1992). Abalone are primitive molluscs, dioecious, and reach maturity at about three to four years of age. Greenlip and blacklip abalone have restricted spawning seasons, and eggs begin to develop in response to changing sea temperatures and the seasonal abundance of algal food. When sperm is released into the water from the holes of the male shell, the nearby females are induced to spawn. The fertilized eggs divide to form larvae which drift in the plankton. Within a day or so, trochophore larvae grow minute shells and form veliger larvae. After another one to several days the larvae settle on crustose coralline algae, and move about on an enlarged foot searching for microscopic algae. Adult abalone graze plants from rock surfaces and also catch drifting algae (Fig. B2.4.1).

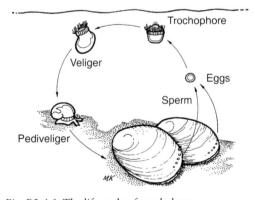

Fig. B2.4.1 The life cycle of an abalone.

not caught locally, but are imported from countries such as the Philippines, whose waters contain the greatest diversity of many marine species.

Several species with shells which have a pearly inner nacreous layer including abalone, *Haliotis* spp., green snail, *Turbo marmoratus*, and top shells, *Trochus* spp., are used in the manufacture of jewellery, ornaments and fashion accessories. The top shell, *T. niloticus*, is collected in the Indo-Pacific and used in the manufacture of buttons for European and Asian markets. Holes are drilled around the spiral shell to produce the circular pieces of shell which are polished to make buttons (Fig. 2.7). Australia, Indonesia and the Pacific islands supply about 7000 tonnes of trochus shell annually and concerns over depleted stocks have prompted studies on culture of the species and the reseeding of depleted reefs (Lee & Lynch, 1997).

Some commercially valuable species have been introduced to areas where they did not exist previously. The top shell, *Trochus niloticus*, for example, which has a natural distribution as far west as Yap in the Pacific Ocean, has been introduced into several island countries further to the west, including the Cook Islands, French Polynesia, and Tokelau (Gillett, 1988).

Fig. 2.7 Two top shells, *Trochus niloticus*, one of which has been drilled to remove the circular pieces of shell used to make buttons.

Cephalopods – squids and octopuses

The class Cephalopoda contains less than 1000 living species of octopuses, squids, cuttlefishes, and chambered nautiluses (Fig. 2.8). Cephalopods, particularly squids, support the largest catches of all molluscan fisheries, with an annual world landing approaching two million tonnes, comprising mainly of open-sea squid. A FAO guide to cephalopods of interest to fisheries has been prepared by Roper et al. (1984). Cephalopods

---

Box 2.5
The first recorded case of overfishing?

In the Mediterranean, murex snails have been traditionally collected for making dye since the days of the Phoenicians. The dye murex (Fig. B2.5.1) secretes a yellowish fluid which is treated and boiled to make the permanent crimson/purple dye known as Royal Tyrian purple. Such was the demand for the dye, that many towns in the western Mediterranean were settled by the Phoenicians as they roamed the seas in search of new beds of murex snails to exploit; the overexploitation of marine species is not a new thing. During the Roman Empire only senators and emperors were permitted to wear cloth dyed with Royal Tyrian purple. After the fall of the empire, cloth dyed in the same way was used by officials of the Christian church, and crimson came to be used as the official colours of a cardinal.

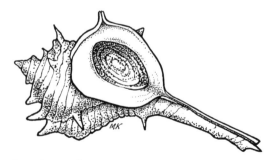

Fig. B2.5.1 The purple dye murex, *Murex brandaris* (8 cm).

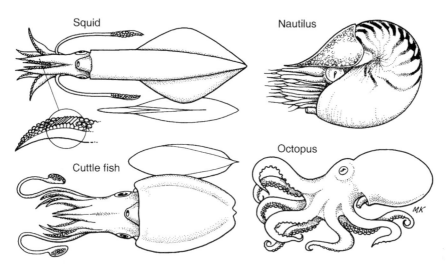

Fig. 2.8 Cephalopod molluscs, including the squid (*Loligo*), cuttlefish (*Sepia*), chambered nautilus (*Nautilus*), and octopus (*Octopus*). Ventral views of the squid and cuttlefish are shown with their internal shells. The inset shows the hectocotylus of the squid.

include the largest of all invertebrates. Some species of deep-water squid weigh over one tonne, although the average size of commercially caught squid is less than one kilogram.

Only *Nautilus* has a completely developed external shell. Squids and cuttlefish have internal shells, reduced in squids to a transparent 'pen' of chitinous material, and in cuttlefish, to a calcified 'cuttlebone' of light cellular material which aids in buoyancy. In octopuses the shell is absent. Squids and cuttlefish have evolved for a pelagic existence in the open ocean, and the mantle is often extended into two stabilizing fins. All cephalopods, including the octopus which is adapted for life on the sea floor, move through the water using a jet-like propulsion, in which water is drawn into the mantle and expelled through a funnel.

The cephalopod head is well developed with the mouth surrounded by arms, which, except for *Nautilus*, bears either suckers or hooks, or both. Squids and cuttlefish have eight arms and two longer tentacles, the octopus has eight arms, and the chambered nautilus has many slender arms without suckers. All cephalopods possess a pair of beak-like jaws to hold prey, and a radula which can be used to rasp at flesh.

All cephalopods are dioecious. In most species, the male has one or more modified arms (the hectocotylus) which is used to transfer spermatophores from the male's mantle cavity to the female. In the squid, *Loligo*, the fourth (ventral) left arm of the male is hectocotylized; on part of the arm, the suckers are replaced by transverse ridges. The eggs are fertilized as they are laid, and are often encased in a gelatinous material. In oceanic species, such as some squids, a large egg mass is left to drift in the sea, but in shallow water species, eggs are attached to the sea floor. Development is direct, and the eggs hatch into miniatures of the adults (see Box 2.6 'The life cycle of a squid').

Octopuses, which are usually nocturnal hunters on coral reefs and in rocky areas, are caught using traps, spears and hooks. They are an important food item in many coastal communities, and stocks have been depleted in some areas of high population. There are only five recognized species of chambered nautiluses (Saunders & Landman, 1987), and all have a limited distribution in the central Indo-Pacific. The most widely distributed species, *Nautilus pompilius*, is caught in traps set in deep water for snapper and caridean shrimps, and it is the shell of this species

---

Box 2.6
The life cycle of a squid

Squid life cycles (Fig. B2.6.1) are not well known, no doubt due to the difficulty of studying such seasonal and pelagic species. In general, spawning may occur in both spring and autumn. The broods of immature squid grow rapidly (1–2 cm mantle length per month), feeding on planktonic

crustaceans, fish and other cephalopods. Approximately one year later, squid move into shallow water to reproduce. Mating may be preceded by a courtship display involving arm movements and colour changes by the male. During mating, the male uses a modified arm (the hectocotylus) to transfer bunches of spermatophores from its own mantle cavity to that of the female. The eggs are fertilized by sperm from rupturing spermatophores as they are released by the female, and are often encased in a gelatinous material. The eggs hatch into miniatures of the adults. During the spawning period, large numbers of squid mass in particular areas, and these congregations are the bases of seasonal but highly productive fisheries. During the fishing season, larger individuals disappear from the catches, suggesting that squid either die after spawning or migrate elsewhere.

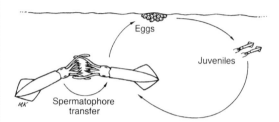

Fig. B2.6.1 The life cycle of a squid.

---

which is usually marketed in the ornamental shell trade. In the Middle Ages when small numbers of nautilus shells were obtained by European travellers in the East Indies and beyond, the shells were regarded as great treasures. Another family of cephalopods, confusingly referred to as paper nautiluses, are the argonauts (family Argonautidae) in which the females produce fragile shell-like cases that contain the eggs (see Box 2.7 'The argonauts').

Over 70% of the world total catch of cephalopods consists of squids (mainly *Todarodes*, *Loligo*, and *Illex* spp.), and the remainder is divided between octopuses (*Octopus* and *Eledone* spp.) and cuttlefish (*Sepia* and *Sepiella* spp.). Other than Japan and other Southeast Asian countries, several European countries, including Portugal, Italy, Spain and Greece have a tradition of cephalopod consumption.

Squids are divided into two suborders, the myopsid (covered-eyed) coastal squids, and the oegopsid (open-eyed) oceanic squids. The former have eyes covered by a transparent membrane with an eye pore anterior to the eye. Myopsid coastal squids such as species of calamari, *Loligo*,

and *Sepioteuthis* of the commercially important family Loliginidae (inshore squids) are often caught with baited jigs, purse seines and trawl nets. Oegopsid oceanic squid, including species such as *Illex*, *Todarodes*, *Nototodarus*, and *Ommastrephes*, of the family Ommastrephidae (flying squids) are taken by gill nets and jigging machines. Jigging machines have unbaited lures on droplines which are wound onto elliptical drums; the machine lowers each line to a preset depth, and retrieves it with a jerking or jigging motion.

Most of the catch of squid is taken by sea-going vessels from many different countries, Japan in particular. Important squid stocks include *Loligo gahi* and *Illex argentinus* in the Falkland Islands, *Loligo vulgaris reynaudi* in South Africa and *Todarodes pacificus* in the northern Pacific. The demand for squid has increased worldwide and after the decline of the *Todarodes pacificus* fishery in the 1970s, the search for new stocks has widened and assessment methods are being refined (Pierce and Guerra, 1994; Boyle & Rodhouse, 2005; Payne et al., 2006). With the decline in finfish stocks across the globe, the importance of squid fisheries is increasing. In the English

Box 2.7

The argonauts

Argonauts (sometimes called paper nautiluses) are abundant in all warm seas and are remarkable in that the female produces a fragile and shell-like case in which she lays her eggs. The male is much smaller than the female and does not make a 'shell'. The female has two specialized arms, which have membranous flaps that secrete and envelop the delicate egg case (Fig. B2.7.1). Argonauts are found in markets in India and Japan where they are used as food and the empty egg cases are prized by beachcombers and shell collectors.

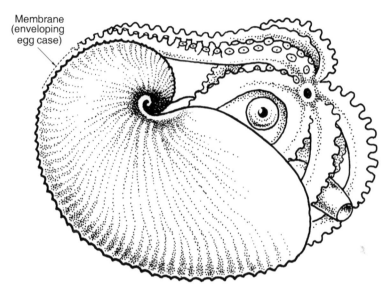

Membrane
(enveloping
egg case)

Fig. B2.7.1 A female argonaut, *Argonauta*, with her egg incubation chamber (up to about 25 cm in diameter) secreted by a membranous flap.

Channel, for example, total catches of finfish in 2003 were 25% lower than in 1983, but catches of cephalopods (mostly cuttlefish) increased by almost 300% from 8000 to 23 000 tonnes over the same period (ICES data in Payne et al., 2006).

Many molluscs, particularly bivalves and gastropods that are sessile and live in shallow water, are vulnerable to overexploitation. Wild stocks of clams, mussels, oysters, scallops and abalone have all been depleted in various parts of the world. Management regulations applied have included size limits, gear restrictions, periodic harvesting, closed areas, licensing systems and catch quotas. Stocks of some clams adjacent to areas of high population density have been overfished and drastic management measures have been imposed.

Many problems are related to the catching methods. In many clam fisheries the use of mechanical catching methods such as water jets or suction pumps are disallowed and in some cases commercial fishing has been totally banned. On the coast of California, the well-known Pismo clam, *Tivela stultorum*, is protected by an armoury of regulations including annual licence fees, a daily fee, a size limit, and a quota of 10 clams per person per day. The highly-prized toheroa, *Paphies ventricosum*, a bivalve mollusc endemic to New Zealand is so threatened that collecting the species has been totally banned except for the customary Maori catch.

A minimum size limit is a widely used regulation and has been applied to many commercial species including pearl oysters, abalone and top shells. A maximum size limit has also been applied to some species to provide protection for larger and more fecund, but less commercially valuable individuals; the shells of older trochus, for example, are often badly affected by a lifetime of attacks from parasites and borers. In the Torres Strait trochus fishery in Australia, a minimum size limit of 80 mm and maximum size limit of 125 mm is applied to all non-traditional fishing.

In commercial fisheries for temperate-water scallops (*Pecten* and related genera), minimum sizes, periodic harvesting seasons, licensing systems and catch quotas have all been applied. A major concern in scallop fisheries is the use of heavy and destructive dredges which cause damage to scallop beds and the sea floor (see Chapter 3 'Fishing gear and methods'). Stocks of the small Atlantic calico scallop, *Argopecten gibbus*, for example, taken by dredges and bottom trawls in North America have been heavily depleted.

In tropical scallops, which are faster growing, pre-season surveys have been used to determine the time of opening of the fishing season in order to maximize the weight and value of the catch (to prevent growth overfishing). In many fisheries for valuable species, including scallops and abalone, commercial fishers are required to hold licences which often reach a very high value.

Stocks of intertidal bivalves, gastropods and reef-dwelling octopuses may be protected by the use of reserves or marine protected areas. An area where fishing is prohibited may allow larger bivalves to reproduce undisturbed and produce larvae that drift and settle in nearby fished areas (see Chapter 6 'Fisheries management'). The establishment of protected marine areas appears to be one of the more effective ways of protecting the large numbers of gastropod species collected for the ornamental shell trade. The ease with which many shelled molluscs can be induced to spawn in simple hatcheries also allows the possibility of restocking depleted areas from artificially produced seed.

Management of pelagic resources is always difficult, and fisheries based on squids are particularly so. Squids generally have a fast individual growth, often over 20 mm per month, a short lifespan and recruitment appears to vary greatly from year to year. Pre-season surveys to determine allowable catches to prevent overfishing, and controlling fishing seasons to maximize the value of the catch are two management possibilities.

2.2.2 Crustaceans

About 35 000 marine animals including prawns, shrimps, lobsters and crabs, together with such animals as sea lice and barnacles, are crustaceans. Crustaceans are predominantly aquatic and are as ubiquitous in the sea as the insects are on the land. Most crustaceans are small and many species, such as copepods, remain planktonic for their entire lifespan. Valuable fisheries are based on the larger decapod crustaceans and these are emphasized in the abbreviated classification shown in Fig. 2.9.

Although superficially mollusc-looking, the acorn barnacle and the gooseneck barnacle shown in Fig. 2.9 are crustaceans in which the 'shell' is made up of plates just like those covering a shrimp or a lobster. Mantis shrimps, or stomatopods, have legs that are similar to those of the insect known as a preying mantis and they use their raptorial claws to hunt smaller creatures. Mantis shrimps are found in many tropical markets and large ones, greater than 20 cm in length, are sold alive from seawater tanks in Southeast Asian restaurants. The beautifully sculptured claws are sometimes used by jewellers, who coat the dried objects in gold to make earrings and brooches.

Amphipods, commonly called beach fleas or sand hoppers, are important scavengers and some species lead an almost terrestrial existence on beaches where they feed on plants and animals washed up by the waves. An even more terrestrial crustacean is the isopod known as a sow bug or slater, which is found under rotting wood and damp vegetation. Most isopods, however, are truly aquatic and are referred to as sea lice or sand lice because of a superficial resemblance to the insects that infest places of careless human habitation. Many species of marine isopods are

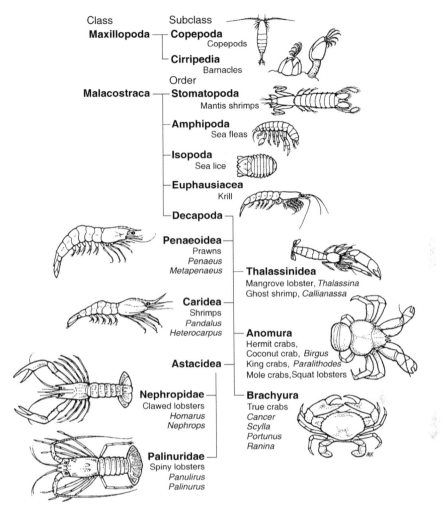

Fig. 2.9 An abbreviated classification of the subphylum Crustacea, accentuating species referred to in the text and commercially important species. A full classification is given in Martin and Davis (2001).

scavengers and are often so numerous and voracious that they are capable of stripping bait from fishing lines and traps in a short time; deep-water species of *Bathynomus* can reach sizes of 35 cm in length. Some specialized isopods, notably those of the family Limnoriidae, sometimes called gribble, are wood borers and cause massive destruction in timber wharves. Others are parasites on molluscs and other crustaceans as well as fish, in which they attach to the gills, tongues and fins, and burrow into flesh (Fig. 2.10).

The 80 or more shrimp-like species that make up the euphausids or krill, form dense shoals in Antarctic waters where they feed on diatoms.

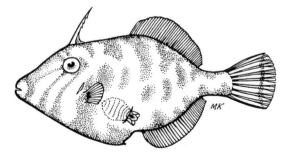

Fig. 2.10 The parasitic isopod, *Ourozeuktes owenii*, embedded in the side of a leatherjacket, has grown to such a size that it is unable to leave its host. Redrawn from Hale (1976).

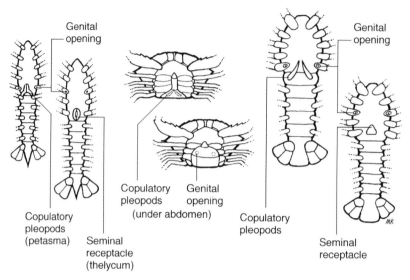

Fig. 2.11 Diagrammatic representations of the reproductive structures of a penaeid prawn (left), a crab (middle) and a lobster (right). Males are shown at the top, and females below.

They are important in southern food webs and are eaten by fish and birds as well as baleen whales; the blue whale, for example, can sieve as much as 8 tonnes of krill from the water each day. The high densities of krill populations, often over 20 kg/m^3 of seawater, have encouraged fishing by Japanese and Polish fleets.

Crustaceans typically have jointed legs and other appendages attached to a segmented body covered with a hard chitinous shell (or exoskeleton) hardened with calcium carbonate. Decapods have a carapace which encloses a fused head and thorax (the cephalothorax), and an abdomen which consists of six segments. In prawns and lobsters, the cephalothorax is referred to commercially as the 'head', and the abdomen as the 'tail'.

In all decapods, the first three pairs of thoracic appendages are modified as maxillipeds, and the name Decapoda refers to the remaining five pairs of thoracic appendages which are legs, or pereiopods. The abdomen bears five pairs of appendages, or pleopods. In prawns and shrimps the abdomen is generally well developed with five pairs of pleopods adapted for swimming. In lobsters and crabs, the cephalothorax is relatively large and bears five pairs of legs adapted

for crawling, or modified to include claws. In true crabs (Brachyura), the abdomen is usually greatly reduced and curved tightly under the cephalothorax.

During mating, the males release sperm from a genital opening at the base of the fifth pair of pereiopods. Appendages of the male's first pair of pleopods may be modified for the transfer of sperm clusters (spermatophores) to an organ or depression on the ventral carapace of the female (Fig. 2.11). In most decapods, the female has paired oviducts which open to the exterior at the base of the third pair of pereiopods, although in most true crabs, the oviducts lead directly into a seminal receptacle. When spawning occurs, the eggs are released and fertilized as they pass over the stored sperm. In all commercial crustaceans except penaeid prawns, fertilized eggs become attached to the pleopods beneath the female's abdomen. In penaeids, fertilized eggs are released directly into the sea, and hatch to nauplius larvae. In other crustaceans which carry broods, the eggs hatch into more advanced larval stages.

The possession of a hard exoskeleton prevents crustaceans from growing with a gradual increase in size as other animals do. In order to grow in size, a crustacean must periodically cast off its restrictive exoskeleton, and expand in the short time before a new and larger shell hardens. This process, called moulting, involves several stages (see Box 2.8 'New shells for old – the crustacean moult cycle'). The moult cycle of crustaceans is under hormonal control. A hormone which initiates moulting is secreted from the Y-glands in the ventral part of the head. However, a hormone which inhibits both moulting and ovary development is produced by the X-organ within each eye stalk. Although the moulting process is under hormonal control, environmental factors may provide the stimulus which triggers moulting, presumably by interrupting the supply of moult-inhibiting hormone. In commercial catches of penaeid prawns, for example, the number of moulting (soft-shelled) individuals is often greatest at the time of full moon.

In theory, two different patterns of step-wise growth are possible in crustaceans. Moulting may occur at regular time intervals, and the amount of growth occurring (the growth increment) at each succeeding moult decreases with increasing age (Fig. 2.12a). On the other hand, the growth increment at each successive moult may be the same, but the inter-moult period increases with increasing age (Fig. 2.12b). Although either strategy would result in a step-wise growth pattern, the growth of most species of crustacean appears to be a combination of both decreasing growth increments and increasing inter-moult periods. If the step-wise growth of individuals in the population is unsynchronized, the average growth of the population approximates the usual smooth growth curve of animals without a hard exoskeleton. A further

---

Box 2.8

New shells for old – the crustacean moult cycle

The process of crustaceans casting off their shells is called moulting, and involves several stages (Fig. B2.8.1). In the pre-moult stage, a new soft exoskeleton forms beneath the older, hard exoskeleton. Calcium from the old exoskeleton is absorbed into the blood, and the new one becomes tinged with red carotene. In the moult, or ecdysis stage, the old shell splits at fuse lines, one of which is usually at the junction of the cephalothorax and the abdomen, and is shed by the soft-shelled crustacean. In the post-moult stage, the animal expands, assisted by the uptake of water, before the new, larger shell gradually hardens. Calcium salts, in temporary storage in the blood, are deposited in the hardening shell, and as soon as the mandibles are sufficiently hard, the animal feeds voraciously. The animal is particularly vulnerable until its shell fully hardens, and often hides or burrows (depending on the species) to avoid predators; these behavioural changes may result in a reduced vulnerability to fishing gear such as traps

and trawls. Each moulting event is separated by an inter-moult period which may be relatively long, and during which the animal may increase in weight (particularly in the early part of the period), but not in length.

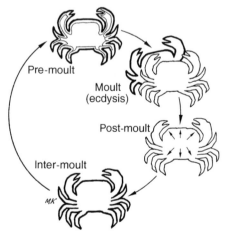

Fig. B2.8.1 Stages of the crustacean moult cycle: pre-moult, moult (ecdysis), post-moult, and inter-moult. The heavy outline represents the hard shell, and the light outline represents the soft shell.

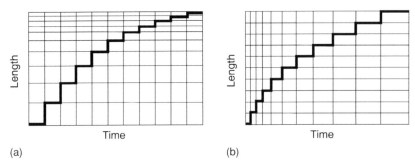

(a)   (b)

Fig. 2.12 Crustacean growth. (a) Moulting at regular time intervals with decreasing growth increments; (b) moulting at increasing time intervals with similar growth increments.

possibility is that, as crustaceans are particularly vulnerable to predation during moulting, a survivorship curve for a crustacean with a synchronized moult cycle will decrease in a step-wise pattern, rather than in the smoothly decreasing curve of other groups (see Chapter 4.3).

Penaeids and carideans – prawns and shrimps

The two groups Penaeoidea and Caridea contain species commonly referred to (sometimes interchangeably) as either shrimps or prawns. This text refers to penaeids as prawns, and carideans as shrimps. Carideans differ from penaeids in that the pleuron (covering shell) of the second abdominal segment overlaps the pleura of both the first and the third segments, and the third pair of thoracic (walking) legs does not have pincers.

Unlike penaeids, caridean shrimps carry fertilized eggs externally beneath the abdomen, which is often proportionally smaller than that of penaeids.

Penaeids include the widespread tropical and subtropical exploited prawns of the genera *Penaeus* and *Metapenaeus*. Species of the genus *Hymenopenaeus*, often called royal red prawns commercially, are found in deeper, cooler waters. Most penaeid prawns can be identified by the number of 'teeth' on the rostrum, and the sculpturing of the carapace. *Penaeus* species have one or more teeth on the lower margin of the rostrum, and the smaller *Metapenaeus* species have none (Fig. 2.13). A guide to commercial species of the world is given in Holthuis (1980) and to Australian species in Grey et al. (1983). Many

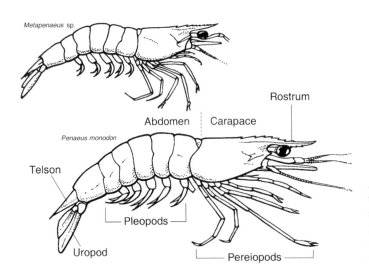

Fig. 2.13 Penaeid prawns. Top: the greasy-back prawn, *Metapenaeus* sp. Bottom: the giant tiger prawn, *Penaeus monodon*. The bottom illustration shows the general external features.

commercial penaeids now have new generic names after a revision by Pérez-Farfante and Kensley (1997), who split the genus *Penaeus* into several genera including *Litopenaeus*, *Farfantepenaeus* and *Fenneropenaeus*.

Prawns are swimming crustaceans with five pairs of swimming legs (pleopods) as well as five pairs of walking legs (pereiopods), the front three of which have claws (chelae). Although prawns are capable of swimming, they spend most time on, or close to, the sea floor. They are usually nocturnal, burrowing into the bottom sediments during the day, and emerging to feed at night. As opportunistic omnivores, they feed on a wide range of plant and animal material. Although a major part of their diet consists of bacteria and microscopic algae growing as a film on particles of sand and mud, prawns will also feed on animal material including faeces and the carcasses of larger animals. They are also able to feed on small mobile animals, including molluscs, polychaete worms, and other smaller crustaceans.

All commercially important penaeid prawns have separate sexes, and the female, which is usually larger than the male, produces several hundred thousand eggs in ovaries, visible through the shell or exoskeleton. Male prawns can be identified by the presence of a sex organ, the petasma, joining the base of the first pair of swimming legs (Fig. 2.11). In some species, mating occurs between a hard-shelled (inter-moult) male and a recently moulted, soft-shelled female. During mating, the petasma is used to transfer spermatophores to the thelycum, located between the last pair of walking legs of the female. During spawning, eggs are released from the female's genital opening, at the base of the third walking leg, and are fertilized as they pass over the spermatophores. The fertilized eggs are not carried beneath the abdomen as in most other crustaceans, including caridean shrimps, but are released directly into the sea.

Penaeid prawns are the bases of valuable trawl fisheries in coastal, warmer waters throughout the world. The distribution of penaeid prawns is generally restricted to tropical and subtropical waters, although some species range as far south as Tasmania and South Africa. In the northern hemisphere, penaeids have entered the Mediterranean Sea through the Suez Canal and the Japanese prawn, *Penaeus japonicus*, is now fished on the coasts of Egypt, Israel, and Lebanon and is farmed in Italy.

Because of the requirement of most penaeids for brackish water, they are found only in areas which have sufficient rainfall to produce rivers and estuaries. There are exceptions, such as the king prawn, *Penaeus latisulcatus*, in which the juveniles grow in high salinity nursery areas (see Box 2.9 'The life cycles of penaeid prawns'). In some tropical areas where prawns do exist, extensive coral formations often preclude the use of towed trawl nets, except in small areas near the mouths of rivers, where brackish water prevents the growth of coral. Even where only small fishing areas exist, these may be sufficient to supply brood stock or juveniles for aquaculture development.

In Southeast Asia, penaeids have been farmed as incidental crops in tidal fishponds for centuries. The giant tiger prawn, also called the jumbo tiger prawn and black tiger prawn, *Penaeus monodon*, is the most commonly marketed prawn in the world, and most are from Southeast Asia, where increasing numbers of prawn farms have resulted in the loss of coastal wetlands.

Penaeid prawns are farmed in many tropical areas, where shallow-water ponds are stocked with juveniles collected from the wild. Modern prawn farming began in the 1930s with the spawning of the Japanese prawn *Penaeus japonicus* by M. Fujinaga in Japan, although intensive outdoor culture did not appear until the 1960s. In the late 1950s, techniques for culturing marine phytoplankton for feeding larval penaeids were developed in the USA. Now penaeid prawns are farmed in many parts of the world including the USA, Latin America, and Australia.

Most commercially important species of caridean shrimp belong to the family Pandalidae. Pandalids include the cold and temperate water shrimps of the genus *Pandalus*, which are the bases of important fisheries in the northern hemisphere, and *Heterocarpus reedi*, which is trawled

Box 2.9
The life cycles of penaeid prawns

There are three types of life cycle in penaeid
prawns – estuarine, oceanic, and 'mixed' estuarine
and oceanic. Some *Metapenaeus* species
complete their life cycle in brackish estuarine
waters while others such as the brown tiger prawn,
Penaeus esculentus, live entirely in oceanic waters.
However, most commercially important species of
Penaeus have a 'mixed' life cycle, in which adult
prawns live and spawn in oceanic waters, and the

juveniles grow in shallow water where the salinity
is usually low (Fig. B2.9.1). In a 'mixed' life cycle
species, the larvae progress through nauplius,
protozoea, and mysis stages before the demersal
post-larvae reach inshore creeks and estuaries
which act as nursery areas for the juveniles.
Prawns in this category include the banana prawn,
Penaeus (Fenneropenaeus) merguiensis, and the
king prawn, *Penaeus latisulcatus*; in the latter,
juveniles grow in shallow water which is often
hypersaline.

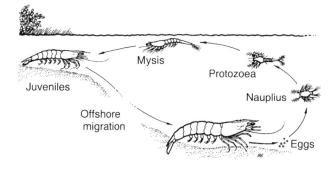

Fig. B2.9.1 The life cycle of
a penaeid prawn.

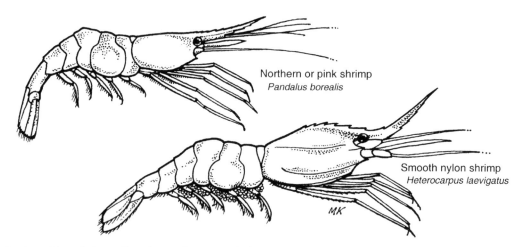

Fig. 2.14 Caridean shrimps. Top: the northern or pink shrimp, *Pandalus borealis*. Bottom: the smooth nylon shrimp,
Heterocarpus laevigatus. Note the egg mass carried on the pleopods of the nylon shrimp.

off the Pacific coast of South America. Species of
both genera are shown in Fig. 2.14.

The northern or pink shrimp, *Pandalus borealis*, is a circumpolar species found in the Atlantic
Ocean in Canada, Greenland, Iceland and as far

south as the English Channel, and in the Pacific
Ocean in Japan, the Bering Strait and as far south
as Washington state in the USA. They are primarily caught with bottom trawls and traps and
are commercially fished in Canada, Greenland,

Norway and Iceland. Some pandalids, such as the spot shrimp, *Pandalus platyceros*, in North America, are also caught in traps. Shrimps of the Pacific coast of Canada are reviewed by Butler (1980). The infraorder Caridea also includes the giant freshwater shrimps of the genus *Macrobrachium*, which are farmed in many tropical countries.

Exploited temperate-water pandalids are protandrous hermaphrodites: after functioning as males for three to four years, they change sex, and spend the remainder of their life cycle as females. Fertilized eggs are carried beneath the female's abdomen, on the hairs of the pleopods, before hatching. During this incubation period, larvae develop through the nauplius stage within the eggs. The eggs hatch directly to advanced protozoea larvae which are planktonic.

Deep-water pandalids, particularly those belonging to the genus *Heterocarpus*, appear to have a worldwide distribution in tropical waters. Unlike temperate-water pandalids, deep-water tropical species have separate sexes. A range of species has been caught in surveys using either baited traps or trawls in depths between 200 m and 800 m off continents and islands in the Caribbean, the Indian Ocean and the Pacific Ocean. In most cases, however, catch rates have been insufficient to offset the high estimated costs of commercial fishing in deep water (King, 1986).

Nephropidae – clawed lobsters

The Nephropidae includes the large clawed lobsters of the genus *Homarus*, which are the bases of valuable fisheries in the northern hemisphere. The European lobster, *H. gammarus*, is caught in traps off rocky shores in northwestern Europe, and is similar to the American lobster, *H. americanus* taken off the Atlantic coast of North America (Fig. 2.15). Most species of lobster are caught in pots or traps. There are many related freshwater species (generally referred to as crayfish) with a worldwide distribution including the largest of all freshwater species, the Tasmanian

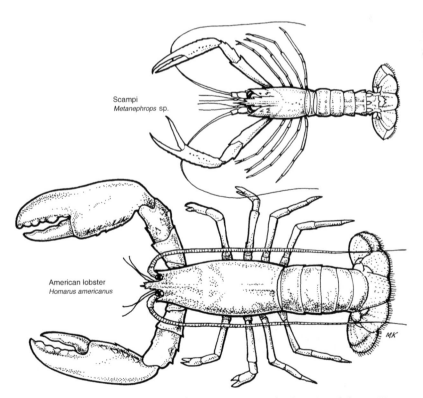

Fig. 2.15 Clawed lobsters. Top: the scampi, *Metanephrops* sp. Bottom: the American lobster, *Homarus americanus*.

Lobster are some of the most valuable of all seafood products. Here, a freshly cooked clawed lobster is displayed in front of a boiling vat at a stall at Fisherman's Wharf, San Francisco.

crayfish, *Astacopsus gouldi*, which grows to a weight of about 6 kg.

The family Nephropidae includes the crustaceans commonly called scampi, Dublin Bay prawns, or Norway lobsters (Fig. 2.15). The best-known commercial scampi species, the Norway lobster, *Nephrops norvegicus*, is caught in the North Atlantic Ocean from northern Norway to the Mediterranean. The Caribbean scampi (or 'lobsterette') is abundant from the Gulf of Mexico to the Caribbean. Several scampi species are widely distributed across the northern and southern Indian Ocean, and *Metanephrops velutinus* (= *andamanicus*) forms the basis of small but valuable trawl fisheries in southeast Africa and northwestern Australia. Scampi may be caught in traps but are more commonly trawled in depths between 300 m and 600 m.

Male lobsters are generally heavier, and have larger claws than females. The first pair of pleopods also differ, in being adapted for transferring spermatophores to the female; the copulatory pleopods, which are hard, tapered, and grooved, are the most conspicuous external differences between males and females (Fig. 2.11). Mating occurs between a hard-shelled male and a recently moulted, soft-shelled female. During mating, the male uses the first pair of pleopods to transfer spermatophores from the genital openings at the base of the fifth pair of walking legs to a triangular seminal receptacle on the ventral side of the female. The spermatophores remain as a gelatinous plug in the seminal receptacle of the female until spawning, which occurs at any time between one month and two years after mating. Eggs are released from the female's genital openings, at the base of the third walking legs, and are fertilized as they pass over the spermatophores. The fertilized eggs become cemented to the hairs on the pleopods where they are carried beneath the female's abdomen for up to a year before hatching.

As lobsters and scampi hatch from eggs at a comparatively late stage of development, the larval transformations are less extensive than those of penaeid prawns. The planktonic larvae pass

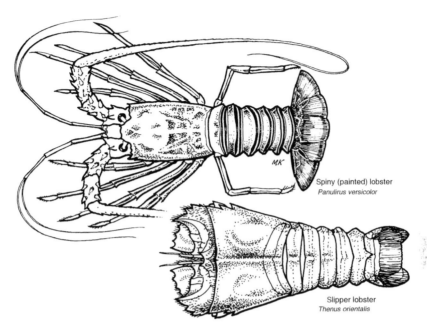

Spiny (painted) lobster
Panulirus versicolor

Slipper lobster
Thenus orientalis

Fig. 2.16 Palinurid lobsters. Top: the spiny lobster, *Panulirus versicolor*. Bottom: the slipper lobster, *Thenus orientalis*.

through three stages, which are of a similar appearance, within three weeks before settling on the sea floor. The reproduction and life cycle of clawed lobsters is similar to those of spiny lobsters, except that the larval transformations and periods are less extensive.

Palinuridae – slipper lobsters and spiny lobsters
The Palinuridae includes two distinct types of clawless lobsters, the slipper or shovel-nosed lobsters, and the spiny lobsters (Fig. 2.16). Slipper lobsters of the family Scyllaridae are so-named for the flat carapace, and the broad, flattened antennal plates at the front of the head. Some species are caught in shallow water, such as the queen slipper lobster, *Scyllarides squamosus*, which is found on coral reefs, but most are taken in deeper waters down to depths of 500 m. Some species are caught incidentally in spiny lobster fisheries, and as a bycatch in trawling operations for penaeid prawns and saucer scallops.

Spiny or rock lobsters of the family Palinuridae lack the claws of true lobsters, such as *Homarus*, but have spine-covered carapaces and antennae which are used in defence. Reproduction is similar to that of clawed lobsters, except that females do not have a seminal receptacle, and spermatophores are deposited as waxy masses on the ventral surface of the female's thorax. The most conspicuous external differences between males and females are the two terminal branches on the female's pleopods (to aid in egg carrying), and the presence of a small claw at the end of the female's fifth pair of thoracic legs. The life cycle of a spiny lobster is shown in Box 2.10 'The life cycle of the spiny lobster'.

Species of the genus *Palinurus* are the bases of small but important fisheries in western Europe and the Mediterranean. The spiny lobster or langouste, *P. elephas*, occurs from as far north as the Hebrides off northern Scotland, to the coast of northwest Africa and into the Mediterranean. Species of the genus *Jasus* are distributed in temperate waters of the southern hemisphere, and are caught in baited traps, or pots, in southern regions of Africa, Australia, and New Zealand.

Spiny lobsters of the genus *Panulirus* are more tropical than those of the genera *Palinurus* and *Jasus*. Their distribution ranges from the California spiny lobster, *Panulirus interruptus*, and the

Box 2.10
The life cycle of a spiny lobster

The Western Australian rock lobster, *Panulirus cygnus*, is the basis of one of the world's largest spiny lobster fisheries, and is used here as an example of a life cycle (Fig. B2.10.1). Unlike many other *Panulirus* species, the western rock lobster is readily caught in baited traps. Female rock lobsters reach sexual maturity within seven years and in a pre-spawning moult around June, the fringing hairs, or setae, of the female's pleopods become longer. During mating, in July and August, the male releases spermatophores, which are cemented as a 'tar spot' on the ventral side of the female. Within the female's cephalothorax, up to a million orange eggs become mature, and are eventually released through pores at the base of the third pair of legs.

The fixed claws on the fifth pair of legs are used to scratch the tar spot and release the stored sperm. The eggs are fertilized as they are drawn across the tar spot by currents created by the beating of the pleopods. The eggs become attached to the long setae on the pleopods. A female may carry the eggs (remain 'berried') for about three months before they all hatch over a period of several hours. Leaf-like phyllosoma larvae are released, and drift in the plankton for up to eleven months, during which time they may make a journey of at least 1500 km out into the Indian Ocean before returning to concentrate at the edge of the continental shelf on the Western Australian coast. Between August and December, the phyllosoma larvae metamorphose into puerulus larvae capable of swimming towards the coast, and eventually settling in algal growth on reef areas.

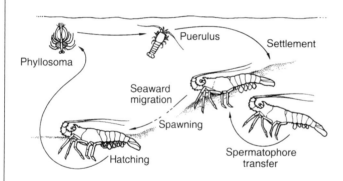

Fig. B2.10.1 The life cycle of the rock lobster, *Panulirus cygnus*. Adapted from Phillips et al. (1980).

Caribbean spiny lobster, *P. argus*, in northern latitudes to the Western Australian rock lobster, *Panulirus cygnus*, in southern latitudes. Some species of *Panulirus*, including *P. cygnus* and *P. argus*, are readily caught in baited traps and hoop nets, and these species are the bases of the largest spiny lobster fisheries in the world. However, many species of the genus *Panulirus* appear not to be attracted to bait, and are caught by other means.

The ornate lobster, *Panulirus ornatus* and the double-spined lobster, *Panulirus penicillatus*, in particular are widely distributed in the Indo-Pacific, and a review of these species is given in Prescott (1988) and Pitcher (1993). The preferred habitat for *P. penicillatus* is on the exposed sides

of coral reefs down to a depth of 10 m, where they are caught by divers. During the night, the species forages over the reef tops where they are taken by groups fishing with spears and lights. The ornate lobster, *P. ornatus*, is usually taken by divers or by using tangle nets and is fished commercially in many countries bordering the Indian Ocean. Ornate lobsters are known to make an annual migration from the coast of northern Australia across the Torres Strait to breeding grounds in Papua New Guinea (Bell et al., 1987). During this migration, large aggregations moving across the sandy sea floor have been targeted by vessels towing trawl nets. Catches were often greater than three tonnes per haul until 1984 when trawling was made illegal to protect breeding stocks. At present, the species is locally important as a dive and tangle net fishery.

Brachyuran crabs

With approximately 4500 species, the infraorder Brachyura, which includes the true crabs, are the most successful of all decapods. Species are distributed over a great range of latitudes and depths, and some crabs live in freshwater and even terrestrial habitats. The abdomen of true crabs is usually greatly reduced, and curved tightly up under the cephalothorax. The pleopods, or swimming legs, are retained in the female for carrying eggs, but only the anterior pleopods, which are adapted for copulation, are retained in the male. The width of the abdomen is often narrower in the male than the female (Fig. 2.11). The

walking legs are well developed with the first pair of legs (the chelipeds) often bearing large claws or chelae. The infraorder includes the Japanese spider crab, *Macrocheira kaempferi*, which is possibly the world's largest crustacean, with a maximum length from claw to claw exceeding 3.5 m. A wide range of crabs is collected for food including many small species that are used as flavouring in soups and rice dishes. Most crabs are edible, although a number of xanthid or pebble crabs, which live on coral reefs of the Indo-Pacific are poisonous (see Box 2.11 'Toxic crabs'). Some crabs of commercial importance are shown in Fig. 2.17.

The cancer crabs represent the typical crab known from markets in Europe. The European crab, *Cancer pagurus*, occurs in northwest Europe, from Scandinavia to Portugal, and is

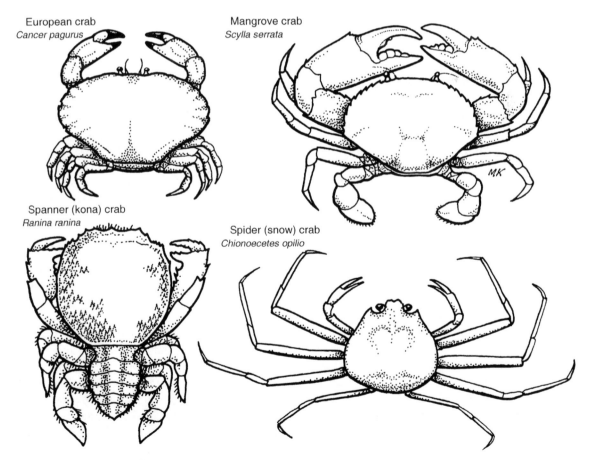

European crab
Cancer pagurus

Mangrove crab
Scylla serrata

Spanner (kona) crab
Ranina ranina

Spider (snow) crab
Chionoecetes opilio

Fig. 2.17 Brachyuran crabs.

Box 2.11
Toxic crabs

Most, but not quite all, crabs are edible. Some xanthid or pebble crabs (family Xanthidae) which live on coral reefs are poisonous. Xanthid crabs contain saxitoxin, one of the most powerful of all neurotoxins, which results in paralytic shellfish poisoning (PSP). Symptoms of PSP include tingling and burning sensations followed by numbness and eventual paralysis. The large and eye-catching toxic reef crab, *Zosimus aenus*, with its brown or red pattern of blotches on a pale cream background is sometimes common on coral rubble and reefs (Fig. B2.11.1). The flesh of this meaty crab, even when cooked, is toxic, and can result in severe sickness and death from respiratory failure.

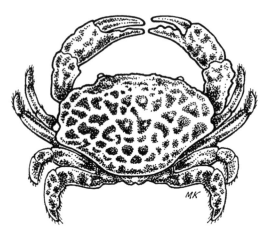

Fig. B2.11.1 The toxic reef crab, *Zosimus aenus* (10 cm carapace width).

taken in traps and baited nets. Other species of the genus, including the jonah crab, *C. borealis*, and the rock crab, *C. irroratus*, are caught on the Atlantic coast of North America, and the Dungeness crab, *C. magister*, is caught on the Pacific coast of North America.

A primitive group of burrowing crabs includes the spanner or frog crab, *Ranina ranina*. The spanner crab, which takes its name from the unusual spanner-like shape of its claws, is believed to be present in deeper water from Africa across the Indo-Pacific to Hawaii (where it is known as the Kona crab). The species is the basis of commercial trap fisheries in Hawaii and northeastern Australia.

Although most crabs have evolved for walking, members of the family Portunidae are powerful swimmers. Members of this family, which contains such commercially important species as the blue crab, *Callinectes sapidus*, the mangrove crab, *Scylla serrata*, and swimming crabs of the genus *Portunus*, have flattened paddles at the end of the last pair of legs. On the Atlantic coast of America, the blue crab is distributed from Cape Cod in the north to as far south as Uruguay, and is the basis of a renowned fishery around Chesapeake Bay. Catching methods include the

use of traps (pots) and dredges. A gourmet market for soft-shelled blue crabs is met by catching pre-moult crabs, which are held in floating cages until just after ecdysis. The blue crab is used as an example of a brachyuran life cycle in Box 2.12. Mangrove crabs have been farmed using juveniles captured from the natural environment, but methods of producing larvae have been discovered providing the potential for closed-circuit aquaculture (Keenan & Blackshaw, 1999).

The world demand for crab meat is high, and large stocks of spider crabs are heavily fished. Exploited species include *Chioneocetes opilio*, (commercially called the snow crab) in North America and *Chioneocetes opilio elongatus* (labelled as snow crab and zuwai crab for export) in Japan.

Many species of crabs are caught for food merely by hand collection, or by spearing, on shallow-water reefs and in estuaries, and some artisanal fisheries may involve the use of traps and hoop nets. Large commercial fisheries may be based on the use of large baited traps, trawl nets, and either baited or unbaited tangle nets. Dredges are sometimes used to collect crabs, such as the American blue crab, when these are buried in the sea floor. Some species, particularly the

> **Box 2.12**
> **The life cycle of a blue crab**
>
> The commercial blue crab, *Callinectes sapidus*, is the basis of a valuable fishery around Chesapeake Bay on the Atlantic coast of North America and its life cycle is illustrated in Fig. B2.12.1. Prior to mating, the hard-shelled male blue crab locates a pre-moult female, which he carries below him until she moults. After the female moults, the male transfers spermatophores via the modified first pair of pleopods to seminal receptacles on the ventral side of the female's thorax (see Fig. 2.11). Within a period of nine months after mating, females spawn and up to 2 million eggs are released and fertilized by the stored sperm. The fertilized eggs form a spongy mass attached to setae on the female's pleopods. At this stage, females living in brackish water must migrate out to sea, where water salinity is sufficiently high to permit the eggs to hatch. The eggs hatch to planktonic zoea larval stages, which develop through several moults into megalopa larvae, before drifting back to inshore areas.
>
>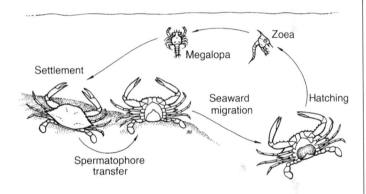
>
> *Fig. B2.12.1* The life cycle of the blue swimming crab, *Callinectes sapidus*.

swimming portunid crabs, are caught as a bycatch in prawn and fish trawling operations.

Anomuran crabs

Although superficially crab-like, hermit crabs, porcelain crabs, king crabs, and horseshoe crabs belong to the infraorder Anomuran. Unlike true brachyuran crabs, the anomuran abdomen possesses uropods and a telson that in some species is used for either burrowing (as in hippoid mole crabs) or holding the body inside a gastropod shell (as in hermit crabs). The last pair of legs are reduced and often hidden under the carapace so that only four pairs of thoracic legs are visible.

In most hermit crabs, the abdomen is not curved under the cephalothorax as in brachyuran crabs, but is modified to fit within the spiral chamber of empty gastropod shells. The crabs use empty shells for shelter, never killing the original occupants, and seek larger shells to accommodate their growth. Two groups of anomurans, the lithodid or stone crabs and the coconut crabs (Fig. 2.18), have evolved away from a total dependence on housing their abdomens in gastropod shells.

In lithodid crabs, the abdomen is flexed beneath the thorax as in true (brachyuran) crabs, but is not symmetrical, and bears pleopods only on the left side. Lithodid crabs include the red king crab, *Paralithodes camtschaticus*, whose common name leads to confusion with the horseshoe crab, *Limulus*, also called a king crab but which is more closely related to spiders than to crabs. The lithodid king crab (Fig. 2.18), which occurs across the northern Pacific from Korea through Kamchatka to the coast of Canada, reaches a maximum weight of over 12 kg and is a major source of canned and frozen crab meat. The king crab and related species are caught in

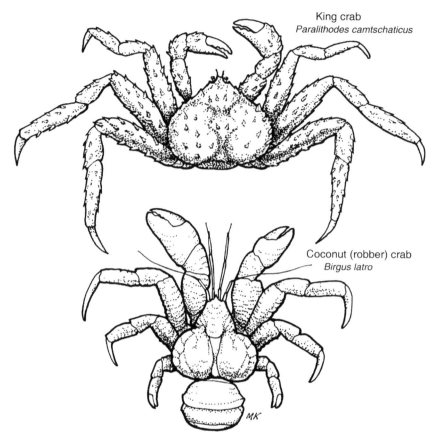

King crab
Paralithodes camtschaticus

Coconut (robber) crab
Birgus latro

Fig. 2.18 Exploited anomuran crabs. Top: the red king crab, *Paralithodes camtschaticus* (12 kg). Bottom: the coconut, or robber crab, *Birgus latro* (4 kg).

large baited traps or pots, and catches of up to one tonne per haul have been made. However, due to declining stocks, the fishing season has been repeatedly reduced and in the four-day season permitted in 2005/06, 250 boats caught over 6000 tonnes of red king crab.

The coconut, or robber crab, *Birgus latro*, (Fig. 2.18) which spends most of its life cycle on land, reaches weights of over 4 kg. Although the juveniles are housed within shells as other hermit crabs, the large adults have evolved away from requiring this protection. The adults possess massive crushing claws, and long legs which enable them to climb trees. The diet includes carrion, rotting leaves, fruit, and coconut flesh, and large adults are capable of husking and breaking through the shell of a fallen coconut.

The coconut crab is distributed in tropical islands from the Seychelles in the Indian Ocean to the Tuamotu Archipelago in the Pacific Ocean, and is a highly valued food item in many parts of its range. Its vulnerability to domestic and feral animals as well as the destruction of its coastal habitat has probably accounted for its disappearance on continents. Coconut crabs are a highly prized food item and its ease of capture has resulted in its depletion in many island countries. In the tropical Pacific, the species has been the subject of an intensive research programme designed to examine the conservation and management of its populations, and to investigate the possibility of farming the species. Reviews are presented in Brown and Fielder (1991), Fletcher (1993) and Fletcher and Amos (1994), from which details of the life cycle given in Box 2.13 'The life cycle of the coconut crab' are derived.

Box 2.13
The life cycle of the coconut crab

Coconut crabs are sexually mature at approximately five years of age. Mating occurs during the summer, and spermatophores are deposited on the female. About three weeks after copulation, the female extrudes her eggs, which are fertilized as they pass over the spermatophores to form a spongy, orange egg mass attached to the pleopods. As the embryos mature, and their eye spots develop, the eggs darken to a dark brown-purple colour. When the embryos are fully developed, the female moves to the shoreline and into shallow water where the eggs hatch on contact with the water to zoea larvae less than 3 mm in length. The larvae develop through four zoeal larval stages within three weeks, and then metamorphose into benthic glaucothoe stages about 4 mm long. After finding and entering a

suitable gastropod shell, glaucothoe emerge from the water and live among coral rubble in the upper intertidal zone. Glaucothoe metamorphose into juveniles at a size of approximately 5 mm, and discard the gastropod shell soon after. Juveniles adopt a terrestrial habit, and move further inland as they grow (Fig. B2.13.1).

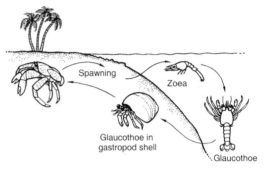

Fig. B2.13.1 The life cycle of the coconut crab, *Birgus latro*.

Dungeness crabs found off the west coast of North America are targeted in one of the world's most famous crab fisheries.

Trawl nets used in prawn and shrimp fisheries are blamed for being unselective (indiscriminately catching many different species) and causing damage to marine ecosystems. Management measures include the compulsory fitting of devices that reduce the bycatch of other species (see Chapter 3).

In many respects, penaeid prawns with their high growth rates, high fecundity, and high commercial value represent an ideal fisheries resource. Because most penaeids reach sexual maturity in less than one year, and produce large numbers of eggs, stocks of these species can generally recover after high levels of exploitation. However, penaeid populations fluctuate greatly from year to year and a major threat to the sustainability of many fisheries is the destruction of inshore nursery areas through coastal development.

Restrictions of the mesh size in prawn trawl nets is a common regulation directed at reducing the capture of small prawns, although the survival prospects of small individuals passing through the meshes are not generally known. The use of closed fishing seasons at the time of recruitment, when young prawns are migrating out to the fishing grounds, may be a preferable way of delaying the age at capture. Closures are also imposed in some prawn fisheries at the time of full moon, when many prawns are soft-shelled, and therefore of low commercial value. Large commercial fisheries are often subject to limited entry through a licensing system, and the imposition of catch quotas (see Chapter 6).

Trap fisheries, for caridean shrimps and lobsters, are less damaging to the environment than trawls, and a minimum mesh size in the traps is usually imposed to reduce the catch of small individuals. There have been some concerns that lines from traps to buoys have resulted in gear entanglements with endangered North Atlantic right whales. Fisheries regulations applied to trap fisheries may include catch quotas, licensing systems, and restrictions on the number of traps carried per boat. The traps used may be required to include an escape gap, a rectangular or round hole worked into the side of the trap, to allow the escape of small individuals.

Regulations to protect females or spawning females are common in trap fisheries based on crustaceans such as crabs and lobsters, in which the different sexes are easy to distinguish. Some regulations require the return of egg-bearing, or berried, females to the sea and others require the return of all females, irrespective of reproductive condition. In some trawl fisheries, fishing is banned during spawning seasons.

In spear or dive fisheries on spiny lobster species, some localized areas have been depleted by artisanal fishers. However, there is some doubt that restrictions on local fishing would be beneficial, because of the long larval life of the species and the possibility that recruitment results from spawning populations remote from the area fished. In some artisanal fisheries, the use of diving gear (scuba) to collect live spiny lobsters has been banned in the interests of conservation. However, the use of scuba gear does allow fishers to hand-collect live lobsters, which have a higher value than speared ones and to reject both small individuals and females with eggs.

2.2.3 Other invertebrates

A small number of invertebrates other than molluscs and crustaceans are collected for food, including jellyfishes and worms. The phylum Echinodermata, however, contains animals that are important in catches in many parts of the world.

The phylum Echinodermata contains approximately 5300 species, including sea lilies, sea stars, sea cucumbers and sea urchins. Its members are exclusively marine, and have an internal skeleton of calcium carbonate. The skeleton is in the form of an enveloping test of interlocking plates covered by a thin layer of tissue in sea urchins, and is reduced to small spikes or spicules embedded in a thick body wall in sea cucumbers. The most distinctive feature of echinoderms is the water vascular system – an internal system of water filled tubes from which blind-ended sacs, called tube feet, project through the body wall. Echinoderms have separate sexes, and eggs are fertilized externally before developing into planktonic larvae. Two of the five classes of the phylum Echinodermata (Fig. 2.19), the holothurians and echinoderms, include species which are the bases of valuable export fisheries, and are widely used artisanally. Echinoderm fisheries of the world are discussed in Sloan (1985).

Holothurians – sea cucumbers

Holothurians, or sea cucumbers, include approximately 1200 species, of which perhaps 15 species are used as food. The biology of, and fisheries for, sea cucumbers has been reviewed in Conand (1989) and in Preston (1993), from which much of the following is derived. Sea cucumbers typically have an endoskeleton reduced to microscopic spicules embedded in the thick muscular body wall, which forms the edible part of the animal. The spicules, which vary in shape, are an important characteristic in taxonomy.

Sea cucumbers move across the sea floor by means of rows of tube feet on the ventral surface. Elsewhere on the body the tube feet may be reduced, and those around the mouth are modified to form tentacles. Mobile sea cucumbers are

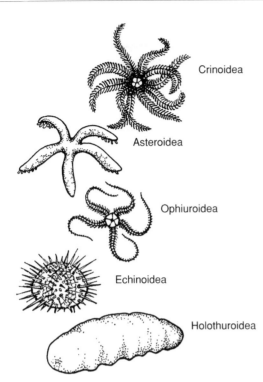

Fig. 2.19 The five classes of the phylum Echinodermata. The classes Echinoidea and Holothuroidea contain species that are commercially important.

deposit feeders, and move across the bottom ingesting sand from which the organic material is removed before being expelled as castings. When under stress, or under attack by a predator, sea cucumbers are able to eviscerate their intestinal tract and respiratory tree, possibly as a defence, and regrow these organs. Sea cucumbers have separate sexes, and a typical life cycle is shown in Box 2.14.

Holothurians are found throughout the world's oceans, and in a great range of latitudes and depths. However, most commercially important species of sea cucumbers are found in the tropical Indian and Pacific oceans, where they have been used traditionally as food in many coastal areas. Some species of sea cucumbers occur in very large densities, several hundred individuals per hectare, in shallow-water tropical lagoons. The white teatfish, *Holothuria fuscogilva*, which reaches a length of over 40 cm and a live weight of over 1 kg, is one of the most valuable species and has a wide distribution in the Indo-Pacific. The black teatfish, *H. nobilis*, the sandfish, *H. scabra*, and the prickly red fish, *Thelenota ananas*, are also commercially valuable. In Japan, the sea cucumber, *Stichopus japonicus*, is produced in hatcheries for release into the wild.

Valuable export industries are based on the production of beche-de-mer, or trepang, in which selected species of sea cucumbers are boiled, smoked and dried for markets in Southeast Asia. During processing, shrinkage in length is approximately 50%, and weight loss about 90%,

---

Box 2.14
The life cycle of a sea cucumber

Sea cucumbers have separate sexes, and release eggs and sperm directly into the surrounding water (Fig. B2.14.1). Males and females rear up on their posterior podia and exhibit a swaying motion as eggs and sperm are released. For a given species, individuals spawn simultaneously in synchrony with particular lunar and tidal conditions. Fertilized eggs hatch and develop through two planktonic larval phases, the auricularia and the barrel-shaped doliolaria, which drift with ocean currents before becoming benthic. The juveniles of most species are cryptic, and growth rates vary greatly between species, with probable lifespans of between five and ten years in the tropics.

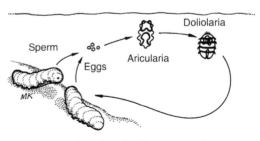

Fig. B2.14.1 The life cycle of a sea cucumber.

depending on the species. Although it is usually the thick body walls of sea cucumbers that are eaten, there are some places in which the internal organs are consumed with relish. In Japan, the organs of sea cucumbers are fermented to form the expensive delicacy known as *konowata*. In several Pacific islands, including Palau, Pohnpei, Samoa, Tonga and the Cook Islands, the guts and gonads are eaten raw or partially fermented in sea water. Usually women collect sea cucumbers (often the curryfish, *Stichopus variegatus*) and make a small hole or slit in the body wall from which they remove the internal organs. The animal is then returned to the sea where it is believed to regenerate its internal organs.

Echinoids – sea urchins

The class Echinoidea includes approximately 800 animals commonly called sea urchins, heart urchins, and sand dollars. Sea urchins have globular bodies with the mouth directed downwards towards the substrate, and the anus directed upwards. The body is arranged radially in five-part symmetry around the mouth–anus axis, with five ambulacral areas (containing tube feet) alternating with five inter-ambulacral areas. The tube

feet, which end in suckers, are arranged in double rows on each ambulacral area, and are operated by an internal water vascular system (Fig. 2.20).

Sea urchins feed on algae and sessile animals using a specialized apparatus called Aristotle's lantern in honour of the Greek naturalist and philosopher. This apparatus includes five calcareous plates (pyramids) that support five band-like teeth. The upper (growing) ends of the bands are encased in dental sacs, and the lower ends protrude beyond the mouth to form a set of five pointed teeth.

The movable spines characteristic of sea urchins are distributed over the entire test, and are interspersed with pedicellariae. The pedicellariae, which terminate in jaws of various kinds, are used to remove debris and settling larvae from the test, although some contain poison glands and are used in defence. Spines vary in size and form, but in most species are round in section and taper to a point. The needle-like spines of the tropical hat-pin urchin, *Diadema*, reach over 30 cm in length, and contain toxins capable of inflicting a painful wound. The spines of the slate-pencil urchin, *Heterocentrotus*, on the other hand, are heavy and blunt, and are adapted for

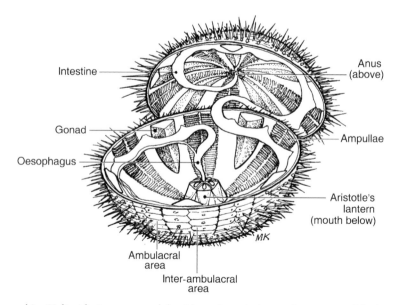

Fig. 2.20 The sea urchin, *Heliocidaris*, cut around the side to show the internal structures of the lower (oral) half and the upper (aboral) half.

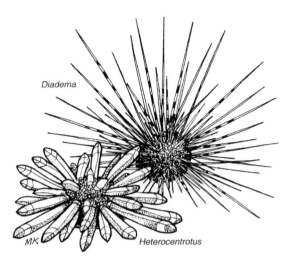

Fig. 2.21 The tropical hat-pin urchin, *Diadema*, and the slate-pencil urchin, *Heterocentrotus*.

wedging the animal into crevices on coral reefs (Fig. 2.21).

Sea urchins are dioecious and have five gonads suspended on the inside of the test, one on each ambulacral area. Each gonad is connected via a short duct to a pore in one of the five genital plates arranged around the anus. Sperm and eggs are released into the water. After fertilization, the

zygote develops into a distinctive, armed echino-pluteus larval stage which may remain in the plankton for up to several months. Newly-settled juveniles are believed to gain protection by living initially under the canopy of spines of adult sea urchins.

Sea urchins are found in both temperate and tropical waters, and are used as food by many coastal communities. The reproductive organs or roe of sea urchins are regarded as delicacies, and small-scale fisheries are based on collecting species, such as the temperate-water species of *Strongylocentrotus* and *Erythrogramma*. The roe is placed in shallow trays, or small wooden boxes, for export to Japan where it is known as *uni*, and is consumed raw. Catches of the red sea urchin, *Strongylocentrotus franciscanus*, in California have exceeded 11 000 tonnes per year (Kato & Schroeter, 1985). Sea urchins including the smaller heart urchins (see Box 2.15 'Heart urchins and sand dollars') are collected as subsistence food by the coastal communities of many tropical countries.

Because of their sedentary nature and shallow-water habitat, commercially exploited sea cucumbers are subject to depletion. In Fiji, for

---

Box 2.15
Heart urchins and sand dollars

In addition to sea urchins the class Echinoidea contains the irregularly-shaped urchins commonly called heart urchins and sand dollars (Fig. B2.15.1). The dried tests of these urchins, with their attractive patterns of perforations, are often found washed ashore. Heart urchins, which are sometimes collected for food by subsistence fishers, have elongated bodies adapted for burrowing and moving through sand with bristle-like spines ending in paddles. Sand dollars have flat bodies and are found on sandy substrates.

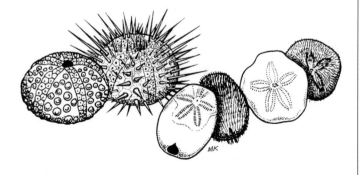

Fig. B2.15.1 From the left: a sea urchin, heart urchin and sand dollar. The living animals are shown behind their empty tests.

example, catches of sea cucumbers reached a peak in 1988 when approximately 7000 tonnes live-weight of sea cucumber were landed, a weight equivalent to the catch of all other seafood, including the cannery catch of tuna, combined for that year. Such huge production levels from relatively slow-growing species are impossible to sustain, and fishery management regulations are imperative.

Size limits on individuals taken are the main fisheries regulations applied to sea cucumber fisheries, although the sizes applied appear to be abitrarily chosen: for example, a minimum size of 75 mm has been imposed on the sandfish, *Holothuria scabra* in India, and 150 mm for the same species in Australia. Other regulations which could be imposed include placing a limit on the number of commercial fishermen, banning the use of scuba gear which is used in some fisheries, and restricting fishing to certain areas or seasons. Opening different lagoons and reefs to fishing on a rotating basis may allow periods for the stock to recover and reproduce. Difficulties in policing any regulations applied to artisanal fisheries are related to the fact that the village-level operations are usually geographically scattered; in the case of sea cucumbers, controls are more practically applied to the buyers and exporting companies, which are generally few in number (see Chapter 6 'Fisheries management').

Commercial fisheries on sea urchins are usually seasonal as sea urchins have to be collected when the roe is well developed prior to spawning. As with holothurians, sea urchins have a sedentary nature and shallow-water habitat, and exploited species are therefore vulnerable to overexploitation. Minimum size limits, and the restriction of fishing effort by licensing commercial divers are possible management measures. As spawning animals are deliberately targeted in commercial fisheries, closed areas may be used to protect part of the reproducing stock (see Chapter 6).

2.3 Fishes

Almost half of all species of animals with backbones, the vertebrates, are fishes. Different species are distributed over environments from high mountain pools to the deepest parts of the ocean, and from the warm waters of coral reefs to the cold waters of Antarctica. Fishes are a diverse group that include over 25 000 living species (Gill & Mooi, 2002) of which about 60% are marine.

The first animals with backbones, the ancestors of modern-day fishes, entered the seas from fresh water 500 million years ago. The first fishes that appeared in the sea were heavily armoured, were too heavy to float and, most debilitating of all, had no jaws. The relatives of some of these jawless fishes survive today as the sea lampreys, eel-like fish that attach themselves to larger fishes from which they suck blood. Although most jawless fish died out, the remnant lampreys did relatively well without jaws. After being introduced into the Great Lakes in the early 1900s they nearly destroyed the existing populations of all large fish species.

In other ancient fishes, rudimentary jaws evolved from the skeletal supports of the gill slits, and the earliest fossils of jawed fish are found in rocks about 450 million years old. With jaws that could evolve for using different food and fins that could be adapted for different habitats, fishes were poised to be most successful of all marine animals. Some adapted to eat plants, others to eat meat and they evolved to fill all available niches in the marine environment. Some evolved to hunt in caves and some to swim in the open sea. Some became camouflaged for protection and capturing their prey and others relied on their ability to swim at speed for the same purposes. Modern fish have two sets of paired fins, the side, or pectoral, fins and the pelvic fins. These fins, which are the evolutionary precursors to the four limbs of terrestrial vertebrates, allowed a great variety of swimming behaviour. Single fins include the dorsal or back fin, the anal fin, and the caudal or tail fin (Fig. 2.22).

Because tissue consists of water together with various other heavier substances, aquatic animals are generally more dense than the water that surrounds them. Thus, without some aid to flotation, they will sink to the bottom of the

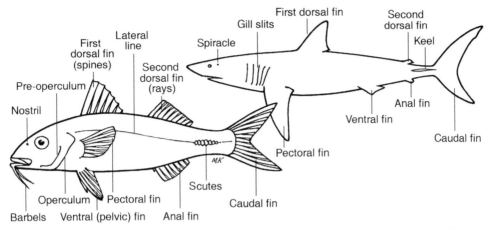

Fig. 2.22 A bony fish (left) and a shark (right) showing external features referred to in the text. Features such as barbels and scutes are found in only a few species.

sea. Ancient and heavily armoured fishes evolved along two main evolutionary lines, each of which solved the problem of buoyancy in different ways.

In one group, the ancestors of modern sharks and rays, the heavy calcified skeleton was replaced by lighter cartilage, a type of connective tissue consisting of living cells embedded in a rubbery matrix stiffened by fibres of collagen. Buoyancy in modern sharks is also aided by having a large liver which is rich in the low density oil, squalene, and by having fixed pectoral fins which act as paravanes. As the shark moves forward through the water, pressure on the underside of the pectoral fins provides uplift and this prevents the animal from sinking. Thus many, but not all, species of sharks have to swim continually to stay afloat.

The other evolutionary group, the ancestors of present-day bony fishes or teleosts, retained the heavy calcified bones in their skeleton, but solved the problem of remaining buoyant in a different way. The primitive lung acquired by ancient fishes living in shallow, deoxygenated water evolved into the swim bladder of modern bony fishes, most of which obtain oxygen through their gills. The lungs of ancient fishes therefore became hydrostatic organs which allowed some fishes to remain at a constant depth with neutral buoyancy. The evolution of the swim bladder allowed fish to evolve away from speed as a way

of life. Pectoral fins, no longer required for aiding flotation, could evolve to allow a greater range of movements. Present-day bony fishes use pectoral fins to hover, to swerve, to swim backwards, and even, in the case of flying fish, to glide through the air. The ability to take advantage of a variety of ecological niches, to be either bottom-dwelling or pelagic, has allowed modern bony fishes to dominate the waters of the world. Representative teleost fish families are shown in Fig. 2.23.

Demersal fishes, those that live near the sea floor, have evolved a much greater variety of shapes. Some have large filmy pectoral fins to manoeuvre around coral heads and into caves, and some are adapted for burrowing in the substrate. Some bottom-living fish, such as the boxfishes have re-evolved a hard covering for protection, some scorpionfishes have developed poisonous spines and pufferfishes contain potent toxins (see Box 2.16 'Poisonous puffers'). Other fishes retained a reliance on speed to obtain prey and escape predators and evolved shapes that allowed higher speeds by reducing drag (see Box 2.22 'Speed and disguise' in Section 2.3.3). Modern fast-moving fishes have many other drag-reducing adaptations, such as fins that retract into depressions in the body.

Fishes have adapted to many different forms of feeding, including sifting plankton from the water, scraping algae from rocks, eating coral

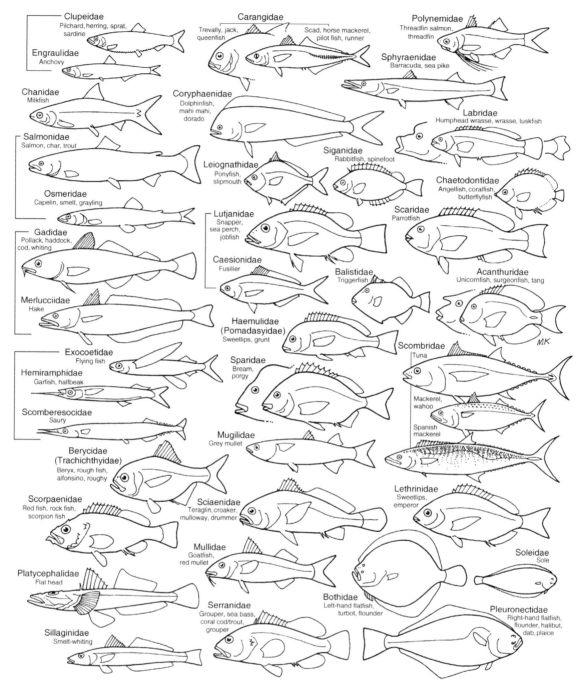

Fig. 2.23 Some representative fish families.

and catching a wide range of molluscs, crustaceans and other fish. The mouth of a fish may be turned either upwards or downwards, or may be terminal, depending on whether the fish is a surface-feeding, a bottom-feeding, or a mid-water-feeding species. The type of teeth and the spacing of gill rakers are related to the type of food taken: the gill rakers, comb-like structures

Box 2.16
Poisonous puffers

Although a number of fishes are venomous, very few are toxic. Some fish become toxic through food chains as in the case of high mercury levels and ciguatera (Chapter 1) but not many are permanently so. The few fishes that are always toxic include the pufferfishes, toadfishes or blowfishes belonging to the family Tetraodontidae. Pufferfishes are related to the boxfishes which are encased in a rigid armour of bony plates (Fig. B2.16.1). Pufferfishes get their common names from their ability to inflate themselves by taking in large quantities of water or air. They have strong fused teeth which can impart a strong bite which is not venomous. However,

the flesh of puffers contains a potent toxin called tetrodotoxin which appears to act as a respiratory block in all animals including humans. Pufferfish are distributed in all oceans and approximately 80 species are known to contain tetrodotoxin (Kantha, 1987).

Poisoning is relatively common in Japan where pufferfish flesh, or *fugu*, is considered a delicacy. *Fugu* chefs in Japan are required to hold a national licence and cases of poisoning have occurred when the delicacy is prepared by unqualified individuals. Symptoms of poisoning include numbness and tingling of lips, tongue and inner surfaces of the mouth. This is followed by weakness, paralysis of limb and chest muscles, decreased blood pressure, quickened and weakened pulse and sometimes death.

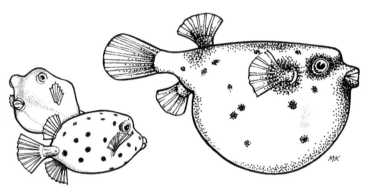

Fig. B2.16.1 Left: the blue-spotted yellow boxfish, *Ostracion tuberculatus*. Right: the large (28 cm) black-spotted puffer fish, *Arothron nigropunctatus*.

attached to the inside curve of the gill arches, sift particles of food from the water which enters the mouth and flows out through the gill slits. The digestive system (Fig. 2.24) includes an S-shaped stomach leading to an intestine which is often longer in herbivores than in carnivores. At the junction of the stomach and the intestine, there are often finger-like sacs, the pyloric caeca, whose function may include aiding food absorption.

The development of senses in fishes depends on their habitat. Fish eyes are spherical and have given their name to the photographer's fisheye lens used to take in a field of vision covering up to 180 degrees. The eyes of many predatory fishes that live in clear water are often large, and

presumably sight in these species plays an important role in locating prey. The eyes of many fishes appear to be capable of distinguishing colours. The sense of touch is particularly well developed in some species, with sense organs distributed over the entire body, and in the feelers or barbels of species such as catfish and goatfish. Goatfish, sometimes called red mullets, move across the sea floor feeling and tasting the substrate with their barbels to discover food.

As in vertebrates, the ear of a fish is a paired organ of both equilibrium and hearing. The ears, which are internal with no external connection, consist of a left and a right ear, one on each side of the brain (Fig. 2.25). Each ear consists of a

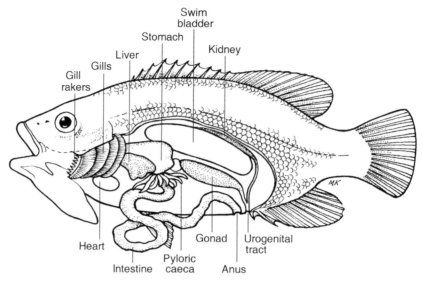

Fig. 2.24 The internal organs of a teleost fish.

system of three semicircular canals arising from a chamber called the utriculus. The utriculus is connected to a second chamber, the sacculus. Each of these chambers contains a sensory spot (macula) with hair-like projections which are associated with dense calcium carbonate 'ear-stones' or otoliths which float in the fluid contained in the inner ear. As a fish grows, deposits of calcium carbonate are laid down on the otoliths. Movement of the fish is detected, and equilibrium is maintained by the otoliths pressing on the sensory hairs, and by the movement of

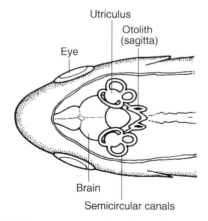

Fig. 2.25 The ear structure of a teleost fish.

fluid in the semicircular canals. Sound is perceived when sound waves, travelling through the water and the head, strike the hard otoliths, which vibrate against the sensory hairs. A teleost fish may have six otoliths, three in each inner ear, and the largest of the these, the saccular otolith (or sagitta), is usually the one in which the concentric rings of deposited material is used by scientists to estimate age (see Chapter 4).

Many fishes produce sounds and this is often reflected in their common names – drums or croakers (sciaenids) and grunts (haemulids). Croakers may use well-developed soniferous muscles associated with the swim bladder as part of reproductive behaviour. The highly-prized giant yellow croaker, *Bahaba taipingensis*, was hunted almost to extinction in China because fishers could locate sparse individuals by the sound that they made. Unsurprisingly, otoliths appear to be larger in fishes that specialize in sound production, and smaller in fishes such as wrasses (labrids) that rely on bright or contrasted colour patterns for visual communication and recognition (Cruz & Lombarte, 2004). Although the sense of hearing in fish in general may be poor, the lateral line is also believed to be capable of detecting low-frequency vibrations, as well as pressure changes. The lateral line, which runs

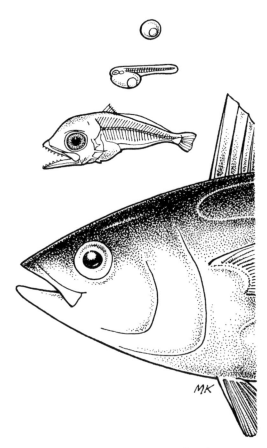

Fig. 2.26 The egg, larvae, and adult of the big-eye tuna, *Thunnus obesus*, which reaches a length of over 1.8 m in a lifespan of about eight years. Larval stages redrawn from Matsumoto (1961).

longitudinally down each side of a fish, consists of a row of pores, connected by a canal running beneath the skin.

Female fishes have ovaries which are usually paired, and occupy a large part of the body cavity when fully developed. In most fishes, eggs enter a hollow central portion of the ovary before being released into the sea where they are fertilized externally by sperm from males. External fertilization occurs in about 96% of all fish species and eggs usually measure about 1–5 mm in diameter irrespective of the size of adults. The fertilized eggs hatch to small larvae (often about 5 mm in length) which are often transparent, with prominent eyes and a large yolk sac (Fig. 2.26). Most larvae are planktonic and drift with ocean

currents. In many species, recently-hatched fish larvae move to the surface by night to feed on phytoplankton, and when larger, feed on other zooplankton. After a period which varies from species to species, the larvae metamorphose, and benthic species settle on the sea floor. The juveniles of many fish species grow in nursery areas which include reefs, banks, bays, and estuaries.

In the few viviparous species (many sharks and rays, for example), fertilization is internal, and part of the pelvic or anal fin of the male is adapted for the transfer of sperm to females. In viviparous females, the ovary itself, or an area of the duct, is adapted to contain the developing young.

Some fishes are sequential or simultaneous hermaphrodites. Sequential hermaphrodites may change sex from male to female (protandry) or female to male (protogyny) or both ways. Simultaneous hermaphrodites produce both eggs and sperm at the same time. Protogyny, in which an individual starts life as a female before changing to a male, is the most common form of hermaphroditism and is found in groupers (Serranidae), parrotfishes (Scaridae) and many wrasses (Labridae). This sex change sequence may give advantages to a species in which there is strong competition among males for access to females: here there may be advantages in being a small female until reaching a size at which it can successfully compete with other males. When dominant males are experimentally removed from the area, the largest female may change sex within days. Less common is protandry in which an individual starts life as a male before changing to a female. Here, advantages are associated with the greater egg-carrying capacity of larger females. A review of the behavioural ecology of reproduction is given in Forsgren et al. (2002).

Although most fishes release eggs and sperm into the sea, a few species provide parental care to fertilized eggs, larvae and even their young. Salmon excavate nests or redds in which to lay their eggs, male seahorses carry eggs in pouches, and some cardinal fishes (Fig. 2.27) are mouth brooders, carrying their eggs and eventually the young in their mouths. A review of breeding behaviours in fishes is given in Moyle and Cech (2004).

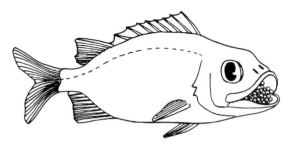

Fig. 2.27 A mouth-brooding cardinal fish (Apogonidae) carrying eggs within the mouth.

Daily migrations of fishes are usually associated with gaining food and avoiding predators. Many coastal species, for example, move into estuaries or mangrove areas on a daily basis to hunt food in these highly productive areas. Seasonal movements, however, are more often associated with spawning and some fishes make spectacular migrations as part of their life cycle. Well-known examples include the anadromous species that spend most of their lives at sea but move into rivers to release their eggs and the catadromous species that grow in rivers and migrate out to sea to breed.

The best-known anadromous fishes are the various species of salmon (family Salmonidae). Species of Pacific salmon (*Onchorhyncus*) return to the rivers of their birth to lay their eggs in shallow depressions and die. The hatchlings migrate to sea in order to grow to maturity before seeking out their home rivers to complete the cycle (Fig. 2.28a). The most celebrated examples of catadromous fishes are the anguillid eels of North America, *Anguilla rostrata*, and Europe, *A. anguilla*, although there has been some suggestion that not all anguillids are catadromous (Tsukamoto et al., 1998). The adults of both species leave rivers and make long migrations to breeding grounds in the Sargasso Sea where the adults spawn and die. The eggs hatch into leaf-like leptocephalus larvae that eventually reach rivers. After reaching the rivers, the larvae metamorphose into juvenile eels, or elvers, that grow to adults before migrating offshore (Fig. 2.28b).

Some strictly oceanic fishes, notably some species of tuna, migrate across large tracts of ocean to reach spawning areas. The southern bluefin tuna, *Thunnus maccoyii*, which swims

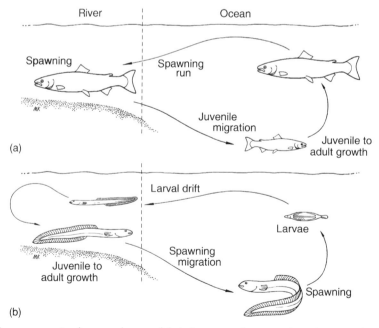

Fig. 2.28 (a) Anadromous species that spend most of their lives at sea but move into rivers to release their eggs include species of salmon (family Salmonidae). (b) Catadromous species that grow in rivers and migrate out to sea to breed include anguillid eels.

Box 2.17

Life cycles of fishes

Although life cycles in fishes vary greatly, about 96% of all species are external fertilizers. In most cases, released sperm fertilize eggs drifting in the sea. Floating eggs hatch to planktonic larvae that are eventually recruited as juveniles back into the adult populations. Variations include the adults of some species migrating to particular spawning areas and juveniles growing in nursery areas which may include bays and estuaries (Fig. B2.17.1).

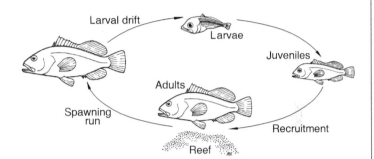

Fig. B2.17.1 A generalized life cycle of fishes. Some species have distant spawning and nursery areas.

south of Australia and into the southern Pacific Ocean spawns in a relatively small area of ocean off Java. Many coastal species make less spectacular migrations in order to breed. Young mullet, *Mugil cephalus*, for example, feed and grow in estuaries and then, as adults, move out to spawn in ocean water.

Some species distributed on coasts and reefs migrate to specific spawning locations to form dense aggregations at particular times of the year (see Box 4.12 'Spawning aggregations' in Chapter 4). These spawning aggregations involve species such as groupers and snappers, many of which do not normally live in schools. Gathering in one place and at one time for spawning has the advantage of maximizing the confluence of eggs and sperm and hence increasing the production of larvae. In addition, many of the aggregation sites appear to be in locations where currents maximize the distribution of the resulting larvae. Although fish in dense gatherings may be particularly vulnerable to predation, the spawning aggregation event is usually sufficiently short to prevent predators becoming accustomed to the presence of super-abundant food. However, human predators have learned, sometimes through traditional knowledge gained over many generations, the location and timing of the aggregations and

heavily exploit the species. In the Caribbean, the Nassau grouper, *Epinephelus striatus*, is just one of many species that are particularly threatened by exploitation of spawning aggregation sites.

In spite of the vulnerability of dense gatherings, the existence of distinct spawning locations and seasons presents opportunities for fisheries managers to protect fish stocks by establishing areas and periods in which fishing is prohibited (see Section 6.5.3 'Controls to protect ecosystems'). In any case, knowledge of the general life cycle of a targeted species, including its migratory movements if any, as well as the key habitats on which it depends, is necessary in order to manage a fishery. A generalized life cycle of fishes is shown in Box 2.17 'Life cycles of fishes'.

On a world scale, the diversity and abundance of fishes vary with both longitude and latitude. From east to west, the largest number of fish species (the greatest diversity) is found in what is believed to be a centre of evolution in the Southeast Asian region of the Indo-Pacific (although doubt has been cast on centre of origin hypotheses – see Section 1.3). The highest number of fish species is found in an area which includes Indonesia, New Guinea and the Philippines and is home to about 2800 species of inshore marine fishes (Mooi & Gill, 2002). Reduction in species diversity away from this area is suggested by the number of species on the Great Barrier Reef (about 1600 species), Samoa (900), Society Islands (630), and Easter Island (126). The trend in endemism is usually the reverse of that in diversity, and the highest endemism of 22% is in Easter Island (Randall et al., 1998).

Along a north to south axis, the greatest diversity of species is generally found in low tropical latitudes and decreases towards polar regions. Although there is a lower species diversity in cooler waters, some species are present in such high abundance that they are the bases of the world's great commercial fisheries. Many large temperate-water fisheries are based on a single species, such the cod, *Gadus morhua*. The opposite is often true in tropical waters, where large catches from a coral reef fishery, for example, are made up of many different species, but relatively low numbers of each. A similar generalization may be made between inshore and offshore ecosystems, in that the latter is characterized by a low diversity of specialized pelagic fish species, which reach high levels of abundance in areas of high productivity such as upwellings.

2.3.1 Demersal fishes of cooler waters – cods, hakes and haddocks

Large demersal fisheries are located on the continental shelves, where primary productivity is higher than in the open sea, and the depths are sufficiently shallow to be accessible to towed trawl nets. The continental shelf consists of a gradual gradient down to depths of less than 400 m, where the steeper continental slope begins. The most productive demersal fishing areas occur where the shelf is very wide, such as in the North Sea, and on the Grand Banks of Newfoundland, or where there are large shallow gulfs.

Up to 19% of the total world catch has consisted of cods, haddocks and hakes of the order Gadiformes (Fig. 2.29). The cod, *Gadus morhua*, and the haddock, *Melanogrammus aeglefinus*, are found across the North Atlantic from Europe to North America. The cod has been exploited for centuries, and Europeans were fishing for this

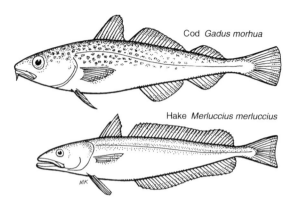

Fig. 2.29 Temperate-water demersal fishes: the cod and hake.

species on the Grand Banks of Newfoundland before the arrival of European colonists. In the days of sailing fishing vessels, the catch was salted for the return trip to Europe, and is still prepared this way for some traditional markets in the Mediterranean. Most of the catch from modern trawlers is sold fresh or frozen. The cod migrates to a number of different spawning grounds, and releases its eggs in depths of about 200 m. The eggs are widely spread by currents, and the cod becomes established as several discrete stocks. Hakes of the family Merlucciidae have a wider temperate water distribution than cods, and different species are the bases of fisheries in both the northern and southern hemispheres. The European hake, *Merluccius merluccius*, is trawled in depths of up to 600 m from Iceland to the northwest coast of Africa. Spawning occurs in the summer months, and larvae drift inshore, where young fish may be caught by hook and line during their first year of life.

2.3.2 Demersal and reef fishes of warmer waters

There are several large tropical trawl fisheries, including those in gulfs, such as the Gulf of Mexico, the Persian Gulf and the Gulf of Thailand, and on continental shelves, such as those off India and China. Demersal species on muddy areas are often dominated by sciaenids (drums and croakers) and polynemids (threadfins).

Box 2.18
The biggest and the fastest

The so-called whale shark is the world's largest living fish, growing to a length of 18 m and a weight of over 40 tonnes (Fig. B2.18.1). This clownish and gentle giant is dark green with yellowish spots and has three longitudinal ridges on each of its sides. In spite of its large size, it is not aggressive and filters its food, plankton and small fishes, from the water. Because of its lack of shyness, the whale shark is a favourite with divers, many of whom have been filmed hitching rides on the shark's massive dorsal fin. The smaller basking shark, which can reach lengths of over 12 m, is hunted for its huge liver that, in common with other sharks, contains large quantities of low-density oil.

Sailfish, which can grow to reach 100 kg, have large sail-like dorsal fins that are over twice as high as the body is deep (Fig. B2.18.1). They appear to hunt in groups, and their tall blue dorsal fins, cutting through the surface of the sea, are used to herd prey species into a tight ball. The sailfish then move through the ball, slashing from side to side with their long bills to kill or maim the smaller fishes. With a timed short-burst swimming speed of 110 km/hr, the sailfish appears to be the fastest animal on the planet with a top speed greater than the fastest of all land animals, the cheetah.

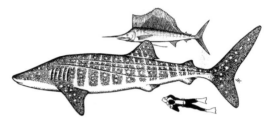

Fig. B2.18.1 The whale shark, *Rhincodon typus* (18 m) and the sailfish, *Istiophorus platypterus* (3.2 m). A human diver is shown as a size comparison. Drawing of whale shark by Jeremy King.

On sandy bottoms, sparids (sea breams) and pomydasids (grunts) are common. Rockier areas include species of groupers, snappers and emperors.

The primary production in coastal areas is often much higher because of the nutrients derived from terrestrial sources (see Section 1.4 'Marine ecosystems'). Several species of fishes are adapted to take advantage of the inshore productivity indirectly through food webs. Mullets (Mugilidae) are able to 'telescope' the food chains by feeding directly at lower trophic levels. Ponyfishes, *Leiognathus* spp., have mouths which are protractile, and ideally suited for feeding on diatoms, filamentous algae and detritus. Goatfish (Mullidae) have sensitive barbels which aid in locating food in the organic-rich substrates of inshore areas.

Although some species of fish are permanent inhabitants of sheltered inshore areas, a much larger number of species use such highly-productive waters as either spawning grounds, or nursery areas, or both. The mullet is typical of fishes which use estuaries and coastal areas for some part of their life cycle. Large schools of mullet make spawning runs along coastlines, and sometimes move offshore as far as the edge of the continental shelf to spawn. In tropical waters, fish such the barramundi, or sea bass, *Lates calcarifer*, spawn near the mouths of rivers, and species such as the mangrove red snapper, *Lutjanus argentimaculatus*, produces young which grow and live in coastal areas before moving out to deeper water.

Inshore species in temperate waters are important in recreational fisheries. In tropical areas, inshore fish are more diverse, and include species of ponyfish, garfish, rabbitfish and tropical clupeids (Fig. 2.30). Such species provide an easily accessible source of protein for subsistence fishers; although such catches are not included in official catch statistics, collectively they are believed to be large. In addition to food requirements, a number of species in Southeast Asia are collected for use in traditional medicines (see Box 2.19 'Fish in traditional medicines').

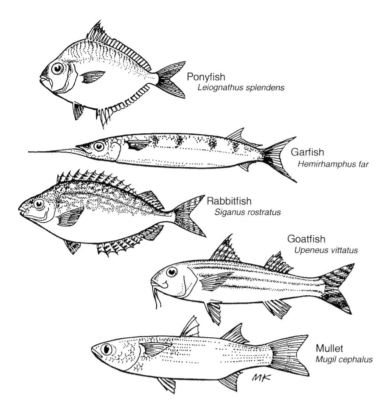

Ponyfish
Leiognathus splendens

Garfish
Hemirhamphus far

Rabbitfish
Siganus rostratus

Goatfish
Upeneus vittatus

Mullet
Mugil cephalus

Fig. 2.30 Some fishes of tropical inshore areas.

Box 2.19
Fish in traditional medicines

Dried seahorses are used in Japanese *kanpo*, Korean *hanyak*, and Indonesian *jamu* medicines and the rarer they become, the higher the price they command. In Hong Kong the best quality seahorses are valued at over US$1000 per kg and such high prices are encouraging more fishers to seek them out. About one-sixth of the world's population depend on traditional medicines in some form or other, and dried seahorses are touted as a cure for baldness, rabies, infertility and impotence. However, any global bans placed on the seahorse trade by the world community would result in claims of cultural imperialism as many fishing communities in Thailand, India, Vietnam and the Philippines rely on the income from seahorse fishing.

Beside the large number of species which use inshore waters during parts of their life cycle, either as spawning adults or as juveniles, there are many more species which regularly visit inshore areas seeking food. The movement of species to and from inshore areas is an important mechanism whereby the high productivity of such areas is transferred, through the food chains, to offshore areas which are often less productive.

Coral reefs provide a substrate for the growth of coralline algae, and the coral polyps, with their symbiotic algae (zooxanthellae), contribute to the primary productivity. Coral reefs provide food and shelter for a greater variety of living things than most other natural areas in the world. Although the diversity of fish species on coral reefs is high, the numbers of individuals within each species is relatively low.

Many coral reef fishes, particularly trevallies, groupers, and surgeonfish, produce larvae which

A Malaysian seafood restaurant where patrons can chose their meal from the live fishes held in a bank of aquaria. The live food-fish trade has created opportunities in many communities but has resulted in the overexploitation of several reef species.

drift far out in the ocean before returning and settling on coral reefs as adults. Generally, coral reef fishes live within a small home range on the reef, reproduce repeatedly over a lifespan of less than ten years, and produce pelagic larvae. The species with longest larval phases have the greatest potential for dispersal and are widely distributed in isolated reefs across the ocean. After metamorphosis, many coral reef fish become established in a particular home position on the reef and remain there throughout their adult life. The recruitment to coral reefs appears to be variable over time and space, and new recruits may be derived from adults living on distant reefs. The periodicity of spawning may be linked to the need to produce larvae at times when favourable currents take the larvae offshore, away from reef-dwelling plankti-vores. Similarly, the length of time that the larvae spend in the plankton may be influenced by the need to utilize currents to return to the reef systems.

Although there are thousands of different fishes on coral reefs, many of these belong to the relatively few families shown in Fig. 2.31. The larger fishes including emperors, groupers and snappers are particularly important in artisanal fisheries and the smaller species are caught for local food by subsistence fishers. Some of the larger wrasses and groupers are targeted by commercial vessels for the live food-fish trade in Southeast Asia (see Box 2.20 'The live food-fish trade'). The most conspicuous of the smaller fishes are often the many species of damselfishes that form large plankton-feeding schools above the corals. Some of the various species flash

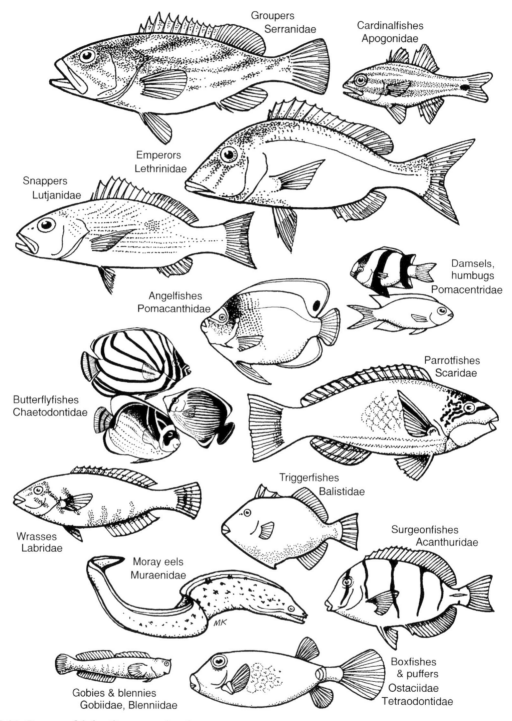

Fig. 2.31 Common fish families on coral reefs.

Box 2.20
The live food-fish trade

It has been customary in countries in Southeast Asia, for those that can afford it, to consume fishes that have been kept alive in tanks until just before being ordered, cooked and eaten. By the 1970s, large fishes presented alive to patrons of expensive restaurants in Hong Kong were commanding very high prices. The increasing demand for live fish resulted in fishers extending their operations further into the South China Sea. As nearer stocks of fish became depleted, boats supplying the

market moved into the Philippines and then Indonesia. By the early 1990s, the live fish trade had moved into islands in the western Pacific Ocean. The fishes most commonly imported into Hong Kong include the humphead or Napoleon wrasse, *Cheilinus undulatus*, (Fig. B2.20.1) and several species of groupers including the leopard grouper, *Plectropomus leopardus*, and the spotted coral trout, *P. aroelatus*. The average retail price paid for the humphead wrasse in Hong Kong was over US$60 per kg in 2003 and there have been reports of prices reaching US$180 per kg.

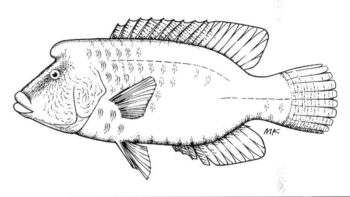

Fig. B2.20.1 The humphead wrasse, *Cheilinus undulatus*, which reaches 2 m in length.

between opal colours of blue and green and the small humbugs have zebra stripes of black and white. The neon damsel has been described as the bluest thing on earth.

All of the fishes shown in Fig. 2.31, even the small ones with the possible exception of gobies and blennies, are used as food by coastal peoples in various parts of the world. Some of the most important food fishes in many Pacific islands, for example, are the various species of surgeonfishes. Even the tiny damselfishes may be caught and used to make soup. Unfortunately, the catching of damselfishes often involves surrounding small coral heads with a fine net and breaking the coral with sticks.

Deep-water snappers are caught by artisanal fishers using hook and line and traps in depths of about 200 m on the steep slopes beyond reefs and on seamounts in tropical countries around the world. The main species caught belong to

the genera *Etelis*, *Pristipomoides*, and *Aphareus* (Fig. 2.32). These large species are particularly valuable in tropical areas as their distance from coral reef ecosystems means that they are unlikely to be affected by ciguatera.

2.3.3 Coastal pelagic fishes – clupeoids

Coastal pelagic fish, particularly the clupeoids, make up two-thirds of world fish landings. Most of the world's largest fisheries resources are located near the coast, and are taken off coasts with wide continental shelves, on shallow banks, and in areas of upwelling. Clupeids (sardines, herrings, sprats, pilchards) and engraulids (anchovies) are pelagic fish typical of upwelling areas, and some species are the bases of large fisheries. Although some of the catch is eaten fresh, canned and pickled, most is ground into fish meal, a protein additive used in agriculture

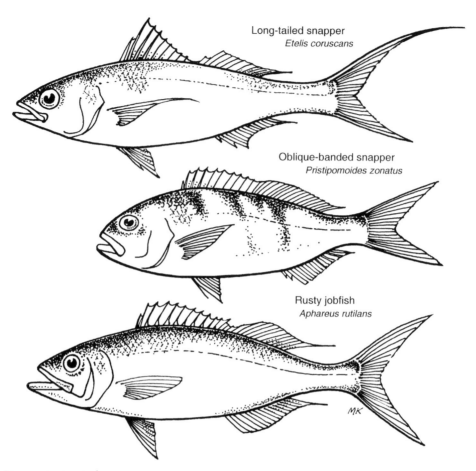

Long-tailed snapper
Etelis coruscans

Oblique-banded snapper
Pristipomoides zonatus

Rusty jobfish
Aphareus rutilans

Fig. 2.32 Deep-water tropical snappers.

Box 2.21

Fatty fish – the good oil

Populations that traditionally eat a lot of fish, such as the Eskimos and the Japanese, have relatively low rates of heart disease. Fish contain omega-3 fatty acids, types of polyunsaturated fats that appear to promote cardiovascular health. Since the human body cannot manufacture its own supply of these fatty acids, it has to obtain them from food – mainly from fish but also from such plants as flax, soybeans and walnuts. Omega-3 fatty acids are especially abundant in oily fishes such as sardines, mackerel, herring and to a lesser degree in tuna.

---

for poultry and pigs, and in the aquaculture industry in food pellets for salmon and shrimp. Clupeoids and other oily fishes are sources of polyunsaturated omega-3 fatty acid, an important human nutrient (see Box 2.21 'Fatty fish – the good oil').

Some of the world's largest single-species fisheries are based on clupeoids such as the Japanese pilchard, *Sardinops melanostictus*, the South American pilchard, *S. sagax*, the Peruvian anchovy, *Engraulis ringens*, and the Atlantic herring, *Clupea harengus* (Fig. 2.33). The fishery for the Peruvian anchovy, *Engraulis ringens*, off the northwest coast of South America collapsed in 1972, due to overexploitation and the effects of a natural phenomenon known as El Niño (see Section 1.8).

Many clupeid species, generally less than 40 cm in length, move inshore, or into estuaries

The eggs from several species of fish are regarded as delicacies in many cultures. The most famous is caviar, eggs from sturgeon, sold here at a specialist caviar cafe in San Francisco, USA.

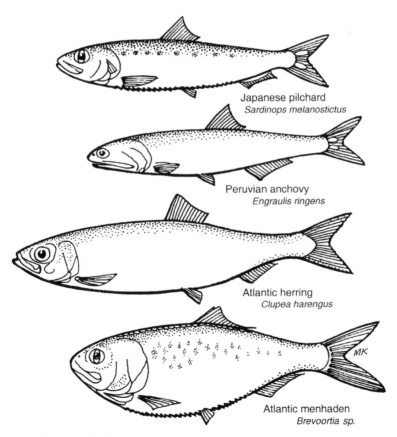

Japanese pilchard
Sardinops melanostictus

Peruvian anchovy
Engraulis ringens

Atlantic herring
Clupea harengus

Atlantic menhaden
Brevoortia sp.

Fig. 2.33 Clupeoid fish of commercial importance.

Box 2.22
Speed and disguise

Predominantly, pelagic fishes rely on speed to catch their prey and to avoid predators. As water is 800 times as dense as air, any part of the body which creates friction or turbulence causes a large amount of drag. Fast fish have bodies that are fusiform or spindle-shaped, as it is this shape which offers the least resistance when moving through the water (Fig. B2.22.1a). The caudal fin, which provides the propulsion, may be shaped like a scythe, with both a long leading edge, and a small surface area (a high aspect ratio). In many fast fishes, the pectoral fins are used as brakes and rudders, and fit into depressions in the body when the fish is swimming at speed. Independently, this fusiform shape has evolved in aquatic mammals such as dolphins and whales. Not so independently perhaps, marine architects have used the shape in designing vessels. The fin keel of a modern racing yacht is blunt at the leading edge, has its greatest width about one-third from the front, and tapers to a fine edge at the stern or back. It seems as though the shape of the front edge is less important than the back edge in reducing drag.

In addition to their speed, most pelagic fish have a very subtle form of camouflage called counter-shading to avoid predators (Fig. B2.22.1b). Fish that habitually swim near the surface often have dark backs that shade to lighter underparts. To a predator swimming below such a fish, the lighter underparts appear the same shade as the sky and the bright surface of the sea. But to a predator such as a sea bird flying above, the dark back of the fish merges in with the deep blue shades of the sea.

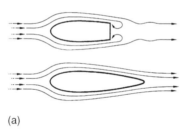

(a)

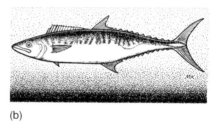

(b)

Fig. B2.22.1 (a) Laminar flow of water past a blunt-ended shape (top) that creates turbulence and drag, and flow past a fusiform shape (bottom) which minimizes drag. (b) Counter-shading in a pelagic fish.

to spawn. The alewife, *Alosa pseudoharengus*, returns from the northwest Atlantic to spawn in estuaries in North America, where they are caught in large quantities by gill nets. Other species of *Alosa*, known as shad in Britain, spawn in estuaries of northeastern Europe. In spite of the small size of most clupeids, the total catch weights of some coastal species are very high. Their schooling behaviour makes them highly vulnerable to surrounding nets, and their coastal migrations expose them to gill nets and fence traps set at right angles to shorelines (see Chapter 3). Annual catches of menhadens (*Brevoortia* spp.), which form dense schools close to estuaries on the Atlantic coast of North America, have been over one million tonnes – the highest catch of any species caught in the United States in some years. Guides to clupeoid fish of the world are given in Whitehead (1985) and Whitehead et al. (1988).

Clupeoids, in common with many other fish that swim near the surface of the sea, are counter-shaded with dark backs that shade to lighter underparts to make them less visible to predators from below and above (see Box 2.22 'Speed and disguise').

Fishes of the family Carangidae (along with mullets and sauries) make up an important group in world catch statistics. Large carangids, the jacks or trevallies, include *Caranx* spp., which form hunting schools off many tropical coasts. The horse mackerels or scads (*Decapterus*,

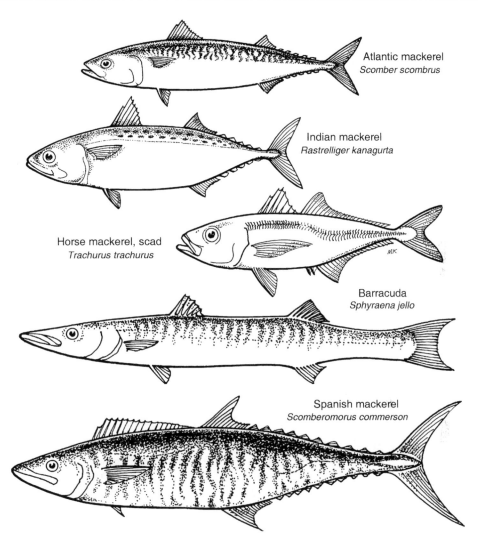

Atlantic mackerel
Scomber scombrus

Indian mackerel
Rastrelliger kanagurta

Horse mackerel, scad
Trachurus trachurus

Barracuda
Sphyraena jello

Spanish mackerel
Scomberomorus commerson

Fig. 2.34 Coastal pelagic fishes.

Trachurus, Salar) are also found mainly in tropical and warm temperate seas, but some species, such as the Atlantic scad or horse mackerel, *Trachurus trachurus*, ranges from Scandinavia to South Africa. Scads occur in great abundance, and are often used for reduction to fish meal, although some species are important food fishes in tropical areas. Because of their schooling behaviour, and hence vulnerability to surrounding nets, the numbers of many coastal pelagics, including the scads, have decreased dramatically in some areas.

Scombrids are found in tropical and warm temperate seas worldwide. Some, such as the Atlantic mackerel, *Scomber scombrus*, moves well into high-latitude coastal areas of the North Atlantic during summer, and takes large quantities of planktonic crustaceans, sprats, herrings and sand eels. The chub mackerel, *Scomber japonicus*, has a worldwide distribution in the tropical and warm temperate seas, and moves into coastal areas to feed during the summer months. The Indian mackerel, *Rastrelliger kanagurta*, is confined to tropical waters, and is commonly caught on coasts of the Indian and Western Pacific oceans (Fig. 2.34). Another group of scombrids is the Spanish mackerels, which includes many species of *Scomberomorus*, and the large wahoo (or seerfish), *Acanthocybium solandri*, which can attain weights of 40 kg. Spanish mackerels and barracudas (*Sphyraena*) are common and voracious predators in the vicinity of coral reefs.

2.3.4 Offshore pelagic fishes – tunas and sharks

Offshore pelagics include tunas, billfishes and sharks. In the open sea, primary productivity is generally low and this is particularly so in tropical waters, where nutrient-rich deeper water is effectively locked below the thermocline. However, in areas where nutrients are carried up into surface waters by upwellings, food for large pelagic predators such as tunas is abundant.

The best-known open sea fishes are the species of tuna, which are distributed over large areas of ocean where they hunt smaller planktivorous fish. Tunas are fast-swimming pelagic fish related to marlins and sailfish, and a guide to scombrids of the world is given in Collette and Nauen

(1983). Unlike most other fish, tunas are warm-blooded and keep their bodies at higher temperatures than the surrounding water. A higher body temperature allows increases in muscle power and may account for a tuna's ability to swim at speeds of over 50 km/hr to catch smaller fish (Fig. 2.35).

There has been an artisanal net fishery for the Atlantic bluefin tuna, *Thunnus thynnus*, in the Mediterranean Sea since the days of the Roman Empire. Since the late 1990s industrial fishing began with the installation of 'tuna ranches', circular floating cages about 50 m in diameter. In the cages, the tuna are fattened on smaller fish for months before being slaughtered and shipped to Japan. In 2006, the catch was limited to a quota of 32 000 tonnes and it was estimated that this would be exceeded by at least 50%. The stock is believed to be threatened by this level of exploitation as well as by pollution in the Mediterranean.

Tuna are caught artisanally by local fishers in many subtropical and tropical countries, usually by trolling lures behind small boats. Distant-water fishing vessels fish for tuna using longlines

Tsukiji fish market in Tokyo covers 56 acres and employs over 50 000 people. Many thousands of tonnes of seafood from all parts of the globe pass through this huge market every day. Large, high-quality bluefin tuna can be sold here for over US$60 000 for a single fish and can eventually cost over US$200 a plate at a local sushi bar.

to catch albacore, big-eye tuna and adult yellowfin tuna. Pole-and-lining and purse seining is used to catch surface fish including skipjack tuna and juvenile yellowfin tuna (see Chapter 3 'Fishing and fishers'). Some species of tuna, including albacore, move across large areas of the ocean, either to reach new feeding grounds or to reach spawning areas, whereas some species, such as skipjack tuna, may stay in one area for their whole life.

Catches of tuna made by fishing vessels of many nations are particularly high in the Western Pacific, where two parallel currents (the North and South Equatorial currents) move apart, creating an equatorial upwelling. This upwelling results in the large stocks of tuna in waters of Papua New Guinea, the Solomon Islands and Kiribati. Poorly-handled tunas and some other pelagic species are responsible for histamine poisoning, one of the most common forms of seafood poisoning (see Box 2.24 'Scombroid or histamine poisoning').

Sharks are fished in many parts of the world for their flesh, fins and liver oil. In addition to being targeted in some fisheries, they are also commonly taken as an incidental catch in trawl and longline fisheries. An evaluation of the status of shark species throughout the world (Castro et al., 1999) concluded that sharks are threatened by overexploitation in many parts of the world. Indeed, the only reported case of driving a marine fish species close to extinction involved another cartilaginous fish, the skate, *Dipturus (Raja) batis*, in the Irish Sea (Brander, 1981).

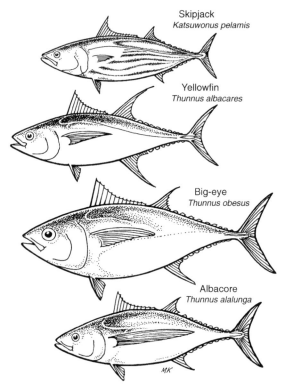

Skipjack
Katsuwonus pelamis

Yellowfin
Thunnus albacares

Big-eye
Thunnus obesus

Albacore
Thunnus alalunga

Fig. 2.35 Tunas.

Because of their low fecundity and trophic position as top predator, sharks are unlikely to support large sustainable fisheries. The demand for shark fins is a serious threat to many shark populations and sharks are targeted for this market even in countries where shark flesh is not eaten for religious or cultural reasons. In places where they are caught for human food, sharks are bled immediately after capture as the blood contains urea which can taint the flesh. Otherwise shark flesh is good white protein and free of bones. Shark finning is wasteful and FAO (1999) has recommended the minimization of waste in shark catches by requiring the retention of sharks from which fins are removed (see Box 2.25 'Shark fin soup').

Box 2.24
Scombroid or histamine poisoning

Scombroid or histamine poisoning is one of the most common forms of seafood poisoning. The fishes involved in scombroid poisoning include tunas and less commonly, mahi mahi and mackerels, all of which have naturally high levels of the amino acid histidine. Post-mortem changes include the enzymatic conversion of histidine to histamine. Fish kept at low temperatures for long periods, even to the point of bacterial spoilage, are unlikely to have high histamine levels. But at room temperatures, say above 10°C, toxic levels of histamine can be reached within hours. High levels of histamine are indications of poor fish handing procedures, particularly a failure to chill fish immediately after capture.

The symptoms of scombroid poisoning are allergic responses to high levels of histamine and can occur within minutes of eating affected fish. Symptoms include a metallic or peppery taste, nausea, vomiting, abdominal cramps, diarrhoea, facial flushing, dizziness, palpitations and a weak pulse. Immediate relief can be obtained by the administration of antihistamines and full recovery usually occurs within 24 hours. Many countries set limits for the presence of histamines in seafood and typical maximum levels are 500 mg/kg for canned tuna and 200 mg/kg for fresh and frozen fish. Histamine levels are not reduced by either cooking, freezing or smoking fish and the only way of avoiding scombroid poisoning is to make sure that fish have been chilled soon after capture and stored at low temperatures. Properly-handled fishes have eyes that are clear and bright, scales or skin that is shiny, and gills that smell seaweed fresh; when raw, the flesh is firm and does not separate easily; when cooked, it does not have a honeycombed appearance.

**Box 2.25
Shark fin soup**

Shark fins are used to make a soup that is considered a delicacy in many parts of Asia. Fins are salted and dried before being treated with hot water to extract the gelatinous 'needles' between the rays of the fins. It is estimated that millions of sharks are killed each year by fishers who slice off the fins and return the sharks, often still alive, to the sea. Feelings of compassion aside, hunting sharks for their fins alone is a waste of good protein and many countries have now banned the landing of shark fins without the accompanying carcass. Sharks have slow growth rates and low fecundity, and if taken in the large numbers required for the soup industry, are easily overexploited. If another reason is needed not to buy shark fin soup, there have been recent claims that shark fins contain mercury levels many times higher than the recommended safe limit for humans. With only about 20 shark attacks recorded on humans each year, this makes dead sharks potentially more dangerous than live ones.

Exercises

Exercise 2.1

Prepare a resource status report on an exploited marine species that you are familiar with locally. The report should address: the biology of the species; the history of the fishery; status of the resource; current management measures and recommendations.

Exercise 2.2

Conduct a brief survey of a local fish market. Make a list of all species offered for sale with an estimate of weights available, price per kg and place of origin. Interview sellers to find out where each species comes from and how the availability of the marketed species varies seasonally.

3 Fishing and fishers

3.1 Introduction

Fishers vary in their motives for fishing. Some fish for profit, some fish for recreation and some fish for food. Intermingled with these basic motives are intangibles such as lifestyle, independence and a love of the sea. Types of fishery operations range from daily food gathering expeditions in the intertidal zone by subsistence fishers, to extended fishing trips using large sophisticated vessels by commercial fishers. Large industrial fisheries target a relatively small number of species that are present in sufficient numbers to make fishing profitable. Small-scale, and particularly subsistence, fisheries, on the other hand, target a much larger range of species in order to satisfy the protein requirements of coastal communities. Many of the exploited species have been introduced in Chapter 2. This chapter provides a brief description of some of the fishing gear used, the fishers involved and the effects of fishing on target species, non-target species and the marine environment.

3.2 Fishing gear and methods

Fishing gear and methods used depend on the species fished. Techniques vary from very simple ones, such as the hand collection, or gleaning, of shoreline invertebrates, to complex and expensive operations such as purse seining for tuna. A large range of fishing gear is used by commercial and artisanal fishers and some basic types are described in the following subsections.

Some of the fishing gear and techniques described, such as traps and gill nets, are classed as passive: that is, the gear remains stationary, and relies on fish moving to the gear. Fishing gear regarded as active, such as seine nets and trawl nets, are designed to be dragged or towed in order to catch fish. The distinction between the two types is important from a consideration of fishing costs and risks of damage to the environment. As they do not require towing, passive gears are relatively inexpensive to operate and have a lower potential to do physical damage to the sea floor. However, active fishing gear, particularly trawl and seine nets, are more efficient in terms of landed catch weight, and account for a major part of the world's catch. The most efficient fishing methods take into account the behaviour of the intended target species.

3.2.1 Gleaning, spears and traps

The gathering or gleaning of marine animals and algae from shallow and intertidal areas of the sea is a common and often social activity in many parts of the world, particularly where seafood is an important part of the local diet. Gleaned species include many molluscs either attached to rocks or buried beneath the substrate. In many developing countries, gleaning is often practised by women who need to remain close to home and who collect sea urchins, sea cucumbers and octopuses from nearby reef flats. Although individual catches are often small, the collective catch made by a large number of subsistence fishers is large and the combined impact on the intertidal areas can be substantial.

Traps are devices designed to encourage the entry of animals, which are then prevented from escaping either by particular aspects of their behaviour, or by the design of the trap itself. The principle of baited traps is that animals, attracted

A traditional Maldivian fishing *dhoni* on Buruni Island in the Maldives, undergoing maintenance, with a small *bokkura* dinghy in the foreground. Buruni is famous throughout the Maldives for its fish paste made from skipjack tuna. Photograph courtesy of Kelvin Passfield, fisheries biologist.

to the bait, enter the trap through tapered openings from which it is difficult to escape. Baited traps or pots are used to catch various carnivorous species of crustaceans, molluscs and fishes and the trap shapes as well as the design of their entrances vary greatly. In crustacean traps, tapered ramps set in the side of the trap, or conical entrances set in the top or sides of the trap are common. The trap shown in Fig. 3.1, used to catch temperate-water spiny lobsters, is made from a steel rod frame covered with wire or twine mesh and a traditional wicker top with a conical plastic entrance.

Fish traps are often designed to take advantage of the species' behaviour. Fishes approaching the trap shown in Fig. 3.2, which is designed to catch breams, emperors and snappers, tend to swim

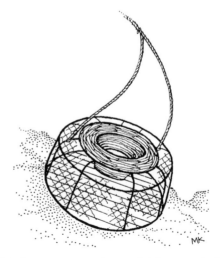

Fig. 3.1 A pot or trap used to catch lobsters and other crustaceans.

Land-based fisheries extend as far south as the southern tip of South America. Here, a wooden fishing vessel is moored on the outside of a steel vessel in Punta Arenus, the world's most southern city. Note the conical crab traps on the foredeck of the steel vessel.

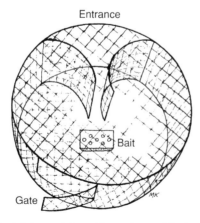

Fig. 3.2 A fish trap (viewed from above). The bait is placed (often in a perforated container) in the centre of the trap and the catch is removed from a gate in the trap's side.

around the outer circumference of the trap, and are directed into the entrance by the curved sides. Once inside the trap, fishes swimming around the inside perimeter are deflected away from the opening by its inwardly curved shape. The number and position of the openings, and the size of meshes used in constructing traps are often controlled by fisheries regulations. In addition, traps used in some fisheries are required to have an escape gap, or an opening of a prescribed size, through which small individuals can escape. To prevent 'ghost fishing' by lost traps, some authorities insist that traps include a biodegradable panel which rots away during prolonged immersion in water.

Some animals may be induced to enter unbaited traps designed to resemble a refuge for the species.

Fig. 3.3 A Pacific island fence, or maze trap used to catch migrating coastal fish.

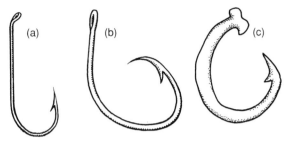

Fig. 3.4 Fish hooks: (a) a common J-shaped hook; (b) a modern circle hook; (c) a Pacific island traditional bone hook.

Octopus traps are often made up simply from clay pots or short lengths of plastic pipe hung from a line set along the sea floor. As the animal entering such a trap is often territorial, and prevents the entry of other individuals, a large number of small traps must be set in order to make a commercially viable catch.

The use of barrier and fence traps is perhaps one of the oldest ways of communal fishing. The simplest traditional traps use the ebbing tide to strand fish in areas hollowed out on reefs and sandbanks, and contained by V-shaped or semicircular walls of stone or coral. Fence traps are built at right angles from shorelines and reefs to guide migrating coastal fish such as mullet into a large retaining area (Fig. 3.3). Fish may be either isolated in the retaining area by the retreating tide, or prevented from escaping by a complicated design or maze. Designs are often traditional, and vary between regions. Although originally made from stone or coral blocks, such traps are now usually made from modern materials such as wire-mesh netting. Their ease of construction and the demand for their use by increasing populations have resulted in authorities limiting the number of fence traps used in some areas.

3.2.2 Hooks and lines

Hook and line gear is used in a wide range of configurations, the simplest being hand-held lines with one or more baited hooks. In some areas, hooks and lines are the only type of gear permitted for use by recreational fishers. The most familiar type of manufactured steel hook is J-shaped, with the pointed part of the hook more or less parallel to the shank (Fig. 3.4). In many commercial fisheries, however, the hooks used are more circular in shape. Circle hooks are similar in design to the bone or shell hooks which have been used since prehistoric times by many coastal peoples (see Box 3.1 'Fish hooks as symbols'). When a fish strikes a circle hook, the point rotates around the jawbone, ensuring that the fish remains caught without the fisher having to maintain pressure on the line. Steel circle hooks are now used to catch tuna, sharks, halibut and deep-water snappers. Hooks are manufactured in a wide range of sizes, and the gap between the point and the shank appears to be the dimension which most determines the size range of fish caught by a particular hook.

Handlining gear consists simply of one or more baited hooks attached to a line, which is weighted at the bottom in the case of demersal target species. If more than one hook is used, these are connected to the main fishing line by short sidelines or snoods. When handlines (or droplines) are used to catch demersal fish in deep water, mechanical means may be used to haul in the line. Many small artisanal vessels used to catch deep-water snappers in tropical areas use simple wooden hand reels to retrieve lines from depths of about 200 m (Fig. 3.5).

A longline is made up of a mainline to which branch lines with baited hooks are attached at intervals. Bottom longlines are used in the North

Box 3.1
Fish hooks as symbols

Fish hooks carved from shells and bones are artefacts common to many coastal societies and some of these came to have a symbolic as well as a functional significance (Fig. B3.1.1). Besides being used for fishing, carved fish hooks were worn as personal ornaments and to assure the wearer of the successful outcome of fishing trips. The fish hook became a symbol of abundance and plenty and some were used to show kinship with cultural groups and family ties. In one Polynesian legend, a demigod fished all over the Pacific with a hook made from his grandmother's jaw bone, symbolizing the continuation of the supply of seafood from one generation to the next.

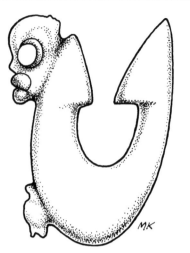

Fig. B3.1.1 A centuries-old Maori fish hook, about 30 mm long, carved from whalebone. The fishing line is tied around the groove on the neck of the upper mask and bait is tied to the lower mask.

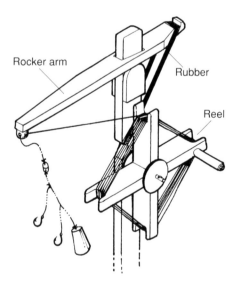

Rocker arm

Rubber

Reel

Fig. 3.5 The FAO-sponsored Samoan hand reel used to retrieve droplines set to catch deep-water snappers in depths of about 200 m.

Atlantic to catch cod and halibut and in depths of up to 2500 m in the fishery for the Patagonian toothfish in Antarctic waters. Surface longlines, which may be over 100 km in length, are set near the surface for pelagic fishes such as tuna (Fig. 3.6). A vertical longline, with hooks set from snoods along its lower section, may be set perpendicularly in the water column in areas where the sea floor is steep or rugged.

Natural or artificial lures attached to lines may be towed (or trolled) behind boats to catch pelagic species, such as mackerels, tunas and billfishes. There are many different ways to rig trolling lines, and the design of lures varies greatly, depending perhaps more on social custom than on the species sought. In general, lures are designed to attract fish by having one or more of the following attributes – an erratic movement when towed through the water (to resemble an injured prey), a bright or reflective surface, and fluttering appendages of feather, plastic, rubber or cloth (Fig. 3.7). Instead of artificial lures, small silver fish such as garfish and flying fish, or pieces of larger fish, may be threaded onto one or a series of hooks. Many countries set floating fish aggregating devices (FADs) off the coast to attract and concentrate pelagic fish for trolling operations (see Box 3.2 'Fish aggregating devices').

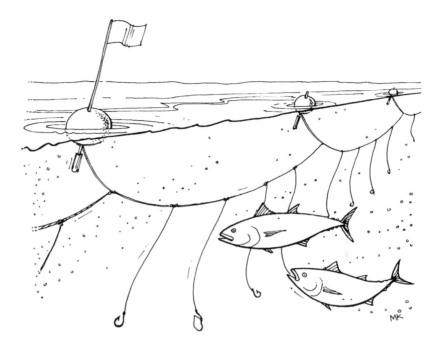

Fig. 3.6 A tuna longline.

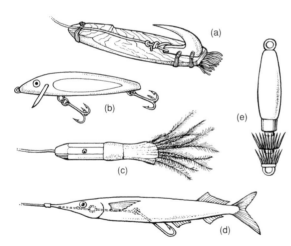

Fig. 3.7 Fish lures: (a) a traditional pearl-shell lure from the Pacific island of Kiribati; (b) a manufactured 'hard' fish lure; (c) a manufactured 'soft' fish lure; (d) a fish hook baited with a garfish; (e) a squid jig.

Tunas may be caught commercially by pole-and-lining, involving the use of barbless, unbaited hooks on short lines attached to poles. The fishes are often encouraged to strike the bright metal hooks by 'chumming' the water with live baitfish to induce a feeding frenzy. Squid are also caught commercially on barbless lures, or jigs (Fig. 3.7) attached either to handlines, or to automatic jigging machines. The machine automatically lowers the line to a set depth, and an elliptical drum retrieves the line with a jerking or jigging movement. During night fishing, bright lights are used to attract squid into the fishing area.

3.2.3 Stationary nets

Stationary nets are those that are passively left in the sea and include gill nets, tangle nets and trammel nets. Gill nets and trammel nets are panels of netting held vertically in the water column by a series of floats attached to their upper edge (the floatline, or corkline), and weights attached to their lower edge (the footrope, or leadline). These nets may be either anchored in shallow water, or set to drift in the open ocean. As passive gear, their catching ability relies on the movement or migration of fishes through the area where the nets are set. The nets are more efficient at night when they are less visible, although most nets are now made from almost invisible monofilament

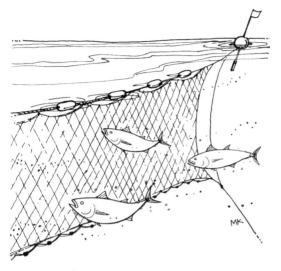

Fig. 3.8 A gill net.

nylon, which locks behind the gill covers (opercula) of bony fish or the gill slits of sharks.

Gill nets (Fig. 3.8) are used in shallow water to catch species such as mullet and mackerel. In deeper water they may be set on the sea floor for demersal species such as sharks, or near the surface for pelagic fish such as tuna. There have been concerns regarding the use of very long gill nets, or drift nets, left in the open sea. Drift nets may be 50 km in length and catch non-target species including dolphins. Due to international concern and political pressure, the number of vessels using drift nets has been dramatically reduced over recent years. The Wellington (New Zealand) convention of the South Pacific Forum recommended a maximum length of 2.5 km for drift nets.

A trammel net is constructed from a panel of small-mesh net sandwiched loosely between panels of larger-mesh net. The nets are set in the same way as gill nets, but catch a much larger size range of fish by entangling rather than gilling them. Fish coming into contact with the middle panel of small-mesh netting, are prevented from breaking free by the outer panels of larger-mesh netting. There have been concerns regarding 'ghost fishing' by lost gill nets, which are made of synthetic rot-proof material and will continue fishing for many years.

3.2.4 Towed nets and dredges

Trawls and dredges are gears which are towed through the water to sieve out fish and marine invertebrates. Trawling is one of the most widely used methods of catching fish in the world, and there are many variations in design and towing methods. Most methods, however, rely on the ability to tow nets at faster speeds than the fish are able to sustain.

An otter trawl consists of a conical or funnel-shaped net leading into a bag, or codend, in which the fishes are retained (Fig. 3.9). The net may be extended laterally by panels of netting

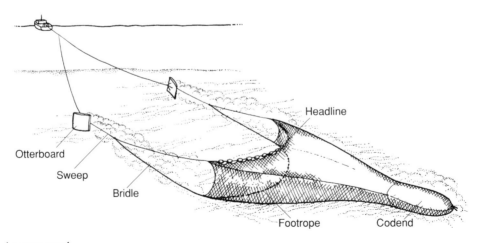

Fig. 3.9 An otter trawl.

A cobble beach is one of the most difficult of all intertidal areas for marine organisms to become established in. This one in Naxos, Greece, has cobbles that range from small to large with increasing distance up the beach slope.

Highly productive inshore areas include shallow bays, inlets, and sand flats that receive nutrients from adjacent coasts. Hill Inlet, Whitsunday Islands in northeastern Australia. Photograph courtesy of Jeremy King.

The Campo de Hielo Sur glacier in Patagonia is one of many that are rapidly being reduced in size by global warming. Bare and scoured rock is becoming visible at the glacier's sides.

Seaweed is common in the diets of many local communities. Its iodine content guards against goitre. Here, several species of algae are offered for sale in a market in Borneo.

Small clupeoids are caught by small purse-seiners in the coastal waters of Southeast Asia. These small fish are dried as *ikan bili*, which is an ubiquitous flavouring in Southeast Asian food.

Lost or discarded netting and other synthetic material washed ashore on a beach south of Calais, France. Such material on the sea floor is responsible for entangling and killing many marine species.

A triangular push net being used in the surf zone near Cap Gris Nez on the Atlantic coast of France.

Local purse seine fishing vessels in harbour, Borneo. These small vessels use purse seine nets to make large catches of coastal fishes, particularly sardines and anchovies.

Small longline vessels are used to fish in the rich Humboldt Current off the coast of Chile. These boats, in Valparaiso, are launched by crane from a nearby pier.

Atlantic coast seafood offered for sale in an Amsterdam market.

Species of tuna are the most ubiquitous of pelagic resources and provide food security and export earnings for many coastal people including those on Pacific islands. Photograph courtesy of Bob Gillett.

A fishing *dhoni* in the Maldives unloading skipjack tuna which will be boiled and smoke dried to make 'Maldives fish' which has been a staple in Maldives for hundreds of years. Photograph courtesy of Kelvin Passfield, fisheries biologist.

The big and the small. Modern tuna longline vessels moored behind an artisanal fishing vessel with a gill net draped over its cabin top. Nuku'alofa harbour in the Kingdom of Tonga.

A traditional sailing *bokkura* trolling for small tuna in Shaviyani Atoll, Northern Maldives. Although small sailing boats are still common, most of the larger fishing *dhonis* have converted from sails to diesel engines. Photograph courtesy of Kelvin Passfield, fisheries biologist.

In countries where a wide range of seafood is traditionally consumed, there is very little that is discarded from the nets of local fishers. This is reflected in the variety of species offered for sale in a fish market in Kota Kinabalu, Malaysia.

A seine net being used on a gently sloping beach in Puerto Madryn, Argentina. The net is hauled out to sea and dragged around in a circle before being hauled back onto the beach.

(wings) at its opening, and long cables (sweeps) which help to herd individuals into the mouth of the net. Otter trawls derive their name from the otterboards, or doors, which act as paravanes to hold open the mouth of the net when towed. During towing, the sweeps and the clouds of sand swept up by the doors help to herd individuals into the path of the approaching net.

Demersal otter trawls are designed for bottom trawling, in which case the target species include a large variety of demersal fish species such as flatfish in shallow water, cod and haddock in depths of about 250 m, and grenadiers and orange roughies in depths of over 1000 m. Trawls are also used for invertebrates including northern shrimps, penaeid prawns and tropical scallops. In invertebrate fisheries, the herding effect of sweeps is unnecessary, and the trawl net is attached

directly to the otterboards without the use of sweeps. A pelagic trawl consists of a bag or codend towed behind a single boat or by two boats working together. In the case of a single-vessel operation the net is kept open by doors in a similar way to otter trawls. Herrings, mackerels, horse mackerels, sardines and sprats are some of the fishes caught by pelagic trawls in the Atlantic and the Alaskan pollack is targeted in a large pelagic trawl fishery in the northern Pacific.

Beam trawls are similar in design to otter trawls, but the net is spread open laterally by a horizontal beam instead of the boards (Fig. 3.10). Beam trawls are used commercially to catch sole and plaice and with their fixed openings are relatively easy to set from smaller vessels. The known working width of a beam trawl makes it an ideal sampling gear for scientific survey work, as the

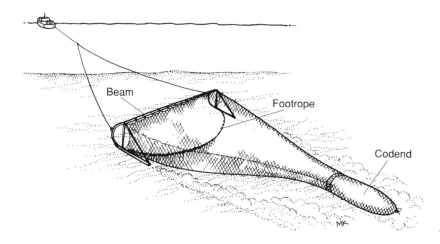

Fig. 3.10 A beam trawl.

area of the sea floor covered, or the volume of water sieved during a tow of the net can be easily calculated.

Dredges are usually made from steel frames and steel mesh, and are towed behind vessels fishing for molluscs such as scallops and oysters (Fig. 3.11). The leading edge of the dredge usually has either a heavy chain, an inclined flat bar, or a toothed bar, which scrapes, or digs, the molluscs from the substrate. Such heavy gear is expensive to tow and environmentally destructive: in addition to causing damage to the sea floor, a scallop dredge damages the uncaught scallops over which it passes.

Surrounding nets are designed to be dragged or towed in an arc around fish. The schooling behaviour of coastal and oceanic clupeids (sardines and anchovies) and tuna make them particularly vulnerable to such gear. A beach seine in its simplest configuration consists of a long panel of netting which is dragged around shoreline schools of fish. The net is weighted to keep the lower side of the panel in contact with the sea floor, and has floats to keep its upper side at the sea's surface. Some beach seines have a central panel of loose netting which forms a bunt or codend to retain fish. Ways of employing beach seines vary, although in many cases, one end of the net is anchored on the shore, and a boat is used to drag the other end seawards in a large arc and back to the shore before hauling. In deeper water, large boat seines may be used from one or more fishing vessels. In a method known as anchor (or Danish) seining, the net is anchored and buoyed at one end, and the other end towed round in a large circle before hauling.

Fig. 3.11 A scallop dredge.

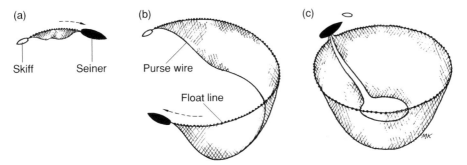

Fig. 3.12 The sequence of operations in purse seining. (a) The end of the net attached to a drifting buoy or a skiff (outline shape) is released. (b) The seiner (solid shape) releases more net as it moves away in an arc. (c) The two ends of the net, and the purse wire at the lower edge of the net, are hauled in. Movement of the seiner is shown by dashed arrows.

Purse seining is one of the main industrial fishing methods with vessels ranging in size from small boats used to catch sprats to large superseiners used to catch tuna in the open sea. A purse seine is a net which is set in a circle around a school of pelagic fish such as tuna, mackerel and pilchards. In terms of total catch weight from a single operation, a purse seine is one of the most efficient types of fishing gear in the world, and over 100 tonnes of fish may be captured in a single shot of the net. The net is essentially a long panel of netting (up to 2 km in length) with floats fastened to its upper edge and weights and purse rings attached to its lower edge.

Once a school is located, sometimes with the aid of sonar and a ship-based helicopter, the end of the net is attached to a buoy or a skiff, which is cast off (Fig. 3.12a). The vessel releases more and more of the net as it moves around in a large circle, which is completed when the two ends of the net are brought together (Fig. 3.12b). After the net ends are retrieved, the purse wire, which runs through the rings around the lower weighted edge of the net, is hauled in to close off the bottom of the net, and prevent fish from escaping downwards (Fig. 3.12c). The net is hauled in by passing it over a large motorized drum (a power block) suspended over the aft deck of the vessel. When the net has been hauled alongside the vessel, a large dip net (or brail) is used to transfer smaller portions of the catch into the hold. In the case of small pelagic species, a submersible fish pump may be used to pump a mixture of fish

and sea water on board. Lampara nets are similar to purse seines, but do not have the purse line to prevent fish escaping below the net. The lower (ground) line, which is shorter than the surface line, is pulled under the school of fish as the net is hauled. When the targets are smaller fish, such as sprats, sardines and herrings, floating lights may be used to attract fish before surrounding them with the seine net. Many species of fish that inhabit the open sea are attracted to floating objects, and some purse seine fishers seeking tunas will leave rafts or logs to drift in the sea and attract fish (see Box 3.2 'Fish aggregating devices').

3.3 Fishers

Fishers and fisheries are sometimes classified as either industrial, artisanal or subsistence. An industrial fishery involves large commercial vessels harvesting large numbers of just a few key species. Catches from just ten species collectively account for about 30% of the world catch in terms of quantity (see Fig. 1.38 in Chapter 1). However, up to 30% of the world total output of fish or about half the landings for human consumption come from small-scale or artisanal fisheries which involve up to 90% of all fishers (various authors in Misund, 2002). An artisanal fishery is a small-scale, low-cost, and labour intensive one in which the catch consists of many different species that are sold and consumed locally.

In many parts of the world, seafood is a vital source of protein, and fishing is undertaken for

Box 3.2
Fish aggregating devices

Many species of fish that inhabit the open sea are attracted to floating objects. This behaviour has been used in the deployment of fish aggregating devices (FADs): floating rafts anchored offshore to attract pelagic fish. A range of materials, including coconut logs, bamboo, and aluminium pontoons have been used in the construction of rafts. Material such as old fish nets, palm leaves, and car tyres are suspended beneath the rafts in the belief that this increases the raft's effectiveness as a habitat for fish. Mooring FADs in depths of over 1000 m, off coasts with no continental shelf, poses particular problems. Mooring systems consist of a sinking (nylon) rope spliced to a buoyant (polypropylene) rope attached to a chain and concrete block anchor (Fig. B3.2.1). This combination of ropes induces a mid-water S-bend, or an inverse catenary, which contains the slack line in the system and prevents chafing (Boy & Smith, 1984).

The benefits of FADs are that they reduce the search time involved in a fishing trip, and therefore reduce fuel costs, as well as increase fish catches. However, the costs of building and setting FADs are high (approximately US$5000), and, because of

storms, wear and vandalism, their average lifespan is less than 12 months. Data on any increase in fish production where FADs have been deployed are generally not available. Because of this, it is difficult to complete cost–benefit analyses, and even more difficult to devise methods whereby the fishers benefiting from the deployment of FADs contribute to their cost.

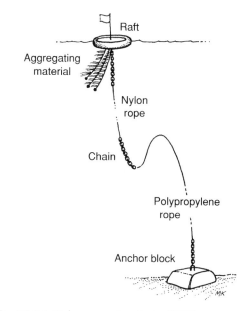

Fig. B3.2.1 Fish aggregating device (FAD).

food rather than for profit. In subsistence fisheries, a major part of the catch is used by the fishers and their families for food although a lesser part of the catch is often sold. The distinction between these categories, however, is often blurred; in many subsistence fisheries, for example, increasing consumerism is resulting in an increasing proportion of daily catches being sold to purchase other essentials.

3.3.1 Fishing for food

For many coastal communities, particularly those with limited productive land for agriculture, the sea has traditionally provided the major source of protein. The traditional reliance on seafood

for nutrition has persisted for so long in many communities that a strong cultural relationship with the sea and the marine environment has developed. In many cases, various customs and traditions have evolved for the purpose of protecting marine resources. Often coastal communities have assumed traditional or customary control over adjacent marine areas and their resources. This form of property right is referred to as customary marine tenure (CMT) and usually involves members or leaders of the community using traditional knowledge, including taboos, to protect fish stocks (see Box 3.3 'Customary marine tenure').

In some developing countries, virtually all people in a coastal community spend some part

Box 3.3
Customary marine tenure

Customary marine tenure (CMT) is the term used to denote legal, traditional or *de facto* control and use of marine areas and resources by indigenous people. Villages, clans, or communities have the right of ownership and use of the land or coastal areas where they live or over which they have ancestral claims. CMT involves property rights that often depend on hierarchies of community structure; these can include *use rights* for people to catch fish and *control rights* for community leaders to make decisions on how traditional fishing areas can be used. Usually, there are also *transfer rights* under which rights are conveyed to others including to heirs through inheritance.

In many communities, conservation rules have been set by traditional leaders for many generations and these have included placing bans or restrictions (tapus, tabus or taboos) on fishing in certain areas or on particular species. Customary marine tenure is being revitalized by communities and encouraged by governments and NGOs as it is seen to be a way of ensuring the conservation of subsistence fisheries resources and marine environments. However, most communities are experiencing rapid population growth, increasing consumerism and a resulting need to exploit natural resources, not just for food but for profit. A study of the association between social and economic characteristics and marine tenure systems in Indonesia and Papua New Guinea is given in Cinner (2005). An analysis of the Fiji Environment Management Act and its recognition of the centuries-old traditions and customs of natural resource management is provided in Sutton (2005).

Customary control of marine resources does not necessarily ensure sustainability. Many traditional fishing methods are destructive and examples include the use of plant-derived fish poisons and communal fish drives across vulnerable reef areas. The effects of some destructive fishing methods have been exacerbated by increasing population numbers. In the past, the marine environment was able to sustain occasional, localized damage because the frequency of the activity was low and fewer people were involved (King & Lambeth, 2000).

National governments in most Pacific islands have control of resources below the high water mark and this is counter to the customary ownership of marine areas by clans and communities. However, governments often make formal or informal concessions to CMT by allowing some community control. In the Cook Islands, the Ra'ui system, under which community leaders can place a tapu or ban on fishing in particular areas and on catching threatened species, is being reinstituted and supported by the government. Communities in Samoa can set their own conservation bylaws, chiefs in Fiji can approve licences for outsiders to fish in their customary areas, village councils in Tuvalu can ban fishing in certain areas and communities in Tonga can control 'special management areas' designated as such by the government. Some other traditional practices act indirectly to conserve marine resources. Some communities in Samoa, for example, place a ban on commoners eating turtle meat. Some communities in Fiji ban catching small baitfish (in the belief that these attract larger fish inshore) and ban all fishing for 100 nights in mourning for the death of a high chief.

of each day collecting seafood or catching fish. Fishing methods used are generally low-cost and labour intensive, including gleaning and using simple gear such as lines with hooks, spears and traps. The relative inefficiency of some of these methods has allowed seafood resources to be shared between large numbers of people. Where daily catches of seafood are vital, the amount of energy gained from food has to be higher than that expended on collecting it. Consequently a

The sea and fishing are entwined in the culture of many coastal communities. This stone carving is of the sea spirit, Kezoko, a bird with a man's body, from the western Province of the Solomon Islands. If the land spirit, usually depicted as a carved man's head placed on the bow of fishing boats, blinks, the sea spirit will kill him.

wide range of abundant and accessible invertebrates and fishes are harvested, including jellyfish and small reef fishes which may be used to flavour soups.

Some seafoods have medicinal, social or cultural significances beyond their worth as energy-providing food items. In the Polynesian island of Samoa, for example, the guts and reproductive organs of a species of sea cucumber are partially fermented in seawater before consumption. The concoction is sought after, even by urbanized Samoans, and is sold from roadside stalls. Also, in many Pacific islands, the harvesting of palolo worms that appear on just one or two nights a year in large reproductive swarms (see Fig. 4.32 in Chapter 4) is a long-standing tradition and the basis of traditional feasts.

Divisions in the labour involved in fishing may be based on sex, rank and tradition. In many subsistence fishing communities, for example, fishing offshore from canoes and boats has been the domain of men whereas collecting seafood species from inshore areas has traditionally been the task of women. Separate fishing roles and areas for men and women reflect the community and family obligations of both. The gleaning and shoreline fishing activities of women allow them to stay close to their homes and address other necessary tasks such as gardening, food preparation and caring for young children. Seafood collected from inshore areas, such as bivalve molluscs, also provide emergency food when the weather is too rough to go to sea (King & Lambeth, 2000).

Some traditional fishing methods involve an entire community and sometimes have great cultural significance. Communal fishing methods include the use of fish drives and palm frond sweeps for fish across shallow reef areas. In many communities, people seldom collect seafood or go fishing alone and fishing activities perform an important social function in establishing and strengthening relationships between individuals and families. Excess catches are distributed throughout the community and may be based on customary obligation towards certain families and high-ranking individuals. Some cultures have evolved strict rules on which members of the community are permitted to engage in certain types of fishing. In the Pacific island of Palau, offshore fishing for several species of shark has been practised only by a few prestigious specialists – with the prestige being related to the danger of the fishing method rather than an appreciation of the shark as food. Other cultures place restrictions on who can catch, cook or consume certain species; turtle, for example, is only eaten by high chiefs or royalty in some parts of Polynesia.

The increasing trend to cash economies is affecting many communities in which fishers were subsistence food gatherers. Even in many isolated communities, there is an increasing requirement for cash to pay for facilities such as power supply, health and schooling for the young. As a consequence, many subsistence fishers are fishing longer and harder in order that part of their catches can be sold to pay for other necessities. Dalzell et al. (1996) estimate that 80% of the catch from inshore fisheries in the South Pacific, whether from reefs, estuaries or fresh water, is taken for subsistence purposes with the remainder going to commercial markets. The cash economy has also resulted in communities that previously relied entirely on locally-collected seafood now being able to buy cheap imported protein. In Pacific islands, for example, imported meats include off-cuts such as frozen lamb ribs and turkey tails exported from more developed countries. The consumption of such low-quality, fat-laden imports is believed to be responsible for the increasing incidence of diet-related diseases including diabetes, hypertension, stroke and heart disease in many Pacific countries. Dietary deficiencies, particularly of vitamin A, which is abundant in fish protein, are adversely affecting mothers and infants (King & Lambeth, 2000).

In many developing countries, the collective subsistence catches from many coastal communities total more than that obtained from commercial fishing. In spite of their importance in providing food, these often widely-scattered fisheries are rarely assessed and even less often managed by centralized management authorities. Governments are often slow to recognize the importance of subsistence fisheries which provide nutrition and a livelihood to people who might otherwise be destitute. Some government and non-government organizations (NGOs) are encouraging fishing communities to reassert customary fisheries management controls that have been used in the past, including closing areas and seasons to some types of fishing. This form of management, referred to as community-based fisheries management, is discussed in Chapter 6.

Although management of large-scale fisheries often involves placing restrictions on the amount of fishing or the quantity of fish caught, such controls are unsuitable for many subsistence fisheries where most of the daily catch is required as food. In addition, nationally-imposed fisheries regulations are all but impossible to enforce in subsistence fisheries in which many thousands of fishers operate along extensive coastlines or in many different islands. Fisheries regulations and their enforcement are discussed in Section 6.5.

3.3.2 Fishing for income

Industrial fisheries involve relatively large vessels with mechanized fishing gear and advanced fish-finding equipment seeking catches from a relatively few key species. Many larger commercial fisheries are referred to by the area fished, the target species and the fishing gear used, such as the North Sea herring purse seine fishery, the Gulf of Mexico shrimp trawl fishery, and the Southern Ocean Patagonian toothfish longline fishery. Industrial fisheries are capital-intensive with a high production capacity and the catch per unit effort is normally relatively high. In some areas of the world, industrial fisheries target species such as sandeels and clupeoids for reduction to fish meal and fish oil.

Competition in large industrial fisheries has led to fishers using larger vessels and nets to improve fishing power and efficiency and this excessive fishing capacity and overinvestment has been responsible for the depletion of many fish stocks. The FAO Code of Conduct for Responsible Fisheries (FAO, 1995a) recognizes that excessive fishing capacity threatens the world's fishery resources and their ability to provide sustainable catches and benefits to fishers and consumers. In Article 6.3 it is recommended that: 'States should prevent overfishing and excess fishing capacity and should implement management measures to ensure that fishing effort is commensurate with the productive capacity of the fishery resources and their sustainable utilization'. In industrial fisheries, key management objectives include reducing the number of fishing vessels, improving selectivity of the fishing gear and reducing the proportion of non-target species in catches.

Illegal fishing is of concern to legitimate fishers as well as fisheries managers. Legal fishers adhering to conservation regulations (such as respecting closed areas and seasons) and contributing to management costs resent illegal fishers who have

Fishing centres in many parts of the world have become famous for seafood. Fisherman's Wharf in San Francisco is well known as a fishing port, a seafood market place and a mecca for tourists.

no such restrictions and costs. Such is the concern over illegal fishing activities on a worldwide scale that countries have formed a monitoring control and surveillance (MCS) network (see Section 6.5.4 'Compliance and enforcement').

Modern technology, including the use of global positioning systems (GPS) and sonar on hulls and nets to locate fish have made fishing more efficient. If resources were unlimited, increases in efficiency would be welcome as they reduce fishing costs by increasing the catch for a given amount of effort. However, making fishing effort more effective is detrimental in the many fisheries based on stocks that are already over-exploited (see Box 6.7 'Technology creep' in Chapter 6). An ideal situation would be one in which increases in fishing vessel efficiency were counterbalanced by a reduction in the total number of vessels in the fishing fleet (say by employing buy-back schemes described in Chapter 6, Box 6.8 'Licence buy-back schemes'). In some fisheries, each vessel is required to install a vessel monitoring system (VMS) which allows a central management authority to be aware of its position, course and speed. The use of VMS has the potential to allow more complex, and hopefully more efficient, management systems including the banning of fishing in particular areas and at particular times (see Section 6.5.4).

Fishers and vessels involved in industrial fisheries are greatly outnumbered by the many millions of fishers engaged in small-scale or artisanal fisheries. Artisanal fisheries usually involve low-cost operations from relatively small, often family-owned, vessels that make short fishing trips. In practice, the definition varies between countries, from a single fisher in a canoe in a developing country to a larger family-owned gill-netter in a more developed country. With about half of the world's total fish landings for human consumption coming from small-scale fishing operations, the importance of artisanal fisheries is often not recognized (Misund et al., 2002). Artisanal fisheries employ about 90% of all fishers, over 12 million people, using a wide variety of gears in less energy consuming fishing operations. In small-scale fisheries, the involvement of fishing communities in fisheries management is perhaps the only way of ensuring that the marine resources and the rural livelihoods that are dependent upon them are sustainable (see Section 6.3.3 'Community-based fisheries management').

Some artisanal fisheries are quite large. In Senegal, for example, small purse seines and surrounding gill nets are used to catch mainly sardines (*Sardinella* spp.) and catches are over 200 000 tonnes per year. Similar large artisanal fisheries can be found in South America and Southeast Asia. If the seafood products from undeveloped countries are in demand on international markets, their profitability will almost inevitably lead to overexploitation of the resource. In fisheries in which artisanal fishers collectively catch quantities of fish that rival industrial fisheries operating on the same stock, conflicts are inevitable. Much fisheries management in these cases involves allocating resources between artisanal and industrial fishers.

Commercial fishing is undertaken in the expectation that catch rates will be sufficient to provide monetary returns that cover the often high costs of fishing and provide enough excess profit to make fishing worthwhile. The economic viability of a commercial operation may be viewed from the possible effects of sustainable catch rates and product price on a vessel's net return. The following example is based on financial information collected during surveys on deep-water caridean shrimps on a plateau to the north of Mauritius. The preliminary estimated costs of fishing using a 17 m vessel are shown in Table 3.1. The initial investment includes the fully-equipped vessel and fishing gear. Fixed costs, including loan repayments, depreciation, repairs, maintenance, and insurance, are those which are associated with the vessel, and are incurred whether or not the vessel goes fishing. Depreciation allows for eventual vessel replacement. Running costs are those associated with the fishing operation, and include crew payments, fishing gear replacement, fuel, food, bait, and ice.

If catch per unit effort (CPUE) is in terms of kg per trap per day, the annual returns from a fishing operation will be:

$$\text{price} \times \text{CPUE} \times \text{traps} \times \text{days fished} \qquad (3.1)$$

where 'price' is the value per kg and 'traps' is the

Table 3.1 Preliminary estimated costs (in Mauritian rupees) of a proposed commercial shrimp trapping venture using a 17 m vessel. Adapted from King (1990).

Number of traps used	90
Estimated number of days worked per year	150
Initial investment	
Vessel cost	800 000
Traps (at R1200 each)	108 000
Floats and lines	60 000
Total initial investment	968 000
Fixed costs per year	
Loan repayments/return to capital (10%)	96 800
Depreciation (10%)	80 000
Vessel repairs (10%)	80 000
Insurance (3%)	24 000
Total fixed costs per year	280 800
Running costs per day	
Crew payments (R250 per crew per day)*	1 250
Fishing gear replacement**	560
Fuel and food	1 000
Bait (R6 per trap)	540
Ice (R5 per trap)	450
Total running costs (per day)	3 800
Total running costs (per year)	570 000
Total annual costs	850 800

* based on five crew members
** based on replacement of 50% of traps and lines annually due to wear and loss

number of traps used. The annual cost of fishing is the sum of the daily running costs, RC, and the annual fixed costs, FC, i.e.:

$$(\text{RC} \times \text{days fished}) + \text{FC} \qquad (3.2)$$

At the 'break-even' point, where returns to fishermen just balance the total cost of fishing (where no profits are made), Equation 3.1 is equal to Equation 3.2:

$$\text{price} \times \text{CPUE} \times \text{traps} \times \text{days fished} = (\text{RC} \times \text{days fished}) + \text{FC}$$

Making price the subject of this equation gives the market price required to cover the cost of fishing as a function of catch per unit effort:

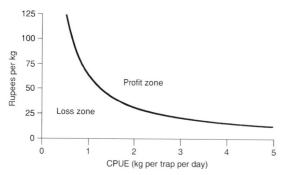

Fig. 3.13 A preliminary break-even curve for a proposed deep-water shrimp fishing operation in Mauritius using a 17 m vessel. The curve is based on a static analysis of one vessel's costs without a consideration of the timing of returns.

price = [(RC × days fished) + FC]/[CPUE × traps × days fished] . . .   (3.3)

The break-even curve relating the cost-recovery price to catch per unit effort in the example of the proposed trap fishery in Mauritius is shown in Fig. 3.13. The area above the curve represents the profit zone, and the area below the curve represents the loss zone. For a given level of catch per unit effort, the graph may be used to indicate the price required to cover fishing costs. Alternatively, if market price is known, the graph may be used to suggest the long-term catch per unit effort which must be maintained to cover costs. If the expected price received by a fisher for shrimp caught was 75 rupees per kg, for example, from the curve, long-term mean catch rates from a commercial vessel would have to be greater than about 0.8 kg per trap to cover fishing costs.

There are, however, some caveats required when relating catch rates obtained during fishery surveys on undeveloped resources to those expected in subsequent commercial operations. Initial catch rates on a previously unexploited resource are higher than they will be when the fishery is fully exploited. In a fully exploited fishery, catch rates are likely to be about half of those originally obtained when fishing the virgin stock (see Chapter 5). In addition, the economic performance of a fishing operation is affected by various unpredictable factors including changes in the cost of key inputs such as increased fuel costs, falling demand for the product and regulatory restrictions on fishing.

The collection of financial data is relatively easy to incorporate into fisheries survey work, and their value is in their use in determining the viability of commercial exploitation. Collecting such data is a recognition that the exploitation of a potential resource may be biologically feasible, but not financially so; the exploitation of deep-water species, where fishing costs are high, is one example. The financial analysis described in this section is based on a simple static analysis of one vessel's costs without a consideration of the timing of returns. Nevertheless, the construction of a break-even curve represents a simple way of assessing the economic viability of a proposed commercial operation.

Theory described in Section 6.2.2 suggests that in an open-access, unregulated fishery, more fishers will enter the fishery until the point at which total revenue equals total costs. At this point, when no profits are made, it is supposed that fishers will leave the fishery and allow stocks to recover. However, fishers may continue fishing even in the face of marginal earnings for many reasons including the unreasonable expectation that catch rates will improve. A FAO survey of fishing operations suggested that, in spite of fully and sometimes overexploited fisheries resources, most marine capture fisheries are financially viable undertakings (Lery et al., 1999). Many fisheries, however, are surviving in the face of low catch rates only because governments are providing subsidies for boatbuilding, cheaper fuel, low-interest loans and other incentives.

3.3.3 Fishing for recreation

Recreational fishers usually engage in their sport for entertainment rather than for obtaining food. In developed countries, many recreational fishers are willing to accept that catching a fish is more expensive than it would be to purchase the same fish on the open market. Recreational fisheries pose particular problems for fisheries management.

Although individual catches may be small, the total recreational catch is often large and represents a large proportion of the total fishing mortality on a stock that is shared with commercial fishers. From the viewpoint of the stock, it makes no difference whether fishing mortality is caused by commercial fishing or recreational fishing and in managing a shared stock, managers are obliged to consider the effects of recreational fishing.

However, recreational fisheries are usually difficult to assess as they involve a large number of fishers who are often spread out over large geographical areas. Methods of surveying recreational fisheries are reviewed by Cowx (1996). Event recall methods, involving the use of log-books, mail questionnaires, or phone surveys, require the respondent to recall past fishing events. On-site intercept methods involve collecting information at the point of access to fishing areas including interviews at boat launching ramps. The latter method provides more reliable responses but at a higher cost.

If the process of assessing a fishery includes economic aspects, it may be found that there are more benefits to society, or a particular community, in allowing recreational rather than commercial fishers access to a resource. This would be the case if recreational fishers were prepared to spend large sums of money on accommodation, boats, fishing gear, food and bait in the community to secure relatively small individual catches. In some fisheries, this has been recognized in the application of management objectives to maximize recreational expenditure and returns to the community. An example is the case in which game fishers are required to pay a high licence fee to catch fishes such as a marlin.

Most people pursue recreational fishing for sport or relaxation and the quantity of fish caught is a secondary consideration. However, some non-professional fishers take fishing more seriously and seek larger catches; there is a maxim among fisheries managers that 90% of the total recreational catch is taken by 10% of recreational fishers. Most often, these are the amateur fishers that are prepared to work harder and fish the tides irrespective of the time of day or night. However, the few amateur fishers, sometimes referred to as 'shamateurs', that sell their catch illegally on the black market cause great resentment from commercial fishers and additional enforcement efforts.

It is generally difficult to enforce rules and regulations on recreational fishers for the same reasons that it is difficult to assess them. Recreational fishers usually fish in many widespread locations that are often isolated and this makes enforcement expensive and not very effective. In order to keep individual catches relatively small and maximise both participation and enjoyment, recreational fisheries are usually subject to strict limitations on gear and catch. The use of commercial types of gear, such as gill nets, is generally forbidden and often only hooks and lines are permitted. In some cases, only one hook is permitted to be used on each line and a limited number of lines are allowed to be used on each boat. Minimum sizes and bag limits (quotas) are also commonly applied in recreational fisheries. Due to the difficulty of enforcement in recreational fisheries, a better option is a publicity campaign that results in participants appreciating the need for restrictions. If people are sympathetic to the aims of management, either personal conscience or peer pressure will result in compliance with most reasonable management measures.

Like its commercial counterpart, recreational fishing is being affected by habitat loss as well as overexploitation (Cowx, 2002) and this is particularly so in enclosed and sheltered waters. Most recreational fisheries are based inshore where their demands compete with other user groups including sailors and birdwatchers, and other uses including wastewater and sewerage disposal.

3.4 The effects of fishing

In addition to its effect on targeted species, all fishing affects the marine environment and ecosystems to a greater or lesser degree. Some fishing gears, particularly those that scrape the sea floor are just plain damaging from a physical viewpoint. But all fishing removes particular target species from an ecosystem and this is unlikely to leave both predator and prey species unaffected. Most fishing operations are unselective and, as they catch species other than those targeted, have much wider and devastating consequences on ecosystems. The adverse effects of fishing are introduced in this section.

3.4.1 Effects on target species

All fish stocks are capable of being overfished and depleted but some are more susceptible than others. Reynolds et al. (2002) selected five key attributes of fishes and fisheries that result in stocks being highly susceptible to overfishing (Table 3.2). It is interesting that at least three of these features give doubt to the rather optimistic belief that when catch rates drop below some

cost-recovery level, fishing will cease and stocks will recover.

Highly valuable species will continue to be hunted and caught even when remaining stocks are at dangerously low levels as long as the small catches cover costs. Examples include the Southern bluefin tuna, *Thunnus maccoyii*, which can be sold for over US$60 000 for a single fish and the giant yellow croaker, *Bahaba taipingensis*, which is hunted for its highly esteemed swim bladder (see Box 3.4 'Yellow croaker in the red').

Other species at risk include those in which catchability remains high as stock sizes decrease; these include schooling and aggregating species that remain vulnerable even if the overall abundance is low. Particularly at risk are the species that form spawning aggregations targeted by fishers: an example is the Nassau grouper, *Epinephelus striatus*, which was previously

Table 3.2 Key features of fish stocks and fisheries that render species susceptible to overfishing. Adapted from Reynolds et al. (2002).

Fish are of high value	Highly valuable species that remain profitable to catch even when abundance and catch rates are low
Catchability of fish remains high as stock size decreases	Species that school or form spawning aggregations and so remain vulnerable even if the overall abundance of stocks are low
Fish are susceptible to capture as a non-target species	Incidentally-caught species that continue to be caught, even at very low abundances, as long as the target species remains profitable to catch
Life history of fish results in low productivity	Species that have long lifespans, slow growth rates, low fecundity or other characteristics that result in low productivity (*K*-selected species)
Per capita recruitment of fish decreases as stock size decreases	Species in which reproduction is less successful at low stock sizes; i.e. recruitment is insufficient despite the presence of some spawning stock (the Allee effect)

Box 3.4
Yellow croaker in the red

The giant yellow croaker, *Bahaba taipingensis*, reaches over 2 m in length and is endemic to coastal waters of China where it once concentrated in large aggregations close to major estuaries during its spawning season each year. Even after numbers had been heavily reduced by fishing,

individual fish could be easily located by the loud drumming noises they make using their swim bladders. Its large size and highly esteemed swim bladder resulted in high market prices and ensured that it continued to be targeted even at very low stock densities. The species was listed as Critically Endangered on the 2006 Red List due to overfishing and is now under protective legislation in China.

distributed widely over the Caribbean before the disappearance of entire regional stocks (see Chapter 4, Box 4.12).

Non-target species under threat include those that are caught incidentally by a fishing operation (as discussed later in Section 3.4.2). Species that are particularly under threat are those that exist at a lower abundance or biomass than the target species. These species will continue to be caught, even at very low abundances, as long as the target species remains economical to catch (see Fig. 5.6 in Chapter 5).

Other target species that are susceptible are those that have life histories that result in low productivity and those in which per capita recruitment decreases as stock size decreases.

Species with low productivity generally have long lifespans, low growth rates and low fecundity (see *K*-selected species in Section 1.3.3) and an extreme example would be species of sharks, on which large fisheries are unlikely to be sustainable.

In some species, recruitment is greatly reduced or even fails at some low level of stock size. The requirement for a minimum stock size for successful reproduction, called the Allee effect, may be common (see Box 3.5 'The Allee effect'). Examples include overfished sessile animals that have become so sparsely distributed that drifting eggs have a low chance of being fertilized.

All fishing depletes stocks of the target species. In sustainable fisheries, numbers lost are replaced by recruitment (the addition of young) into the

Box 3.5
The Allee effect

The Allee effect (named after W.C. Allee) describes the situation in which the per capita growth rate decreases at low population levels. In Fig. B3.5.1, the straight line shows a population in which the per capita growth rate increases as the population biomass decreases. In other words, the model shows a compensatory increase in growth rate that allows the population to recover from low stock levels. The lower curve in the figure illustrates the

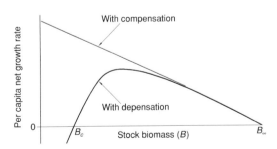

Fig. B3.5.1 Per capita net growth rate as a function of biomass under conditions of compensation (straight line) and depensation (curve). B_∞ is the maximum population biomass and B_c is the critical biomass level, below which the stock cannot drop without proceeding to extinction.

Allee effect, or depensation, in which the per capita growth decreases at low biomass levels. The Allee effect refers to the situation in which reproduction or survival is reduced at low population densities. This could occur if individuals in low densities suffered from a reduced ability to find mates, to resist predators or to withstand the results of random fluctuations in the environment.

Fertilization is external in most marine species and, if individuals are few and scattered, the chances of drifting sperm and eggs meeting is greatly reduced. This is particularly so in sessile species or those with limited mobility, such as abalone and sea urchins. In schooling species, defence against predators may be less effective when the size of schools has been reduced. All animal and plant populations driven to low population densities are likely to be more susceptible to random catastrophic events as well as fluctuations in the numbers of prey and predator species. In a strong Allee effect, there is a critical population size (B_c in Fig. B3.5.1) below which the population cannot decline on average without extinction becoming inevitable. Because of this alarming prospect, the Allee effect is attracting the renewed interest of biologists working on endangered species, including fish stocks that have been driven to very low levels.

stock. Theory suggests that fishing depletes the target resource in order to create a surplus production that can then be harvested. However, overfishing occurs when stocks are reduced to levels in which numbers produced are less than the numbers caught.

In the face of remorseless and heavy fishing by many nations on earth, it would be reasonable to assume that many marine species have become extinct. However, there is very little evidence to suggest that this is so. A major problem is the difficulty of conducting censuses on marine species. If a terrestrial animal, particularly a larger, charismatic furry one, went missing, many people would be aware of it. But there are just too few people studying and counting marine species, and efforts are biased towards the ones of economic importance.

Several marine mammals, such as Steller's sea cow and the Caribbean monk seal, have disappeared off the face of this earth because of hunting. There are some species of fish, such as particular groupers, that are known only from preserved specimens collected for museums over a century ago, and have not been seen since. The Banggai cardinalfish, *Pterapogon kauderni*, was originally thought to be extinct before it was rediscovered in 1994. This rediscovery resulted in aquarium fish suppliers rushing to collect the rare fish, and it is believed that this finally caused its disappearance from the wild.

There are a few species, particularly those with limited distributions, in which extinction has been narrowly averted. The white abalone, *Haliotis sorenseni*, which inhabits waters from Southern California to northern Baja California, was listed by the USA National Marine Fisheries Service as an endangered species in 2001. The species was fished down to an estimated 1600 individuals remaining in the wild and it was claimed that these would disappear without some form of human intervention. A captive-breeding programme has been suggested as numbers in the wild were below the Allee effect's critical density required for successful reproduction. In addition, there have been many extinctions, sometimes called extirpations, of local populations. In

temperate waters, the Pacific salmon has been fished out of some river systems and is unlikely to return. Giant clams live in clear shallow water that makes them particularly vulnerable and some species have disappeared from several tropical island countries within their range.

The Convention on International Trade in Endangered Species (CITES) has established a worldwide system of controls on international trade in threatened wildlife and wildlife products. Protection of endangered species is provided by member countries banning trade in the listed species (listed in CITES Appendix I, which includes all those threatened with extinction and which are, or may be, affected by trade) and by regulating and monitoring trade in others that might become endangered (listed in CITES appendices II and III). Marine species currently listed by CITES include all species of giant clams (*Tridacna* spp.), the Caribbean queen conch (*Strombus gigas*), sturgeons (*Acipenser* spp., from which caviar is obtained), the giant humphead wrasse (*Cheilinus undulatus*), the basking shark (*Cetorhinus maximus*), the whale shark (*Rhincodon typus*), as well as whales and sea turtles. Problems include the listing of species that are endangered in the wild but farmed successfully (such as giant clams) and species that may be threatened in some areas but are abundant in others.

Although there are very few verified examples of overfishing leading to the extinction of a species, many fisheries have collapsed. In many cases, but not all, commercial fishing will cease when catch rates become too low to cover fishing costs. That is, fishing on a target species will stop before the resource is driven all the way to extinction. Fishing on the great whales, for example, stopped when it became uneconomical to do so, not because of any concern for the species. However, there are several circumstances in which fishing on depleted stocks will continue in spite of very low catches. This can happen when the target species is very valuable (Table 3.2) or where there are government subsidies to prop up commercial fisheries. It can also happen in some recreational and game fisheries, where the sport is more important than costs, and fishing will continue even when catch rates are very low.

An additional risk is that, if a stock of fish is reduced down to very low numbers, there is a chance that the vacant niche will be opportunistically filled by another, perhaps hardier or unfished, species. The collapse of the world's largest fishery in which stocks of the Peruvian anchovy were decimated in 1971/72 allowed sardines to increase in numbers and become the dominant species for many years (Section 1.8).

Fisheries management objectives include addressing the problem of fishing gears that indiscriminately catch many non-target species and small individuals of the target species. Incidental catches of non-target species is discussed in the following section (3.4.2). Reducing the catch of small individuals of the target species has been a long-standing aim of fisheries management. Minimum size limits and minimum mesh sizes in nets are conventional fisheries management tools used when the aim is to allow individuals to reach a size at which they can reproduce at least once before capture.

It is unlikely that any fishing technique used will be equally efficient in catching all sizes of fish; that is, all fishing gear is selective to some degree. In a trawl net, for example, very small fish are less likely to be retained than large ones, whereas in a gill net, both very small and very large fish are unlikely to be caught – the distinction between large and small being related to mesh size.

Gill nets have a mesh size designed to catch fishes of a specific size range, and do not gill very small and very large fish. The main determinant of the range of lengths of fish caught by gill nets is the hanging ratio, which is defined as the ratio of the length of the headline to the length of the stretched net. If the hanging ratio is low, say less than 0.5, the net will hang slack in the water rather than taut. In this case the net becomes less selective, as it will entangle fish as well as gill them. Some nets, referred to as tangle nets, are deliberately made this way.

Fish hooks of a given size are designed to catch fish within a particular size range, and are less

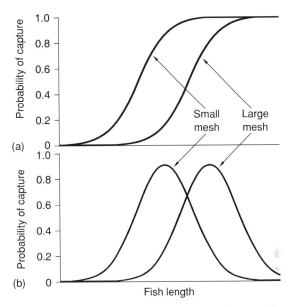

Fig. 3.14 Mesh selectivity curves for (a) trawl nets; and (b) gill nets. In each case, two curves are shown – one for a smaller mesh size (left curve), and one for a larger mesh size (right curve).

likely to catch very small and very large fish. Even explosives, which are destructively and illegally used for fishing in some countries, may be selective! – at least to the extent that, at the periphery of the explosion, larvae and small individuals may be more susceptible to the blast than larger individuals. The selectivity of fishing gear in relation to a particular species is often examined by means of a graph which relates the probability of capture to fish size (Fig. 3.14).

Data used to determine the selectivity of fishing gear may be obtained by several different methods. The examples given here are based on covered codend experiments (related to the selectivity of trawl nets), and on alternate haul experiments, which were originally based on trawl nets, but have applicability to other fishing gear.

Covered codend experiments involve fitting the codend of a trawl net with a cover made of smaller-mesh netting (Fig. 3.15). Fish escaping through the meshes of the codend are retained in the cover net. Data from a covered codend experiment are straightforward to analyse as suggested in Table 3.3. The proportion retained may be

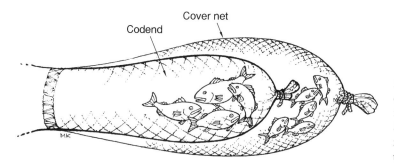

Fig. 3.15 The codend of a commercial trawl net with a cover net made of smaller-mesh netting. Small fish escaping through the meshes of the codend are retained in the cover net.

Table 3.3 Fish length (column 1), the numbers of fish caught in a codend cover (column 2) and in the codend of a commercial net under test (column 3), total fish caught (column 4), and the proportion retained in the commercial net (column 5). Data used to construct a logistic curve are given in column 6 (see text).

1 L (cm)	2 F (Cover)	3 F (Codend)	4 F Total	5 Proportion retained (P)	6 $\ln[(1-P)/P]$
10	2	0	2	0.00	not defined
11	24	3	27	0.11	2.09
12	27	8	35	0.23	1.21
13	26	11	37	0.30	0.85
14	35	26	61	0.43	0.28
15	22	47	69	0.68	−0.75
16	12	72	84	0.86	−1.82
17	5	67	72	0.93	−2.58
18	0	48	48	1.00	not defined

calculated as the number of fish in the codend divided by the total number caught for each length class (column 5 in Table 3.3).

The proportion retained may be plotted against length, and an S-shaped curve drawn through the points (Fig. 3.16a). The important point on the graph is the mean length at first capture (L_c), at which a fish has a 50% chance of being retained by the net (or a 0.5 probability of being caught). Although the curve in Fig. 3.16a can be drawn by hand, it is preferable to fit the data with an S-shaped or logistic curve of the form:

$$P = 1/(1 + \exp[-r(L - L_c)]) \qquad (3.4)$$

where r is a constant with a value which increases with the steepness of the selection curve, and L_c is the mean length at first capture. Equation 3.4 may be transformed as follows:

$$1 = P + P \exp[-r(L - L_c)]$$
$$(1 - P)/P = \exp[-r(L - L_c)]$$
$$\ln[(1 - P)/P] = rL_c - rL \qquad (3.5)$$

which is of the form of a straight line where the slope, $b = -r$ and the intercept, $a = rL_c$. The values of $\ln[(1 - P)/P]$ in column 6 of Table 3.3 may be plotted against L as shown in Fig. 3.16b to provide an estimate of r and L_c as:

$$r = -(b), \text{ and,}$$

$$L_c = \text{intercept}/r$$

In the example, $r = 0.775$, and $L_c = 10.736/0.775 = 13.9$.

The mean length at capture could also be estimated more directly by non-linear least squares using a computer-based search for the parameter values (L_c and r in this case) that result

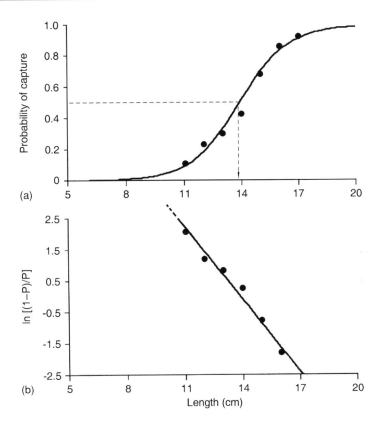

Fig. 3.16 (a) A selectivity curve fitted through the data given in Table 3.3. The mean length at first capture, L_c, is shown by the broken line as 13.9 cm. (b) A plot of $\ln[(1 - P)/P]$ against length using data from column 6 of Table 3.3. The slope is –0.775 and, by extrapolation, the intercept on the y-axis is 10.746.

in a logistic curve fitting the observed data. Fitting a logistic curve by least squares is described in Appendix A4.5 in which a model relating maturity to length is presented.

The alternate haul method is particularly useful because it can be applied to some other fishing gear as well as trawl nets. However, the data require some manipulation to allow for the possible different catches in each of the alternate 'hauls'. Length–frequency data for the deep-water shrimp, *Heterocarpus ensifer*, caught in fine-mesh traps and coarse-mesh traps are shown in Table 3.4.

As there was a larger number of coarse-mesh traps used than fine-mesh traps, the number of shrimps caught in the fine-mesh traps have to be adjusted before the catches of the two types of traps can be compared. To do this, a correction factor must be applied to the fine-mesh catches. Individuals with a length greater than say, 25 mm carapace length, may be regarded as fully selected by both mesh sizes, and the numbers of shrimp

above this size can be used to obtain a correction factor. The number of shrimps above 25 mm in the coarse-mesh traps (B) divided by the number of similar sized shrimps in the fine-mesh traps (A) is the correction factor (C):

$$\sum_{26}^{35} A = 415 \qquad \sum_{26}^{35} B = 670 \qquad C = \sum B / \sum A = 1.6$$

Multiplying the frequencies for the fine-mesh traps by the correction factor results in an adjusted frequency (A × C) which allows for the larger sample obtained from coarse-mesh traps (B). As the two samples may now be compared, the proportion caught by the coarse-mesh traps is estimated by B/(A × C).

Trawl nets are usually subject to minimum mesh size regulations and sometimes on the geometry of the meshes; in some fisheries the requirement to use nets with square meshes (rather than the more conventional diamond-shaped ones) has reduced the capture of small individuals. These

Table 3.4 Length–frequency data for the deep-water shrimp, *Heterocarpus ensifer*, caught in fine-mesh traps (A) and coarse-mesh traps (B). A×C represents the adjusted frequency for the fine-mesh traps, and (B/A×C) is the proportion of shrimps caught in the coarse-mesh nets (samples of 5 or less are excluded).

Carapace length (mm)	A Frequency (fine)	B Frequency (coarse)	A×C Frequency (adjusted)	B/(A×C) Proportion caught
15	6	1	9.6	0.10
16	7	1	11.2	0.09
17	3	1	4.8	—
18	8	4	12.8	0.31
19	8	5	12.8	0.39
20	14	8	22.4	0.36
21	15	11	24.0	0.46
22	15	17	24.0	0.71
23	26	35	41.6	0.84
24	25	34	40.0	0.85
25	29	38	46.4	0.82
26	33	49	52.9	0.93
27	35	57	56.1	1.02
28	62	98	99.3	0.99
29	64	98	102.5	0.96
30	67	106	107.3	0.99
31	51	88	81.7	1.08
32	38	68	60.9	1.12
33	32	52	51.2	1.01
34	23	39	36.8	1.06
35	10	15	16.0	0.94

and other technical modifications to reduce bycatches and the impacts of bottom fishing gears are discussed in van Marlen (2000).

3.4.2 Effects on non-target species

In addition to the target species, most types of fishing gear catch some other species, collectively known as non-target species. These effects are introduced in this section. Non-target species are also threatened less directly by fishing effects on habitats, food chains and ecosystems; these effects are discussed in the next section (3.4.3).

Non-target species can be divided into those species that have some commercial value (referred to as byproduct) and those species that are unwanted (referred to as bycatch) because they either have no commercial value, are threatened, or are an endangered species protected by

law. As an example, Fig. 3.17 shows the target species and some non-target species in Pacific

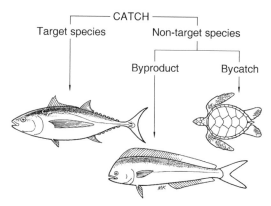

Fig. 3.17 The catch is made up of target species and non-target species. Non-target species can be further divided into byproduct species (those with some value) and bycatch species, which either have no value or are a prohibited catch.

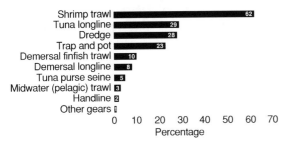

Fig. 3.18 Discards expressed as percentages of the total catch for different fishing methods. FAO (2003).

tuna longline fisheries. Byproduct species include sharks, marlin, sailfish, mahi mahi, wahoo and opah. Bycatch species include species such as snake mackerel and pelagic rays (that have no value) and species such as sea turtles (that are endangered and protected by law).

In many fisheries, the quantity of marine species discarded from catches is very high. An analysis by FAO (Fig. 3.18) suggests that shrimp trawls produce the largest amount of discarded bycatch. In general, shrimp and prawn trawls are relatively fine-meshed and catches of non-target species are often very high. In tropical prawn fisheries, non-target to target species ratios in trawls may be as high as 20 : 1 (Misund et al., 2002). It should be noted, however, that in the mixed catches from trawl fisheries on the soft bottom continental shelves in the tropics, very little of the catch is discarded. In developing countries in particular, a large variety of marine species are utilized and can be found in local markets. It has been pointed out that in many artisanal fisheries only about 5% of the total catch is discarded (Misund et al., 2002). This may be because many artisanal fishers work in countries where a much wider range of seafood is traditionally consumed.

The incidental catch of species that are threatened (such as turtles) or charismatic (such as dolphins) often results in public outrage. Individuals and conservation groups are justifiably concerned about anything that reduces an endangered species' chance of survival. Actions taken include the boycotting of products from the fishery and prohibitions through law courts, and these have affected tuna fisheries in which dol-phins, seabirds and turtles are caught incidentally as part of fishing operations.

In the eastern tropical Pacific, schools of tuna are accompanied by dolphins, some of which were accidentally caught in the large purse seine nets once used to catch the tuna in the area. A public campaign persuaded canneries to buy only 'dolphin-safe' tuna and this resulted in drastic changes in the fishery. In the 1990s there was growing public concern about accidental catches of a small number of leatherback turtles in the longline fishery for swordfish in Hawaii. In 1999, conservation groups initiated legal proceedings against fishing interests and the US National Marine Fisheries Service and this resulted in the court ordering the closure of the fishery.

The Secretariat of the Pacific Community in New Caledonia has published material suggesting actions that tuna longline fishers can take to reduce the incidental catch of turtles (King, 2004). Suggested actions include setting longlines deeper (below 100 m), setting the longline at least 12 nautical miles from reefs, not using squid as bait, and using large circle hooks rather than J-shaped hooks. Actions are necessary, not only to reduce the catch of endangered species but to avoid negative public reaction to buying tuna from tuna longline fisheries in the South Pacific (Fig. 3.19).

Seabirds are often associated with fishing operations and some common species are shown in Fig. 3.20. Some seabirds are attracted in to fishing operations, such as the retrieval of trawl nets, to scavenge small incidentally-caught fishes spilling from the nets. In surface fisheries, the presence of seabirds is often used by fishers to indicate the whereabouts of pelagic fish. For tuna fishers, a 'bird pile' of seabirds actively striking the water to feed on baitfish usually suggests that there is a school of tuna feeding from below. In some fishing methods involving baits, hooks, lures and lines, seabirds are accidentally caught. The incidental catch of albatrosses, for example, by longline fishing vessels working in high latitudes has been widely publicized. To reduce the catch of seabirds, longline fishers use baits that have been dyed blue (and are therefore less visible

Fig. 3.19 A sticker (produced by the Secretariat of the Pacific Community) designed to be placed in the wheelhouses of tuna longliners in the South Pacific as a reminder to take steps to avoid catching turtles.

to birds) as well as a chute or a funnel that feeds the longline beneath the surface of the sea during shooting, quickly placing the baits beyond the reach of seabirds.

Public concern is not restricted to the incidental catches of threatened or rare species but is extending more and more to considerations of waste and ecological well-being. Concern is particularly directed at trawl fishing operations, which release large quantities of small fishes and invertebrates that have little or no chance of survival. Often the public will only become aware of the quantity of discards when, perhaps due to unusual weather or currents, these are washed ashore. When the flotsam of dead fish includes species that are also sought by recreational fishers the public reaction is often severe. Fig. 3.21 shows some local press headlines reflecting a public outcry after tonnes of fish discarded from prawn trawlers were found washed up on a beach in Queensland, Australia in March 2006. By law, prawn fishers are allowed to retain only certain species and must discard other species even if edible.

Much development in gear technology is aimed at reducing the catch of small individuals, and

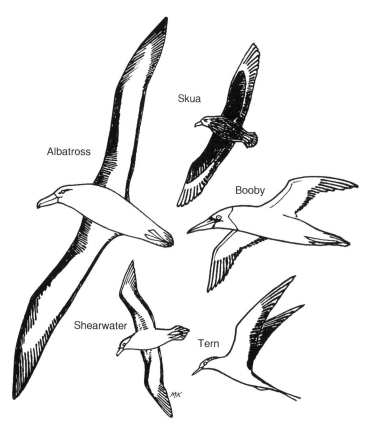

Fig. 3.20 Some seabirds associated with schools of tuna and longline fishing operations. From King (2004).

Fig. 3.21 Local press headlines reflecting a public outcry after tonnes of fish discarded from prawn trawlers were found washed up on a beach in Queensland, Australia.

non-target (bycatch) species. This is being treated as urgent in cases where endangered species are included in bycatches. The skate, *Dipturus (Raja) batis*, was fished close to extinction in the Irish Sea as a bycatch, not as a target species. Its

juveniles were caught as a bycatch and the species does not reach sexual maturity until it is eleven years old (Brander, 1981) – see Box 3.6 'The fate of bycatch species'.

In some countries or areas, unselective fishing gears such as trammel nets are banned. In many cases, existing gear is being modified to protect particular non-target species. Fishers setting pots in the spiny lobster fishery in Western Australia, for example, have agreed to change the design of the baited pots used along part of the coast where sea lion cubs enter and become trapped in the pots. A rod extending up through the entrance to the pot prevents access by sea lion cubs.

To reduce the number of bycatch species caught, special devices have been incorporated into net designs. Such devices which funnel unwanted or threatened bycatch species out of trawl nets are often referred to as TEDs (trawl efficiency devices, trash eradication devices, or even turtle exclusion devices). Bycatch reduction devices are required to be fitted in many trawl fisheries. These include the turtle exclusion devices in the northern prawn fishery of Australia (Salini et al., 2000) and the Nordemore grids that sort fish from retained shrimps in northern shrimp trawl fisheries (Fig. 3.22).

According to FAO data, the quantity of marine fish caught and discarded has fallen by several million tonnes since the mid-1990s (FAO, 2004). This is claimed to be due to improved gear selectivity, fishing practices that reduce bycatch, decreased access to some stocks and no-discard policies that require all fish caught to be landed.

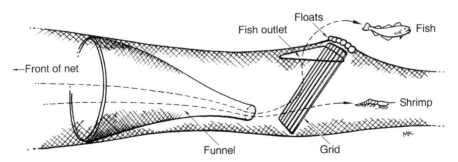

Fig. 3.22 A Norwegian-designed fish exclusion device fitted ahead of the codend of a shrimp trawl. The catch of fish and shrimp is forced through a funnel against a grid which deflects fishes (small cod and haddock) out of the net through a triangular outlet. Shrimps pass through the grid into the codend. Isaksen et al. (1992).

Box 3.6
The fate of bycatch species

Bycatch species are at risk, particularly if they are less abundant or more vulnerable than the target species. Consider the case in which a target species and a bycatch species are equally susceptible to the fishing gear, but the bycatch species is present in a lower initial biomass. If the target species is fished at its maximum sustainable yield, the bycatch species could be extirpated (Fig. B3.6.1) as was the skate, *Dipturus batis*, in the Irish Sea. The development of selective fishing gear, which excludes much of the bycatch is an aim of many management policies.

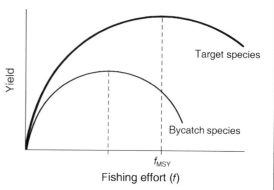

Fig. B3.6.1 The yield of target species and a less abundant bycatch species. At the fishing effort required to secure the maximum sustainable yield of the target species (f_{MSY}), the bycatch species would be drastically overfished.

3.4.3 Effects on the environment and ecosystems

In the broadest sense, all fishing is environmentally damaging to a greater or lesser degree. Adverse effects include physical damage to habitats caused by the fishing gear and the ecological effects of removing target and non-target species from food chains and ecosystems. The effects of fishing in relation to non-target species and habitats are the subject of a collection of papers in Kaiser and de Groot (2000).

Particular types of active fishing gear, such as trawl nets with heavy ground chains, and steel dredges, are known to be highly destructive to the sea floor and its epifauna. Even the continual use of light trawling gear may prevent the settlement of marine benthos. In many fisheries, steps are being taken to replace destructive gears with those that are less environmentally damaging. New types of lighter trawling gear are being designed to skim the sea floor rather than dig it up.

Steel dredges (Fig. 3.11), which are towed to catch species of scallops, *Pecten*, are known to cause severe environmental effects including degradation of the sea floor, sediment suspension, and destruction of benthic communities (NMFS, 2002). In some scallop fisheries, dredges

are being replaced by trawl nets. Several types of fishing gear have been implicated in 'ghost fishing': even without bait, some traps continue to trap organisms and lost gill nets can continue fishing for many years after they are lost. In Norway, there is a government programme to dredge for gill nets that have been reported missing (Misund et al., 2002).

In some countries, the use of explosives and poisons to disable and capture fish represents a serious threat to marine ecosystems and the long-term viability of fisheries. These destructive fishing methods include the use of toxic plants, commercially available poisons such as cyanide and bleaches (sodium hypochlorite) as well as explosives. Poisonous plant material is traditionally used to catch fish in many tropical countries. In many Pacific islands, for example, fish poisons are derived from the roots of the climbing vine, *Derris elliptica*, and the nut of the coastal tree, *Barringtonia asiatica*, which is ground into a paste and wrapped in small parcels made of leaves. Fishers drive fishes into the shelter of a pre-selected coral head where two or three parcels of poisonous material are placed. More seriously, commercial poisons, including bleaches, are poured into pools isolated at low tide to capture small coral fishes. Explosives are either

When you want a coconut
you don't chop down the whole tree

**So, when you want a fish,
DON'T kill the whole reef**

People who use dynamite and chemicals
to kill fish are destroying our reefs.
They are also destroying our future.

Fig. 3.23 A cartoon used in Samoa to raise public awareness of the damaging effects of using explosives and chemicals to catch fish.

thrown from a canoe into a school of fish such as mullet, or set on coral where fish have been encouraged to gather by setting bait. Explosives and severe poisons are many times more damaging to small animals, such as fish larvae and coral polyps, than they are to large fish. Although the use of chemicals and explosives is illegal, the practice may be tolerated in isolated communities in which the illegally-caught fish are shared. A publicity campaign was used in Samoa to make the public aware of the long-term damage done by the illegal fishers (Fig. 3.23). Destroyed coral reefs result in low fish production, and may not recover for many years; dynamited reefs in the Philippines, for example, took an average of 38 years to recover (Alcala & Gomez, 1979).

Many stocks that have been heavily depleted take an unexpectedly long time to recover, even after fishing has stopped. In addition to the Allee effect of low reproductive success at small stock sizes, there may be broader ecological reasons for slow recovery. These can include changes in predator–prey relationships, the emergence and dominance of a previous competitor, and the disturbance of the habitat in some way. The effects of fishing on ecosystems are described by Kaiser and Jennings (2002).

The removal of a target species from an ecosystem, where it is likely to be a predator, prey and competitor to a range of other species is unlikely to be inconsequential. The loss of a predatory species may allow its prey to increase in numbers and the removal of a prey species may induce facultative predators to prey on other species. This flow-on of the effects when removing a species from an ecosystem has been referred to as a 'trophic cascade' (Pace et al., 1999).

Many large commercial fisheries target and remove particular species from an ecosystem and this may result in an imbalance. Artisanal fishers, on the other hand, use many different fishing methods and catch a wide range of species within an ecosystem, and it has been suggested that this is less damaging (Misund et al., 2002). This is based on the view that an ecosystem is less affected by the removal of sizes and species from many different trophic levels rather than one. The size composition and species in artisanal catches often resembles the structure found in ecosystems.

Ecosystem effects on non-target species are generally not known. It is not known, for instance, what effect fishing for krill has on stocks of baleen whales in Antarctica, or what effect fishing for capelin (the major food of cod) has on stocks of cod off Newfoundland. Wherever humans go fishing they compete with other animals for seafood, and of all human competitors, marine mammals attract the most attention (see Box 3.7 'Competing with marine mammals'). Several stocks of marine mammals are increasing in numbers due to conservation actions, market resistance and simply the increasing costs of hunting them. Actions by conservationists and an outcry against cruel hunting practices resulted in the collapse of the seal skin market in Europe and elsewhere in the early 1980s. Since these developments, populations of many marine mammals have been growing. Seal populations off Canada, the United States, South America's west coast, Namibia, Scotland and Norway are reportedly increasing, sometimes at rates of up to 13% per year. In the Mediterranean, Black, and the Baltic seas, however, seal populations are still endangered, mainly due to pollution and diseases.

Even if more selective fishing gear is used and the amount of fishing is kept within sustainable

---

Box 3.7
Competing with marine mammals

In some places where marine mammal populations are increasing, fishers are concerned about competition for dwindling stocks of fish. Dolphins, toothed whales and the various seals feed on fishes and squid. In some cases they take fish already caught by humans – an action sometimes referred to as depredation. In the Pacific Ocean, pilot whales swim along baited surface lines and eat the hooked tuna leaving fishers with a line full of fish heads. In the eastern Mediterranean, dolphins bite pieces of netting from the codends of trawl nets. In Alaska, killer whales pluck sablefish, a deep-water species and thus not their normal prey, from hooks as lines are being hauled in.

Harp seal numbers in the northwest Atlantic have been increasing and the total population size is now estimated to be about 5 million animals. A large population of seals off the east coast of Scotland, where there are calls for a culling programme, is estimated to consume over 450 tonnes of fish per day and is suspected to be the cause of the sharp decline in local cod catches.

Action by campaigners in the 1970s resulted in Canada cutting its harvest of seals. However, in response to the resulting growth in seal populations the cull was reinstated with a limit of 300 000 in 2004.

On one hand, there is a wish to protect the mammals and on the other there is the possibility that their increasing populations are threatening the recovery of depleted stocks of fish. In addition, there is a chance that humans have somehow altered the balance of nature and allowed the populations of some protected marine mammals, such as seals, to increase beyond their previous natural levels. This would be so if their food supply had been increased by humans fishing, or if the numbers of their predators had been reduced. These reasons are being used to make a case for controlled culling not only to conserve fish stocks but to prevent starvation in seal populations. In the past, when fish catches were small, sharing fish with marine mammals such as seals was less of a problem. But today, human population numbers and requirements are such that people will tolerate no competition from the other animals with which they share this earth.

---

limits for the target species, marine ecosystems are not safe from harm. Food chains and predator–prey relationships are almost certainly affected by the removal of a particular target species from the environment. In response to this, fisheries management has moved away from controlling the harvesting of the target species to include the effects of fishing on bycatch species. Management has now been extended further to consider the effects of fishing on habitats and marine ecosystems. More recently, fisheries managers are required to take into account non-fishing activities that affect populations of marine species.

If a fish stock has been depleted by excessive fishing and the environmental degradation of its habitat, a reduction in fishing effort alone may not allow to stock to recover. Either or both could cause a stock to become locally extinct. If this occurs, the question of whether overfishing or environmental effects caused the extirpation of the species is merely an academic one. The management of species and marine ecosystems is discussed in Chapter 6.

Exercises

Answers to selected exercises are available at www.blackwellpublishing.com/king

Exercise 3.1

Choose a particular type of fishing gear that is used locally and discuss how it is used, how selective it is, how much bycatch is caught in a typical operation, and how damaging it is to the environment.

Exercise 3.2

An experiment involving covering the codend of a trawl net with a cover made of smaller mesh netting produced the following small sample. Make an initial estimate of the mean length at first capture, L_c.

Length (cm)	Frequency (codend)	Frequency (cover)
9	0	8
10	1	9
12	3	11
14	3	13
16	12	8
18	14	4
20	13	2
22	25	1
24	19	0

Exercise 3.3

Design a computer spreadsheet program that estimates the mean length at capture from the data in Table 3.3 using non-linear least squares as described in Appendix 4. See the example used to estimate mean length at maturity for the spreadsheet design.

Exercise 3.4

Graph the proportion caught against carapace length for the deep-water shrimp data given in Table 3.4. Draw a selectivity curve through the data by eye, and estimate the mean length at first capture, L_c. Fit a logistic curve by log transformation or by designing a spreadsheet program as in Appendix 4.

Exercise 3.5

Lobster pots or traps are often fitted with an escape gap, a rectangular or round hole worked into their side, to allow the escape of small individuals. Design a field experiment to examine the selectivity of such traps, and to estimate the mean length at first capture.

Exercise 3.6

During a survey on a potential fisheries resource, the likely commercial fishing costs were estimated. Fixed costs were estimated to be US$30 000 per year, and the daily running costs US$800 per day. Meteorological information suggests that approximately 180 days per year are suitable for fishing. Construct a break-even curve for a proposed commercial operation of one vessel.

Exercise 3.7

Build a computer spreadsheet model to record fishing costs and construct a break-even curve. Base your model on the example given in Table 3.1.

4 Stock structure and abundance

4.1 Introduction

Fisheries biologists contribute to fisheries science in two main areas: first, by studying the basic biology and distribution of resource species; and second, by studying the dynamics of fish stocks. The first will lead to the elucidation of the life cycle of the species, including its place in marine ecosystems and the habitats that are crucial for its survival. The second will provide the basis for stock assessment – for the quantitative studies that lead to predictions of how stocks will respond to management actions. The ultimate purpose of fisheries science is to provide fisheries managers with advice on the relative merits of alternative management strategies.

It is instructive to consider a fish population or stock as a simple biological system in which the forces acting to control stock numbers and weight are shown (Fig. 4.1). Ignoring environmental factors that impose an upper limit, stock numbers are being increased by reproduction and the eventual addition, or recruitment, of small fish. In addition, the weight, or biomass, of the fish stock is increased by the growth of individuals – in the figure, three consecutive age groups,

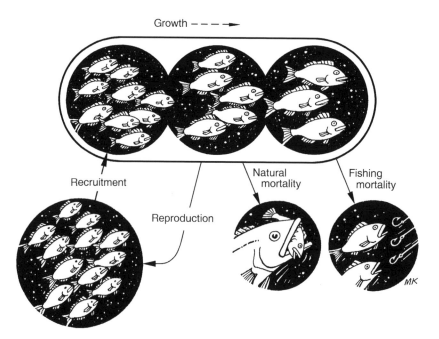

Fig. 4.1 An exploited fish stock viewed as a simple biological system. The stock biomass (top elongate shape) is increased by recruitment and growth (three age groups are shown), and is reduced by natural mortality and fishing mortality.

Box 4.1
Sampling the catch

Besides collecting catch and effort data, sampling the catch either on vessels, at landing places or in processing factories can be used to obtain the following types of information.

• **Morphometrics and condition indices.** Relationships between length and weight can be used to assess the well being of individuals and to determine possible differences between separate unit stocks of the same species. Electrophoretic and DNA analyses may also be used for the latter purpose.

• **Length–frequency data**. The number or frequency of individuals in different size classes may be suitable for the determination of stock structure and the estimation of growth.

• **Tagging or marking**. Tagging or marking individuals, in the expectation that a small number of these are later recaptured, has the potential to be used, with some assumptions, in the estimation of abundance, movement, growth and mortality.

• **Hard part analyses**. Hard parts including shells of molluscs, scales and otoliths of fish and the vertebrae of sharks have been used to assess the age of individuals.

• **Reproductive condition**. The development of gonads (usually ovaries) can be used to trace development and suggest spawning times.

• **Stomach contents**. The analysis of stomach contents may be possible in some species and those caught by some fishing methods (fish brought up from deeper water often disgorge their stomach contents). There is renewed interest in such analyses and species interactions because of multispecies and ecosystem-based stock assessments.

or year classes, are shown. At the same time, the stock is being reduced by natural mortality and, in exploited species, by fishing mortality as well. Fishing mortality refers to fish caught by fishers, and natural mortality refers to fish which die by other means, most commonly by predation.

In a species that is unexploited or fished at a low level, losses due to mortality are balanced, on the average, by gains through recruitment, and stock abundance will fluctuate, often greatly, around a mean level. If exploitation is high, the number of mature fish may be reduced to a level where reproduction and recruitment are unable to replace losses and stock numbers will decrease. Excessive levels of fishing, referred to as overfishing or overexploitation, as well as damage to marine ecosystems and the destruction of fish habitats have all been implicated in the poor condition of fish stocks around the world. Methods of estimating the population parameters implied in Fig. 4.1, namely stock abundance, growth, recruitment and mortality are presented in this chapter.

4.2 Structure and abundance

Questions of abundance and geographical limits of stocks arise at the start of all fisheries. At this stage, both fishers and fisheries managers require some order of magnitude assessment of the stock. Fishers and investors need some idea of the extent of the resource to justify some level of investment in boats, gear and shore facilities. Fisheries managers, well aware of the difficulty of having to reduce fishing effort in an overcapitalized or overexploited fishery, have to make some judgement on the level of fishing that should be encouraged or permitted within the fishery. In this case, initial estimates of stock abundance and spatial limits are required to gain some idea on whether the fishery will be able to support 10, 20 or 200 fishers.

In fisheries studies, the estimation of stock abundance is important in determining the effects of fishing and environmental disturbances as well as in estimating parameters such as mortality. Although estimates of absolute abundance (the

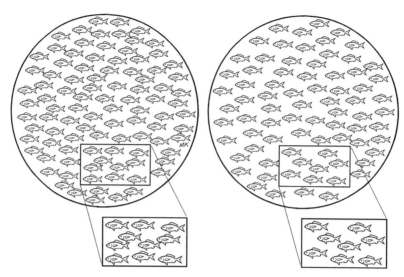

Fig. 4.2 The use of CPUE as an index of abundance. Each circle represents a different area of stock and each square represents the catch taken by one unit of fishing effort.

total number of individuals) are sometimes required, it is often sufficient to obtain estimates of relative abundance – the number of individuals in one area in relation to the numbers present in another area, or in the same area at another time. Ways of estimating total stock size include sampling surveys, mark–recapture studies and depletion methods. Besides these methods, which are discussed in the present section, population sizes are also estimated by back-calculations from historical catch data in virtual population analyses (see Chapter 5).

4.2.1 Relative abundance

The most commonly used index of relative abundance in fisheries studies is catch per unit effort. Catch (C) and fishing effort (f) data are usually collected in all managed fisheries, either from research surveys or from commercial operations (see Box 4.2 'Fishery-dependent and fishery-independent data').

Catch per unit effort (C/f or CPUE) may be recorded in many ways: for example, as the number or weight of fish caught per hook per hour, of lobsters caught per trap per day, or of demersal fish caught per hour of trawling. Within a particular fishery, fishing effort may be measured in

units of increasing refinement. In a gill net fishery, for example, a gross measure of effort is the number of fishers using gill nets, in which case CPUE would be recorded as kg per fisher. If the number of days fished was also known, a more precise measure of CPUE would be kg per day. Ideally, the length of the gill nets used would also be known, and CPUE could be recorded in units such as kg per 100 m of net per day.

The concept of using CPUE as an index of abundance is illustrated in Fig. 4.2. If a fisher catches 10 fish per hour from the circular area on the left and 8 fish per hour from the area on the right, the inference is that the right-hand area contains 20% less fish than the left-hand area. How reasonable this inference is, depends on the answers to several questions. Did the fishers (the samplers) in both areas have the same fishing skills? Was the fishing gear used in each area the same? Are the fish randomly distributed and equally vulnerable to the fishing gear in both places? Consider the effects of non-random distribution in the case of fish species that form schools.

In commercial fisheries, catch, effort and fishing area data may be collected through fishers' logbooks, port sampling or observer programmes. In many licensed fisheries, the

Box 4.2
Fishery-dependent and fishery-independent data

The data may be collected from fishing vessels and fish processors (*fishery-dependent data*), and from the activities of fisheries researchers, in some cases working from a fisheries research vessel (*fishery-independent data*). The merits of each method can be discussed in terms of quality, quantity and cost.

In terms of quality, concerns are often expressed about the accuracy of data collected from commercial operations (on how well fishers fill in logbooks or otherwise provide information). The quality of data supplied by fishers sympathetic to the aims of fisheries management, or better still, actively involved in management, can be high. However, the data may biased in several ways – commercial fishers selectively fish in areas of high stock density and CPUEs may suggest that the stock abundance is continually high. Alternatively, as high-density areas become depleted and fishers move to low-density areas, decreasing CPUEs will suggest that stock abundance is rapidly falling.

In terms of quantity, fisheries research vessels are able to collect only a small number of samples compared with commercial operations. In fishery-independent data, the variability between CPUEs is often large because the samples (catches) represent a small fraction of the total stock. However, research vessel surveys may produce data from a wider area of the stock's distribution than commercial fishing, in which effort is often concentrated on a small part of the stock at a given time. In terms of costs, a research vessel is usually the most expensive of all scientific equipment and represents a very costly method of collecting data. In many developing countries, research vessels (often supplied under foreign aid) cost more to run than the value of the fisheries that they are involved in managing.

One solution involves a combination of collecting fishery-dependent data through logbooks and using cooperating or chartered commercial vessels for special surveys. Logbooks are a relatively cheap and comprehensive way of collecting information, and ensuring that samples of the population are large with lower variability. It is relatively easy to involve commercial vessels in occasional survey work if the surveys are perceived to be beneficial to the industry. The use of commercial vessels for surveys is especially effective in cases of cooperative management in which fishers are actively involved in management decisions (see Section 6.3.2).

completion and submission of fishing logs is a legal requirement of participation in the fishery (see Box 4.3 'Logbooks'). Graphs of catch, effort and CPUE are updated regularly, often on a monthly basis, and these are used to decorate the office walls of those doing the stock assessment. The data and graphs are important in that they trace the history of the fishery and provide a record of changes in the amount of fishing and the quantity of fish caught. The difficulty, however, is in accepting that fluctuations and trends in CPUE are a reasonable representation of changes in the abundance of the stock.

In many fisheries, CPUEs fluctuate wildly, making their use as indices of abundance particularly difficult. This is especially so in fisheries on aggregating species, such as a schooling fish, or on benthic species such as abalone and sea urchins. In these cases, the catch rate for a particular fishing trip may range from very high to zero depending on the skill and the luck of the fisher.

Even in a developed fishery, new gear and ways of locating fish are continually being discovered. These improvements in efficiency may be either dramatic or may occur gradually over a longer time period. Dramatic changes may be relatively easy to detect, but a slow 'technology creep' may not be so easily noticed, particularly by those fisheries scientists who have little contact with fishery. This means that each unit of fishing will increase in efficiency and catches will be greater

Box 4.3
Logbooks

In many licensed fisheries, the completion and submission of fishing logs is a legal requirement of participation in the fishery. However, the quality of information supplied by fishers depends on their empathy with the aims of fishery management and, often, their respect for the people involved in stock assessment. Fishers actively involved in fisheries management (through co-management) who receive monthly updates on fishery information from the managing agency are more likely to fill in accurate fishing logbooks.

The design of a fishing log is often a compromise between the desire of scientists to obtain as much information as possible, and the desire of fishers to spend as little time filling in logbooks as they can.

Most commercial fishers have more than the usual aversion to filling in forms, and a complicated fishing log, or one demanding too much information, is more likely to be completed carelessly or even fictitiously. To the scientist, a small amount of accurate information is better than having a large amount of suspect information.

Often daily fishing logs are issued in the form of a logbook containing a chart of the fishing area (with grid references), and many duplicated daily log pages (Fig. B4.3.1). Fishers enter the date, the fisher's name, fishing depths and locations, fishing gear used, catch composition, and fishing effort data for each day's fishing. Each daily log page is forwarded to the fisheries authorities, and the duplicate page is retained by the fisher as a personal record.

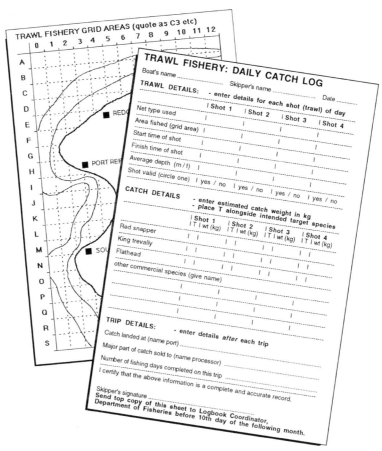

Fig. B4.3.1 A daily page from a simple trawl fishing log, which includes a chart of the fishing area (with grid references).

for the same fishing effort. In these cases, more fish are being caught for the same amount of apparent effort, and CPUEs will give the misleading impression that abundance of the stock is high (or even increasing), when in fact it is decreasing. The same effect would occur if effort was measured in terms of the number of boats, and less efficient boats were gradually replaced by more efficient ones.

In many cases, fishing effort has been measured in relatively coarse units. It is generally easier to determine the number of boats fishing, for example, than the number of hours actually spent fishing. However, coarse measures of effort may make it impossible to detect changes in effort that occur at finer levels. If fishing effort in the fishery was recorded as the number of fishing trips per month, for example, the measure would be unable to detect the fact that fishers may have increased the number of days worked on each trip. In this case, CPUE defined as the catch per trip may be increasing even though stock numbers may be actually falling; CPUE recorded as catch per hour fished would provide a better measure of changes in stock abundance.

The spatial distribution of the stock and fishing effort can also result in CPUE being a poor index of overall abundance. Consider the case of a developing trap or pot fishery for lobsters, in which fishers initially set pots in the more easily accessible inshore grounds. As catches begin to fall, fishers may set their pots at successively greater distances from shore to maintain high catch rates. Even though inshore stocks are being depleted, CPUE will remain high as fishers move to previously unexploited offshore parts of the stock. Unless managers take into account the changing behaviour of fishers, they will be unaware of problems in the fishery until CPUEs begin to fall when pots are being set at the outer limit of the stock's distribution. In this case, where fishers are selectively fishing areas of high density to keep catch rates high, CPUEs will overestimate abundance.

The opposite situation will occur if only part of the stock is accessible. In a trawl fishery, for example, where nets cannot be used in rocky areas, some part of the stock may not be exposed to fishing. The young of some species collected by divers are cryptic and are likewise not exposed to exploitation. In these cases, CPUEs may fall as the fishable areas are exploited even though overall stock abundance remains high; here, collected CPUEs will underestimate abundance and there may be concern when none is warranted.

Problems due to spatial effects can be avoided by stratifying catch and effort data by area, rather than as aggregated or totalled data and analysing data from each area separately. Other problems are related to collecting catch and effort data from fisheries targeting multiple species or using different types of gear on the same target species (see Box 5.4 'Multispecies and multigear fisheries' in Chapter 5).

The above examples give cases in which CPUE data will either overestimate or underestimate stock abundance. What is hoped for is the situation in which CPUEs will be a reasonable index of abundance. This may be close to being achieved in a highly mobile and well-mixed stock in which individuals removed from one area will be offset by redistribution of the stock. In this case, CPUE will be directly proportional to abundance – that is, there is a straight-line relationship between CPUE and abundance. One unit of fishing effort will catch a constant proportion (referred to as the catchability coefficient, q) of a total homogeneous stock:

$$CPUE = qN \qquad (4.1)$$

In most cases the value of q is not known and therefore absolute abundance cannot be estimated from the index. This defect is unimportant if information on abundance is required in relative terms. That is, if the questions to be answered are – is the relative abundance of fish on ground A different from that on ground B? – or, is the relative abundance of fish in one month different from that of the previous month?

It should be noted, however, that in many cases the catchability of a species is not constant, and may vary in response to behavioural changes associated with the time of day, the lunar monthly cycle, and the season of the year. If

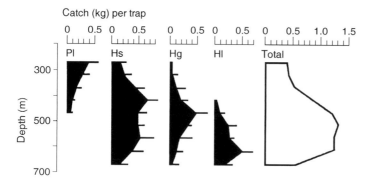

Catch (kg) per trap

Fig. 4.3 Relative abundance by depth (kg per trap per night) of deep-water pandalid shrimp in Fiji. Species indicated are: Pl, *Plesionika logirostris*; Hs, *Heterocarpus sibogae*; Hg, *Heterocarpus gibbosus*; Hl, *Heterocarpus laevigatus*. From King (1984).

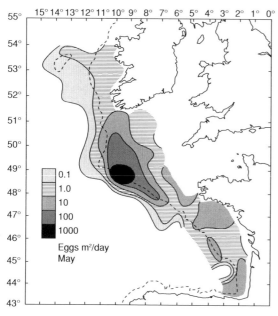

Fig. 4.4 The distribution of mackerel (*Scomber scombrus*) eggs over the main western spawning grounds off the British Isles. From Lockwood (1988).

fishing is conducted over the total area of a fish stock, and over a range of depths, CPUE data may be used as indices of abundance to suggest a species distribution by depth and area. Fig. 4.3 shows the depth distribution of pandalid shrimps using catch per trap per night as an index, and Fig. 4.4 shows the distribution of mackerel (*Scomber scombrus*) eggs using the number of eggs per m^2 per day caught in towed sampling nets as an index.

4.2.2 Sampling surveys

Absolute abundance refers to the actual number of individuals in an area or total fish stock. In a very few cases, such as the counting of salmon climbing 'fish ladders' over weirs as they return up rivers to spawn, it is possible to count every individual in a population. In most cases, it is necessary to estimate absolute abundance by counting the numbers in small samples taken from the total stock.

The most direct way of estimating absolute abundance is by counting the number of individuals in small parts (or quadrats) of the whole population. In the distribution of sea cucumbers suggested in Fig. 4.5, estimates of absolute abundance may be obtained by quadrat and transect sampling. If sampling was carried out by divers operating from a small boat, the boat could be anchored at randomly selected locations within the stock area, and the divers could collect all sea cucumbers within a 20 m radius of the anchor. In this case, a diver collecting sea cucumbers represents the *sampling method*, the random sampling is the *sampling design*, and the 20 m radius quadrat is the *sampling unit*.

Random sampling requires that each sample unit in the stock has an equal probability of being chosen, and random number tables (Appendix 7) may be used to select the position of the sampling units. In practice, biologists often use less time consuming ways of choosing the location of quadrats, such as throwing a quadrat frame (a metal rod in the form of a square or a circle) into the sampling area without looking. In the case of the sea cucumber example, the equivalent is throwing the anchor over at haphazardly chosen locations over the stock area. Such methods are not strictly random, and may result in marginal areas being under-sampled – if the divers avoided sampling near the outer limits of the stock area, for example (see Box 4.4 'Accuracy, precision and bias').

Simple random sampling as described above would provide precise estimates only if the individuals in the stock were distributed randomly. Where individuals are distributed differentially with depth (as they appear to be in Fig. 4.5), sampling along a line (or a transect) at right angles to the depth contours represents a better option. In the sea cucumber example, this could be done by using the centre of the shallow bank as a base point, and using random number tables to select

a compass course (a transect) along which the samples are taken. The boat moving along this transect could be stopped and anchored at regular intervals, and the sea cucumbers counted within the quadrat with the anchor at its centre. Sampling at regular intervals (at a set distance apart, or at regular depth intervals) is an example of *systematic sampling*, and is advantageous when a secondary aim of the survey is to map the species' distribution. Its disadvantage is that the usual formula for variance does not strictly apply. However, if the transect was randomly chosen, the survey should provide an unbiased estimate in which the variance is approximated by the usual formula for random samples as in the procedure given below.

Samples are taken from a population about which we wish to make statements. In the case of

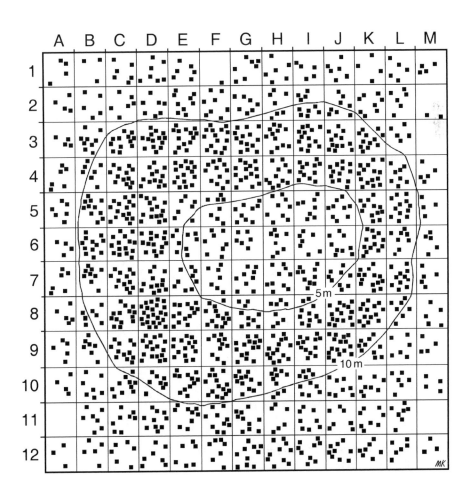

Fig. 4.5 The distribution of sea cucumbers (square black points) in a total area of 15 600 m² around a sandbank. Each square grid (quadrat) is 100 m². Contours are shown at depths of 5 m and 10 m.

Box 4.4

Accuracy, precision and bias

In statistics, the terms accuracy, precision and bias have particular meanings. *Accuracy* is the closeness of a measured value to its true value. If the divers in the sea cucumber example had sampled quadrats only at the centre of the stock's distribution, the stock estimate would be inaccurate or biased (in this case, the stock would be underestimated). *Precision* is the closeness of repeated measurements to the same value, and is usually measured as the spread of values around their mean value. These measures include variance and standard deviation, and precision increases as the variance decreases. Precision is therefore independent of accuracy. A useful measure of precision is the coefficient of variation of the estimated abundance, which is equal to the standard deviation of the estimate divided by the estimate; field surveys usually aim for coefficients of variation of less than 0.2 (20%). Assuming that the values obtained have a normal distribution around their mean, the concepts of accuracy and precision can be illustrated as in Fig. B4.4.1, in which the vertical broken lines represent the true mean.

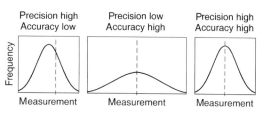

Fig. B4.4.1 The concept of accuracy and precision. The vertical broken lines represent the true mean.

the sea cucumbers on the sandbank in Fig. 4.5, we are using a small number of samples to estimate the total population size. The total population (N) is too large to count and a limited number of samples (from quadrats in this case) is taken. From our samples, the mean number of sea cucumbers in each quadrat is estimated. This sample mean (denoted by $\bar{x}$ and referred to as x-bar) is regarded as an estimate of the population mean (denoted by μ).

For the purpose of the example, it is assumed that sea cucumbers in 7 quadrats have been counted and the results are shown in the second column of Table 4.1. The right-hand column will be used in subsequent calculations (see Appendix 2).

The mean number of sea cucumbers per quadrat, $\bar{x}$, is calculated as the sum (denoted by sigma, Σ) of the number of individuals in each quadrat divided by the number of quadrats sampled, n. In this book, this is abbreviated to:

$$\bar{x} = \Sigma x/n \qquad (4.2)$$

$$= (4 + 15 + 9 + 6 + 7 + 13 + 5)/7 = 59/7 = 8.43$$

The computed sample mean ($\bar{x}$) is regarded as the best estimate of the population mean (μ). An estimate of the total stock, N, can therefore be obtained by multiplying the mean by the ratio of the stock area to the quadrat area; i.e. multiplying by (A/a), where A is the total area occupied by the stock, and a is the quadrat area:

$$N = (A/a) \times \Sigma x/n = (15\ 600/100) \times 8.43 = 1315$$
sea cucumbers $\qquad (4.3)$

As the stock size of 1315 sea cucumbers is only an estimate (and not the *actual* stock number), it is also necessary to complete some calculations to show how precise the estimate really is. For this

Table 4.1 Worksheet for calculations based on a sample taken from the stock of sea cucumbers shown in Fig. 4.5.

Quadrat	x	x^2
6A	4	16
6C	15	225
6E	9	81
6G	6	36
6I	7	49
6K	13	169
6M	5	25
	$\Sigma x = 59$	$\Sigma x^2 = 601$
	$(\Sigma x)^2 = 3481$	

Box 4.5
The size and number of sampling units

The size of a sampling unit has a marked effect on the precision of sample estimates. The choice of size depends on both the size of the organism and its distribution (Fig. B4.5.1). The ratio of the size of the area of the sampled organism to the sampling unit should be small – say less than 1 : 200. In the case of aggregated species, the size of the

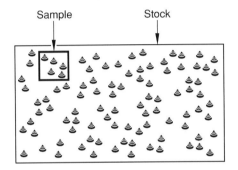

Fig. B4.5.1 The size of sampling units.

sampling unit should be larger than the distance between the aggregations, so that each sampling unit will include at least part of an aggregation.

The choice of how many samples to take depends on the level of precision required. There is usually a trade-off between the requirement for a high level of precision and the cost of taking a large number of samples. For example, if stock abundance estimates of ± 20% are required, with 95% confidence limits, and an initial trial provides a mean of 50 individuals per m^2 with a standard deviation of $s = 18$, the mean and confidence limits are given by $\bar{x} \pm t(s/\sqrt{n})$; that is, $50 \pm t(18/\sqrt{n})$, where $t = 1.96$ if the sample is large (Appendix 7). If confidence limits are to be $\leq 20\%$ of the mean (say, 10) then $10 \leq 1.96 \times 18/\sqrt{n}$, and:

$$n \geq [(1.96 \times 18)/10]^2 \ldots \text{ or } 13 \text{ (rounded up)}$$

As the estimated sample size (n) is small, the value of t will be greater than 1.96, and it is necessary to re-calculate for different values of t and n. By iteration, a minimum sample of 15 is required to obtain an estimate of ±20%.

we need to calculate confidence limits at some chosen level of probability. That is, we want to be (say 95%) confident that the real value lies between certain limits (see Appendix 2). In this example, there is 95% confidence (or 0.95 probability) that the true stock size is 1315 ± 46%, or between 710 and 1920 sea cucumbers. If these limits are regarded as too wide – if a fisheries manager demands a more precise estimate – the standard error and therefore the confidence limits could be reduced by increasing the sample size. However, sampling a larger number of quadrats would increase the time and costs involved in completing the survey. Sampling is discussed in Box 4.5 'The size and number of sampling units'.

For a given amount of sampling effort, a greater degree of precision (narrower confidence limits) can sometimes be attained by concentrating sampling in areas, or strata, of high abundance (Saville, 1977). Stratified sampling involves conducting an initial survey to identify different

strata in the stock. In Fig. 4.5, for example, the stock may conveniently be divided into two areas on the basis of density of individuals – stratum A which includes the densely populated depths between 5 m and 10 m, and stratum B which includes both shallower and deeper areas. The area of each stratum (if based on depth) can usually be obtained from marine charts, and the mean numbers per quadrat in each stratum can be estimated by initial sampling. One set of results from Fig. 4.5 is given in Table 4.2, which shows the stratified sampling ratio as the product of the area ratio and the density ratio. These calculations suggest that the relative sampling effort to be expended in strata A and B should be in the ratio of 290 : 196, or approximately 3 : 2. Thus, if time and funding constraints allowed only 10 quadrats to be sampled, the optimum sampling plan would consist of completing 6 quadrats in depths between 5 m and 10 m (stratum A) and 4 quadrats in other depths (stratum B).

Table 4.2 The calculation of the sampling ratio (area ratio multiplied by density ratio) in stratified sampling based on the data in Fig. 4.5.

Stratum	Area		Density		Sampling ratio
	(m²)	Ratio	Numbers	Ratio	
A	5800	58	14.5	5	290
B	9800	98	5.8	2	196

The stratified mean, $\bar{x}$, may be calculated as:

$$\bar{x} = [\bar{x}_a \,(\text{area A}) + \bar{x}_b \,(\text{area B})]/\text{total area} \qquad (4.4)$$

where $\bar{x}_a$ and $\bar{x}_b$ are the mean numbers per quadrat in strata A and B respectively. The variance of the stratified mean, s^2, may be calculated as:

$$s^2 = (s_a^2/n_a)(\text{area A/total area})^2 + (s_b^2/n_b)(\text{area B/total area})^2 \qquad (4.5)$$

where s_a^2 and s_b^2 are the variances, and n_a and n_b are the numbers of quadrats sampled in strata A and B respectively. Confidence limits may be estimated as $\bar{x} \pm t\sqrt{s^2}$.

Underwater visual census (UVC) techniques involve the counting (and sometimes recording the size class) of fish through underwater observations. Observations may be made by means of a towed underwater video camera or by divers. In a few cases, observations have been made from submersible vessels, as in the case of deep-water caridean shrimps in Hawaii (Gooding et al., 1988). An evaluation of using visual estimates of abundance from submersibles is given in Richards and Schnute (1986). A video camera may be mounted on an underwater vehicle which is towed and controlled to move along a transect at a constant height above the sea floor, and covers a fixed swathe of the sea floor. Fish counts are made either instantaneously by a trained observer viewing a monitor screen on board the towing vessel, or the whole transect is recorded on video tape for later assessment.

Divers may complete a census using either *transect counts* or *stationary (point) counts*. In the case of transects, divers may use a manta board or a diving sled towed by a vessel at a set speed, or may swim over a measured distance, to count fish in a swathe or long rectangular area of perhaps 5 m by 50 m (Fig. 4.6). In point counts, a diver enters the water and counts fish while descending to the sea floor. Features of the sea floor at the periphery of the diver's vision are noted as reference points which are measured after the census to estimate the radius of the sampled area. The width of the transect, or area of the circle, in which fish may be counted depends on the clarity of the water, and the skill of the divers, as well as the density and mobility of the fish population being observed.

In UVC surveys, the numbers of cryptic species, those which live in crevices, or under coral ledges, are likely to be underestimated, as are species which are diver-shy and avoid approaching divers. It is also possible that some species are attracted to divers, in which case the density of fish will be overestimated. However, the technique offers a relatively inexpensive and fishery-independent way to estimate the

Fig. 4.6 Covering a 5 m wide transect in an underwater visual census (UVC) survey.

abundance of coral reef fish. Applications of UVC techniques to spiny lobster populations are given in Smith and van Nierop (1986) and Pitcher et al. (1992). A review of UVC techniques applied to mobile reef fishes is given in Thresher and Gunn (1986) and in McCormick and Choat (1987). UVC techniques are commonly used on coral reef communities and manuals have been produced to encourage the use of a standard methodology (English et al., 1997; Labrosse et al., 2002).

An equivalent to transect sampling involves the use of a trawl net to estimate the mean catch at a number of stations in a fish stock. In the so-called swept-area method, the mean catch (C) taken in the area swept by the trawl (a) is multiplied by the stock area (A) to estimate the stock size, or more usually, the total stock weight or biomass (Fig. 4.7). A towed trawl net in fact samples fish in an area which is equivalent to a long rectangular quadrat with an area, a, estimated as:

$$a = W \times TV \times D \qquad (4.6)$$

where W is the effective width of the trawl, TV is the towing velocity, and D is the duration of the tow. The effective width of the trawl is often taken to be the distance between the otterboards or doors (doorspread), although this assumes that the doors and sweeps are effectively herding fish into the path of the net. An examination of

how the depth of water, towing speed, and other trawl parameters may effect doorspread is given in Hurst and Bagley (1992).

As no trawl net is totally efficient, the catch weight, C_W, made during the tow of the net is usually less than the actual weight of fish in the path of the towed net. The proportion of fish in the path of the net which is retained in the codend is termed the vulnerability, v. The weight or biomass of fish in the path of the trawl is therefore C_W/v. An estimate of the total stock biomass, B, is obtained by multiplying the biomass of the fish in the path of the trawl by the ratio of the stock area to the trawled area:

$$B = C_W/v \times (A/a) \qquad (4.7)$$

where C_w = mean catch weight per tow, v = the vulnerability of the fish, A = the total area occupied by the stock, and a = the area covered by the standard trawl.

Vulnerability is difficult to estimate by conventional means (although underwater video cameras may be useful in this respect), and a 'guessed' value of v = 0.5, which assumes that 50% of the fish in the path of the net escape capture, is sometimes used. The value of v depends to a large degree on the target species' ability either to manoeuvre out of the path of the trawl, or to keep on swimming ahead of the approaching net. The most conservative estimate of stock biomass

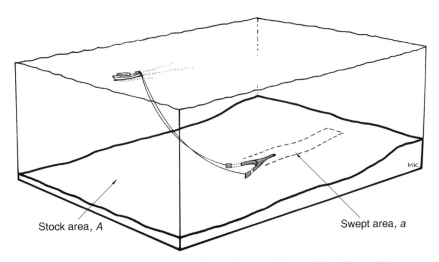

Stock area, A   Swept area, a

Fig. 4.7 The relationship between the swept area (a) of a trawl net and the total stock area (A).

is obtained by assuming that the species is totally vulnerable ($\nu = 1$). Note the distinction between vulnerability, the proportion of fish in the gear's area of influence which is retained, and catchability, the proportion of fish in the stock which is caught by one unit of effort.

Analyses analogous to the swept-area method may be applied to other gears if some estimate of the gear's area of influence is available. In the case of where baited traps are set along a bottom line, some indication of their area of influence may be gained by noting the spacing at which the traps begin to compete with each other; if traps are set too close together, competitive effects will cause catch rates to decrease. When traps are set along a bottom line for deep-water shrimps, those spaced at more than about 50 m apart appear not to compete with each other. From this it may be assumed that a single trap attracts shrimps from within a radius of half this distance (although currents on the sea floor may cause the area to be elliptical). In this case, a rough estimate of biomass may be obtained by a modification of Equation 4.7:

$$B = \text{CPUE} \times (A/a) \qquad (4.8)$$

where A is the total area of the stock, and a is the circular area of the trap's influence.

In the case of synchronous spawners (species in which all individuals collectively release eggs within a discrete period) the concentration of eggs in the surrounding seawater may be used to estimate stock biomass (e.g. Parker, 1980). A plankton net fitted with a flow meter is towed through the spawning stock, and the eggs retained are counted to estimate the concentration per unit volume of water. As long as the sex ratio of the stock and the relationship of fecundity to fish size are known, the stock biomass required to release such a volume of eggs may be estimated. The success or otherwise of this method also depends on knowing the proportion of females that release eggs over the particular spawning season being studied. One equation used to estimate biomass (B) is:

$B = $ daily egg production/(fecundity $\times$ proportion spawning) $\qquad (4.9)$

where fecundity is in terms of eggs per kg of body weight, and the proportion of spawning females (in weight) is estimated from a large sample of fish. Egg production techniques are particularly useful in estimating the stock sizes of fish which form dense spawning aggregations, in which catch per unit effort is highly variable, and therefore a poor indicator of stock size.

In addition to the methods outlined above, a variety of high-technology methods are presently being used to estimate abundance. These methods, which are beyond the scope of this book, range from aerial surveys to computer-enhanced satellite imagery.

Acoustic techniques involve the use of an echo sounder or sonar to estimate the dimensions and packing densities of schools of fish. Pulses of acoustic energy released from an echo sounder strike objects such as the sea floor or fish, and bounce back in the form of echo signals which can be detected by a transducer set in the hull of a vessel. Fish may be counted and their size estimated by the number and amplitude of their echo signals. Alternatively, all of the echo signals may be integrated (by an *echo integrator* attached to the echo sounder) to obtain an estimate of the total biomass of fish in the volume of sea covered by the vessel. Such was the interest shown in the use of acoustics to obtain fishery-independent biomass estimates of fish stocks, that the Food and Agriculture Organization (FAO) of the United Nations produced a manual on the topic in the early 1980s (Johannesson & Mitson, 1983).

4.2.3 Mark–recapture methods

The simplest mark–recapture experiment to estimate stock size is known as the Petersen method. With this method, a known number of marked or tagged fish is released into a fish stock, and the proportion of recaptured tagged fish in subsequent catches is used to estimate the stock size. The large rectangle in Fig. 4.8 shows a fish stock of unknown size, of which 32 tagged fish (solid shapes) were released. At a later time, a catch of 36 fish (in the small rectangle in the lower right-hand corner) was found to include 6 tagged

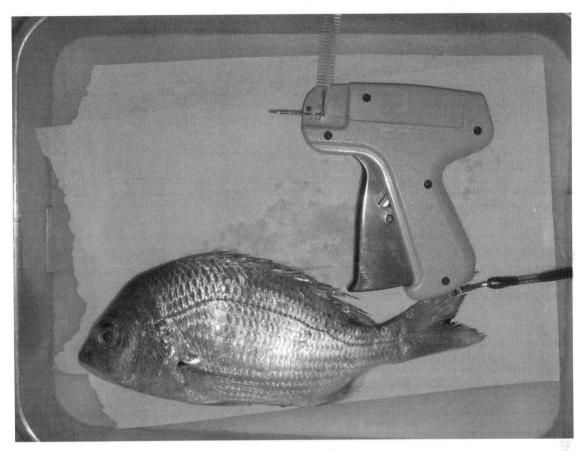

A gun used to inject plastic tags into fish for mark–recapture experiments.

individuals. The stock size may be estimated by assuming that the ratio of tagged fish (T) in the stock (N) is equal to the ratio of recaptured tagged fish (R) in the catch (C):

$$T/N = R/C \qquad (4.10)$$

From this an estimate of the stock size (N) may be obtained as:

$$N = TC/R \qquad (4.11)$$

A large number of replicate catches would produce a set of estimates which are distributed about the population size with a standard error (SE) of $\sqrt{[T^2C(C - R)/R^3]}$. In the case of the example shown in Fig. 4.8:

$$N = (32 \times 36)/6 = 192 \text{ fish}$$

$$SE = \sqrt{[32^2 \times 36 \times (36 - 6)/6^3]} = 71.6$$

The accuracy of the Petersen estimate depends on several assumptions being met. Initially, the tagged individuals must be distributed randomly over the population and, after tagging, there must be no recruitment or migration before sampling the stock. For these reasons, experiments on relatively closed populations, on isolated reefs, in bays or rivers, are ideal. Although short time intervals between marking and recapturing will reduce the possibility of additions or losses, repeated Petersen estimates can be used to monitor long-term population changes. Repeated estimates, for example, have been used to suggest that one of the largest remaining populations of Atlantic sturgeon, *Acipenser oxyrinchus*, in the Hudson River has declined by over 80% in under 20 years (Peterson et al., 2000).

The presence of the tag must not alter the

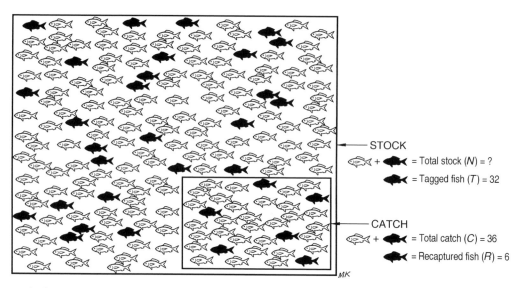

Fig. 4.8 The large rectangle represents a stock of fish which includes 32 tagged fish (solid shapes). The small rectangle at the lower right represents a catch of 36 fish of which 6 have tags.

chance of a fish either surviving or being caught by the fishing gear. If an external plastic tag, for example, resulted in fish being more susceptible to capture by becoming entangled in a gill net, Equation 4.11 suggests that the stock size would be underestimated. In fact, Equation 4.11 will tend to overestimate stock size, particularly for small sample sizes, because of the use of reciprocals in the equation. There are several more sophisticated ways of using mark–recapture data to estimate stock size, and some allow the use of data from multiple releases of tagged fish (Youngs & Robson, 1978) but many of the assumptions are seldom satisfied.

4.2.4 Depletion methods

Depletion methods involve the removal of individuals from a stock and measuring the resulting decrease in relative abundance, usually using CPUE as an abundance index. In its simplest form this involves fishing on a closed population, in which there is no immigration and emigration, and the time interval is short enough to ignore losses due to natural mortality.

The concept can be illustrated by a simple example of a fish survey, in which the initial catch

per unit effort (CPUE) was 50 fish per hour, and, after 3000 fish had been caught, the CPUE was reduced to a CPUE of 30 fish per hour – i.e. by 40%. As the removal of 3000 fish caused CPUE to drop by 40%, it would take a catch of 7500 to decrease the CPUE by 100%, i.e. to the point where the CPUE and the stock size are both zero. As the total cumulative catch is equivalent to the removal of the whole population, the estimate of the initial exploitable stock is 7500. This estimate is likely to be reasonable as long as CPUE is proportional to stock size, and has been reduced over a short time period, during which recruitment, migration, and natural mortality can be ignored.

Although a depletion experiment involves deliberately overfishing a population of fish it is not as drastic as the name implies. Often depletion experiments can be conducted on subpopulations or isolated parts of the overall stock, say on widely separated reefs or bays, so that the overall stock is not threatened. In the early stages of development of a fishery, heavy but short-term fishing on a previously unexploited stock can provide an estimate of virgin stock biomass, and the use of this in management may avoid future commercial overfishing.

After the commencement of fishing, the numbers, N_t, present at time t will be equal to the original stock size, N_∞, less the accumulated catch (ΣC_t) up to time t:

$$N_t = N_\infty - \Sigma C_t \tag{4.12}$$

By definition, the catchability coefficient, q, is the proportion of the total stock caught by one unit of effort, so that, at time t, $CPUE_t = qN_t$ and, therefore:

$$N_t = CPUE_t/q \tag{4.13}$$

Substituting equation 4.13 in Equation 4.12 gives:

$$CPUE_t = qN_\infty - q\Sigma C_t \tag{4.14}$$

Equation 4.14 suggests that if a fish stock is fished heavily, catch per unit effort at time interval t ($CPUE_t$) may be graphed against cumulative catch (ΣC_t) as a straight line. In this graph, referred to as a Leslie plot, the positive value of the slope of the relationship is equal to the catchability coefficient, q. An estimate of the initial stock size, N_∞, which is equivalent to the point where the line crosses the x-axis (where $CPUE = 0$), may be found by dividing the intercept, qN_∞ by the slope, q. The Leslie model and the Delury model can be used to analyse depletion data (Ricker, 1975; Hilborn & Walters, 1992); the latter model is similar to the former, except that logarithms of CPUE are used in the plot.

An ideal experimental situation is where a species exists in a small area which is isolated from the larger part of the stock. Such situations occur naturally where non-migratory species are found in separate bays along a coastline, or on a chain of separated banks or seamounts. In some cases, experimental situations can be created artificially; if fish live in a large estuary, for example, where part of the area can be closed off with small-mesh nets, the enclosed part of the population can be intensively fished.

The following example is based on stocks of deep-water snapper (Lutjanidae) in Samoa in the South Pacific, where commercial fishers began fishing in an area remote from the usual fishing grounds. The area of 3.4 nmi² (11.7 km²) was

Table 4.3 Effort (in line-hours), catch (numbers), CPUE (numbers per line-hour), cumulative catch (numbers) and adjusted cumulative catch (numbers) for the deep-water snapper, *Etelis coruscans*, off Samoa.

1 Period (week)	2 Effort (line-hours)	3 Catch (numbers)	4 CPUE	5 ΣC	6 ΣC (adjusted)
1	170.6	60	0.35	0	30.0
2	453.8	274	0.60	60	197.0
3	513.4	240	0.47	334	454.0
4	714.4	244	0.34	574	696.0
5	679.1	301	0.44	818	968.5
6	419.9	151	0.36	1119	1194.5
7	470.3	127	0.27	1270	1333.5
8	318.4	90	0.28	1397	1442.0
9	136.8	31	0.23	1487	1502.5
10	177.6	7	0.04	1518	1521.5

considered to be sufficiently isolated from the rest of the stock to allow catch and effort data to be used in a depletion analysis (King, 1990). In Table 4.3, fishing effort is recorded as the number of hours each fishing line was used (line-hours). The catch per unit effort (in numbers caught per line-hour) is shown in column 4. The cumulative catch which is conventionally recorded as the number of fish caught up to the beginning of time interval t (column 5), is also shown in the table as the number of fish caught up to the beginning of time interval t, plus one half the catch from time interval t (column 6); this adjustment proposed by Chapman (in Polovina, 1986a) allows for the decline in CPUE during each time interval.

Fig. 4.9 shows a Leslie plot of the data in Table 4.3 with a straight line relating CPUE to ΣC. The stock size, N_∞, is estimated by the intercept divided by the slope (or as the point where the line cuts the cumulative catch axis), and the catchability coefficient, q, as the positive value of the slope:

$$q = -(\text{slope}) = -(-0.0002) = 0.00020 \text{ per line hour} \tag{4.15}$$

$$N = -(\text{intercept/slope}) = qN_\infty/q = 0.524/0.0002 = 2620 \text{ fish} \tag{4.16}$$

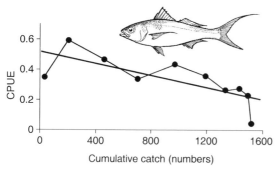

Fig. 4.9 The relationship of CPUE (numbers per line-hour) to adjusted cumulative catch (numbers) for the deep-water snapper, *Etelis coruscans*, off Samoa.

The catchability coefficient, q, applies to the area under consideration, in this case 11.7 km², and suggests that the use of 1 line-hour of fishing effort will catch a proportion of 0.0002 of the total stock in this area. The proportion caught in 1 line-hour per km² will be larger, and therefore obtained by multiplying q by 11.7 to get $q = 0.00234$ per km².

If the initial catch rate (CPUE_∞) is available from another adjacent stock with an area, A km², the numbers of vulnerable fish ($N_{A\infty}$) in this area can be estimated by multiplying the area of habitat by CPUE_∞/q, where q is value of the catchability coefficient per km²:

$$N_{A\infty} = A(\text{CPUE}_\infty)/q \qquad (4.17)$$

One of the most commonly overlooked considerations in running depletion experiments is that fishing effort must be sufficient to reduce the stock in the chosen area over a short time. Several attempts to conduct depletion experiments have failed because either the available fishing effort was too small, or the experimental area was too large to produce a reduction in CPUE over a short time.

An additional problem in the case of multi-species stocks is that the catchability of each of the component species may not remain constant during the course of the depletion experiment. In the case of hook and line fisheries, for example, a more aggressive species (or a species with a greater ability to detect baits) may have a higher catchability than the other species. As the numbers of this species are reduced, other species may gain more access to the baits, and therefore have an increasing catchability as the depletion experiment progresses. A model which allows the catchability of one species to vary inversely with the abundance of competing species has been suggested by Polovina (1986b).

It should be remembered that all of the caveats associated with the use of CPUE data as an index of abundance given earlier in this chapter (Section 4.2.1) apply to the application of depletion methods. Depletion experiments will produce incorrect estimates of catchability and abundance if CPUE is not directly proportional to stock size. If CPUE is recorded from a commercial fishing operation where a large degree of fisher skill is involved, fishers may maintain their catch rates as the abundance drops by targeting the remaining high concentrations of fish. In this case, q would be underestimated, and the initial stock size overestimated. Conversely, if only a proportion of the stock is vulnerable at any one time, recorded CPUEs would decrease at a high rate, and stock abundance would be underestimated; this may occur in the case of cryptic species, or species with spatially separated age groups.

Although depletion models described above apply to closed populations or subpopulations they have been adapted for use in open populations and have incorporated migration rates and even recruitment. Reviews are given in Cowx (1983) and Hilborn and Walters (1992). Depletion methods have been used on a wide range of species and a recent example is their use on squid stocks in Scottish waters (Rasero & Pereira, 2004) as well as in the Falkland Islands (Agnew et al., 2002) where the method has been used for a number of years (Boyle & Rodhouse, 2005).

4.3 Factors that increase biomass

Factors that act to increase the biomass of a fish stock include growth, reproduction and recruitment. Individuals in a stock grow in both size and weight and, within a time that varies between species, reach sexual maturity. A complex array of selective forces determines how fast an animal grows and whether it reproduces once or twice in a short lifespan or reproduces many times over an extended lifespan.

From a fisheries viewpoint, the balance between growth and mortality rates determines the optimal harvesting time – catching individuals in an age class before it reaches a maximum weight is referred to as growth overfishing. More importantly, maintaining numbers of mature adults at a level that assures the continuation of populations is one of the most important functions of fisheries management – fishing stocks down to a level at which recruitment cannot replace the numbers lost is referred to as recruitment overfishing. This section presents methods used to estimate growth parameters as well as describing reproductive cycles and the relationships between the size of spawning stocks and recruitment.

4.3.1 Size and growth

Growth may be described in terms of changes in length, width or any other linear dimension as well as weight. Length is most often used as it is easier to measure, particularly when out at sea, where the vessel's vertical motion affects weighing instruments. The size of marine animals is usually recorded as a 'standard length' which is chosen because it is capable of being easily and accurately measured (Fig. 4.10). Although size is usually measured as length, weight measurements are often required for fisheries work (in calculating yield for example). It is, therefore, useful to determine the relationship between standard length and measurements of weight as well as other length dimensions over a wide size range of individuals.

When animals are growing at the same rate in all linear dimensions, that is, increases in length, width and height are proportional to each other, growth is said to be isometric. In these cases there is a simple linear relationship between the standard length and other dimensions. The following example of the relationship between two length measurements is based on the carapace lengths (mm) and total lengths (mm) for a small sample of the penaeid prawn, *Penaeus latisulcatus*; in practice a much larger sample would be taken across a wide size range.

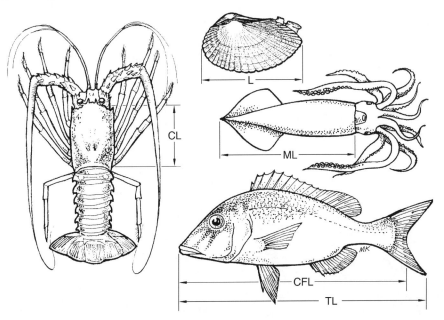

Fig. 4.10 Standard length measurements for various groups of marine species. From left and clockwise: carapace length, CL, in a lobster; shell length, L, in a bivalve mollusc; mantle length, ML, in a squid; and either total length, TL, or caudal fork length, CFL, in a fish.

Table 4.4 Small selection of carapace length and total length data for the penaeid prawn, *Penaeus latisulcatus*.

Carapace length (mm)	Total length (mm)
20	87
26	105
33	135
38	148
44	173
49	187
55	211
58	220

A scatter plot based on the raw data in Table 4.4 is shown in Fig. 4.11. The plot suggests that total length and carapace length are positively related and that the relationship appears linear. As long as the individuals to be measured were chosen at random from the stock, the correlation coefficient (*r*) can be calculated to express the degree of association (or correlation) between the two variables, both of which are subject to random variation. The correlation coefficient takes values between 0 (no correlation) and 1 (perfect correlation) – see Appendix 3. In this case, the correlation coefficient of $r = 0.999$ is very close to 1, indicating that there is a high degree of correlation between total length and carapace length in

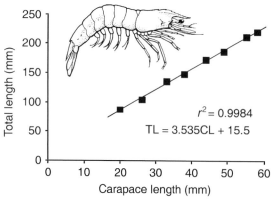

Fig. 4.11 Scatter plot of carapace lengths and total lengths for the penaeid prawn, *Penaeus latisulcatus*. The coefficient of determination r^2 is 0.9984 and the correlation coefficient *r* is therefore 0.999. The regression line is TL = 15.51 + 3.535CL.

the sample; its positive value reflects the fact that the slope is positive.

In the case of this example, it is relatively easy to draw a line by eye through the data points; a better alternative is to use linear regression which minimizes the differences between the data points (the observed data) and the straight line (the expected data) which has the equation:

$$y = a + bx \qquad (4.18)$$

where *y* is the dependent variable, *x* is the independent variable, *a* is the intercept on the *y*-axis and *b* is the slope or gradient of the line. The analytical solution is provided in Appendix 3. If the data are entered onto a computer spreadsheet programme, built-in functions may be used to estimate the slope, gradient and coefficient of determination. In the case of the penaeid prawn, the intercept *a* is 15.5 mm and the slope *b* is 3.535. Total length (TL) is related to carapace length (CL – the standard measurement of a penaeid) by the linear relationship in Equation 4.18:

$$TL = 15.5 + 3.535(CL) \qquad (4.19)$$

The slope, or gradient, of the regression line, which is positive in this case ($b = 3.535$) reflects the fact that total length increases 3.535 times with every unit increase in carapace length. As the value of the intercept ($a = 15.51$) is not equal to zero, the two length measurements are not proportional over the whole size range of prawns, at least not for very small prawns. These analyses are described more fully in Appendix 3.

If an animal is growing isometrically (increasing in all dimensions at the same rate) and doubles in length, its weight will increase in relation to the increase in volume; that is, by 8 (or 2^3) times (Fig. 4.12). Thus there is a cubic relationship between length (*L*) and weight (*W*) which can be represented by the cubic or power curve equation:

$$W = aL^b \qquad (4.20)$$

where *b* is close to three in isometric growth, and *a* is a constant determined empirically.

A power curve, or any non-linear curve, is more difficult to fit than the straight line

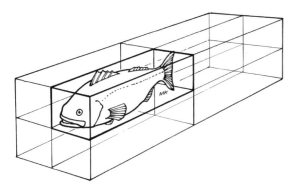

Fig. 4.12 The cubic relationship between length and weight. If all linear dimensions are doubled, volume (and therefore weight) increases by 2^3 or eight times.

considered previously in the relationship between carapace length and total length. In many cases, a more complex curve can be transformed into a linear form by the use of natural logarithms (symbolized in this book as ln). Power curves may be fitted to length and weight data by transforming Equation 4.20 to a linear form using natural logarithms as:

$$\ln[W] = \ln[a] + b\ln[L] \qquad (4.21)$$

This equation is of the same form as the linear equation $y = a + bx$ and the data may be treated as a linear regression by plotting $\ln[W]$ against $\ln[L]$. The example given here is based on a small subsample of measurements on the Pacific oyster, *Crassostrea gigas* (Table 4.5). The table may be set up on a computer spreadsheet with additional columns for the log transformed values.

The log transformed data, shown plotted at the top of Fig. 4.13, provide values for the intercept of $\ln[a] = -5.37$ and a slope of $b = 2.17$. The constant a in the power curve (Equation 4.20) is obtained by back-transforming $\ln[a]$ as $\exp[-5.37] = 0.0047$. This value and the value of $b = 2.17$ can be substituted in $W = aL^b$ for a range of length values and a smooth curve drawn through the points; this is the power curve $W = 0.0047L^{2.17}$ which relates weight to length and is shown at the bottom of Fig. 4.13. The fact that the value of a relates to the condition of a fish (the larger a is, the larger the individual's weight will be for a given length) is sometimes used in condition indices (see Box 4.6 'Condition factors').

Table 4.5 Length and total wet weight (and natural logarithms) for the Pacific oyster, *Crassostrea gigas*. A small sample of data from King (1977). In practice, a much larger sample would be taken across a wide size range.

Length (mm)	ln[length]	Weight (g)	ln[weight]
24	3.178	4	1.386
35	3.555	18	2.890
38	3.638	9	2.197
45	3.807	13	2.565
49	3.892	29	3.367
55	4.007	25	3.219
65	4.174	38	3.638
75	4.317	58	4.060
89	4.489	80	4.382
95	4.554	86	4.454
99	4.595	100	4.605
105	4.654	120	4.787
110	4.700	90	4.500
121	4.796	152	5.024
129	4.860	195	5.273
147	4.990	241	5.485

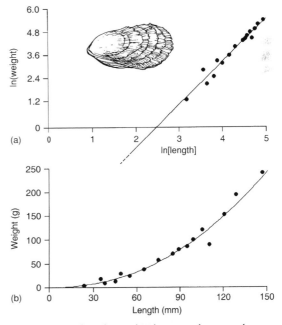

Fig. 4.13 (a) The relationship between the natural logarithms of total wet weight and length in the Pacific oyster (data from Table 4.5). The intercept is −5.37, the slope is 2.17 and the coefficient of determination is 0.964. (b) The curve $W = 0.0047L^{2.17}$ which relates total wet weight to length.

Box 4.6

Condition factors

When considering an individual fish, the value of a in the equation $W = aL^b$ may be used as an index of 'well-being', or a relative condition factor, CF, for the fish. The equation $CF = W/L^b$ suggests that the more a fish weighs for a given length, the greater will be its condition factor. An alternative is to compare the mean weight of fish in a sample with the predicted weight of fish from a generalized length–weight relationship, such as the one shown in Fig. 4.13. If the length–weight relationship had been determined from samples collected over a year, for example, the relationship would be a general one (the value of a in the equation $W = aL^b$ would be an average value). In this case, monthly values of mean weight (W_m) could be compared with the general predicted value (W_p) for fish of the same mean length; ie $CF = W_m/W_p$. Condition factors calculated from monthly samples, for example, may be used to detect seasonal variations in the condition of fish, which may vary with food abundance and the average reproductive stage of the stock.

A length–weight relationship for a species which is of irregular shape, such as the oyster chosen in the example, may show a large amount of variation or scatter around the fitted curve and confidence limits can be estimated as shown in Appendix 2. In this case, the confidence interval for b (from 1.94 to 2.40) does not include 3, the expected value if the relationship between weight and length was cubic.

Of the food taken by a particular species, much of the dietary energy is used for body maintenance, activity and reproduction; only a small part, often less than one-third, is available for growth in size. Growth rates and lifespans vary greatly between species (Fig. 4.14). A lower growth rate reduces energy requirements and an extended lifespan permits repeated spawning events. However, advantages of a fast growth rate may include reaching a size early in life which gives the species some immunity from its predators; generally, larger individuals suffer less predation than smaller ones. In a dangerous environment where survival prospects are low, fast growth rates may allow at least some individuals to reach a size at which they can contribute to the next generation.

Several models have been used to express growth using simple mathematical equations,

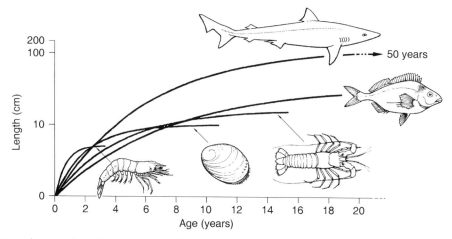

Fig. 4.14 Growth curves for selected commercial species (from left, counter-clockwise): the penaeid prawn, *Penaeus latisulcatus*; abalone, *Haliotis laevigata*; rock lobster, *Jasus edwardsii*; fish, *Nemadactylus macropterus*; and shark, *Galeorhinus australis*. Each curve (except the one for the shark) has been terminated at 95% of its maximum length.

and a review of growth models and rates of development is given in Jobling (2002). The von Bertalanffy growth equation, possibly because of its incorporation into fisheries yield equations by Beverton and Holt (1957), has been the most commonly used in studies on marine species. This model, based on physiological concepts, has been found to fit data from a wide range of species, although the use of any single model is unlikely to represent growth over the entire lifespan. The von Bertalanffy equation, in terms of length, is:

$$L_t = L_\infty(1 - \exp[-K(t - t_0)]) \qquad (4.22)$$

where L_t is the length at age t, L_∞ is the theoretical maximum (or asymptotic) length that the species would reach if it lived indefinitely, and K is a growth coefficient which is a measure of the rate at which maximum size is reached. As an animal is unlikely to grow according to the above equation throughout its whole lifespan, particularly in pre-adult stages, the curve often cuts the x-axis at a value less than zero, hence t_0 (the theoretical age at zero length) often has a small negative value (see Box 4.7 'Age at zero length in the von Bertalanffy growth equation'). If a fish is growing isometrically, weight has a cubic relationship to length and the von Bertalanffy equation in terms of weight is:

$$W_t = W_\infty(1 - \exp[-K(t - t_0)])^3 \qquad (4.23)$$

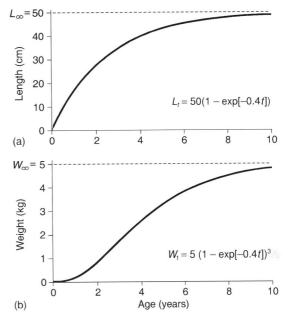

Fig. 4.15 Von Bertalanffy growth curves in terms of (a) length and (b) weight; $K = 0.4 \, \text{yr}^{-1}$, $L_\infty = 50$ cm and $W_\infty = 5$ kg.

where W_∞ is the theoretical maximum (or asymptotic) weight. Von Bertalanffy growth curves in terms of length and weight are shown in Fig. 4.15.

Conventionally, the three most widely used methods of studying growth and estimating the parameters of the von Bertalanffy growth curve are: length–frequency distributions; mark–

Box 4.7

Age at zero length in the von Bertalanffy growth equation

A von Bertalanffy growth curve often cuts the length axis at a value different from zero; hence t_0 (verbalised as t-zero, the theoretical age at zero length) often has a small positive, or more usually, a small negative value. T-zero can be thought of as a scaling factor that positions the growth curve along the age axis. This applies to the mackerel data analysed in Fig. 4.18 in which the exact ages of the first and subsequent modal groups are not known.

Growth curves are fitted through data representing larger (exploited) individuals and it is unreasonable to expect that the same curve extrapolated back to zero length will pass through the age axis at zero. It is unlikely that growth over the entire lifespan can be represented by a single growth curve; pre-adult stages, particularly in species in which juveniles live in nursery areas with different environmental conditions, may grow quite differently from adults. However, if a general curve did represent growth over the lifespan, extrapolation back to the age axis would provide an estimate of a 'birth date', which is generally taken to be the time at which the larvae hatch from eggs.

recapture experiments; and analysis of growth checks formed in hard parts such as scales, otoliths and vertebrae.

4.3.2 Growth from length–frequency data

When a large unbiased sample is taken from a stock, lengths of individuals in the sample may be measured and graphed as a length–frequency diagram as shown in the top graph in Fig. 4.16 (see Box 4.8 'Recording length for length–frequency distributions'). If spawning occurs as discrete events, these will produce different size groups or classes which are evident as 'peaks' or modes in the length–frequency distribution. In addition, if spawning occurs at regular intervals, say one year apart, it is possible to attribute approximate ages to the various size classes.

Considering the single length–frequency sample from year 1 in Fig. 4.16, the largest age class, $N_{1.1}$, consists of the youngest individuals in the stock, and has resulted from a spawning event less than one year ago and is designated the 0+ group. The second mode, $N_{1.2}$, has resulted from a spawning one year before the 0+ group and is 1+ years old; each successive group, therefore, is

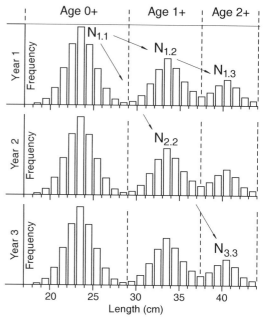

Fig. 4.16 Length–frequency distributions with three different age groups shown for three consecutive years. The single sample (from year 1) shows three pseudo-cohorts $N_{1.1}$, $N_{1.2}$ and $N_{1.3}$, which are assumed to represent groups spawned one year apart. Annual samples show one cohort followed from $N_{1.1}$ to $N_{2.2}$ and to $N_{3.3}$ over years 1, 2 and 3. The numbers in each group are reduced by mortality over time.

Box 4.8

Recording length for length–frequency distributions

Fish lengths are often measured using a graduated board with a block at one end, against which the snout of the fish is placed (Fig. B4.8.1). If fish are to be measured to the nearest cm, for example, lengths may be recorded either as the nearest unit *below* (16 cm caudal fork length in the figure), or as the nearest unit (17 cm in the figure). In the former case, all fish recorded as 16 cm will be between 16 cm and 17 cm in length, and have a length interval mid-point of 16.5 cm. In the latter case, all fish recorded as 17 cm will be between 16.5 cm and 17.5 cm in length, and have a mid-point of 17 cm.

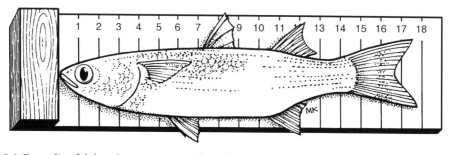

Fig. B4.8.1 Recording fish length on a measuring board.

assumed to be one year older than the one before it. In other words, each annual spawning has produced a normally distributed age class within the polymodal distribution of the total sample. Considering the multiple length–frequency samples in Fig. 4.16, the 0+ age class $N_{1.1}$ in the first year's sample appears as the 1+ age class, $N_{2.2}$ in the second year's sample and the 2+ age class $N_{3.3}$ in the third year's sample.

The arrows in Fig. 4.16 follow the increases in length between pseudo-cohorts (age classes $N_{1.1}$ to $N_{1.2}$ to $N_{1.3}$, with different spawning dates, usually assumed to be one year apart) and the progress of a single 'real' cohort (with the same spawning date) from $N_{1.1}$ to $N_{2.2}$ to $N_{3.3}$ along the length axis.

A growth curve can be estimated from the relative position of the modes in a single length–frequency sample. The crucial assumption is that the modes, $N_{1.1}$ to $N_{1.2}$ to $N_{1.3}$, in Fig. 4.16 have resulted from spawning events that occurred at equal time intervals apart – typically, one year. This would be known if gonad indices were being calculated, say on a monthly basis (see Section 4.3.5).

The length–frequency data shown in Fig. 4.17, based on a species of mackerel, are used here as an example of estimating the von Bertalanffy growth parameters. A commonly used graphical method of estimating the parameters from data representing equal time intervals is a Ford–Walford plot. Here, the mackerel data are analysed using both the traditional graphical method and then by a more direct search for the parameters using non-linear least squares. The derivation of the Ford–Walford plot is based on the growth curve (Equation 4.22) with t_0 equal to zero:

$$L_t = L_\infty(1 - \exp[-Kt]) \tag{4.24}$$

$$L_\infty - L_t = L_\infty\exp[-Kt] \tag{4.25}$$

After substituting L_{t+1} for L_t in Equation 4.24, the difference between this new equation and Equation 4.24 is given by:

$$
\begin{aligned}
L_{t+1} - L_t &= L_\infty(1 - \exp[-K(t+1)]) - L_\infty\exp[-Kt] \\
&= -L_\infty\exp[-K(t+1)] + L_\infty(1 - \exp[-Kt]) \\
&= L_\infty\exp[-Kt](1 - \exp[-K]) \tag{4.26}
\end{aligned}
$$

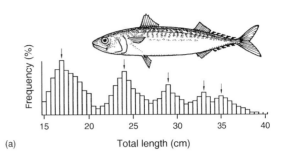

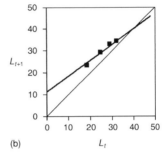

Fig. 4.17 Growth analyses based on the mackerel. (a) A length–frequency distribution with modes at 17, 24, 29, 33, and 35 cm (arrowed); (b) a Ford–Walford plot with a slope of 0.7023 and a y-axis intercept of 12.166.

Substituting Equation 4.25 into 4.26 gives:

$$
\begin{aligned}
L_{t+1} - L_t &= (L_\infty - L_t)(1 - \exp[-K]) \\
L_{t+1} &= L_\infty(1 - \exp[-K]) + L_t\exp[-K] \tag{4.27}
\end{aligned}
$$

This equation is of a linear form, and suggests that length at age t, (L_t), can be plotted against length at age one year later (L_{t+1}). The straight line fitting these data will have a slope of $b = \exp[-K]$ and an intercept on the y-axis of $a = L_\infty(1 - \exp[-K])$. These may be manipulated to estimate K and L_∞ as:

$$K = -\ln[b] \tag{4.28}$$

$$L_\infty = a/(1 - b) \tag{4.29}$$

The length–frequency data in Fig. 4.17 shows the numbers of mackerel in each 0.5 cm length class, and modes are evident at 17, 24, 29, 33 and 35 cm. For the purpose of this exercise, it is assumed that there is an annual spawning period which produces modes representing successive year classes. If the sample is an accurate representation of the population structure, each modal group is one year apart. Although the age of the

Table 4.6 Length in one year (L_t) against length one year later (L_{t+1}).

L_t	L_{t+1}
17	24
24	29
29	33
33	35

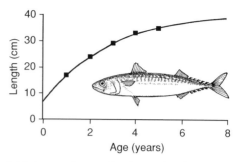

Fig. 4.18 A growth curve for the mackerel produced by the program described in Appendix A4.1.

first modal group (at 17 cm) is not known, the second modal group (at 24 cm) is assumed to be one year older than the first modal group, and so on. However, as growth slows down, and modal groups get closer together with age, there may be several year classes bunched together in the 'group' with a mode at 35 cm.

The first step is to prepare a table of values of L_t and L_{t+1} as these represent the x and y values in the Ford–Walford plot shown in Fig. 4.17 (Table 4.6). Values of the slope and intercept of this plot are used in equations 4.28 and 4.29 to give $K = 0.35$ yr^{-1}, and $L_\infty = 40.87$ cm; alternatively, L_∞ can be read from Fig. 4.17 as the point where the regression line crosses the diagonal line of equality between L_t and L_{t+1} (where there is no growth between one year and the next, or between age t and $t + 1$).

The remaining parameter in the von Bertalanffy growth equation, t_0, can be estimated if length at a particular age is known. In the mackerel example, this could be the case if the timing of the spawning event which produced the first modal group was accurately known. Making t_0 the subject of the von Bertalanffy growth equation (Equation 4.22) gives:

$$t_0 = t + (1/K)(\ln[(L_\infty - L_t)/L_\infty]) \qquad (4.30)$$

If the youngest (17 cm) modal group was known to have resulted from a spawning event 18 months previously, its absolute age would be 1.5 years. Substituting $K = 0.35$ yr^{-1}, and $L_\infty = 40.87$ cm and the age and length of the first modal group ($t = 1.5$ years and $L_t = 17$ cm respectively) provides an estimate of $t_0 = -0.036$ years. A growth curve can be drawn by using the values of K, L_∞ and t_0

in the von Bertalanffy equation to predict fish length for a range of ages from 0 to, say, 10 years.

The growth parameters estimated by the graphical method described above can also be estimated more directly by non-linear least squares. This involves a computer-based search for the optimal parameter values that result in a model fitting the observed data. These are the parameter values that minimize the sum of the squared differences (or residual errors) between the observed data and the expected values (those predicted from the use of the parameters in the model). This measure of fit, called least squares, and the use of the Solver function in Microsoft Excel in completing a directed search for the minimum sum of squared residuals (SSR) and the corresponding parameters is described in Appendix 4.

Fig. 4.18 shows a growth curve using the parameters estimated by non-linear least squares as $K = 0.35$ yr^{-1}, $L_\infty = 41.2$ cm and $t_0 = -0.52$ years with a minimum residual sum of squares of SSR $= 0.163$ (see Appendix A4.1). In this case, the computer search has the difficult task of estimating three unknowns from only five data pairs and there may be different combinations of the parameters that result in a growth curve fitting the data equally well.

The single sample method presents several difficulties. All fishing gear is selective to a greater or lesser degree and samples taken by the gear may not represent the actual stock structure very well – that is, some size classes may not be well represented in the length–frequency data (see Box 4.9 'The effects of mesh selectivity on length–frequency samples').

Box 4.9

The effects of mesh selectivity on length–frequency samples

The use of particular fishing gears results in samples in which some size classes may not be well represented in the length–frequency data. The upper length–frequency histogram in Fig. B4.9.1 shows the structure of the stock consisting of five overlapping year classes with selectivity curves for a gill net and a trawl net superimposed. The lower figure shows the portion of the stock structure (the sample) taken by each type of gear. The trawl net, depending on its mesh size, may not catch smaller individuals (in this case and for this particular mesh size, individuals less than about 25 cm). The gill net will take a very narrow range of sizes from the stock and is the poorest type of sampling gear to collect samples for length–frequency analysis. A method of adjusting length–frequency data for gear selectivity is given in Appendix 5.

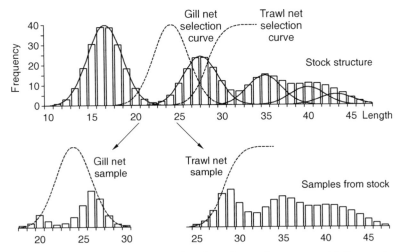

Fig. B4.9.1 Effects of gear selectivity on samples taken from the stock structure with five overlapping year classes. The bottom figures show the part of the distribution obtained using a gill net (left) and a trawl net (right).

In a fast-growing species with a brief annual spawning period, several year classes may be clearly evident in length–frequency distributions. However, the 'spread' or variation within each year class or pseudo-cohort means that there is often a large degree of overlap between them. The two sources of variation in size within a cohort are different growth rates among individuals and different birth dates within a particular spawning season.

If the spawning period is extended, or growth is slow, pseudo-cohorts in the length–frequency distribution may overlap to such a degree that it is not possible to identify separate modes. As growth slows with increasing length, older age classes may 'bunch' together in what is termed the 'pile-up' effect. In addition, if spawning occurs more than once a year, the modes will not be from groups 12 months apart. Some of these difficulties are illustrated in Fig. 4.19, which shows three length–frequency distributions for fish with the same growth rates but with different spawning period lengths.

In Fig. 4.19, the top graph of a species with a brief annual spawning period and widely-separated age groups (with small variances about the means) are relatively straightforward to analyse. In the middle graph of a species with

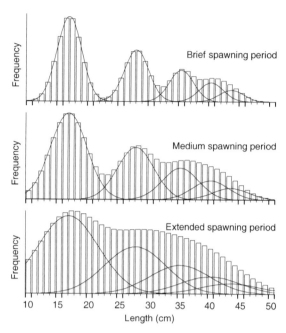

Fig. 4.19 Computer-generated length–frequency histograms of three species with increasingly extended spawning seasons (from the top). Growth and mortality are the same for each species.

The von Bertalanffy plot is a graphical method of analysing length–frequency data, as long as an estimate of L_∞ is available. The von Bertalanffy growth equation (Equation 4.22) may be manipulated as follows:

$$L_t/L_\infty = 1 - \exp[-K(t - t_0)]$$
$$1 - L_t/L_\infty = \exp[-K(t - t_0)]$$
$$-\ln[1 - L_t/L_\infty] = -Kt_0 + Kt \qquad (4.31)$$

Equation 4.31 is of a linear form and its left-hand side may be plotted against age, t: in this plot the slope $= K$ and the intercept $= -Kt_0$. From this:

$$K = \text{slope} \qquad (4.32)$$

$$t_0 = \text{intercept}/(-K) \qquad (4.33)$$

The method requires that the relative ages of the modal groups are known, and that an estimate of L_∞ is available. An example is given using length–frequency data from a stock of banana prawns, *Penaeus merguiensis*, in the Persian Gulf, Iran. Data from Table 4.7 are used to construct the length–frequency histograms in Fig. 4.20.

Several techniques are available for estimating L_∞ independently (including the Wetherall plot presented in Section 4.4). Alternatively, an approximation of L_∞ may be based on the largest fish in catch samples. As the largest fish found (L_{max}) is often about 95% of the estimated L_∞, then $L_\infty = L_{max}/0.95$. Alternatively, L_∞ can be approximated as the mean of the 10 largest individuals found in the catch (Sparre et al., 1989), using this approximation for the prawn data from the Persian Gulf, $L_\infty = 44.6$ mm. This estimate is likely to be reasonable only if the sample on which it was based is large, and from an unexploited or a lightly exploited stock. The largest individuals in a heavily exploited stock are likely to be considerably smaller than L_∞.

The mean carapace lengths have to be estimated for each cohort in the length–frequency data in Fig. 4.20. Note that a cohort of larger and older prawns (cohort B – presumably resulting from a spawning in April 1990) is visible in the length–frequency data in the first several months, and two cohorts are present in June and August.

a more extended spawning period, the three oldest age groups overlap to produce a single and misleading mode at 36 cm. In the bottom graph, from a species with a very long spawning period, the age classes are so close together that no usable modes are evident in the resulting length–frequency diagram, and it is impossible to use the data to estimate growth.

If length data are collected at different times, length–frequency distributions can be arranged sequentially and the progress of a single cohort used to estimate growth. This method, sometimes called modal progression analysis, is illustrated in Fig. 4.16 in which the modes of the cohorts $N_{1.1}$, $N_{2.2}$, $N_{3.3}$ can be traced as they progress along the length axis. If modal progression data were not collected from equal time intervals, as is often the case, the Ford–Walford plot cannot be used. Other suitable graphical methods include the von Bertalanffy plot (described here) and the Gulland–Holt plot which is introduced in Section 4.3.3 on estimating growth from tagging.

Table 4.7 Frequencies of female banana prawns, *P. merguiensis*, by carapace length caught during 10 surveys in the Persian Gulf, Iran, 1991. A spawning peak appears to occur on 30 April each year. Unpublished data, Fisheries Research Centre, Bandar Abbas.

CL (mm)	16 Mar	17 Apr	25 Apr	25 Jun	21 Aug	10 Sep	16 Oct	9 Nov	9 Dec	31 Dec	Total
10.5					3						3
11.5				1	4						5
12.5					2						2
13.5					1						1
14.5					2						2
15.5				2	12	2					16
16.5				2	15	6					23
17.5				6	23	4					33
18.5				13	30	10					53
19.5				20	27	16					63
20.5				33	57	51	1	9			151
21.5				14	45	50	2	3			114
22.5				60	39	72	3	36			210
23.5				21	33	41	2	71			168
24.5				24	31	74	10	123	5		267
25.5				16	20	81	25	202	25		369
26.5				19	2	69	30	219	20	6	365
27.5				6	8	87	57	252	32	12	454
28.5				6	2	46	81	241	22	14	412
29.5	4	7		4		21	68	269	28	30	431
30.5	3	14		3	1	7	59	254	20	42	403
31.5	5	21	2	4		4	31	155	32	51	305
32.5	13	28	1	1		1	12	87	31	65	239
33.5	11	38	2				4	62	17	55	189
34.5	19	43						17	16	35	130
35.5	35	60	3					16	6	40	160
36.5	26	13	2	2			2	6	3	25	79
37.5	21	23	10	5			1	2	1	11	74
38.5	12	16	12	6	2				1	4	53
39.5	6	4	31	5				1	1	4	52
40.5		13	26	7	2		1	1		1	51
41.5	4	4	36						1	1	46
42.5	1	1	16	1	1			2			22
43.5			9	2							11
44.5			8	1	1						10
45.5			1								1

Total = 4967

The small newly-recruited prawns appearing as a new cohort in June (cohort A) result from a more recent spawning in April 1991.

Calculations can be set up as a table or on a computer spreadsheet (Table 4.8) that contains:
• The mean age of each group in the cohort. The relative ages of the modal groups are estimated as the number of weeks after a 'birth date' (based on a spawning peak) which occurred at the end of April in both 1990 and 1991.
• The mean carapace length for each group in the cohort (modes may be used in some data).

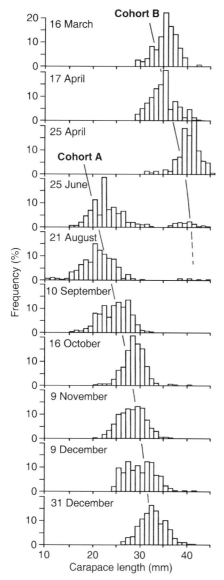

Fig. 4.20 Length–frequency histograms (percentages) for female banana prawns, *P. merguiensis*, in the Persian Gulf, Iran, 1991. The progression of two cohorts (A and B) is traced by the connecting lines.

• The values of $-\ln[1 - Lt/L_\infty]$ required for the von Bertalanffy plot.

A von Bertalanffy plot of $-\ln[1 - Lt/L_\infty]$ against the age, t, is shown in Fig. 4.21. The straight line is fitted through all data except the first point (the first cohort of small prawns from 25 June, which may have been affected by the

selectivity of the trawl nets) and the last point (the second cohort of larger prawns from 21 August, where the sample size is only 6 prawns). The plot has a slope of 0.036, and a y-axis intercept of 0.064. From equations 4.32 and 4.33:

$$K = 0.036 \text{ wk}^{-1}, \text{ or (multiplying by 52) } K = 1.87 \text{ yr}^{-1}$$
$$t_0 = 0.064/(-0.036) = -1.78 \text{ wk, or } -0.034 \text{ yr}$$

Variances for the slope and therefore the value of K, can be estimated (see Appendix 2) for 95% confidence limits as 0.036 ± 0.012. A growth curve can be drawn by using the values of K, L_∞ and t_0 in the von Bertalanffy growth equation (Equation 4.22) to predict fish length for a range of ages from 0 to, say, 10 years. Fig. 4.22 shows a growth curve fitted through all data except the first point (for age 8.1 weeks) as this point may have been affected by the selectivity of the trawl nets and is excluded from the analysis.

The growth parameters estimated using the von Bertalanffy plot can also be estimated by non-linear least squares on a computer spread-sheet (see Appendix A4.2). Non-linear least squares found a minimum residual sum of squares at parameter values of $K = 0.030 \text{ wk}^{-1}$, $L_\infty = 46.04 \text{ mm}$ and $t_0 = -5.75 \text{ wk}$. The results differ from those obtained using the von Bertalanffy plot and this reflects the fact that there are often different combinations of growth parameters that result in a curve that appears to fit the data equally well. Also, the least-squares analysis had the difficult task of estimating three unknowns from the data set; different runs could be tried with L_∞ locked to the previously estim-ated value and leaving Solver to estimate the remaining two parameters.

The use of computer programs also allows the fitting of more elaborate curves which take into account possible variations in growth with sea-son. A series of length–frequency data collected at time intervals smaller than one year (e.g. on a monthly or even weekly basis) allows the pos-sibility of detecting changes in the rates of modal progression with season. In temperate waters, growth rates may be depressed during the cold winter months and, in tropical waters, the

Table 4.8 Worksheet for analysing the data from Table 4.7 by means of a von Bertalanffy plot. The relative age (weeks) is calculated from an estimated 'birth date' of 30 April.

Date of collecting sample:

Day	16	17	25	25	21	10	16	9	9	31
Month	Mar	Apr	Apr	Jun	Aug	Sep	Oct	Nov	Dec	Dec

Mean age (weeks) of prawns in each cohort:

Cohort A	—	—	—	8.1	16.3	19.1	24.3	27.7	32.0	35.1
Cohort B	45.7	50.3	51.4	60.1	68.3	—	—	—	—	—

Mean carapace length (mm) of prawns in each cohort:

Cohort A	—	—	—	22.9	21.0	24.6	28.8	28.4	30.0	32.7
Cohort B	35.5	34.9	40.4	39.8	40.8	—	—	—	—	—

$-\ln[1 - L_t/L_\infty]$, where L_t is mean length, and $L_\infty = 44.6$ mm:

Cohort A	—	—	—	0.72	0.63	0.80	1.04	1.01	1.12	1.32
Cohort B	1.59	1.52	2.36	2.22	2.47	—	—	—	—	—

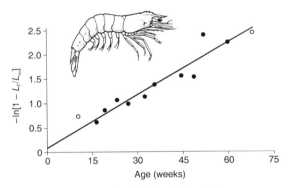

Fig. 4.21 A von Bertalanffy plot for female banana prawns, *P. merguiensis* (data from Table 4.8). The straight line is fitted through all data except the first and last points (see text). The slope is 0.036, and the y-axis intercept is 0.064.

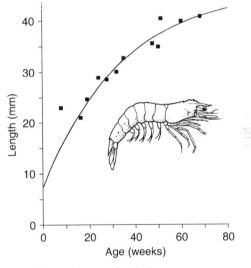

Fig. 4.22 A growth curve for the banana prawn, *P. merguiensis*. Graphical output from the program described in Appendix A4.2.

growth of estuarine species may vary between wet and dry seasons.

Following the cohorts in the length–frequency histograms for the surf clam, *Donax deltoides*, (Fig. 4.23) in southern Australia shows reduced growth during the colder months of the southern winter. A growth curve fitted to these data, say by using a von Bertalanffy plot, would be an average curve that would not reflect changes in seasonal growth and have residuals in a wave form above and below zero. The von Bertalanffy curve may be modified by the inclusion of a sine wave

form (Pitcher & MacDonald, 1973; Haddon, 2001) as:

$$L_t = L_\infty(1 - \exp -[C\sin(2\pi(t - s)/12) + K(t - t_0)]) \tag{4.34}$$

which has two parameters in addition to the usual von Bertalanffy growth model; *C* reflects the amplitude of the oscillation above and below the regular (non-seasonal) growth curve and *s* is the starting time for the sine wave.

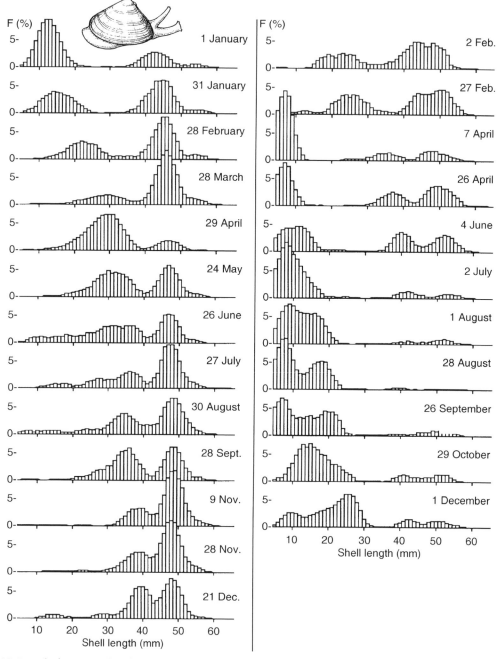

Fig. 4.23 Length–frequency data for the pipi or surf clam, *Donax (Plebidonax) deltoides*, collected from a surf beach in southern Australia. Collection dates are shown on the right of each graph, and sample sizes range from 176 to 712 individuals.

Using the *Donax* data in Fig. 4.23, the major cohort that appears on 1 January at a modal length of 12.5 mm shell length (at which stage it is given the arbitrary age of 8 weeks) can be followed over time. Data for this cohort are entered on a computer spreadsheet with relative ages in one column and modal lengths in another column (see Appendix A4.3). A computer spreadsheet

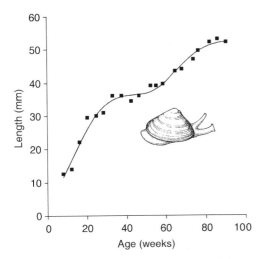

Fig. 4.24 A seasonal growth curve for the surf clam (*Donax*) produced by the spreadsheet model described in Appendix A4.3.

program such as that described in Appendix A4.3 will have difficulty locating a global minimum value when trying to adjust five parameters and some experimentation is necessary. A curve fitted by non-linear least squares is shown in Fig. 4.24.

In summary, growth analyses based on length–frequency data work best on species in which recruitment occurs over a short time period, and in which growth rates are relatively high. If both of these conditions apply, a single length–frequency sample may show several widely-spaced year classes, and a time series of length–frequency samples is likely to produce narrow cohorts with modes which rapidly progress along the length axes. In most cases, single samples can be used only if it can be safely assumed that the component size classes (pseudo-cohorts) are year classes: that is, the size classes are produced in equally-spaced, annual spawning events. The analysis of a time series of length–frequency samples has the advantage of allowing the growth of the *same* group of animals (a cohort) to be traced over time, and no assumption of annual spawning is required. In all cases, the accuracy of length–frequency analyses depends on how well the collected samples reflect the size structure of the actual population.

One of the difficulties in fitting a growth curve is that there are usually very few small and large fish in the samples obtained for analysis. Small fish are often undersampled by the fishing gear, and, if the fishery is heavily exploited, the largest fish in the sample may be much smaller than the actual maximum length for the species. In most length–frequency data, therefore, the smallest fish has a length well above zero, and the largest fish often has a length well below L_∞. Values of L_∞ and t_0, therefore, which relate to the extreme ends of the growth curve, often represent large extrapolations beyond the range of the sample data. Although the estimated growth curve is likely to be an adequate description of growth over the range of lengths in the sample, it may not be so reasonable much beyond this range. Another difficulty is the negative correlation between the values of K and L_∞ in the von Bertalanffy equation. With small data sets, there is often a range of possible curves which will fit the data equally well and the more data points there are to fit a growth curve to, the better.

4.3.3 Growth from tagging information

Tagging or marking fish or invertebrates usually

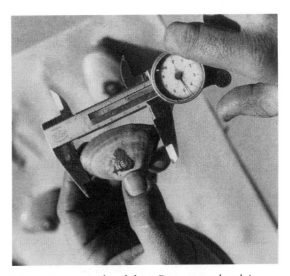

Measuring a tagged surf clam, *Donax*, on a beach in South Australia. The tag is attached to the shell of the clam with underwater-setting epoxy glue.

Box 4.10
Graphical and computer-based analyses of growth

Some graphical and computer-based routines are available for dissecting a length–frequency distribution into its component groups, which are assumed to represent separate age classes. Early methods of dissecting length–frequency data into the component groups which represent separate age classes were graphical (Cassie, 1954; Bhattacharya, 1967; Pauly & Caddy, 1985). Graphical methods have been generally superseded by computer programs which search for the parameters that provide the best goodness of fit. Programs include ELEFAN (Pauly & David, 1980), MIX (Macdonald & Green, 1985), and MULTIFAN (Fournier et al., 1990, 1998). Bhattacharya's method has been incorporated into computer programs, and the use of this method in separating out normal components of length–frequency data is described in Appendix 6. The program ELEFAN is incorporated into a suite of fisheries tools called FISAT distributed by FAO (Pauly & David, 1980).

ELEFAN, in particular, has been widely used to analyse both single and multiple length–frequency distributions. The program is claimed to be more objective than most methods, although some subjectivity may still be involved in interpreting the results from difficult data; however, if L_∞ is

estimated independently (e.g. by the use of a Wetherall plot – see Section 4.4) the program can be used more objectively to estimate K only. ELEFAN may give misleading results when used on a single length–frequency distribution with a large degree of overlap in the age groups. The program MIX, which requires the assumed number of age groups in the sample, is applied to length–frequency data for the ornate lobster, *Panulirus ornatus*, in Fig. B4.10.1. The relative performance of different programs has been examined by Hampton and Majkowski (1987), Basson et al. (1988), Castro and Erzini (1988), and Isaac (1990). A guide to a range of computer packages is given in Pitcher (2002).

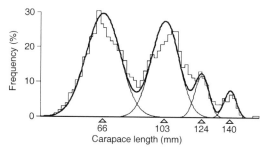

Fig. B4.10.1 Output from the computer program MIX using length–frequency data for the ornate lobster shown as a histogram. Component normal curves are shown as light lines and the summed distribution is shown by the continuous heavy line. Estimated means are shown by arrow heads. Data from Trendall (1988).

involves spending considerable time at sea and the payment of rewards for the return of recaptured tagged animals. In spite of the time and expense involved, tagging experiments are widely used in fisheries biology because of the usefulness of the data collected. In certain circumstances, tagging data can be used to estimate population size, mortality rates, and migration, as well as growth rates.

Much ingenuity has been applied to tag design and some are shown in Fig. 4.25. External tags include the widely used spaghetti tag, which has a barbed shaft that is inserted in muscle and a soft,

brightly coloured plastic tube on which information is printed. The fish shown in Fig. 4.25 has a spaghetti tag (A) inserted between the rays of the base of the dorsal fin. A spaghetti tag is also shown inserted into the muscle of the lobster's abdomen between the covering plates, or pleura; its position is such that the tag remains in place when the lobster moults. A disk tag (B) and an opercular clip tag (C) are attached to the gill cover of the fish.

The tail fan of the lobster is marked by defacement (D in Fig. 4.25) – in this case the uropods have been punched through with holes consisting

Fig. 4.25 Tagged, or marked, marine species (see text for details).

of a sequence of symbols which represent a code for information such as the location, date, and size class at release. As the tissue beneath the exoskeleton is destroyed, the punched holes persist through several moult cycles. The trochus, a gastropod, has a small printed plastic tag (E) glued to its shell with quick-setting epoxy resin. The bivalve molluscs have been marked, at some previous date, by filing a notch (F) on the shell margin. After recapture, the increase in distance from the old shell margin to the new shell margin represents growth during the period between release and recapture.

A mechanized tagging method involves the use of coded wire tags (CWTs) or microtags (G in Fig. 4.25) – fine (0.25 mm diameter) pieces of wire, bearing binary codes (a pattern of grooves),

that can be injected into species ranging from shrimp (Prentice & Rensel, 1977) to salmon (Potter & Russell, 1994). The microtag is magnetized, and its presence in recaptured fish is determined by passing the catch over a detector unit which measures small changes in magnetic fields. Passive integrated transponders (PITs) contain an integrated circuit transmitting a unique code when powered by a magnetic field. They are injected in any part of the flesh thick enough to maintain the tag or in the body cavity and are detected by automated systems in weirs, factories or boats.

Simple body cavity tags (BCTs) of plastic or metal can be placed in the body cavity of individuals to be recovered during gutting and processing. A wide range of internal tags are available and these appear to have higher retention rates and less effect on behaviour and catchability. The incidence of tag loss is often assessed by double tagging experiments in which individuals are tagged with two different types of tags.

In most tagging experiments used to estimate growth, it is necessary to be able to recognize individual fish – that is, each tag used must have a unique number or code. The main assumption in such experiments is that the growth of the individual is not affected by either the tagging process or the presence of the tag. Some tags have been found to cause injuries that reduce the growth of tagged individuals; a comparison of the use of otoliths and mark–recapture studies in a sparid fish for example, showed that external tags attracted hydroids that caused severe lesions in the fish flesh and resulted in underestimated growth rates (Brouwer & Griffiths, 2004). Experiments on captive tagged and untagged (control) individuals are often used to assess the effect of tags on growth and survival.

Tagged animals are recaptured after being at liberty for varying lengths of time. In a heavily exploited fishery, tagged individuals may be recaptured in a very short time. In some species the time between release and recapture may be much longer; internal tags used in the shark *Galeorhinus australis* have been recovered after more than 42 years (Olsen, pers. comm.). If a tagged fish is released at time t_1 when its length is L_1 and recovered at time t_2 when its length is L_2, the growth rate per unit time is equivalent to the change in length $(L_2 - L_1)$ divided by the change in age $(t_2 - t_1)$:

$$\text{growth rate} = (L_2 - L_1)/(t_2 - t_1) \qquad (4.35)$$

During its period of liberty $(t_2 - t_1)$ the tagged fish grows along the curve shown in Fig. 4.26, and the growth rate (the tangent to the curve) is decreasing over this period. The straight line represents the mean growth rate during this period, and is parallel to a tangent to the curve at approximately halfway between t_1 and t_2. If the change in time $(t_2 - t_1)$ is small, this point may be related to the mean length between release and recapture:

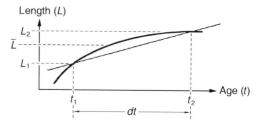

Fig. 4.26 A tagged fish released at time t_1 and recaptured at time t_2 grows from length L_1 to length L_2 along the curve shown. The straight line represents the mean growth rate which may be related to the mean length during this period.

$$\text{mean length} = (L_2 - L_1)/2 \qquad (4.36)$$

If the change in age $(t_2 - t_1)$ is small, the growth rate $(L_2 - L_1)/(t_2 - t_1)$ may be related to the mean length $(L_1 + L_2)/2$ by a straight line called a Gulland–Holt plot (Gulland & Holt, 1959). As growth rate decreases with length, this plot has a negative slope, b, which may be used to estimate the growth parameter K:

$$K = -b \qquad (4.37)$$

The intercept on the x-axis (where the growth rate is zero) is an estimate of L_∞ and may be calculated as the negative of the y-axis intercept, a, divided by the slope:

$$L_\infty = -a/b \qquad (4.38)$$

An example of using a Gulland–Holt plot to estimate growth parameters from mark–recapture information is given in Table 4.9, which is based on selected data for a penaeid prawn. From Table 4.9, the growth rates (in column 4) are plotted against the mean lengths (in column 5) as a Gulland–Holt plot (Fig. 4.27). In this figure, the intercept on the x-axis (where the growth rate is zero) is an estimate of L_∞, and the slope (with sign changed) an estimate of K. From equations 4.37 and 4.38 respectively, $K = 0.017$ wk^{-1} or 0.88 yr^{-1} and $L_\infty = 50.1$ mm carapace length.

In practice, the mean length values used in a Gulland–Holt plot are often bunched together, and a regression line will provide poor estimates of K and L_∞. In this case forcing the line through a

Table 4.9 Mark–recapture data for the penaeid prawn, *Penaeus latisulcatus*.

1 Carapace length at release (mm) (L_1)	2 Carapace length at recapture (mm) (L_2)	3 Time free (weeks) ($t_2 - t_1$)	4 Growth rate (mm/wk) ($L_2 - L_1$)/($t_2 - t_1$)	5 Mean length (mm) ($L_1 + L_2$)/2
28.2	34.0	17.0	0.34	31.1
29.7	35.2	14.7	0.37	32.5
30.6	35.7	16.9	0.30	33.2
32.1	34.7	09.3	0.28	33.4
32.2	36.0	12.1	0.31	34.1
34.1	41.4	35.6	0.21	37.8
35.0	38.2	21.0	0.15	36.6
35.1	37.5	9.3	0.26	36.3
36.5	38.6	16.3	0.13	37.6
37.6	41.5	33.9	0.12	39.6
37.7	39.3	13.9	0.12	38.5
38.7	42.0	18.6	0.18	40.4
39.9	44.5	24.0	0.19	42.2
40.7	46.5	44.6	0.13	43.6
42.1	44.5	36.0	0.07	43.3
45.0	45.8	05.0	0.16	45.4
			Means = 0.208	37.85

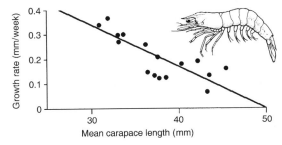

Fig. 4.27 A Gulland–Holt plot between mean length (mm) and growth rate (mm per week) for the penaeid prawn, *Penaeus latisulcatus*. Growth parameters are estimated from the numerical value of the slope and the *x*-axis intercept respectively as $K = 0.017$ wk^{-1} and $L_\infty = 50.1$ mm carapace length.

In the example in Table 4.9, using $L_\infty = 48$ mm, this equation becomes:

$$K = 0.208/(48 - 37.85) = 0.02 \text{ wk}^{-1} \qquad (4.40)$$

A Gulland–Holt plot is a way of dealing with data with variable (and short) time intervals between samples; other methods include those of Fabens (1965) and Munro (1982) as well as the von Bertalanffy plot discussed earlier.

4.3.4 Growth from hard part analyses

Growth is strongly influenced by environmental conditions, and any variation in an individual's environment is likely to affect its growth rate. These changes in growth rate may be visible as growth lines, or abrupt discontinuities in the character of accreting material on the hard parts of animals. Hard parts on which growth lines may be found include the scales, opercular bones, vertebrae, otoliths and spines of fish, the shells of bivalves and gastropods, and the statoliths of squid. In fish, the use of scales is sometimes

fixed value of L_∞ on the *x*-axis may give an approximate value for K. In a so-called 'forced' Gulland–Holt plot, the mean of all the *y*-axis values ($\bar{y}$ = mean growth rate), and the mean of all the *x*-axis values ($\bar{x}$ = mean length) are used as follows:

$$K = \bar{y}/(L_\infty - \bar{x}) \qquad (4.39)$$

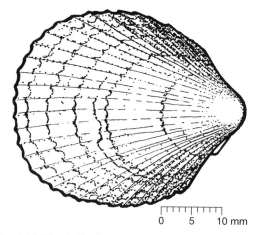

Fig. 4.28 The shell of a coconut-scraper cockle (*Vasticardium*) showing a series of concentric bands, which may have been formed, for example, when wet season rains had diluted the waters in which it was found. The third band from the umbo is likely to be the result of an unsuccessful attack by a predator.

preferred as their removal causes no harm to their owners (scales can be regenerated) but the use of all other hard parts including otoliths, fin rays and gill opercula must follow the death of the fish.

Environmental stimuli which cause an interruption, slowing or cessation of growth, and therefore the formation of growth lines, may be either random or periodic. Random events, such as cyclones or unsuccessful attacks by predators, may cause sufficient stress to produce *disturbance lines*, although the unknown timing of the formation of these lines makes them unsuitable for use in growth analyses (Fig. 4.28).

Regular or periodic events may produce a series of *periodic lines*, which, because they are formed at known time intervals, may be used to estimate growth. Periods most commonly resulting in growth lines are expressions of planetary motion – days, tides, lunar months, and years. Annual events which result in growth lines include decreased water temperatures associated with winter in cooler regions, and the low salinities associated with the wet season in some tropical regions. Lines may also be formed during spawning, which may or may not be an annual

event. Growth lines may also be formed at shorter intervals such as the lunar month and the solar day, in which variations in growth may be associated with changes in behaviour or the availability of food. There has been some interest in the assessment of daily growth rings in the otoliths of tropical fish (Panella, 1974; Gjosaeter et al., 1984). Most directly, all the daily growth increments in the otolith can be counted, but the less tedious method of using the density of increments in several different segments of the otolith has also been used (Baillon, 1988).

Fish have six otoliths, three on each side of the head, and these vary in size between species. Many medium-sized reef species, for example, have otoliths about a centimetre long while some large pelagic species such as marlins have otoliths about the size of a pinhead. Otoliths grow continuously throughout the life of a fish by laying down calcified material at different rates throughout the year. During the warmer summer months, otolith growth is faster and translucent, widely-spaced bands are laid down. In the cooler winter months, otolith growth is slower and closely-spaced, opaque or white bands are formed. Each set of alternating opaque and translucent bands represents one year of growth and these can be counted to estimate age. In some species the bands are not easy to see and the otoliths have to be treated by either cutting, grinding, staining or burning. In some cases, it is necessary to cut a very thin slice of the otolith with a diamond edged saw, polish the thin section and view the bands under a microscope. Reviews of the use of otoliths for age determination are given in Bagenal (1974) and Summerfelt and Hall (1987).

The age of a fish may be determined simply by counting the number of checks, as long as the periodicity of their formation is known. The fish scale shown in Fig. 4.29 is from a temperate-water fish, and has three clear checks corresponding to three winter periods, during which there is reduced growth. In theory, the previous sizes of the fish at the time of each individual check formation can be calculated through a process called 'back-calculation'. The simplest case is where

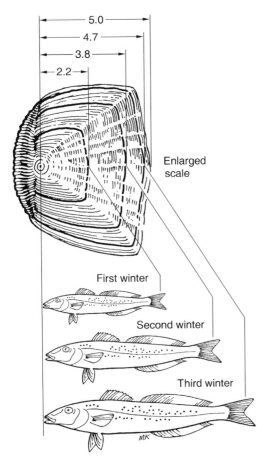

5.0
4.7
3.8
2.2

Enlarged scale

First winter

Second winter

Third winter

MK

Fig. 4.29 The relationship of annual checks to age in the scale of a temperate-water fish, the whiting, *Sillaginoides punctatus*. Distance from the focus (scale centre) to the three winter checks and the posterior edge of the scale are in arbitrary units. Scale redrawn from Anon (1987).

Table 4.10 Back-calculations based on the fish scale shown in Fig. 4.29. Present scale length (S_p) is 5 arbitrary units, and is assumed to be taken from a fish with a length (L_p) of 38 cm.

	Scale length (S_x)	Back-calculated fish length (L_x)
First winter	2.2	16.7
Second winter	3.8	28.9
Third winter	4.7	35.7

38 cm length, the back-calculated fish lengths using Equation 4.41 would be as in Table 4.10.

In theory it is possible (although obviously unwise) to estimate growth parameters from a single fish scale. In the whiting scale example (Fig. 4.29), the calculated fish lengths are from equal time intervals (each check is assumed to be one year apart), and a Ford–Walford plot may be used to estimate K and L_∞. Such a plot for the whiting scale (through only two points!) has an intercept of 19.59 and a slope of 0.56; from equations 4.28 and 4.29 respectively, the growth parameters are $K = 0.58$ yr^{-1} and $L_\infty = 44.5$ cm. However, fisheries biologists are interested in the growth characteristics of the stock as a whole rather than of individual fish. In addition, back-calculated lengths are frequently less than actual fish lengths in older fish, an effect known as Lee's phenomenon (Lee, 1920).

A practical method that avoids the use of back-calculation involves recording the number of rings on either otoliths or scales from a large size range of fish. These data can then be arranged as in Table 4.11, which shows otolith data for a sample of a temperate water fish, the morwong, *Nemodactylus macropterus*. As long as the number of rings on otoliths from each individual fish is assumed to represent age, the mean length at each age can be calculated. The mean lengths at age may be analysed (by using a Ford–Walford plot or a von Bertalanffy plot, for example) to estimate the von Bertalanffy parameters K and L_∞, and produce a growth curve.

The data from Table 4.11 can best be analysed using non-linear least squares on a computer

there is a direct relationship between scale length and body length over the lifespan of the fish – that is, a graph of fish length against scale length passes through the origin of the graph. In this case, the length of fish (L_x) at any previous check formation (x) can be back-calculated using Lea's (1910) formula:

$$L_x = L_p(S_x/S_p) \qquad (4.41)$$

where L_p equals the fish length at present (at the time of capture), S_p equals the present scale length, and S_x equals the scale length at check x. If the scale in Fig. 4.29 was taken from a fish of

Table 4.11 Age–length distribution, sample size (N), mean length in cm, and standard deviation (Sdev) for male morwong, *Nemodactylus macropterus*. Data from Smith (1982).

Length (cm)	Numbers of fish in each age group								
	III	IV	V	VI	VII	VIII	IX	X	XI
23.5	5								
24.5	6								
25.5	15	1							
26.5	27	3							
27.5	21	6							
28.5	14	17							
29.5	12	31	5						
30.5	6	31	14	1					
31.5	2	29	22	3					
32.5		7	26	5	2				
33.5		6	24	12	6				
34.5		3	11	20	6	3	2		
35.5			6	17	7	2			
36.5			1	10	15	8			
37.5			1	1	11	8	4		
38.5					2	5	6	1	
39.5						2	4	1	1
40.5							2	3	
41.5							2		1
42.5						1		2	
43.5									1
44.5									
45.5									1
N	108	134	110	69	49	29	20	7	4
Mean	27.2	30.3	32.6	34.6	35.9	37.3	38.6	40.6	42.5
Sdev	1.8	1.7	1.6	1.4	1.6	1.7	1.9	1.5	2.6

spreadsheet similar to that for modal progression, with age and mean length data entered in the first two columns (see Appendix A4.4). Fig. 4.30 shows a non-linear least squares growth curve with the very small samples from the last two age groups (X and XI) ignored.

Often the analysis of hard parts is more complex than the outline given above. If the relationship between fish length and hard part length is non-linear (or linear without passing through the origin) correction factors must be used. A general review of methods and principles involved is given in Bagenal (1978).

A major criticism of studies involving fish ageing is that many of the techniques used have not been validated (Beamish & McFarlane,

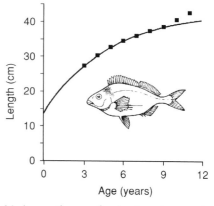

Fig. 4.30 A growth curve fitted to length at age data based on the morwong. Lengths at ages 10 and 11 excluded from the analysis. Graph produced by the spreadsheet program described in Appendix A4.4.

1983). The identification of true annuli or checks should always be validated either indirectly with independent estimates of age, or directly by the examination of hard parts from an animal of known age. It is difficult to distinguish 'true rings' from the 'false rings' or 'disturbance rings' that may reflect periods of stress, starvation, and migration or even an unsuccessful attack by a predator. Repeated examination of large samples by different readers is always wise.

Validation may involve injecting fish with substances such as oxytetracycline (OTC), which causes a visible band to be laid down on the otolith. When the fish are recaptured, the number of zones beyond the OTC mark should correspond with the number of years at liberty. Validated studies have shown that scale ages may be inaccurate, particularly for large fish. The possibility that fish live for many years with little or no scale growth, suggests that scales should not be used to age some species or the older age groups of many species (Beamish & McFarlane, 1987).

In many species, vertebrae and otoliths have been found to be the most reliable method of age determination. In a review of otolith-related research, Campana (2005) determined that most recent studies have been directed at otolith microstructure and determining age and growth. Otolith microstructure studies, involving an examination of the sequence of daily growth increments have been used to describe the early life history of fish species including hatching date, duration of the pelagic larval phase and growth rate. Daily increment analyses have been used on a wide range of fish from anglerfishes (Wright et al., 2002) to deep-water snappers. However, annual growth marks on otoliths continue to be the most common method of determining age and growth in fisheries stock assessments and life history studies (Campana & Thorrold, 2001). Otoliths have also been used been used to study predation (by recovering and identifying otoliths from the stomach contents of known predators – e.g. in seals (Bowen, 2000)). It is also possible to use hard parts to age fish in a way that is independent of ring formation. Many new methods are being investigated including radiometric techniques, which make use of the fact that some radioactive isotopes, such as ^{226}Ra, are accumulated in an individual's hard parts over its lifespan.

In summary, each of the methods used to estimate growth has problems and relies on assumptions that are sometimes hard to accept. Modes in length–frequency samples may not reflect those in the population, tagging may cause injuries that reduce growth, and hard part growth may not reflect somatic growth. Because of the problems associated with different techniques, it is always wise to check growth using as many different methods as possible.

4.3.5 Reproduction

The success of a species is determined by its ability to contribute to the next generation. Most marine species, including 96% of all fish, are external fertilizers; the chances of sperm meeting eggs, high larval mortality and vagaries of currents carrying larvae to suitable settling areas all result in a very low chance of survival. Reproduction and recruitment are two of the major events in the life history of a species. In some cases, these events involve movement between different areas with some species migrating to particular spawning areas and juveniles living and growing in nursery areas that are distant from adult stocks (Fig. 4.31).

As movement requires energy and food, it is remarkable that marine species migrate at all. However, large migrations are relatively common in oceanic species such as tuna, and smaller movements to and from nursery and spawning areas are common in coastal species. Far from being uniform, the sea exhibits variability in illumination, depth, dissolved oxygen, and temperature as well as in the abundance of a species' food and predators. These may vary daily and seasonally as well as spatially. In view of this diversity, it is unlikely that one habitat will be suitable for all stages of a species' life cycle. Many species have evolved migratory life histories and move between different areas, each of which is best for the activities of feeding, growing and spawning

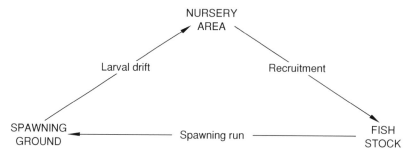

Fig. 4.31 A generalized life-history triangle for marine species. Not all species have geographically separate spawning and nursery areas.

(Metcalfe et al., 2002). Migration and schooling often makes a species particularly vulnerable to exploitation; examples include salmon passing through rivers to spawn and many other species that form spawning aggregations in the sea. Species that migrate to inshore spawning areas or nursery areas may also suffer from the environmental degradation of coastal areas (Reynolds et al., 2002).

Most marine species allow for huge losses by producing millions of eggs during each spawning event. The sunfish, *Mola mola*, is believed to produce up to 300 million eggs (see Box 4.11 'The most fecund vertebrate?'). In many invertebrates and fish, the larvae drift as part of the plankton for periods which vary between species from a few days to many months before metamorphosis into juveniles. The Atlantic cod, *Gadus morhua*, for example, spawns at different grounds over its range of distribution, producing eggs which rise to the surface and hatch to larvae capable of feeding from their yolk sacs for about 2 weeks. After reaching a length of about 4 cm, juveniles assume a benthic habitat. Some species of fish form dense spawning aggregations in particular areas. The Atlantic salmon, *Salmo salar*, makes spectacular migrations up rivers to lay eggs in sand or gravel nests (redds); the eggs hatch to juveniles (fingerlings or fry) in the spring, and, after a period of up to 5 years, the young salmon migrate to the sea, eventually to return to their native rivers to spawn. Most coral reef fish have pelagic larvae which metamorphose into juvenile benthic forms that grow within a restricted home range. In both

temperate and tropical waters, the juveniles of many species become established and grow in inshore nursery areas, such as bays, river estuaries and mangrove areas, before recruitment to the adult stocks.

The term recruitment is often used ambiguously, although broadly it refers to the addition of individuals to an adult unit stock. In benthic species, the term has been used to refer to the process of settlement of metamophosed individuals (e.g. Keough & Downes, 1982), and may be measured by the abundance of age 0+ fish produced (Doherty & Williams, 1988). In fisheries studies, recruitment refers to either the addition of new fish to the *vulnerable* population by growth from among smaller size categories (Ricker, 1975), or the entrance of individuals to the *area* where fishing occurs (Beverton & Holt, 1957). This last definition is perhaps the most useful in fisheries work, as it separates three distinct phases in the life history of exploited species (Table 4.12).

The particular life-history events that are of interest in fisheries studies are the timing of spawning and recruitment (i.e. the time of the year in which they occur), and the mean length or age at which these events happen. Research is devoted to estimating the mean length at recruitment, L_r, the mean length at first capture, L_c, and the mean length at sexual maturity, L_m (the mean length at first capture was discussed in Section 3.4.1). If growth parameters are known, these lengths may be converted to the corresponding mean ages, t_r, t_c and t_m by using the inverse of the growth equation (Equation 4.22):

Box 4.11
The most fecund vertebrate?

The most fecund vertebrate in the world is believed to be the strange-looking sunfish, which produces more than 300 million eggs, each about 2–3 mm in diameter. Sunfishes (six species in the family Molidae) belong to the order Tetraodontiformes, which includes pufferfishes and triggerfishes. The common sunfish, *Mola mola*, which reaches a length of 3 m and a weight of 2 tonnes, is shown in Fig. B4.11.1. Sunfishes have large and flattened bodies with small eyes and a mouth that contains fused teeth. They are found in all oceans and eat a variety of foods including gelatinous zooplankton (jellyfish and salps), squids, crustaceans and fish larvae. In spite of their size, sunfishes appear not to be hunted as food and it is suggested that they may contain the same toxin as pufferfishes. The common sunfish, *Mola mola*, becomes covered with parasites and often moves into reefs where cleaner fish remove them.

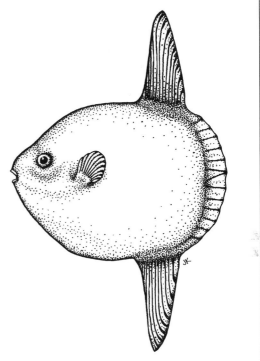

Fig. B4.11.1 The common sunfish, *Mola mola*. Drawing by Jeremy King.

Table 4.12 Three events and three inter-event phases in the life history of exploited species. The mean age at sexual maturity, t_m (of the parental stock), the mean age at recruitment, t_r, and the mean age at first capture, t_c, associated with each event are shown.

Event	SPAWNING		RECRUITMENT		EXPLOITATION	
Phase (between events)		Pre-recruit phase		Post-recruit (pre-exploited) phase		Post-recruit (exploited) phase
Age	t_m		t_r		t_c	

$$t = t_0 - (1/K) \ln[1 - L_t/L_\infty] \quad (4.42)$$

The gonads (both testes and ovaries) of many marine animals appear to have a relatively long inactive or resting phase, after which they grow and develop (become 'ripe'). In fisheries studies, the particular event of interest in the reproductive cycle of a species is the time of spawning when fully developed gametes are released. The spawning event may extend over either a short or long time period, and be at either regular or irregular intervals, but in many species, occurs once every year.

The stimulus (or combination of stimuli) which induces resting gonads to become active may include *endogenous* factors (internal events related to growth and maturity), and *exogenous* events in the surrounding environment. Exogenous stimuli could include light, salinity, food availability, moon phase, and perhaps most commonly, temperature. Although a particular absolute temperature may be required to trigger development, it is also possible that *changes* in temperature (either an increase or a decrease) may be important. The reproductive cycle of temperate-water species is often such that larvae are produced in the spring, so that the larvae can feed on the blooms of phytoplankton that occur in this season.

Because of the role of temperature as a triggering stimulus in many species, it is often claimed that tropical species, which are not subjected to great temperature differences over the year, exhibit continuous, or at least protracted, spawning. However, many tropical species, from corals to giant clams, have short synchronized spawning events. The most spectacular example of a brief synchronized spawning event is in fact displayed in a tropical animal, the palolo worm. These polychaete worms (*Eunice*) live burrowed in the coral, and during one night of each year, release their reproductive tail segments which are present in such abundance that they cover the surface of the sea above shallow-water coral in a writhing mass (Fig. 4.32). The rising of the Pacific palolo in islands such as Samoa is predicted with some accuracy by villagers (often during the third lunar quarter of October or November), and the

segments are traditionally gathered for annual feasts. The Atlantic palolo also forms spectacular swarms in the Dry Tortugas. In addition, many species of fish make migrations to certain areas to form spawning aggregations at particular times of the year (see Box 4.12 'Spawning aggregations').

Spawning may be triggered by stimuli which are quite different from those which induce gonads to develop. The stimuli which trigger spawning (a relatively short process) may be brief environmental events, whereas the stimuli which induce gonads to develop (a relatively long process) may need to have a more sustained influence. Reproductive cycles and the timing of spawning in marine species have been studied by several different methods including:

• The direct observation of spawning, which may be possible during underwater surveys, particularly on sessile species.
• The relative abundance of larvae over time.
• The brooding of eggs (e.g. the proportion of female crustaceans carrying eggs, or in berry) over time.
• The appearance of gonads, according to predefined stages of maturation, over time.
• The relative size or weight of gonads over time.
Of the above, the last two methods are used most often in fisheries studies, and are discussed in some detail here.

Often in fisheries work, only the female ovaries are studied, as these are larger and more easily examined than male testes; it is also assumed that development of both ovaries and testes are synchronous. Ovaries may be examined microscopically and classified into various developmental stages. One such classification is shown in Table 4.13, although it should be noted that gonad development is continuous, and all such staging is somewhat artificial. Whichever classification is used, there is some merit in defining as few stages as possible as the main interest in fisheries studies is in determining the time when the majority of the population is in the final spawning stage. Ovaries taken from samples of females collected at intervals throughout the year may be classified into developmental stages, and

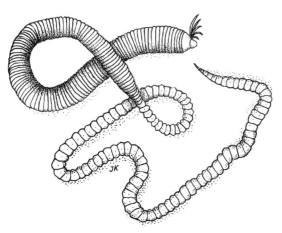

Fig. 4.32 The palolo worm, *Eunice viridis*. Drawing by Jeremy King.

Box 4.12
Spawning aggregations

Some species of fish gather in large aggregations
at particular times of the year and in specific places
to reproduce. These spawning aggregations
involve species such as groupers and snappers,
many of which normally live apart, making
migrations of sometimes hundreds of kilometres to
reach spawning areas. Spawning aggregations may
be triggered by the phase of the moon and last for a
few days or weeks each year. The sites and timing
of aggregations are often well known to traditional
fishers who target the vulnerable fish milling about
in dense aggregations as they release sperm and
eggs. More recently, commercial fishers have
exploited known spawning aggregation sites in
the Caribbean and in Pacific islands.

One well-known and now threatened species is
the Nassau grouper, *Epinephelus striatus*, which
was distributed widely over the Caribbean. About
one-third of the Nassau grouper's aggregation sites
have been destroyed by overfishing and entire

regional stocks are believed to have disappeared.
Many countries are either banning fishing on the
species or protecting the spawning sites. There
is some concern that even moderate fishing may
break up aggregations and thus affect the overall
reproductive success of the spawning stock.
Spawning aggregations of some invertebrates,
notably squids, are also targeted and concerns
have prompted an assessment of the impact of jig
fishing on squids in South African waters (Roel &
Butterworth, 2000).

In some fisheries it is only possible to catch fish
during spawning aggregations when they come
together in dense schools and are more vulnerable.
Large catches of hoki (whiptail, blue hake or blue
grenadier), which is the basis of one of New
Zealand's most important fisheries, are taken from
spawning aggregations during a ten-week period
between June and September. Although fishing on
spawning aggregations is particularly risky, some
of the spawning stock of hoki is within a limited
fishing zone that excludes larger vessels and this
is believed to provide some protection.

Table 4.13 Stages of reproductive development showing the appearance of ovary and eggs.

Stage	Description	Ovary	Eggs
I	Resting	Undeveloped, small, translucent	None visible to naked eye
II	Developing	Opaque, orange colour	Visible and opaque
III	Ripe	Fills body cavity	Translucent, large and round
V	Spawning	Releases eggs when pressed	Large, translucent; some free in ovary
VI	Spent	Shrinking/slack	Some residual eggs

the percentages of each can be plotted by month
in a histogram.

The classification of gonads into at least as
many stages as those shown in Table 4.13 may be
possible with detailed histological studies, but
this research is necessarily laboratory-based and
expensive. In many cases, division into broad
stages (e.g. resting, developing, and ripe) may be
based on simple macroscopic criteria such as
ovary appearance and colour. Such assessments

are useful in field work, although the criteria
should first be validated by histological studies.
Ovary maturation stages based on macroscopic
criteria for a penaeid prawn are shown in
Fig. 4.33a, with analyses summarized in the histogram in Fig. 4.33b. Relatively large numbers
of individuals with highly developed ovaries in
November and March suggest peaks in spawning
activity during these periods.

A gonad or gonosomatic index (GI or GSI) is

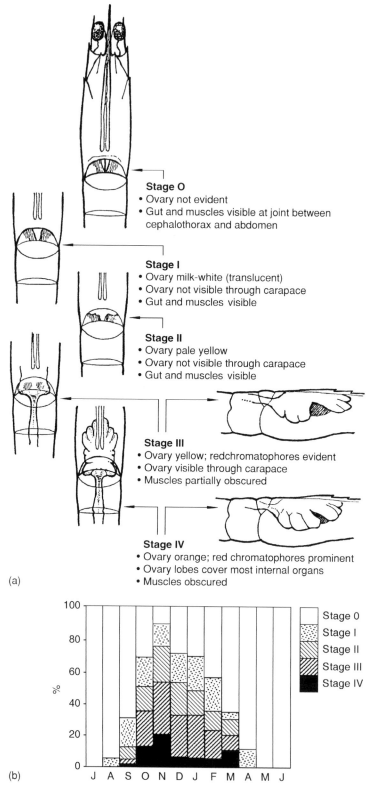

Fig. 4.33 (a) Stages in ovary development based on macroscopic criteria for the penaeid prawn, *Penaeus latisulcatus* (dorsal and lateral views of prawns and ovaries are shown); (b) percentage of female penaeid prawns, *Penaeus latisulcatus*, in five different development stages.

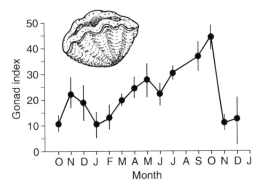

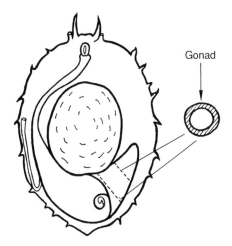

Fig. 4.34 Gonad indices for the giant clam, *Tridacna crocea*. The vertical bars indicate ±1 standard deviation. Adapted from Shelley and Southgate (1988).

Fig. 4.35 Abalone with shell removed showing the visceral coil. In the section taken from within the two dashed lines, the gonad is shown as the shaded area around the digestive gland. From Shepherd and Laws (1974).

often used to follow the reproductive cycle of a species over the year at monthly intervals or less. This index, which assumes that an ovary increases in size with increasing development, compares the mass of the gonad (GM) with the total mass of the animal (TM):

$$GSI = 100 \times (GM/TM) \qquad (4.43)$$

Gonosomatic indices vary greatly and range from less than 1% in many fish to almost 47% in the European eel, *Anguilla anguilla* (Kamler, 1992). However, biologists are usually more interested in how the indices for a single species vary over time. Gonad indices for the giant clam, *Tridacna crocea*, are shown in Fig. 4.34. The sharp decreases in the indices in October and November suggest that the clams spawn during these short time periods.

A variation of a gonad index based on weight is one based on area. This may be useful in species including many molluscs and crustaceans where gonad material is not easily separated from somatic tissue. In the abalone shown in Fig. 4.35, the gonad lies dorsally on the digestive gland and part of the stomach which together form the rudimentary visceral coil. A section through the base of the visceral coil allows the area contained by the perimeters of the gonad and digestive gland to be measured by using a planimeter on tracings or photographs of the section. The area of the gonad divided by the total area of the section and

converted to a percentage becomes a gonad index based on area, and mean values of the index for each month may be plotted on a graph in a similar way to Fig. 4.34.

The above analyses may lead to an estimation of the time of spawning in a marine species by a decrease in the number of individuals with active or ripe ovaries (from graphs such as Fig. 4.33), or by a decrease in gonad size (from graphs such as Fig. 4.34).

Although gonad development and subsequent spawning may depend on various environmental stimuli, individuals must reach a certain age or size before they are capable of reproduction. The mean length at first reproduction, or mean length at sexual maturity (L_m), may be defined as the length at which 50% of all individuals are sexually mature; e.g., as the length at which 50% of all females in a stock of crustaceans are ovigerous, or the length at which 50% of all female fish have ovaries in an advanced stage of development.

In practice, large samples are collected during the spawning season, to estimate the proportion of sexually mature individuals in different length or age classes. The percentage of sexually mature mackerel, *Scomber scombrus*, found in the spawning populations off the British Isles by both

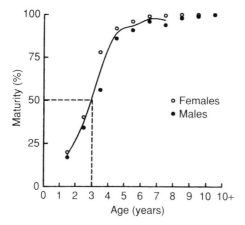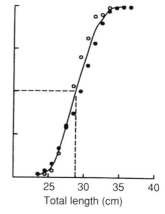

Fig. 4.36 The percentage of sexually mature mackerel, *Scomber scombrus*, found in spawning populations off the British Isles, by both age and total length. From Lockwood (1988).

age and length is shown in Fig. 4.36. The 50% points suggest that the average age (t_m) for a first time spawner is 3 years, and at a total length (L_m) under 30 cm.

The following worked example is based on female tiger prawns (*Penaeus semisulcatus*) in the Persian Gulf. Table 4.14 shows the number of female prawns with ripe ovaries in samples in a range of size classes. The proportion of sexually mature individuals by length class is given in column 4. In this column, the maximum proportion of sexually mature individuals in the samples is 0.62. It is often the case that, even in larger length classes, the proportion of mature individuals is less than one; that is, not all mature individuals are in reproductive condition at the same time. Assuming that only 0.62 or 62% of the mature population are reproducing, the proportions have to be adjusted to represent the number of sexually mature individuals expressed as a proportion of the *reproductive* population in each size class. The data in column 4 have been multiplied by a correction factor (1/0.62 = 1.6) to give the adjusted proportions in column 5.

Table 4.14 Numbers of female prawns (*Penaeus semisulcatus*) with ripe ovaries expressed as a proportion of the total sample. Column 5 contains proportions multiplied by (1/0.62 =) 1.6 to allow for the fact that not all mature females were ripe at the same time (see text). Data from Bushehr Fisheries Research Centre, Iran.

1 Total length (cm)	2 Total number in sample	3 Numbers ripe	4 Proportion ripe	5 Adjusted proportion ripe
12.5	109	5	0.05	0.07
13.5	73	7	0.10	0.15
14.5	42	10	0.24	0.38
15.5	48	21	0.44	0.70
16.5	321	158	0.49	0.79
17.5	458	215	0.47	0.75
18.5	771	396	0.51	0.82
19.5	535	280	0.52	0.84
20.5	180	85	0.47	0.76
21.5	29	18	0.62	0.99
22.5*	1	1	—	—

*excluded from analysis (sample size < 10).

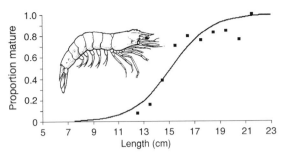

Fig. 4.37 A logistic curve relating the numbers of mature individuals to length for *Penaeus semisulcatus* in the Persian Gulf. Graph produced using the spreadsheet model described in Appendix A4.5.

A logistic curve may be fitted to the proportion (*P*) of sexually mature individuals by length (*L*), in a similar graph to that presented for trawl mesh selectivity (Fig. 3.16 in Chapter 3). Adapting Equation 3.4 in Chapter 3 gives:

$$P = 1/(1 + \exp[-r(L - L_m)]) \qquad (4.44)$$

where *r* is the slope of the curve, and L_m is the mean length at sexual maturity, or the length which corresponds to a proportion of 0.5 (or 50%) in reproductive condition. As in the case of mesh selectivity this equation can be transformed to linear form as:

$$\ln[(1 - P)/P] = rL_m - rL \qquad (4.45)$$

which can be treated as a linear regression and solved analytically in the same way as the trawl mesh selectivity data in Chapter 3 (see Table 3.3 and Fig. 3.16). From Equation 4.45, the intercept is rL_m and the mean length at sexual maturity, L_m = (intercept)/*r*.

An alternative approach is to minimize the sum of squared residuals (SSR) on a computer spreadsheet described in Appendix A4.5. The program in Appendix A4.5 produced the curve shown in Fig. 4.37 and located an optimal value for the mean length at sexual maturity (L_m) of 15.17 cm; an unreasonably high estimation of L_m would have been obtained if the logistic curve had been fitted to unadjusted proportions.

4.3.6 Recruitment

Recruitment into the adult stock often takes place at particular times of the year, and when juveniles have reached a certain age or size. In some species, recruitment may take the form of a migration from established nursery areas. A simple method of examining the timing of recruitment is by plotting the percentage of small individuals (less than an abitrarily chosen size) in samples taken at regular intervals from the stock. The mean length at recruitment (L_r) may be estimated by using a logistic curve similar to that used to estimate L_m. The graph may be constructed by comparing length–frequency data collected from nursery areas with those from the adult stock (in a method similar to alternate haul analysis; Section 3.4).

Some of the least convincing graphs produced by fisheries biologists are those attempting to relate the number of new recruits to the size of the spawning stock. The interest in this relationship is due to its potential to provide insight into what happens to recruitment when a fish stock is reduced by fishing. Data have been used to fit curves as imaginative as that shown in Fig. 4.38. Except for a single, influential, right-hand point in this graph, there is in fact little evidence that recruitment decreases at high stock levels as the curve suggests.

The graph suggests that even at the same stock level, recruitment varies greatly. One hypothesis used to explain large difference in recruitment at similar stock levels is the match/mismatch hypothesis (Cushing, 1990). If the timing of the

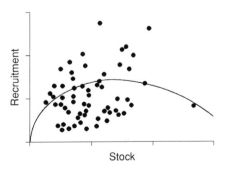

Fig. 4.38 The relationship of recruitment to parental stock size in the sockeye salmon. Rounsfell, from Pitcher and Hart (1982).

production of larvae matches the period of high abundance of their zooplankton prey, larval survival will be low. Alternatively, if the larval production and zooplankton abundance is mismatched many larvae will die from starvation or grow slowly and therefore have an extended exposure to predation.

Intuitively, one would expect the correlation between the numbers of recruits and stock size to be reasonable: that is, the larger the number of reproductive individuals in a stock, the larger the number of resulting offspring. However, even for a given stock size, the variation in recruitment is often large, suggesting that recruitment is related to factors other than stock size. Varying recruitment strengths could result from year to year variations in reproductive effort or success, as well as variations in the survival or settlement of larvae and juveniles. The problem is that recruitment is often separated from the spawning event by a long time period, in some cases several years. This period usually includes a planktonic phase, during which the larvae are exposed to highly variable environmental and hydrological conditions, as well as high mortality rates. Because of this time lag, there is likely to be a poor relationship between the number of eggs released and the strength of recruitment.

The population structure of many commercial species often show strong year classes which persist in catches over many years, and appear unrelated to parental numbers. Perhaps strong recruitments occur periodically when several favourable aspects of the environment coincide. In a review of recruitment in species of coral reef fish, Doherty and Williams (1988) suggested that the intensity of recruitment on reefs is neither uniform or predictable, and that the patchiness of recruitment is most likely due to differences in the concentrations of pre-settlement fish passing over reefs. A good review of recruitment in a fisheries context is given in Myers (2002).

Ultimately, of course, recruitment *does* depend on stock size in all species and, at some low level of stock size, recruitment will fail. The Allee effect (see Box 3.5 in Chapter 3) suggests that for some species recruitment may fail at stock sizes

well above zero – that is, a graph relating recruitment to stock size may not go through the origin.

Several models have been used to suggest the relationship between recruitment and stock size. The Beverton and Holt (1957) equation suggests that recruitment approaches an asymptote at high stock densities:

$$R = aS/(b + S) \qquad (4.46)$$

where R is the number of recruits, S is the number of individuals in the spawning stock and a and b are the parameters of the curve. Fig. 4.39a shows a series of Beverton and Holt curves for different values of a. Species with a short lifespan and high

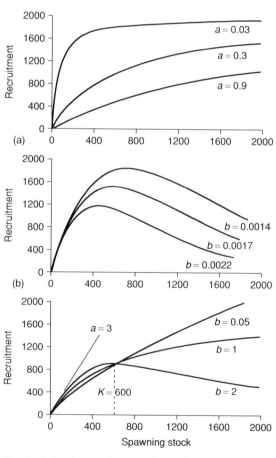

Fig. 4.39 Stock–recruitment relationships: (a) Beverton and Holt curves for $a = 0.03$, 0.3 and 0.9 with $b = 0.0005$; (b) Ricker curves for $b = 0.0014$, 0.0017, and 0.0022 with $a = 7$; (c) Shepherd curves for $b = 0.5$, 1, and 2, with $a = 3$ and $K = 600$.

fecundity are likely to have stock–recruitment relationships with low values of this constant and show no reduction in recruitment even when stocks are reduced heavily by fishing. Recruitment in fisheries based on such species, however, is likely to fail suddenly at low levels of stock abundance.

The Ricker (1954, 1975) model describes the situation where recruitment reaches a maximum before decreasing at higher levels of stock abundance:

$$R = aS \exp[-bS] \tag{4.47}$$

Fig. 4.39b shows a series of Ricker curves for different values of b. This model suggests the presence of density-dependent mechanisms; perhaps adults compete more successfully for the same resources as juveniles, or adults prey on young of the same species.

Other models including those of Shepherd (1982) and Schnute (1985) appear to be more flexible in the shape of their graphs and can mimic the previous two models, but at the expense of introducing a third parameter. Shepherd's model has the equation:

$$R = aS/[1 + (S/K)^b] \tag{4.48}$$

The parameter a is the initial slope at the origin of the curve (the line in Fig. 4.39c), and is a measure of the maximum recruitment per unit stock; this is attainable only at low stock sizes where the density-dependent mortality of pre-recruits is least. At higher stock sizes, recruitment is below this line because of density-dependent effects. The parameter K represents the threshold stock size above which density-dependent effects dominate density-independent effects. Specifically, K is the stock size at which recruitment is reduced to half the level it would have had under density-independent processes only (the vertical line in Fig. 4.39c). If the stock size falls below this threshold, the stability conferred by density-dependent processes becomes less effective, and the likelihood of collapse is high. The compensation parameter b is the degree to which density-independent effects compensate for changes in stock size. For values of $b < 1$, density-dependent

effects are minimal, and the curve rises without limit. For values of $b = 1$ (when the curve is identical to that of Beverton and Holt) density-dependent effects at high stock densities compensate exactly for increases in stock sizes, and recruitment is asymptotically constant. For values of $b > 1$ (when the curve is domed like that of Ricker), density-dependent effects at high stock densities are so strong that they over-compensate for increases in stock size, and recruitment decreases. High values of b may reflect cannibalism and competition for limited resources. Cannibalism in fish and crustaceans appears relatively common, and has been recorded in at least 36 of the 410 teleost fish families reviewed by Smith and Reay (1991). In the case where there is no spatial separation between adults and young fish, it is possible that cannibalism reduces, and even controls numbers of new recruits. However, if the degree of cannibalism is density-dependent, i.e. its occurrence is greater at higher stock densities when food is scarce, heavy fishing may reduce its incidence.

The constants in both the Ricker model and the Beverton and Holt model can be estimated by transforming the equations using natural logarithms and using either graphical methods or fitting the curve by least squares. Ricker's equation, for example, can be transformed by using natural logarithms to give:

$$\ln[R/S] = \ln[a] - bS \tag{4.49}$$

Data such as that given in Table 4.15 can be analysed by a regression of $\ln[R/S]$ against S. From Equation 4.49, the regression line will have an intercept of $\ln[a]$ and a slope of $-b$. The values of a (= exp[intercept]) and b (= −slope) can then be substituted in Equation 4.47 to draw the stock–recruitment curve.

As an alternative to the above method, a nonlinear least squares analysis can be completed on a computer spreadsheet, a design for which is given in Appendix A4.6. The model is used to find values for a and b in the Ricker model and then, separately, to find the values of a and b in the Beverton and Holt model (remembering that a and b in each model are unrelated). Fig. 4.40 shows both curves generated by the model.

Table 4.15 Stock and recruitment indices.

Stock	Recruitment
900	4000
2000	4800
2100	7900
2200	8300
2500	6000
2800	9000
3200	6000
4000	8000
4200	5000
4400	7000
5300	8000
5800	5800
6300	7200
6800	7300
7300	6900

Judging by the closeness of the values of the sum of squared residuals for each model (see Appendix A4.6) it appears that either could be reasonably used to describe the data. In fact, a straight line fitted through the data (without being forced to pass through zero) would fit the data almost as well with a slight positive slope!

The often large amount of scatter around the curves reflects the stochastic nature of recruit-

ment, and the degree of year to year variation attaches a great deal of uncertainty to the form of stock–recruitment relationships. These relationships are however of great interest to fisheries scientists and managers – the shape of a stock–recruitment curve provides an indication of the level below which a stock cannot be driven without causing recruitment to collapse.

4.4 Factors that decrease biomass

Many factors in the marine environment act to reduce the chances of survival of individuals in a population. These include adverse conditions, lack of food, competition and, perhaps most important of all in marine species, predation.

Most marine species produce large numbers of small planktonic larvae, in which mortality rates must be extremely high. Hjort (1914) suggested that a critical period in the early life history of fish occurs at the time when pelagic larvae have exhausted their yolk-sac reserves, and their survival depends on the availability of surrounding food. In addition to predators, environmental hazards including storms and adverse currents can act to prevent larvae from reaching suitable settlement areas or nursery grounds. Although mortality rates of juveniles may decrease with increasing size (as larger animals generally have fewer predators), one of the assumptions often made is that, after recruitment, adult mortality rates are constant over the remainder of the life cycle (Fig. 4.41).

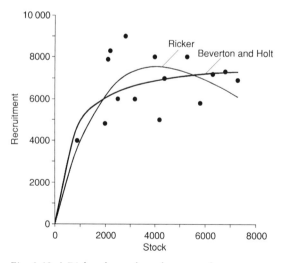

Fig. 4.40 A Ricker dome-shaped curve and a continually rising Beverton and Holt curve relating recruitment to stock size. A spreadsheet program to fit stock–recruitment curves is given in Appendix A4.6.

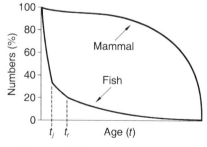

Fig. 4.41 The survival of marine species (lower curve) over lifespan: t_j represents the mean age at transformation from larvae to juveniles, and t_r is the mean age at recruitment. The likely survival curve for a mammal (upper curve) is shown for comparison.

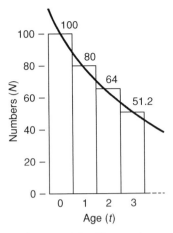

Fig. 4.42 Numbers surviving (figures above columns) over three years, beginning with an initial number (N_0) of 100 fish, when a constant 20% mortality per year is applied. A negative exponential curve of the form exp[−0.223] is drawn through the top of each column.

The loss of individuals in a population through death can be discussed in terms of the percentage of individuals that survive (survival rate) over a particular time interval, or the percentage that die (mortality rate). Fig. 4.42 shows what happens to a cohort of 100 fish over three years when the mortality is a constant 20% per year. Numbers decrease from 100 to 80 to 64 to 51.2 over the three years in a decreasing curve (a negative exponential curve). If the initial population number is N_0, the number of individuals, N_t, in the population after t years is given by the exponential decay equation:

$$N_t = N_0 \exp[-Zt] \tag{4.50}$$

where Z is the conventional symbol for the instantaneous rate of total mortality. The fraction surviving after one year is called the survival rate, S:

$$S = N_{t+1}/N_t = \exp[-Z] \tag{4.51}$$

From this, $Z = \ln[N_{t+1}] - \ln[N_t]$ and using the reduction of 100 to 80 individuals in Fig. 4.42, $Z = \ln[100] - \ln[80] = 0.223 \text{ yr}^{-1}$. The cohort of 100 fish thus suffers an instantaneous total mortality rate of 0.223 yr^{-1}. Instantaneous rates may be converted to percentages by:

$$\text{Survival} (\%) = 100 \exp[-Z] \tag{4.52}$$

and, as mortality is the complementary value of survival:

$$\text{Mortality} (\%) = 100(1 - \exp[-Z]) \tag{4.53}$$

In Fig. 4.42, the cohort of 100 fish suffers a constant mortality of 20% per year or an instantaneous total mortality of $Z = 0.223 \text{ yr}^{-1}$. Unlike percentages, instantaneous rates are additive. The total mortality over the three years, therefore, can be found by multiplying the annual rate by three as in Equation 4.53 as $S = 100 \exp[-Zt] = 100 \exp[-0.223 \times 3] = 100 \times 0.512 = 51.2\%$ as shown in Fig. 4.42.

If the cohort of fish shown in Fig. 4.42 was from an exploited stock, it is necessary to distinguish between mortalities caused by fishing and those caused by natural phenomena. The total mortality rate (Z) is the sum of the instantaneous rate of fishing mortality (F), which is caused by the fishing operation, and the instantaneous rate of natural mortality (M), which includes deaths caused by all other factors:

$$Z = F + M \tag{4.54}$$

In an exploited fish stock, numbers surviving will tend to decline exponentially with time or age according to the sum of the instantaneous rate of fishing mortality (F) and the instantaneous rate of natural mortality (M), so that Equation 4.50 becomes:

$$N_t = N_0 \exp[-(M + F)] \tag{4.55}$$

The survival of fish subjected to different rates of mortality is shown in Fig. 4.43, in which the number surviving decreases with age under an instantaneous rate of natural mortality (M). As individual fish reach a size when they become susceptible to the fishing gear (after the age of first capture, t_c), the effects of the instantaneous rate of fishing mortality (F) have to be added to give the total mortality ($Z = F + M$). In an unexploited stock, $F = 0$ and, therefore, $Z = M$.

The most common method of estimating mortality rates involves transforming the curves shown in Fig. 4.43 and by plotting the natural

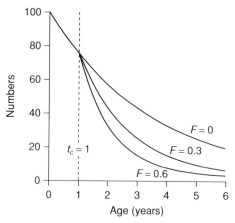

Fig. 4.43 Stock numbers decreasing under the effects of natural mortality ($M = 0.3$ yr^{-1}). After the age at first capture (t_c), the additional effects of three different rates of fishing mortality ($F = 0$, $F = 0.3$ and $F = 0.6$ yr^{-1}) are shown.

logarithms of the numbers of fish surviving by age as a catch curve. Under the assumption of a constant rate of mortality, numbers surviving (N_t) will tend to decline exponentially with time or age (t) as suggested by Equation 4.50. This equation can be log transformed to give the linear catch curve:

$$\ln[N_t] = \ln[N_0] - Zt \qquad (4.56)$$

Graphing the natural logarithms of numbers surviving over successive years will therefore produce a straight-line relationship referred to as a catch curve (Beverton & Holt, 1957; Ricker, 1975). A line of best fit through these data will have a slope numerically equal to the instantaneous mortality rate Z.

4.4.1 Age-based catch curves

An age-based catch curve is a graph of relative numbers against age. Constructing a catch curve usually involves determining the ages of individuals in a large sample of fish (by hard part analysis, for example), and assuming that the age composition of the sample represents the age composition of the stock. This assumption depends on all age groups being equally vulnerable to the fishing gear used to collect the sample. The use of these data in a catch curve also assumes that all age groups have had a similar abundance at recruitment, and have been subjected to the same total mortality rate after recruitment. The assumptions of constant recruitment and constant mortality are illustrated by the hypothetical age-composition data presented in Table 4.16. This table shows the numbers of fish in three age classes for three successive years, during which mortality is a constant 60% per year. Consequently, for each 1000 individuals at age 1, there will be 400 at age 2, and 160 at age 3.

Table 4.16a shows the situation (which must be very rare) in which recruitment is constant; that is, the same number of 1 year old fish are added to the stock each year. Here, the number of individuals in each age class (in each pseudo-cohort across the table) decreases according to a constant mortality rate. In this case, a single large sample in which the age composition is known may be used to construct a catch curve, and estimate mortality. Graphing the natural logarithms of the number of individuals (1000, 400, and 160) against age provides a total mortality rate of 0.92 yr^{-1}, which, by using Equation 4.53, is equivalent to 60%.

Table 4.16 Numbers of fish by age over three successive years under conditions of (a) constant recruitment and (b) variable recruitment. A constant annual mortality of 60% is applied.

(a) Constant recruitment	Age 1	Age 2	Age 3	(b) Variable recruitment	Age 1	Age 2	Age 3
Year 1	1000	400	160	Year 1	1000	2000	100
Year 2	1000	400	160	Year 2	2000	400	800
Year 3	1000	400	160	Year 3	900	800	160

Table 4.16b shows the more usual situation in which recruitment is variable from year to year. In this case, the numbers at different ages will vary according to the strength of recruitment, and the reduction in numbers between age classes in a single year cannot be used to estimate mortality. However, a cohort may be followed diagonally from upper left to lower right, and, in the case of the example, there are 1000 individuals at age 1 (in the first year), 400 at age 2 (in the second year), and 160 at age 3 (in the third year).

Often, pseudo-cohorts in a single sample are used to construct a catch curve, even though recruitment which has produced the age groups is variable. This is less of a problem in a species with a long lifespan, and in which recruitment varies in a random fashion. In this case, there will be a large number of data points (one from each age group), and even though the data are scattered (due to variable recruitment), a line of best fit will provide a reasonable estimate of mortality. An example of a catch curve for a relatively long-lived, temperate-water fish, the jackass morwong, *Nemadactylus macropterus*, is provided here as an example. As the age structure of the morwong stock had been determined from otolith studies (Smith, 1982), it was possible to calculate the relative numbers in each of the twelve age groups in the commercial trawl catch. The first step in constructing a catch curve is to set out the relative numbers and natural logarithms of the numbers by age as shown in Table 4.17 (note that Roman numerals are often used to denote ages based on annuli).

The catch curve for morwong (Fig. 4.44) consists of a regression fitted through the descending data points representing older age groups. The initial ascending data points are not included in the regression – these points represent younger

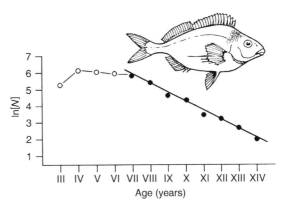

Fig. 4.44 A catch curve for the jackass morwong, *Nemadactylus macropterus*. Redrawn from Smith (1983).

age groups, which are subjected to a lower fishing mortality because they are either not fully recruited or are not fully vulnerable to the fishing gear used. Perhaps only some of each of the younger age classes (those less than 7 years of age, in this case) have moved from nursery areas and have been recruited to the adult stock on the fishing grounds.

The methods described above are based on age composition data, i.e. the abundance of individuals in each different age class. As discussed earlier in this chapter, catch per unit effort (CPUE) is commonly used as an index of abundance. This suggests that the natural logarithms of the CPUE values (in *numbers* per unit effort) of a single cohort may be plotted against age as a catch curve. However, this requires that, in addition to a time series of CPUE data, details on age composition are available.

If information on age composition is not available, CPUE data may still be used if recruitment occurs during well-defined periods. If this is the case, the decrease in total CPUE over the

Table 4.17 Relative numbers by age for the jackass morwong from Smith (1982). Natural logarithms are shown in the third row.

Age	III	IV	V	VI	VII	VIII	IX	X	XI	XII	XIII	XIV
Number	200	403	384	365	305	193	100	71	33	21	12	7
ln	5.30	6.00	5.95	5.90	5.72	5.26	4.61	4.26	3.50	3.05	2.49	1.95

period between one recruitment and the following recruitment will reflect the *average* mortality of all age groups combined. The assumptions are, of course, that all age groups, including the newly-recruited group, are equally vulnerable to the catching method, and that mortality is constant. Other sources of error may be due to migration and changes in catchability. If fish gradually move out of the fishing area, for example, the measured decrease in CPUE will be due to the combined effects of mortality and migration. Similarly, if catchability changes with season (if fish are less susceptible to capture during winter, for example), the changes in CPUE over time will not be due to the effects of mortality alone.

The following example is based on a purse seine fishery for a species of sardine (*Sardinella*), in which the growth parameters are known, and monthly catch and effort data from selected vessels are available (Table 4.18). The basic data include catch (column 3) and effort (column 4). These data enable CPUE (column 5) to be calculated in units of weight (kg per haul of the net). In order to estimate mortality, CPUE in terms of weight must be converted to numbers per unit effort (column 8); this conversion is relatively simple if basic biological information on growth

is available. Newly-recruited sardines are 5 months of age in January, and recruitment appears to be complete by March. Growth in length (column 6) conforms to the von Bertalanffy equation (Equation 4.22) with $K = 0.08$ m^{-1} and $L_\infty = 232$ mm, and weight (g) is related to length (mm) by the equation $W = 0.00002L^3$ (column 7).

The natural logarithms of CPUE (numbers per haul) from Table 4.18 are plotted as a catch curve in Fig. 4.45 as line A. The initial two points (from January and February) are not included in the regression as either recruitment is incomplete or the vulnerability of small fish to the gear is low in these months. The slope of the line (with sign changed) estimates Z as 0.19 per month or 2.3 yr^{-1}. Regression line A will estimate mortality correctly only if is there is no change in the catchability, no recruitment and no migration. If the fish became more vulnerable to the fishing method with increasing size or age, for example, plotted data may be similar to those shown in catch curve B which has less of a slope and therefore underestimates mortality. Alternatively, if larger fish tended to move away from the area fished, for example, plotted decreases in catch rates would not only reflect mortality but also losses by migration – that is, catch curve

Table 4.18 Catch (t), effort (number of hauls) and CPUE (kg/haul) by month for a sardine. Individual mean lengths (mm) and weights (g) are used to calculate catch per unit effort as numbers caught per haul. The natural logarithms of these values are shown in column 9.

1 Month	2 Age (m)	3 Catch (t)	4 Effort (hauls)	5 CPUE (kg/haul)	6 Length (mm)	7 Weight (g)	8 CPUE (nos/haul)	9 ln[CPUE]
Jan	5	3	18	167	76.49	8.95	18 659	9.83
Feb	6	6	15	400	88.44	13.84	28 902	10.27
Mar	7	18	10	1800	99.48	19.69	91 417	11.42
Apr	8	22	12	1833	109.67	26.38	69 485	11.15
May	9	23	14	1643	119.07	33.77	48 653	10.79
Jun	10	18	9	2000	127.76	41.70	47 961	10.78
Jul	11	17	10	1700	135.77	50.05	33 966	10.43
Aug	12	17	11	1545	143.17	58.69	26 325	10.18
Sep	13	17	9	1889	150.00	67.50	27 985	10.24
Oct	14	16	10	1600	156.30	76.37	20 951	9.95
Nov	15	14	8	1750	162.12	85.22	20 535	9.93
Dec	16	10	7	1429	167.50	93.98	15 205	9.63

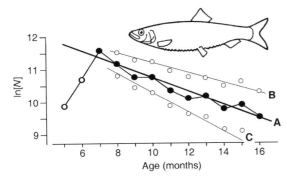

Fig. 4.45 Catch curves for the sardine. (A) A plot of the natural logarithms of CPUE (numbers per haul) by month without error; (B) hypothetical data with increasing susceptibility of larger fish; (C) hypothetical data with the increasing loss of larger fish by migration. Actual mortality is shown by the heavy line.

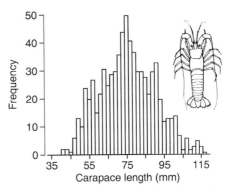

Fig. 4.46 Length–frequency distribution for the female spiny lobster, *Panulirus penicillatus*, in Samoa. Unpublished data, M. King and L. Bell (1990).

C would have a steeper slope and overestimate mortality.

The methods described above are based on either age composition data, or CPUE data. In some cases, it is possible to infer age composition directly from length–frequency data. In the length–frequency data shown in Fig. 4.17 for the mackerel, for example, five different age classes are distinguishable. The numbers in each year class (equivalent to the area under each normal curve) can be counted, and their natural logarithms plotted as a catch curve, the slope of which would provide an estimate of Z. This would assume a constant recruitment and that all age classes are fully vulnerable to the fishing gear used to collect the sample.

This method of constructing a catch curve from length–frequency data is only possible in cases where different age classes can be easily separated from each other. In most length–frequency data, unlike the data shown in Fig. 4.17, it is not easy to distinguish different age groups. In these cases, length–frequency distributions may be converted to age–frequency distributions by means of length-based catch curves.

4.4.2 Length-based catch curves

In most length–frequency data, it is difficult to identify the component age groups and estimate the numbers at each age. The spiny lobster, *Panulirus penicillatus*, for which length–frequency data are shown in Fig. 4.46 has a lifespan of about 7 years and the presence of this number of component curves makes separation into age classes difficult. However, as long as growth data are available for the species, a length–frequency distribution can be converted to an age–frequency distribution by means of a length-converted catch curve (Pauly, 1983; Sparre et al., 1989).

A length-converted catch curve is based on Equation 4.56 in which N_t is replaced by the frequency, F, between L_1 and L_2, (the upper and lower limits of the length class) and t is the age corresponding to the length:

$$\ln[F_{(L_1-L_2)}/\Delta t] = \text{constant} - Z[(t_{L_1} + t_{L_2})/2] \quad (4.57)$$

in which Δt is the time taken for the species to grow through a particular length class, and allows for the fact that as growth slows down with increasing size, older length classes contain more age classes than do younger groups. The value of t is estimated using the inverse of the von Bertalanffy growth formula (Equation 4.42):

$$t = t_0 - (1/K) \ln[1 - L_t/L_\infty] \quad (4.58)$$

The value of t_0 can be set to zero as age in relative rather than absolute terms will not affect the value of the slope and estimates of mortality. Table 4.19 shows the length–frequency data from Fig. 4.46 in a form to construct a length-converted catch curve. The von Bertalanffy

Table 4.19 The use of length–frequency data (from the female spiny lobster in Fig. 4.46) to construct a length-converted catch curve. The von Bertalanffy growth parameter used are $K = 0.39$ yr^{-1} and $L_\infty = 121$ mm carapace length. The asterisks * and ** indicate the first and the last points respectively of those used in the regression (Fig. 4.47).

1 L1	2 L2	3 Mid L	4 F	5 Age at L1	6 Age change (Δt)	7 Age at mid L	8 ln[$F/\Delta t$]
40	42	41	2	1.029	0.064	1.061	3.44
42	44	43	2	1.093	0.066	1.126	3.42
44	46	45	1	1.159	0.067	1.192	2.70
46	48	47	6	1.226	0.069	1.261	4.46
48	50	49	13	1.296	0.071	1.331	5.21
50	52	51	10	1.367	0.073	1.403	4.92
52	54	53	24	1.440	0.075	1.478	5.76
54	56	55	20	1.516	0.078	1.554	5.55
56	58	57	27	1.593	0.080	1.633	5.82
58	60	59	15	1.673	0.083	1.715	5.20
60	62	61	22	1.756	0.085	1.799	5.55
62	64	63	31	1.842	0.088	1.886	5.86
64	66	65	25	1.930	0.092	1.975	5.61
66	68	67	29	2.022	0.095	2.069	5.72
68	70	69	30	2.117	0.099	2.166	5.72
70	72	71	36	2.215	0.103	2.266	5.86
72	74	73	44	2.318	0.107	2.371	6.02
74	76	75	50	2.425	0.112	2.480	6.11
76	78	77	41	2.536	0.117	2.594	5.86*
78	80	79	37	2.653	0.122	2.713	5.71
80	82	81	30	2.775	0.128	2.838	5.46
82	84	83	30	2.903	0.135	2.970	5.40
84	86	85	28	3.038	0.142	3.108	5.28
86	88	87	22	3.181	0.151	3.255	4.98
88	90	89	30	3.331	0.160	3.410	5.23
90	92	91	34	3.492	0.171	3.576	5.29
92	94	93	20	3.663	0.183	3.753	4.69
94	96	95	12	3.846	0.197	3.943	4.11
96	98	97	13	4.043	0.214	4.148	4.11
98	100	99	13	4.257	0.233	4.371	4.02
100	102	101	14	4.490	0.257	4.616	4.00**
102	104	103	7	4.747	0.285	4.886	3.20
104	106	105	2	5.032	0.321	5.188	1.83
106	108	107	6	5.353	0.367	5.530	2.79
108	110	109	2	5.720	0.428	5.925	1.54
110	112	111	3	6.148	0.515	6.393	1.76
112	114	113	5	6.663	0.644	6.965	2.05
114	116	115	3	7.307	0.863	7.703	1.25
116	118	117	1	8.170	1.310	8.742	−0.27

growth parameters used are $K = 0.39$ yr^{-1} and L_∞ = 121 mm carapace length. Columns 1, 2 and 3 contain the lower limit ($L1$), upper limit ($L2$) and midpoint (mid L) respectively of each length class interval. The corresponding frequency (F) is shown in column 4. In column 5 the relative age is calculated using Equation 4.58 with t_0 set to zero. In column 6, the value of Δt, the time taken for the species to grow through a length class from L_1 to L_2 is estimated from Equation 4.58 as $\ln[1 - L_1/L_\infty]/K - \ln[1 - L_2/L_\infty]/K$ which gives:

$$\Delta t = \ln[(L_\infty - L_1)/(L_\infty - L_2)]/K \qquad (4.59)$$

In column 7, the age at the midpoint of the class interval (mid L) is calculated from Equation 4.58. Finally, the natural logarithms of the frequency, F, divided by the change in age, Δt, is calculated in column 8. The length-converted catch curve, a plot of $\ln[F_{(L1-L2)}/\Delta t]$ against mean age, is shown in Fig. 4.47 which is produced by the spreadsheet program presented in Appendix A4.7. The regression line is fitted through the data which exclude:

• the initial ascending data points representing groups of individuals which are either not fully recruited or are too small to be totally vulnerable to the fishing gear;
• data points from mean ages with very small sample sizes; and
• data points close to L_∞, where the relationship between length and age becomes uncertain.

The value of total mortality estimated from the slope of the length-converted catch curve shown in Fig. 4.47 is $Z = 0.99$ yr^{-1}.

As a fish stock becomes more heavily exploited, larger fish are removed from the stock, and, as long as recruitment is not affected, smaller ones are continually added to the exploitable part of the population. This results in a decrease in the mean length of fish in the catch. Fig. 4.48 shows two length–frequency distributions, one from a heavily exploited stock and one from a lightly exploited stock. The mean length of heavily exploited fish (the shaded part of the histogram) is lower than that of the lightly

exploited stock. Beverton and Holt (1956) suggested the following relationship between total mortality, Z, and mean length:

$$Z = K[(L_\infty - \bar{L})/(\bar{L} - L')] \qquad (4.60)$$

where $\bar{L}$ is the mean length of fish equal to, or longer than, length L', which is the lower limit of the first length interval of fish fully vulnerable to the fishing gear. When the length at which fish are fully vulnerable is not known, but the mean length at first capture, L_c has been found from selectivity studies (see Chapter 3), the following version of the Beverton and Holt equation may be used:

$$Z = K[(L_\infty - \bar{L}_c)/(\bar{L}_c - L_c)] \qquad (4.61)$$

where L_c is the mean length at first capture, and $\bar{L}_c$ is the mean length of fish in the catch. If the species represented in Fig. 4.48 has a growth coefficient of $K = 0.12$ yr^{-1} and an asymptotic length of $L_\infty = 39$ cm, then Equation 4.60 estimates Z as 0.42 yr^{-1} for the lightly exploited stock and 0.62 yr^{-1} for the heavily exploited stock.

The Beverton and Holt Z-formula (Equation 4.60) may also be used in a regression plot to estimate both Z/K and L_∞ (Wetherall, 1986; Wetherall et al., 1987). A modified version of this equation (Pauly, 1986) suggests that values of $(\bar{L} - L')$ can be plotted against a series of cutoff points, L', as a straight line:

$$\bar{L} - L' = a + bL' \qquad (4.62)$$

As before, $\bar{L}$ is the mean length of all fish equal to, or longer than, length L', which now becomes a series of lower limits for the length intervals of fully vulnerable fish.

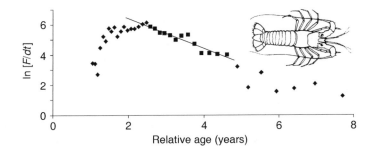

Fig. 4.47 A length-converted catch curve fitted through the data points shown as large squares (see text) for the spiny lobster, *Panulirus penicillatus*, in Samoa. Produced by the program described in Appendix 4.7.

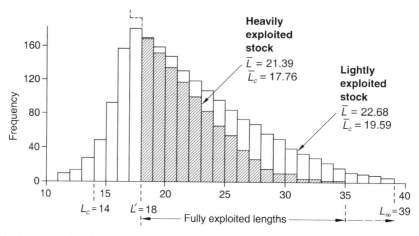

Fig. 4.48 Length–frequency distributions for a heavily exploited stock (shaded area) and a lightly exploited stock (unshaded area).

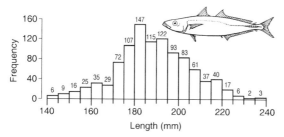

Fig. 4.49 Length–frequency data for the scad, *Decapterus macrosoma*, in Tonga. Data collected by V. Langi and S. Malimali.

The combined length–frequency data from almost a year of sampling on the scad, *Decapterus macrosoma*, in the Vav'au Islands of Tonga (Fig. 4.49) are used here in an example of a Wetherall plot and the worksheet is shown in Table 4.20. Columns 1, 2 and 3 contain the lower size limit ($L1$), upper size limit ($L2$) and the corresponding frequency (F) respectively of each length class interval.

Column 4 contains the midpoint of the length class (($L1 + L2$)/2) multiplied by the corresponding frequency, F. In column 5, the mean length, $\bar{L}$, above the lowest limit of the length interval is estimated as $\Sigma(FL)/n$ where L is the midpoint of the class interval and n is the number of fish above L'. Column 6 contains $\bar{L} - L'$, which is plotted against L' in the Wetherall plot shown in Fig. 4.50. The regression line in the plot is fitted through all data representing the fully exploited

Table 4.20 Wetherall plot calculations for the scad, *Decapterus macrosoma*, in Tonga. Data collected by V. Langi and S. Malimali. The asterisks * and ** indicate the first and the last points respectively of those used in the regression (Fig. 4.50).

1 L1 (L')	2 L2	3 F	4 ΣFL	5 $\bar{L}$	6 $\bar{L} - L'$
140	145	6	855.0	188.40	48.40
145	150	9	1327.5	188.67	43.67
150	155	16	2440.0	189.04	39.04
155	160	25	3937.5	189.63	34.63
160	165	35	5687.5	190.46	30.46
165	170	29	4857.5	191.50	26.50
170	175	72	12 420.0	192.27	22.27
175	180	107	18 992.5	193.98	18.98
180	185	147	26 827.5	196.41	16.41
185*	190	115	21 562.5	199.94	14.94
190	195	122	23 485.0	203.03	13.03
195	200	93	18 367.5	206.78	11.78
200	205	83	16 807.5	210.25	10.25
205	210	61	12 657.5	214.13	9.13
210	215	37	7862.5	217.98	7.98
215	220	40	8700.0	220.96	5.96
220**	225	17	3782.5	225.89	5.89
225	230	6	1365.0	231.14	6.14
230	235	2	465.0	235.50	5.50
235	240	3	712.5	237.50	2.50

part of the sample, often from one length interval to the right of the highest mode in the length–frequency data (185 mm in this case), but excluding length classes with small sample sizes (in this

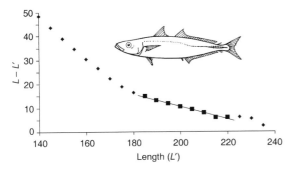

Fig. 4.50 A Wetherall plot produced by the spreadsheet program described in Appendix 4.8. The regression line is fitted through the data points shown as large squares (see text).

example, the final three length classes, which contain less than ten individuals).

A spreadsheet program (using a von Bertalanffy growth coefficient of $K = 1.2$ yr^{-1}) to obtain the regression of $(\bar{L} - L')$ against the cutoff points, L' is described in Appendix A4.8. From the regression line, the value of Z/K is estimated from the slope as:

$$Z/K = -(1 + \text{slope})/\text{slope} = -(1 - 0.265)/(-0.265)$$
$$= 2.8 \qquad (4.63)$$

As the value of the growth coefficient was estimated as $K = 1.2$ yr^{-1} from modal progression analysis, then $Z = 3.3$ yr^{-1}. A useful additional result of the analysis is that L_∞ may be estimated from the intercept with the x-axis as:

$$L_\infty = -a/b = -63.515/(-0.2649) = 239.8 \text{ mm} \qquad (4.64)$$

4.4.3 Mortality from mark–recapture data

Under certain conditions, mark–recapture data may be used to estimate survival and mortality. These conditions are usually more rigorous than in using tagging data to study growth and migration. The types of error which can influence mortality estimates from tagging include the additional mortality caused by the tagging operation itself and by the longer-term effects of the tags on the fish after release. The simplest situation is where a single, and often necessarily massive, number of marked fish is released into the

studied population. If mortality is constant the reduction in numbers of tagged fish returned will follow an exponential curve similar to that shown in Fig. 4.51. That is, the survival rate (Equation 4.50) for unit time, $N_t = N_0 \exp[-Z]$ also applies to tagged fish. If N_0 is the number of fish initially tagged, then the number of tagged fish, N_r, at the beginning of any time interval r, would be:

$$N_r = N_0 \exp[-Zt] \qquad (4.65)$$

The number of fish dying is the complementary value of those surviving – that is, $N_t(1 - \exp[-Z])$. The catch, C, is the proportion of the fish dying due to fishing and is calculated by the catch equation:

$$C_t = (F/Z)N_t(1 - \exp[-Zt]) \qquad (4.66)$$

From the catch equation the number of tagged fish caught (C_r) during time interval r is:

$$C_r = (F/Z)N_r(1 - \exp[-Zt]) \qquad (4.67)$$

Substituting Equation 4.65 in 4.67 gives:

$$C_r = (F/Z)N_0 \exp[-Zt] \, (1 - \exp[-Zt]) \qquad (4.68)$$

Taking natural logs of both sides of the equation, this becomes:

$$\ln[C_r] = \ln[N_0 F/Z] + \ln[1 - \exp[-Z]] - Zt \qquad (4.69)$$

This equation is of the form of a straight line, and suggests that a regression of the natural logarithms of the numbers recaptured per unit effort against time will have a slope which is an

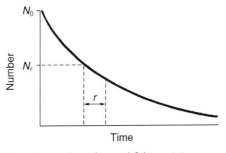

Fig. 4.51 The number of tagged fish surviving over time. N_0 is the number initially tagged, and N_r is the number of tagged fish at the beginning of any time interval r.

estimate of the instantaneous mortality coefficient Z. Under certain assumptions, the intercept on the y-axis can also be used to estimate the fishing mortality, F, from Equation 4.69 (Gulland, 1983). In this equation the intercept, a, is:

$$a = \ln[N_0 F/Z] + \ln[1 - \exp[-Zt]] \quad (4.70)$$

Manipulating this, $\exp[a] = N_0 F(1 - \exp[-Zt])/Z$, and:

$$F = Z \exp[a]/(N_0(1 - \exp[-Zt])) \quad (4.71)$$

where a is the intercept on the y-axis, and N_0 is the number of fish tagged. A hypothetical example of using mark–recapture data to estimate mortality is demonstrated in Table 4.21. A regression of $\ln[C_r]$ against the midpoint of the corresponding time interval is shown in Fig. 4.52. The total mortality coefficient, Z, estimated from the slope, is 0.96 yr^{-1}, and the instantaneous fishing mortality, estimated from Equation 4.71, is 0.24 yr^{-1}.

In most fisheries studies, tagging is carried out over a series of field operations. There are several methods of using data from multiple tagging operations, and these have been reviewed by

Table 4.21 Tagged fish caught (C_r) over 4 years of constant fishing effort, after an initial 400 fish tagged.

Months	0–12	12–24	24–36	36–48
Midpoint (years)	0.5	1.5	2.5	3.5
Tagged fish caught (C_r)	36	15	6	2
$\ln[C_r]$	3.58	2.71	1.79	0.69

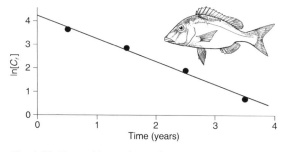

Fig. 4.52 Natural logarithms of tag returns against time. The slope is -0.96 and the intercept is 4.11.

Jones (1976). Robson's method, or developments of it (Youngs & Robson, 1978) are most useful in fisheries studies where recaptures are made continuously, which is usually the case where tagged animals are returned from a commercial fishing operation.

The estimation of mortality from mark–recapture experiments assumes either that there is no additional mortality caused by the tagging operation and the tag itself, or that the additional mortality caused is known. It also assumes that the loss of tagged fish due to migration out of the fished area either does not occur, or is known and can be corrected for. The types of errors which can occur are summarized in Box 4.13 'Tagging errors'.

4.4.4 Natural mortality

In most cases it is easier to estimate total mortality than it is to partition it into its fishing mortality and natural mortality components. Surveys on unexploited (virgin) stocks, of course, may provide data from which natural mortality can be estimated (ignoring the small amount of fishing mortality imposed by the survey). In a stock in which catch values for a range of different levels of fishing effort, and a series of (annual) total mortality estimates are available (Table 4.22), natural mortality may be estimated as follows. As $Z = F + M$, and $F = qf$, where q is the catchability coefficient, and f is fishing effort:

$$Z = M + qf \quad (4.72)$$

This suggests a linear relationship between total mortality and fishing effort as shown in Fig. 4.53, where M is equivalent to the intercept on the y-axis. In this case, natural mortality is 0.36, although the confidence limits for the intercept, and therefore M, are wide (± 0.35).

The natural mortality rate of a species is likely to be related to environmental factors as well as its evolved life-history pattern. Pauly (1980) analysed data from a large number of fish species in an attempt to obtain a general relationship to predict natural mortality from the von Bertalanffy growth parameters, K (per year) and

Box 4.13

Tagging errors

The types of errors which can occur have been classified by Ricker (1975) and include:

• *Type A errors* which affect the value of the intercept (and therefore the estimate of fishing mortality) but not the slope.

• *Type B errors* which affect the value of the slope (and therefore the estimate of total mortality) but not the intercept.

• *Type C errors* which affect the initial part of the regression line, but do not affect either the value of the intercept or the slope as long as initial data are excluded from the analysis.

Type A errors include the additional mortality caused during the tagging operation, and immediately after release of the tagged fish. A loss of tags from fish or tag-induced death shortly after the tagging operation, or a consistent and long-term incomplete recording of tag returns will cause this type of error (Fig. B4.13.1). Type B errors

include additional tag-induced mortality, loss of tags, and emigration of fish which occur at a steady instantaneous rate over the run of the experiment. Type C errors include the effects of short-term changes in fish behaviour due to the presence of the tags; e.g. where the catchability of tagged fish is temporarily different from untagged fish.

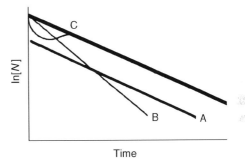

Fig. B4.13.1 Types of errors affecting estimates of total and fishing mortality from tagging experiments. The heavy line represents the regression in the absence of error, and the lighter lines (A and B) represent regression with error.

Table 4.22 Fishing effort and estimates of total mortality by year.

Year	Effort (hr)	Z (yr^{-1})
1	1360	0.56
2	2800	0.54
3	3640	0.79
4	4120	0.87
5	4680	0.85

L_∞ (cm), and mean annual surface temperature, T (°C). The derived relationship, which is sometimes used to obtain a first approximation of natural mortality is:

$$\ln[M] = -0.0152 - 0.279 \ln[L_\infty] + 0.6543 \ln[K] + 0.463 \ln[T] \qquad (4.73)$$

Another rough approximation may be obtained by considering the natural mortality rate required to reduce the numbers of fish in a cohort to close to zero over the period of its lifespan. The oldest

individuals in an unexploited stock are often about 95% of their asymptotic length. If mean lifespan (t_{max}) is defined as the time required for a fish to reach 95% of its asymptotic length, L_∞, then from the inverse of the von Bertalanffy growth equation (Equation 4.22):

$$t_{max} = (-1/K) \ln[1 - (0.95L_\infty)/L_\infty] \text{ or}$$
approximately, $t_{max} = 3/K \qquad (4.74)$

From the exponential decay equation (Equation 4.50) the survival of an unexploited species over its lifespan is $N_t/N_0 = \exp[-Mt_{max}]$, where N_0 is the initial number of fish in the cohort, and N_t is the number of fish remaining towards the end of the lifespan. If N_t is say 1% of the initial number, N_0, then:

$$1/100 = \exp[-Mt_{max}], \text{ and,}$$
$$M = -\ln[0.01]/t_{max} \qquad (4.75)$$

From these equations, for example, a fish with growth parameters of $K = 0.4$ yr^{-1} and $L_\infty = 45$ cm would have an *approximate* lifespan of 7.5 years,

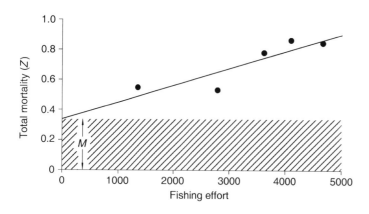

Fig. 4.53 The relationship between total mortality and fishing effort. M is estimated from the intercept on the y-axis as $0.36 \ yr^{-1}$.

and an *approximate* natural mortality rate of 0.6 yr^{-1}. These methods may provide initial values of the mean natural mortality over the lifespan of the species. However, natural mortality is likely to decrease with age, as larger fish have fewer predators than smaller ones.

Exercises

Answers to selected exercises are available at www.blackwellpublishing.com/king

Exercise 4.1

(a) Estimate the numbers of sea cucumbers in the stock shown in Fig. 4.5 by counting the number of individuals in 6 randomly selected quadrats (use the random number/letter table in Appendix 7). Calculate the 95% confidence limits for the estimate.
(b) Repeat the above by sampling every second quadrat along a transect from grid reference G1 to G11 – i.e. use quadrats G1, G3, G5, G7, G9 and G11. Calculate the 95% confidence limits for the estimate. How does the precision of this estimate compare with that of the estimate obtained by random sampling?

Exercise 4.2

Estimate the numbers of sea cucumbers, with 95% confidence limits, in the stock shown in Fig. 4.5 by counting the number of individuals in a total of 10 quadrats by stratified sampling. Either

select your own, or use the following random quadrats in strata A and B.

Stratum A (> 5 m and < 10 m): C6, E3, F4, G9, J3, and J8
Stratum B (< 5 m and > 10 m): A3, G6, K11, and M10

Exercise 4.3

A research vessel completes a standard trawl at 41 stations on a unit stock distributed over an area of 360 km². The trawl net, which has an effective fishing width of 20 m, was towed at a velocity of 8 km/hr for 20 minutes at each station. The mean catch per trawl was 64 kg, and the standard error of the mean ($s/\sqrt{n}$) was 18. Assuming that vulnerability of the fish to the trawl net is 50% ($v = 0.5$), use the swept-area method to estimate the total stock size with 95% confidence limits.

Exercise 4.4

(a) A depletion experiment using traps on an isolated 24 km² stock of crabs was run over four weeks. The number of crabs caught and the number of traps used per week are shown in the following table. Estimate the catchability

Week	Catch	Traps
1	2074	140
2	2376	183
3	2534	235
4	1836	204

coefficient per km^2, and the initial exploitable stock size. List the assumptions that have to be made.

(b) If a survey of an adjacent larger stock of 150 km^2 resulted in an initial mean CPUE of 15.6 crabs per trap, use the catchability coefficient obtained in the above depletion experiment to estimate the exploitable stock size in this larger area. What additional assumptions have to be made?

Exercise 4.5

Determine the relationship of weight to length for a selected marine species. Obtain a random sample (which includes a wide size range) of approximately 50 individuals of the same sex of either a crustacean, mollusc or fish. Construct a table similar to Table 4.5, estimate a and b in the equation $W = aL^b$ and draw a graph as in Fig. 4.13b.

Exercise 4.6

Using Appendix 4 as a guide, construct an Excel spreadsheet and use Solver to fit a curve to the length–weight data for the oyster (Table 4.5) by minimizing the sum of squared residuals.

Exercise 4.7

(a) Based on the four modal lengths visible in the length–frequency histogram for the hypothetical species in Fig. 4.19 (top graph), use a Ford–Walford plot to estimate the von Bertalanffy growth parameters.

(b) Use the estimated parameters K and L_∞ (and assume $t_0 = 0$) in the von Bertalanffy equation to predict fish length for ages from 0 to 8 years

(an age–length key) in steps of 0.5 years and draw a growth curve.

(c) Estimate the approximate mean lifespan of the fish as the age at which the species reaches 95% of the value of L_∞; use the inverse of the von Bertalanffy formula (i.e. make age, t, the subject of Equation 4.22).

Exercise 4.8

The numbers below are the lengths (cm) of 221 fish caught in a single tow of a trawl net.

39	13	12	10	16	13	14	45	13	29	10	12	15
11	8	39	26	12	42	42	43	30	12	28	27	12
32	28	13	9	36	11	12	26	11	30	13	31	14
14	28	11	46	28	11	14	25	13	25	46	38	42
12	31	50	29	34	10	46	46	47	45	47	47	11
48	48	41	49	49	51	10	27	40	30	13	12	11
39	16	39	39	7	39	40	40	27	40	40	41	41
11	13	29	25	16	26	33	15	26	30	26	13	13
24	10	12	50	13	31	13	14	26	29	9	29	10
13	31	12	29	11	26	14	38	38	42	23	15	27
27	8	14	27	10	43	43	11	44	27	14	11	28
14	25	27	32	14	15	12	28	25	29	11	26	11
30	29	12	15	10	16	12	12	29	12	30	28	12
28	14	15	10	14	15	13	17	24	11	13	30	13
35	35	48	36	12	36	37	11	37	37	37	41	45
16	19	22	18	17	21	46	18	23	9	28	11	12
24	9	27	38	28	15	44	38	14	10	17	9	28

(a) Organize the above data as a length–frequency 'tally sheet' by completing a table similar to the sample shown below.

Length (cm)	Frequency	Total (numbers)
7	\	1
8	\\	2
9	₩	5
10	₩ ₩	10
etc.	etc.	etc.

(b) Display the length–frequency data as a histogram similar to that shown in Fig. 4.17. From the left, label the four modal (highest) points on the graph as 0+ years, 1+ years, 2+ years, and 3+ years.

(c) Use the modal values to draw a graph of length at age $t+1$ against length at age t (a Ford–Walford plot with three points, similar to that shown in Fig. 4.17) in order to estimate K and L_∞.

(d) Use the values of K and L_∞ to predict fish length for a range of ages (assume $t_0 = 0$). Enter the estimated lengths in an age–length key and draw a von Bertalanffy growth curve.

(e) Examine the above procedure carefully. What aspects of the biology of this hypothetical fish species could result in the estimates of K and L_∞ being quite wrong?

Exercise 4.9

The heights of a large and random sample of humans could be plotted as a length–frequency graph as used to analyse growth in some fish populations. But it is impossible to use such a graph to estimate the growth of humans from the relative frequencies in size classes. Why is this so?

Exercise 4.10

Re-analyse the penaeid prawn data in Table 4.9 to estimate K using a 'forced' Gulland–Holt plot with $L_\infty = 46$ mm carapace length (ignore negative growth increments).

Exercise 4.11

Analyse the results of the four marked and recaptured bivalves shown in the lower left corner of Fig. 4.25; assume that nine months have passed between marking and recapturing. Measure the shell height along a straight line from the umbo (the pointed tip of the shell) to the shell margin, and which passes through the filed notches. Estimate K and L_∞ by using a graphical method.

Exercise 4.12

Analyse the mean lengths at age of the male morwong, *Nemodactylus macropterus*, (Table 4.11) by means of a von Bertalanffy plot to estimate

the parameters K and t_0, and draw a growth curve. Use $L_\infty = 43.2$ cm, and ignore the samples for ages X and XI which contain less than 10 fish. Compare these estimates with the least-squares results in Appendix A4.4.

Exercise 4.13

(a) Use the drawing of the bivalve, *Vasticardium*, in Fig. 4.28 to estimate K and L_∞ in the von Bertalanffy growth equation.

(b) If the bivalve was collected in mid-June, and the species is known to reach a spawning peak in mid-September, estimate t_0.

(c) Draw the growth curve for the species based on this one individual. Comment on whether or not this is reasonable, and consider the assumptions that the above analyses rely on.

Exercise 4.14

The table below shows the numbers of fish caught per unit of effort in each of three age classes. Fish more than 1 year old are sexually mature and immature fish are recruited to the adult stock before they are 12 months of age. Use these data to estimate a stock–recruitment relationship.

Year	0+	1+	2+
1985	425	260	142
1986	210	120	63
1987	255	154	86
1988	270	95	46
1989	195	92	38
1990	180	86	33

Exercise 4.15

Prepare a graph relating instantaneous mortality rates to percentage mortality rates (place percentage rates from 0 to 90 on the x-axis).

Exercise 4.16

Estimate the growth parameters, K and L_∞, and the total mortality, Z, from the length–frequency data shown in the figure. List the separate assumptions that estimating growth and mortality in this way depend on.

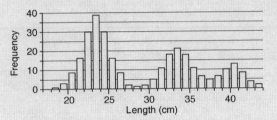

Length–frequency data.

Exercise 4.17

A biologist has access to records from a processing factory as shown below. The factory grades the prawn catch as the number of prawns per kg and the table shows the catches and grade landed in the four months immediately following recruitment.

Month	Numbers (per kg)	Catch weight (kg)
1	35	5690
2	28	5260
3	24	4550
4	22	3670

Make an initial estimate of the total instantaneous mortality rate, Z, and list the important assumptions that must apply for the estimate to be a reasonable one.

Exercise 4.18

A clupeid fish species is recruited at an age of 5 months when it is fully vulnerable to the fishing gear. Mean catch rates per month, starting at the month of recruitment, are 145, 215, 295, 380,

and 466 kg/hr. Growth is described by the von Bertalanffy growth equation with $K = 0.05\ m^{-1}$, and $L_\infty = 35$ cm. Weight (g) is related to length (cm) by a power curve with $a = 0.03$ and $b = 3$. Estimate the total mortality rate (Z) for the species. List the assumptions that must be accepted for the estimate to be a reasonable one.

Exercise 4.19

Catch (t) and effort (hours trawled) by month for a penaeid prawn, *Penaeus latisulcatus*, trawl fishery are shown in the table below. Recruitment is assumed to be complete by March when the newly-recruited prawns are 6 months of age. The von Bertalanffy equation applies to growth with $K = 0.15\ m^{-1}$ and $L_\infty = 55$ mm. Weight (g) is related to carapace length (mm) by the equation $W = 0.0005 L^3$.

Month	Age (m)	Catch (tonnes)	Effort (hr)
Jan	4	169.1	3068
Feb	5	252.9	4738
Mar	6	298.6	4146
Apr	7	314.2	5171
May	8	173.9	3974
Jun	9	79.5	2388
Jul	10	45.8	1693
Aug	11	45.8	1858
Sep	12	62.5	2383
Oct	13	69.4	2738
Nov	14	164.7	4431
Dec	15	59.8	2024

Examine the use of the above data in a catch curve to estimate mortality. There is some evidence that prawns spend more time under the substrate in colder months (southern hemisphere winter months), and may be less accessible to the trawl gear.

Exercise 4.20

When analysing mark–recapture data from a single release of tagged fish, what would be the effects on estimates of total and natural mortality if: (a) a constant 10% of tagged fish caught are unreported by fishers; and (b) fishers gradually lose interest in returning tagged fish due to poor feedback from researchers?

Exercise 4.21

The length–frequency diagram shown in the figure is from a sample of scad, *Decapterus macrosoma*, caught in a purse seine net during surveys. Growth is described by the von Bertalanffy growth equation

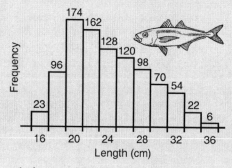

Length–frequency data for a sample of scad.

with $K = 0.8$ yr^{-1}, and $L_\infty = 38.3$ cm. Estimate the total mortality rate (Z) for the species. Accepting that the growth parameters are correct, list assumptions that must be made for the estimate to be a reasonable one.

Exercise 4.22

A fish species grows according to the von Bertalanffy growth equation with $K = 0.4$ yr^{-1} and $L_\infty = 60$ cm, and the mean length at first capture is 18 cm. Over several years of increasing effort, the mean size of fish in the catch has been decreasing as shown in the table below. Estimate the natural mortality rate, M.

Year	Effort (hours)	Mean length (cm)
1	1760	34.8
2	2800	33.2
3	3640	32.1
4	4120	31.7
5	4680	31.2

5 Stock assessment

5.1 Introduction

The overall purpose of fisheries science is to provide decision-makers with advice on the relative merits of alternative management strategies (Punt & Hilborn, 1997). This advice may include predictions of the reactions of a stock and fishers to varying levels of fishing effort and, conventionally, includes an estimate of the level of fishing effort required to obtain the maximum weight or yield that may be taken from a stock on a sustainable basis. However, it is not the task of stock assessors to recommend a single best level of exploitation for a fishery as this will depend on policies that include social, economic and political considerations.

The concept of maximum sustainable yield (MSY), the largest annual catch or yield that may be taken from a stock continuously without affecting the catch of future years, has had a fluctuating history of favour and scorn in fisheries management (Punt & Smith, 2001). The concept of MSY originated in the 1930s and mathematical models that related yield to fishing mortality began to appear in the 1950s. The first model that is most associated with MSY is the surplus production model of Schaefer (1954, 1957) and the second is the yield per recruit model of Beverton and Holt (1957). The increasing availability of age data in the 1960s resulted in the widespread use of age-structured models that addressed the contribution of different age classes to total yield (Megrey, 1989).

The increasing availability of computers has allowed construction of models that use large numbers of more meaningful difference equations that track stock changes in a series of discrete time steps. Computers have also allowed the development of stochastic models, in which one or more factors are allowed to vary randomly, to match the natural fluctuations of populations more realistically. The development of stochastic simulation models provides representations of how a complex system behaves in various circumstances and may be used to predict the possible outcomes of various management strategies. All models, either directly or indirectly, are an acknowledgement of the basic processes that affect a fish stock, including additions by recruitment and individual growth in weight as well as decreases through natural and fishing mortality. This view of stock dynamics, formulated by Russell in 1931, is summarized in Fig. 4.1 in Chapter 4.

Fluctuating environmental factors result in variations in reproductive success, recruitment, and eventually, the strength of year classes moving through the stock and overall abundance. Thus it is hardly surprising that a constant long-term MSY and a corresponding optimal fishing effort are not realities in most fisheries. However, it is evident that annual catches do indeed initially rise with increasing fishing effort and reach a maximum before falling when numbers removed from the stock cannot be replaced by reproduction and recruitment. The difficulty is in knowing the level of fishing effort at which this maximum occurs and how it varies from year to year (Fig. 5.1).

Stock assessment is not a one-off activity. Because of the dynamic nature of fish stocks, fluctuating populations and changes in the amount and efficiency of fishing, it is necessary to continually monitor fisheries. Catch and fishing

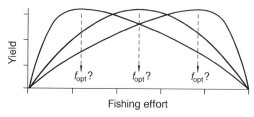

Fig. 5.1 Relationship of catch, or yield, to fishing effort. Three of an infinite number of possible maxima are shown, each related to very different levels of fishing effort (f_{opt}).

effort data are commonly collected for all commercial fisheries, as these are used to calculate catch rates, or catch per unit effort (CPUE). Catch rates are used, sometimes uncritically, as an index of stock abundance and to monitor the 'health' of a fish stock; in commercial fisheries, catch rates are directly related to profitability (see Box 5.1 'Catch rates revisited').

The models described in this chapter are presented in increasing order of complexity from surplus production models that require only a time series of catch and effort data, to models that explicitly include the opposing effects of growth and mortality, to age-structured models that include the contributions of different age classes. A guide to choosing a suitable model is given in Sparre and Hart (2002). The purpose of doing these assessments is to provide advice to fisheries managers whose tasks are much broader and the subject of Chapter 6. Although the sustainability of fish stocks is paramount, fisheries managers have to ensure that benefits to fishers and communities are maximized. In doing this, managers have to take into account political, economic, sociological and environmental factors in addition to the predictions of fisheries biologists.

5.2 Stock abundance and catches – dynamic production models

Surplus production models, sometimes called surplus yield models or biomass dynamic models, are based on the assumption that a fish stock produces an excess or surplus abundance (or biomass) that can harvested. In reference to

Fig. 4.1 in Chapter 4, factors that increase stock biomass are individual growth in weight and the addition, or recruitment, of new individuals. Factors that decrease stock biomass include natural and fishing mortalities.

The biomass in any year equals the previous year's biomass plus recruitment and growth, minus natural mortality and the catch. If recruitment and growth are regarded as production and this is greater than mortality, biomass will increase. Biomass produced in excess of that required to replace losses is regarded as surplus production which can be harvested without adversely affecting the stock. In this respect, the maximum sustainable yield refers to the point at which the rate of surplus production is maximized.

Surplus production models rely on having a long time series of accurate catch and effort data and the assumption that catch rates are a reliable index of biomass (see Box 5.1 'Catch rates revisited'). Tracking and graphing catch and fishing effort is done in all managed fisheries and, in many ways, surplus production models are a natural extension of this. Despite the shortcomings of the models, they serve to exemplify the principle of sustainable harvesting and the effects of overexploitation. Both equilibrium and non-equilibrium methods can be used to fit surplus production models and various techniques are presented in the following sections.

5.2.1 Equilibrium models

In the Schaefer model, the growth of a population or stock is assumed to increase with time in the manner of a logistic or S-shaped curve as discussed in Section 1.3 'Population growth and regulation'. The total weight or biomass of the stock will increase until it reaches a maximum (B_∞) at a carrying capacity imposed by density-dependent factors such as the availability of space or food; the Schaefer model assumes that the increase in stock biomass follows a logistic curve that is symmetrical about the point of inflexion (Fig. 5.2a). In this symmetrical curve, biomass is increasing most rapidly when the stock is at one half of its maximum biomass (Fig. 5.2b). This suggests

Box 5.1
Catch rates revisited

In Chapter 4, catch rates, or catch per unit effort (CPUE), were used as indices of abundance to estimate parameters such as mortality. Catch rates are calculated from catch and fishing effort data for most commercial fisheries, where maintaining high catch rates is the key to profitability. Falling catch rates on the other hand, suggest decreasing profits and the possibility that the target stock is being depleted. Using catch per unit effort (CPUE) to make inferences about stock size (N) assumes that there is a relationship between the two. The formula given in Section 4.2.1 was CPUE = qN, where q is the catchability coefficient, the small proportion of the total stock caught by one unit of fishing effort (say in one hour's trawling). Thus CPUE can reasonably be used as an index of abundance as long as q remains constant. However, there are many circumstances in which q will vary; any improvements in gear technology and fishing methods, for example, will result in an increased value of q. In schooling species, in particular, CPUEs are not a good indication of stock abundance. After each catch made by fishers, the remaining fish gather together in dense schools and are thus just as available as ever, even though the overall abundance has decreased. Because of this regrouping of remaining fish, CPUEs in schooling species may remain high right up to the point of the stock's collapse.

A hypothetical example, in which fishers begin to fish a previously unexploited stock of a non-schooling species that is unfamiliar to them, is summarized in Fig. B5.1.1. The developmental phases of the fishery are shown between vertical broken lines and described below.
- A to B: As fishers become familiar with the species, refine their fishing methods, and locate areas of high stock density, fishing becomes more efficient (the value of q increases).
- B to C: After fishing reaches a peak in efficiency,

catch rates decrease as the more accessible part of the stock becomes depleted.
- C to D: To maintain catch rates, fishers start to exploit parts of the stocks further from the port of landing or in progressively deeper water. Targeting distant areas of higher densities results in catch rates remaining steady even though overall stock abundance is low (as there are less areas of high density).
- D to E: Catch rates fall steeply as the more distant parts of the stock now also become depleted.

A fisheries scientist observing the steady catch rates between C and D in Fig. B5.1.1 might assume that stock abundance is relatively stable when it is, in fact, continuing to decrease to dangerously low levels. The eventual crash may come without sufficient warning for fisheries managers to take remedial actions. This emphasizes the need for fisheries biologists to be familiar with the dynamics of the fishery being managed. In the above example, a scientist working closely with fishers would have been aware of changes in fishing efficiency and the spatial distribution of fishing. Someone working remotely from a fishery, on the other hand, is unlikely to be aware of changes that are occurring but not being reflected in the data collected.

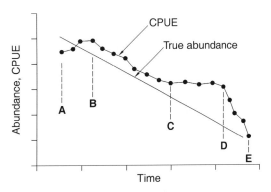

Fig. B5.1.1 Decrease in CPUE (joined points) and trend in actual abundance (straight line) over time.

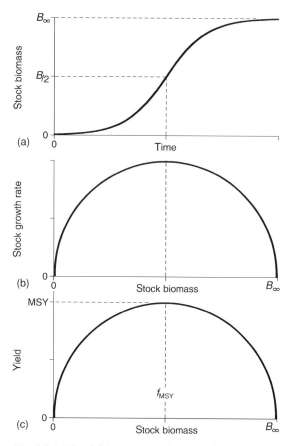

Fig. 5.2 (a) Stock biomass growth along a logistic curve that reaches a maximum biomass, B_{max}. (b) The stock grows most quickly at half its maximum biomass. (c) The maximum sustainable yield is taken at half the maximum biomass.

that, if the stock was exploited, surplus production and therefore yield would be maximized when the stock is reduced to 50% of the unexploited maximum biomass value (Fig. 5.2c). The mathematics behind surplus production models is detailed in Schnute and Richards (2002).

The following year's biomass B_{t+1} will equal the present year's biomass (B_t) plus the surplus production minus the catch (C_t). That is, $B_{t+1} = B_t$ + (surplus production$_t$) – C_t, or:

$$B_{t+1} = B_t + rB_t(1 - B_t/B_\infty) - C_t \qquad (5.1)$$

where r is the intrinsic rate of increase, and B_∞ is the unfished equilibrium stock biomass (sometimes referred to as the carrying capacity of the

environment, K). Under equilibrium conditions, removals in the form of the catch will be balanced exactly by biomass growth, and B_{t+1} will equal B_t so that:

$$C = rB(1 - B/B_\infty) \qquad (5.2)$$

Catch per unit effort (CPUE) is related to biomass by the catchability coefficient, q, so that:

$$CPUE = qB \qquad (5.3)$$

Substituting CPUE/q for B and CPUE$_\infty$/q for B_∞ in equation 5.2 gives:

$$C = rCPUE/q(1 - CPUE/CPUE_\infty)$$

where CPUE$_\infty$ is the catch per unit effort at the maximum biomass, B_∞, of the stock. As CPUE is C/f, this equation can be solved for CPUE as follows:

$$CPUE = CPUE_\infty - (CPUE_\infty q/r)f \qquad (5.4)$$

which is a straight line of the form:

$$CPUE = a + bf \qquad (5.5)$$

A graph of Equation 5.5 is a straight line with an intercept $a = CPUE_\infty$ and a slope $b = -(CPUE_\infty q/r)$. Values for a and b can thus be obtained from a linear regression of CPUE against fishing effort, f. It should be noted that both the variables include fishing effort and this interdependence may influence the correlation between the two. As CPUE is equivalent to C/f, multiplying 5.5 by the fishing effort, f, gives:

$$C = af + bf^2 \qquad (5.6)$$

which is the equation for the symmetrical parabola in Schaefer's model. Differentiating this equation with respect to f gives ($a - 2bf_{MSY}$) which can be equated to zero to give the fishing effort (f_{MSY}) at which yield is maximized. Therefore:

$$f_{MSY} = -a/2b \ldots \text{ or } r/2q \qquad (5.7)$$

Substituting this equation in 5.6 and simplifying gives:

$$MSY = -a^2/4b \ldots \text{ or } rB_\infty/4 \qquad (5.8)$$

Other versions of the surplus yield model have incorporated different forms of biomass growth, which directly influences the estimate of MSY. The model of Fox (1970) uses an asymmetric growth curve in place of Schaefer's symmetrical one. The model of Pella and Tomlinson (1969) adds another parameter that allows the biomass growth curve to be skewed either to the left or right of the symmetrical one. In the case of the Fox model, CPUE decreases in a curve with increasing fishing effort, rather than as the straight line assumed under the Schaefer model. The curve may be converted to a straight line by using natural logarithms:

$$\ln[CPUE] = \ln[C/f] = a + bf \qquad (5.9)$$

Equation 5.9 gives the Fox non-parabolic yield equation:

$$C = f\exp[a + bf] \qquad (5.10)$$

in which MSY $= (-1/b)\exp[a - 1]$, and f_{MSY} $= -1/b$

The conventional approach to both models is to use a long time series of annual catch and effort data, assume equilibrium conditions and graph either CPUE (for the Schaefer model) or the natural logarithm of CPUE (for the Fox model) against fishing effort. The values of a (the intercept) and b (the slope) can then be substituted in either Equation 5.9 or 5.10 to construct the respective yield curves. Note that the values of a and b are different in the two models. The following example is based on a fishery for the sea bass, or barramundi, *Lates calcarifer*, in the Northern Territory of Australia, for which annual catch and effort data are shown in Table 5.1.

In Table 5.1, which may be set up on a computer spreadsheet, mean catch per unit effort values for each year (column 4) are calculated by dividing the annual catch or yield (column 2) by the annual fishing effort per year (column 3).

For the Schaefer model, a graph of the annual CPUE values (column 4 in Table 5.1) against annual fishing effort (column 3) has an intercept on the y-axis of $a = 25.6$, and a slope, $b = -0.000205$, with $r^2 = 0.73$ (upper graph in Fig. 5.3). For the Fox model, a graph of the natural logarithms of CPUE (column 5 in Table 5.1) against annual fishing effort (graph not shown)

Table 5.1 Catch and effort data for a fishery on the sea bass, or barramundi, *Lates calcarifer*. Catch weight is given in tonnes and effort in units of 100 m of net used per day (hmnd). Column 4 shows the CPUE values used to estimate the Schaefer yield curve, and column 5 shows the natural logarithms of CPUE used to estimate the Fox yield curve. Data from Anon. (1991), values for 1976 not available.

1 Year	2 Catch or yield (t)	3 Effort (hmnd)	4 CPUE (kg/hmnd)	5 ln[CPUE]
1972	382.0	17 300	22.08	3.09
1973	431.3	21 000	20.54	3.02
1974	656.0	22 800	28.77	3.36
1975	432.0	15 700	27.52	3.31
1977	1054.0	72 000	14.64	2.68
1978	820.0	95 900	8.55	2.15
1979	745.0	100 700	7.40	2.00
1980	531.7	71 400	7.45	2.01
1981	764.1	66 900	11.42	2.44
1982	856.1	95 400	8.97	2.19
1983	607.0	85 500	7.10	1.96
1984	632.1	71 400	8.85	2.18
1985	592.8	66 300	8.94	2.19
1986	533.9	45 000	11.86	2.47
1987	505.4	41 400	12.21	2.50
1988	507.8	34 900	14.55	2.68
1989	593.8	31 300	18.97	2.94

has an intercept, $a = 3.334$ and a slope, $b = 0.00001413$, with $r^2 = 0$ 80.

The lower graph in Fig. 5.3 shows the annual data points of yield (column 2) against fishing effort (column 3). The Schaefer surplus yield curve is drawn through points predicted by substituting the relevant values of $a = 25.6$ and $b = -0.000205$ in the equation $C = af + bf^2$ for a range of arbitrarily chosen values for fishing effort, say for values from $f = 0$ to 100 000 in steps of 20 000. The Fox curve is drawn by substituting $a = 3.334$ and $b = 0.00001413$ in the equation $C = f\exp[a + bf]$.

From equations 5.5 and 5.6, the Schaefer model indicates that the maximum sustainable yield (MSY) is 800 729 kg and the effort required to take it (f_{MSY}) is 62 544 per 100 m of net used per day (hmnd). That is, a sustainable yield of

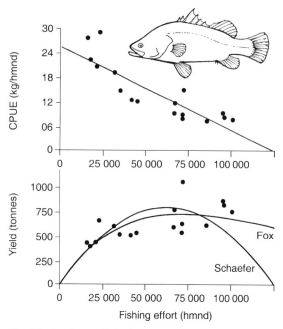

Fig. 5.3 Surplus yield for barramundi. Top: catch per unit effort (kg per 100 m of net per day) against fishing effort (in 100 m net used per day). Bottom: Schaefer and Fox yield curves.

approximately 800 tonnes of barramundi may be taken by, for example, about 42 fishers using 1500 m of gill net, and fishing for an average of 100 days each year. Based on the correlation coefficients of the linear regressions, the Fox model provides a better fit to the barramundi data than the Schaefer model and suggests a lower MSY (730 409 kg) and a slightly higher f_{MSY} (70 783 hmnd). The result of reducing fishing on yield is discussed in Box 5.2 'The effects of reducing fishing effort'.

Equilibrium surplus production models have been used widely in managing fisheries, largely because they are based only on catch and effort data, which are relatively easy to collect. The main disadvantage is in the assumption of an 'equilibrium state' which fails to allow for the fact that the age structure of the stock, and therefore the biological parameters, change with different levels of exploitation. The equilibrium assumption is that a stock reaches a stable balance with fishing pressure and that numbers removed from a stock are replaced by additions. The models ignore the biological processes (growth, recruitment, and mortality) that affect

Box 5.2
The effects of reducing fishing effort

In the barramundi, *Lates calcarifer*, fishery, fishing effort increased dramatically between 1972 and 1979 (Table 5.1). Fisheries managers, recognizing that the barramundi was overfished, took measures including restricting the length of gill net used by each fisher, to reduce fishing effort. In a fishery which has been overexploited, and in which fishing effort has been subsequently reduced over a short time period, equilibrium conditions (already a dangerous assumption) can no longer be assumed; that is, in theory, the new lower rate of fishing needs to be maintained long enough for a new equilibrium to be reached. Fig. B5.2.1 shows lines connecting annual catches for each year. The positions of the data points from 1980 onwards suggests that, following the reduction in fishing

effort, yield remained low, and approached the original yield curve from below as shown by the arrows in Fig. B5.2.1. Time is required for stocks to recover biologically from the effects of overfishing and, in theory, for a new equilibrium to be reached.

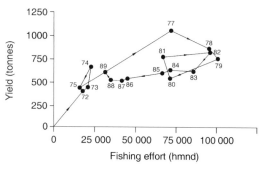

Fig. B5.2.1 Annual yield of barramundi from 1972 to 1989.

stock biomass. In addition, the predictions of the models are linked to the assumed shape of the biomass growth curve – in the case of Schaefer's model, for example, yield is always obtained at half the virgin stock biomass.

Even assuming that CPUE data are accurate and have not been affected by changes in fishing efficiency, surplus production models may produce spurious results. The view that the annual catch is equal to the surplus production at different levels of fishing effort is optimistic and the decreasing catch rates that occur with increasing fishing effort, common in many fisheries, may indicate severely depleted fish stocks.

In production models, the net growth rate is always positive (the curve in Fig. 5.2b is always above the x-axis). This reaction of the stock, responding in a stabilizing way by increasing growth if the biomass declines, referred to as 'pure compensation', cannot be true for all populations at all levels of fishing effort. Many marine species require a minimum density of individuals to ensure that reproduction is successful (see Chapter 3, Box 3.5 'The Allee effect'). Most marine species are broadcast spawners and many that are widely spaced will become increasingly separated at high levels of exploitation. Fig. 5.4 shows a surplus production curve which incorporates one type of depensatory effect in which a widely-spaced species with a low reproductive rate becomes unstable at low stock levels. For such vulnerable stocks there is a particular critical level of biomass (B_x) below which the stock growth rate becomes negative and extinction becomes inevitable.

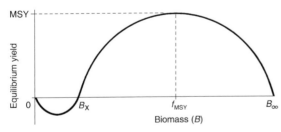

Fig. 5.4 Surplus production curve showing 'critical depensation' in which the stock biomass, once reduced below a critical level B_x cannot recover and extinction is inevitable.

In spite of bad press, surplus production models remain one of the best ways of demonstrating and emphasizing that natural resources are not inexhaustible and that for every fishery there is, on average, some level of fishing effort that should not be exceeded. The models are also useful in examining the costs and returns associated with fisheries; this introduces another management target, the maximum economic yield, which is generally achieved at lower levels of fishing effort and therefore a less risky target than MSY. Maximum economic yield is discussed as a management target in Chapter 6.

5.2.2 Non-equilibrium models

The surplus yield models presented in Section 5.2.1 assume equilibrium conditions and the models have been discredited for producing spurious estimates of maximum sustainable yield and the fishing effort required to take it. Methods of model fitting that do not rely on stock equilibrium assumptions include process-error and observation-error methods (Hilborn & Walters, 1992; Quinn & Deriso, 1999). Process-error methods involve using catch and effort data (assumed to have been measured without error) and that all error is associated with the process (the relationship between stock growth and stock size). Multiple linear regression is then used to estimate the parameters of the Schaefer curve.

The observation-error or time-series fitting method (Pella & Tomlinson, 1969) assumes that all errors are in the observations, in the CPUE values that are used as an index of stock size. Initial data are used to make a preliminary estimate of stock size and trial values of the parameters of the Schaefer model are used to predict catch rates over the period in which data are available. A comparison of observed and predicted catch rates then allows the parameter values to be adjusted to minimize differences between them.

In an assessment of the performance of the three models (equilibrium, process-error and observation-error models) on different fish stocks, Polacheck et al. (1993) found that the observation-error method performed the best and made strong recommendations against the use of equilibrium methods. The following simple example of time-series fitting uses the previously presented barramundi data (Table 5.1) to estimate the usual parameters of the Schaefer model: r, q and B_∞. As in Equation 5.1 in the previous section, next year's biomass B_{t+1} will equal the present year's biomass (B_t) plus the surplus production minus the catch (C_t). As catch is equal to qfB where f is fishing effort:

$$B_{t+1} = B_t + rB_t(1 - B_t/B_\infty) - qf_tB_t \qquad (5.11)$$

If estimates of r, q and B_∞ are available, all that is needed is an initial or starting stock biomass, B_1, and values of fishing effort to enable the calculation of the biomass, catch and CPUE for all subsequent years. As there are already three parameters to estimate, it is best to independently estimate the fourth parameter, B_1. Hilborn and Walters (1992) have suggested using $B_1 = C/fq$ based on an average of the first few data points. In the case of the present data, fishing effort at the beginning of the data series was low enough to regard the biomass as close to its unexploited level; that is $B_1 \approx B_\infty$.

A model based on non-linear curve fitting is described in Appendix A4.9 'Non-equilibrium surplus yield'. Fig. 5.5 is generated by the model described in the appendix which provides least squares estimates of $B_\infty = 4126$, $r = 0.50318$ and $q = 0.0000057$. A non-equilibrium surplus yield curve estimated as $qB_\infty f - (q^2B/r)f^2$ in the model is shown on the upper graph in Fig. 5.5. Note that the MSY is considerably lower, and obtained at a lower fishing effort, than was estimated by the equilibrium Schaefer model.

5.2.3 Multispecies applications

In many fisheries, several species are caught in the same fishing operation. This may be intentional in cases where several valuable species are caught or may be unintentional where unwanted species are caught as bycatch. Some fishing gears, such as trawl nets, are notorious for being unselective, and commonly catch a large number of different

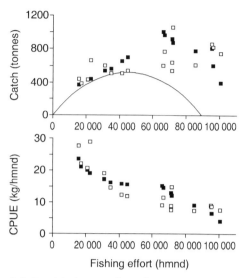

Fig. 5.5 Graphical output from the non-equilibrium surplus production model described in Appendix A4.9. Graphs show observed data (open squares) and predicted data (solid squares) for catch and CPUE. A surplus yield curve is included in the upper graph.

species. Trawl nets used on the soft bottom continental shelves in the tropics, where there is a large diversity of species, catch many different species, very few of which are discarded. Catches of mixed species are the norm in the tropics, where small numbers of a large variety of marine species can be found in local markets.

As all species in the fished area are interlinked by biological interactions including predator–prey relationships and competition for food and space, it is likely that the removal of even some of these species will have far-reaching effects on the ecosystem. These biological interactions are discussed in Section 3.4 and the more direct effects on stocks of the target species will be considered here.

Although equilibrium yield models have been applied to fisheries in which several different species are caught by the fishing operation, problems are associated with the different biological characteristics of the species that make up the catch. The component species are likely to have different levels of abundance and susceptibility to the fishing gear and the effort required to maximize yield in the total fishery of mixed species

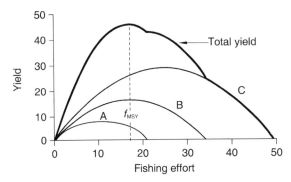

Fig. 5.6 Yields from three species, A, B and C (light curves), and total yield from the mixed stock (heavy curve). The vertical broken line shows f_{MSY} for the total yield of all species.

will overexploit the less productive species. Fig. 5.6 shows Schaefer yield curves for three separate species taken in a fishery and a total yield curve summing the yield from all three species. At the level of effort required to maximize yield from the fishery (f_{MSY} = 17 vessels approximately), species C would be underexploited and catches would be lower than they could be. More seriously, species A would be overexploited, and soon disappear from the fleet's catches. The right hand side of the total yield curve decreases in a step-wise or cascading manner, and an attempt to fit a single curve to the combined data is an approximation at best.

In tropical fisheries, however, where catches are made up of many different species, treating the component species as a single stock and managing it by applying the fishing effort to maximize economic yield may be the only practical option. The fishing effort required to maximize economic yield is lower than that required to maximize MSY, and may offer some degree of protection to species with lower productivity (see Section 6.2). However, although single-species production models suggest that overfishing effects on a single species can be reversed by reducing fishing effort, multispecies models suggest no such protection to component species with lower productivity. That is, even at low levels of fishing effort (say 12 fishing units), species A in Fig. 5.6 would still be at risk of overfishing and eventual local extinction.

(Also see Box 5.4 'Multispecies and multigear fisheries'.)

Surplus production models are usually based on a time series of catch and effort data. However, they can also be applied to data collected from different areas subjected to varying degrees of fishing pressure. An example of such an analysis is provided in Box 5.3 'An area-based, multispecies production model'. In subsistence fisheries, the amount of seafood caught by different coastal communities will be related to the area available for fishing and the number of fishers. Estimates of annual catch by area fished and the numbers of people fishing per km^2 can be used to construct a yield curve that suggests not only the average sustainable catch, but the communities where fisheries resources are presently under pressure. Although the area-based method provides a 'snapshot' in time it does not negate the assumptions of equilibrium and it depends heavily on the different fishing areas having similar ecological characteristics and productivity.

5.2.4 Potential yield – rough estimators

The ideal time to start managing a fishery is at its very beginning. Unfortunately, this has rarely been the case and management actions on most fisheries have been taken when overexploitation is evident. In order to install management at the start of exploitation, it is useful to have some rough idea of the size of resource – that is, whether the potential fishery will support 10, 100 or 200 fishing vessels. There have been several attempts to produce rough estimators to estimate the order of magnitude of yield (e.g. Gulland, 1983; Garcia et al., 1987).

Although more detailed stock assessments can be completed later, future analyses will benefit from research done at this early stage of development. Estimates of natural mortality, for example, are more easily obtained by analyses on stocks in which the age structure is not yet affected by fishing – larger age classes will rapidly disappear from stocks as exploitation increases (see Chapter 4). At this early stage, it is also easier

Box 5.3

An area-based, multispecies production model

A study in American Samoa (Wass, 1982) identified eleven discrete reef areas which were exclusively fished over long periods by people from adjacent villages. Data from this study (Table B5.3.1) were analysed by Munro (1984), and a Fox surplus yield curve of catch (kg per hectare per year) against fishing intensity (people per hectare of reef) is shown in Fig. B5.3.1.

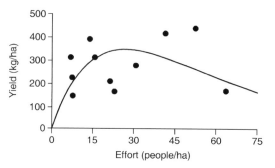

Fig. B5.3.1 A Fox surplus yield curve of catches (kg per hectare per year) against fishing effort (people per hectare of reef) for 11 villages in American Samoa.

Table B5.3.1 Catch (fish and invertebrates combined) from American Samoan villages. Wass (1982) adapted from Munro (1984).

Village	Population	Reef area (ha)	EFFORT People per hectare	YIELD Catch (kg/ha)	CPUE Catch (kg/person/yr)
Faganeanea	191	29.4	6.5	311	47.8
Matu'u	315	44.0	7.2	147	20.4
Faga'ulu	757	36.8	20.6	209	10.2
Utulei	991	19.1	51.9	440	8.5
Leloaloa	789	35.2	22.4	166	7.4
Aua	1471	48.6	30.3	281	9.3
Lauli'i	607	40.0	15.2	315	20.7
Fagaitua	429	31.6	13.6	394	30.0
Masefau	315	44.9	7.0	227	32.4
Fagasa	656	16.0	41.0	418	10.2
Vaitogi	661	10.4	63.6	165	2.6

for managers to provide some sort of insurance against the initial estimates being wrong. It is easier, for example, to close off some areas to fishing to protect against the possible future depletion of spawning stocks. Such closed areas are much harder to impose when the fishery is fully exploited, when catch rates are low and fishers are reluctant to give up any existing fishing areas. These and other aspects are included in Chapter 6 on fisheries management.

Gulland's estimator is based on the assumptions that yield is related to fishing effort by the Schaefer surplus yield model, and that fishing mortality at maximum sustainable yield (MSY) is

approximately equal to natural mortality. If the first of these assumptions is reasonable, the original stock biomass, or virgin biomass (B_∞), will be reduced to half its value at the time the fishery is at its maximum sustainable yield. That is, $B_{MSY} = 0.5B_\infty$. The catch or yield from a stock is $C = (qfB)$, or (FB), therefore:

$$MSY = FB_{MSY} \qquad (5.12)$$

Assuming that fishing mortality at MSY is approximately equal to natural mortality:

$$MSY = 0.5MB_\infty \qquad (5.13)$$

Although Gulland's estimator (Equation 5.13) is

believed to overestimate MSY (Beddington & Cooke, 1983), it may be useful in providing an upper limit to the expected yield from a potential resource. A more general version of Gulland's estimator is:

$$MSY = XMB_\infty \qquad (5.14)$$

where X is a scaling factor which allows for different ratios between B_{MSY} and B_∞ in different species groups. A factor of 0.3 or 0.4 is often used if only because it provides a more conservative value of MSY. Gulland's estimator cannot be used for exploited stocks, in which case MSY may be estimated using the total mortality rate, Z, and the average exploited biomass B_{mean}:

$$MSY = 0.5ZB_{mean} \qquad (5.15)$$

The use of this equation, known as Cadima's estimator, is only reasonable if fishing mortality is close to the value at MSY, and will provide biased estimates at all other levels of exploitation. For exploited stocks, Garcia et al. (1987, 1989) suggested alternative estimators, which are similarly based on surplus yield models, but are unbiased at all levels of fishing effort. In the Fox model (Equation 5.9), catch per unit effort (and therefore biomass, B) reduces exponentially as described by the linear logarithmic equation:

$$\ln[C/F] = \ln[B] = a + bF \qquad (5.16)$$

The Fox yield model is maximized at $F_{MSY} = -1/b$, and:

$$MSY = -(1/b)\exp[a-1] \qquad (5.17)$$

If F_{MSY} is approximately equal to M, then $b = -1/M$, and as $B_{mean} = Y/F$, Equation 5.16 becomes:

$$\ln[Y/F] = \ln[B_{mean}] = a - (1/M)F = a - (1/M)$$
$$(Y/B_{mean}) = a - Y/(MB_{mean})$$

From this, $a = \ln[B_{mean}] + Y/(MB_{mean})$, which may be substituted in Equation 5.17, to give the unbiased Fox estimator:

$$MSY = MB_{mean}\exp[Y/(MB_{mean}) - 1] \qquad (5.18)$$

Thus a rough estimate of MSY may be obtained for a stock as long as the average exploited biomass

(B_{mean}), natural mortality (M), and annual catch (Y) are known. For an unexploited stock (where $Y = 0$), the Fox estimator becomes MSY = $MB_{mean}\exp[-1]$, or:

$$MSY = 0.37MB_{mean} \qquad (5.19)$$

The main advantage of the above rough estimators of sustainable yield is that their use requires only estimates of mortality and stock size. The latter is often difficult to estimate, and may fluctuate from year to year. All three of the rough estimators presented above suffer the same deficiencies as surplus yield models in general. The sustainable yield from a fish stock depends on complex interactions between many biological and environmental parameters, and any attempt at its estimation from just a few parameters, which in any case may be variable, is hazardous. In some species with very low productivity, the sustainable yield is likely to be a much smaller proportion of the total biomass than is suggested by the application of the above models. In stocks of long-lived species in cold or deep water, the sustainable yield may be only a few percent of the virgin biomass. Nevertheless, such estimators may provide an indication of the magnitude of a potential resource, or an un-assessed exploited resource. Knowing whether a potential fisheries resource is able to support either 10, 100, or 200 fishing vessels, for example, is likely to be of great importance in planning research, resource development and management.

5.3 Including growth and mortality

The surplus production models presented in Section 5.2 ignore the biological processes that affect stock biomass. In the present section a selection of models include the opposing processes of growth and mortality. The simplest approach is to examine the effects of growth and mortality on the biomass of a single cohort or year class; a fate of a single cohort of cod is used as an example.

A more practical example uses data for a penaeid prawn, one of the few species in which a

Box 5.4

Multispecies and multigear fisheries

In multispecies fisheries, problems occur when fishing effort is recorded as overall or aggregate effort, and fishers are able to target different component species in the stock. Consider a trawl fishery, for example, where two species, A and B, are taken in the catch, and fishing effort in the fishery is recorded as total hours trawled. Although species A may be the preferred target, if the market price of this species decreases, fishers may switch to target species B. If the hours trawled in the fishery remains the same, the CPUE of species A will decrease dramatically even though its abundance does not. In this case, an uncritical examination of the collected CPUE data will give the impression of imminent disaster for species A, when in fact species B, in which CPUEs are increasing, may be the species under threat. One

of the ways of addressing 'target switching', and ensuring that CPUE is a reasonable index of abundance in multispecies fisheries, is to ensure that fishers record effort directed on the different component species separately (see the fishing log in Chapter 4, Fig. B4.3.1).

Some fisheries, particularly those in tropical regions, are characterized by the use of several different kinds of fishing gear. In theory, it is possible to standardize effort imposed by the different fishing gear. If the gears are used in the same area at any particular time, the mean catch rate of each gear may be used to compare the relative efficiencies of each type of gear. After standardization, it is then possible to combine the CPUE data from different gears. If the gears are used on different parts of the stock, however, standardization is not possible, and each part of the stock should be assessed separately.

fishery may be based entirely on a single large year class. The objective of this exercise is to calculate the time after recruitment when the biomass of a single year class will reach a maximum and to introduce the concept of growth overfishing – a level of fishing in which many small individuals are caught before they grow to a size at which the stock biomass is maximized. Other models include additions to the stock through recruitment, including delay-difference models that allow for a time delay between spawning and recruitment.

5.3.1 The effects of growth and mortality on biomass

For the purposes of examining the effects of growth and mortality on stock biomass, the following example is based on following the fate of just a single age class, or cohort. The stock parameters, based on the cod, *Gadus morhua*, from various sources, including Thurow (1971), are growth, $K = 0.2$ yr^{-1}, $L_\infty = 100$ cm, and the instantaneous rate of natural mortality, $M =$

0.44 yr^{-1}; weight (g) is related to length (cm) by the equation $W = 0.0082L^3$.

The calculations can be set up in tabular form (preferably on a computer spreadsheet) with age in the left-hand column (Table 5.2). In column 2

Table 5.2 Length, weight, relative numbers, and biomass of a single cohort by age, based on the cod, *Gadus morhua*.

1 Age (years)	2 Length (cm)	3 Weight (g)	4 Relative numbers (N)	5 Biomass (g)
2	32.97	294	1000	294
3	45.12	753	644	485
4	55.07	1369	415	568
5	63.21	2071	267	553
6	69.88	2798	172	481
7	75.34	3507	111	389
8	79.81	4169	71	296
9	83.47	4769	46	219
10	86.47	5302	30	159
11	88.92	5765	19	110
12	90.93	6165	12	74

the mean length of individuals can be calculated by using the growth equation (Equation 4.22) and converted to weight (in column 3) by using the length–weight relationship. Assuming that natural mortality rate, M, is constant after 2 years of age, the numbers surviving in the cohort (column 4) may be calculated from the exponential decay equation as $N_{t+1} = N_t \exp[-M]$ where N_t is the number present at the beginning of one year, and N_{t+1} is the number at the beginning of the following year.

As the number of cod recruited is unknown, column 4 lists relative rather than absolute numbers surviving, beginning with an arbitrarily chosen initial number (a recruitment of 1000 young cod at the beginning of the second year is used in this example). The relative biomass, the total relative weight of the cohort (column 5), is calculated by multiplying the individual weight (column 3) by the numbers surviving (column 4). Table 5.2 and the curves in Fig. 5.7 suggest that the weight of each individual is increasing through growth and the number of individuals in the cohort is decreasing through mortality. The total weight or biomass of the cohort therefore increases over time to reach a maximum, the 'critical' point of Ricker (1975) at about four years of age before decreasing as the growth curve approaches an asymptote (Fig. 5.7).

In the case of the cod, the stock is made up of several year classes, all of which contribute to the total stock biomass and analysing a single cohort is of little practical value (although it does suggests a size at which the species should be caught to maximize weight). A model that accounts for the biomass contributions of all age classes is presented later in Section 5.4. As the model now stands, it may be useful in cases in which fishing is based on a single strong year class of fish or on a species with a very short life cycle, such as penaeid prawns and some clupeids. In prawn farming, in particular, where a pond can be harvested in a short time, the biomass curve may be used to suggest the harvesting time that would maximize the resulting weight.

The following example of a biomass model is based on stock parameters for the banana prawn, *Penaeus merguiensis*, given in Chapter 4 on growth. The relevant growth parameters (converted to monthly values) are $K = 0.15 \text{ m}^{-1}$, $L_\infty = 45 \text{ mm}$ carapace length, and $t_0 = -0.44$ months. For the purpose of the example, the instantaneous rate of natural mortality is $M = 0.3 \text{ m}^{-1}$, and weight (g) is related to length (mm) by the equation $W = 0.0008L^3$. Assuming that the prawns are 4 months of age at recruitment, the mean length, individual weight, numbers surviving, and the relative biomass can be calculated,

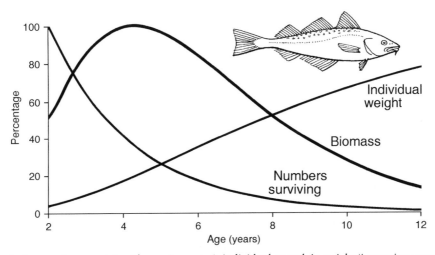

Fig. 5.7 The relative numbers surviving (decreasing curve), individual growth in weight (increasing curve), and cohort biomass (heavy curve), expressed as percentages of maximum values against age for the cod, *Gadus morhua*.

Table 5.3 Biomass and biovalue model for an unexploited fishery, based on the banana prawn, *Penaeus merguiensis*. Unpublished data from Fisheries Research Centre, Bandar Abbas, Iran.

1 Age (m)	2 Length (mm)	3 Weight (g)	4 Relative numbers (N)	5 Relative biomass (kg)	6 Price (US$/kg)	7 Relative biovalue (US$)
4	21.88	8.38	1000.0	8.38	3.00	25.14
5	25.10	12.65	740.8	9.37	3.30	30.93
6	27.87	17.32	548.8	9.51	3.63	34.51
7	30.26	22.16	406.6	9.01	3.99	35.98
8	32.31	26.99	301.2	8.13	4.39	35.70
9	34.08	31.66	223.1	7.07	4.83	34.14
10	35.60	36.10	165.3	5.97	5.31	31.71
11	36.91	40.23	122.5	4.93	5.85	28.80
12	38.04	44.02	90.7	3.99	6.43	25.68

as in the previous example, for each successive month (Table 5.3).

In some species, premium prices are paid for individuals in a particular size range; for example, large prawns usually have a higher market value per kg than small prawns, and in some markets, small 'plate-size' fish and lobsters have a higher value than large individuals. If price per kg varies with size, a curve of the relative biovalue of the cohort would reach a maximum at a different age from that of the biomass curve. For the purpose of the example, prawns are assumed to have a unit value of US$3.00 per kg at the month of recruitment, and increase by 10% in each succeeding month (column 6 of Table 5.3). The relative biovalue of the cohort by month (column 7), is calculated by multiplying the biomass (column 5) by the mean price per kg (column 6).

Table 5.3 and the curves in Fig. 5.8 suggest that the total weight of prawns increases to reach a maximum about 2 months after recruitment (at an age between 5 and 6 months), but the total biovalue of prawns reaches a maximum after about 3 to 4 months (at an age of approximately 7 months). Information such as this can be used to suggest the increase in yield (catch weight) or biovalue (catch value) that would result from protecting or restricting the catch of young individuals (see Chapter 6 on fisheries regulations). Catching the majority of the stock before individ-

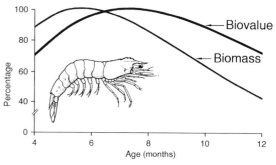

Fig. 5.8 Relative biomass (light curve) and relative biovalue (heavy curve) expressed as percentages of the maximum values against age in months for the banana prawn, *Penaeus merguiensis*, in Iran.

uals grow to an optimum marketing size is referred to as growth overfishing.

The model, however, does not provide information on the timing and intensity of fishing effort required to maximize yield – that is, how many months before the maximum biomass occurs should a fishing season be opened to maximize the catch, and how much fishing effort should be applied. The problem of timing of available fishing effort does not arise in aquaculture operations, where, for example, all prawns could be harvested from ponds over a short time period (equivalent to imposing an instantaneous and massive fishing mortality) to secure the

Table 5.4 Biomass and biovalue model including fishing mortality based on the banana prawn, *Penaeus merguiensis*, data used in Table 5.3.

A	B	C	D	E	F	G	H	I	J	K
Age	M	F	Length (mm)	Weight (g)	Cohort numbers	Numbers dying	Catch numbers	Catch weight (kg)	Unit price (US$/kg)	Catch value
4	0.30	0.30	21.88	8.38	1000.00	451.19	225.59	1.89	3.00	5.67
5	0.30	0.30	25.10	12.65	548.81	247.62	123.81	1.57	3.30	5.17
6	0.30	0.30	27.87	17.32	301.19	135.90	67.95	1.18	3.63	4.27
7	0.30	0.30	30.26	22.16	165.30	74.58	37.29	0.83	3.99	3.30
8	0.30	0.30	32.31	26.99	90.72	40.93	20.47	0.55	4.39	2.43
9	0.30	0.30	34.08	31.66	49.79	22.46	11.23	0.36	4.83	1.72
10	0.30	0.30	35.60	36.10	27.32	12.33	6.16	0.22	5.31	1.18
11	0.30	0.30	36.91	40.23	15.00	6.77	3.38	0.14	5.85	0.80
12	0.30	0.30	38.04	44.02	8.23	3.71	1.86	0.08	6.43	0.53
			Numbers remaining =		4.52		Total weight =	6.81	Total value =	25.06

maximum biovalue. The following yield models are more instructive in wild fisheries.

5.3.2 The effects of fishing mortality on a single cohort

The biomass model may be extended to include exploitation, as in the spreadsheet model shown in Table 5.4 for the Iranian prawn example. A column containing values of fishing mortality, *F*, has been added (a constant value of $F = 0.3$ m^{-1} in this example). However, if fishing mortality varied with age, individual values of *F* could have been added for different ages in the column.

Cohort numbers (column F) are calculated as in the previous tables, but with *F* included: i.e. $N_{t+1} = N_t \exp[-(M_t + F_t)]$. Note that *M* and *F* are the values from the previous month, *t*, not *t* + 1. The numbers dying from both natural and fishing mortality (column G) are not required for the calculations, but are added for interest; they may be estimated as equal to $N_t(1 - \exp[-(M_t + F_t)])$ or simply as $N_t - N_{t+1}$. Catch numbers (column H) are estimated from the catch equation as $C_t = (F/Z)N_t(1 - \exp[-(M_t + F_t)])$, or by (F_t/Z_t) multiplied by the numbers dying (column G). Values in the remaining columns are calculated in the same way as in Table 5.4. The values at the bottom of the table are the relative numbers remaining at the end of the fishing season, the total relative catch weight for the season, and the total relative catch value for the season respectively.

A computer spreadsheet program is described in Appendix A4.10 'Biomass and biovalue model including fishing mortality – prawn example'. This model can be run repeatedly for different levels of exploitation or other scenarios. If effort was increased by 50%, for example, fishing mortality would be increased from $F = 0.30$ m^{-1} to $F = 0.45$ m^{-1} (as long as catchability, *q*, remained the same) and the program can be re-run to compare results with the previous run. Although the values obtained are relative (in units per 1000 recruits in this case), they may be discussed in terms of percentage change as in Table 5.5.

Results in Table 5.5 suggest that a 50% increase in fishing effort would decrease the number of prawns remaining at the end of the season by 74% and increase the landed catch value for the season by only 3.3%. Not many fisheries managers or fishers would be happy with such a marginal increase in value, particularly if the cost of the additional fishing effort was taken into account. In fact, if financial information were available, fishing costs and revenue could be included in the model. The reduction in the

Table 5.5 Runs of the computer spreadsheet model (Appendix 4.10) for two different scenarios. Scenario 1 includes a fishing mortality of $F = 0.30$ m^{-1} and scenario 2, a fishing mortality of $F = 0.45$ m^{-1}. The results of changing from scenario 1 to 2 are shown at the right.

	Scenario 1 ($F = 0.30$ m^{-1})	Scenario 2 ($F = 0.45$ m^{-1})	
Relative numbers remaining =	4.52	1.17	(decrease of 74%)
Relative total catch =	6.81	7.40	(increase of 8.5%)
Relative total value =	25.06	25.91	(increase of 3.3%)

number of prawns remaining at the end of the season would be of concern in a species where these were the sexually mature individuals that provide the following season's recruitment. It is also possible to use the model in Table 5.4 to examine the benefits or otherwise of imposing closures on the fishery, and delaying the opening of the fishing season – that is, by inserting $F = 0$ in the initial months.

The above model is based on the relative yield from a fishery in which a single year class is targeted. It can be expanded to include the contribution of several age classes to the year's catch and this is the basis of the Thompson and Bell model presented in Section 5.4.

Delay-difference models are sometimes called Deriso/Schnute models in reference to those involved in their development (Deriso, 1980; Schnute, 1985) although their first use was by Allen (1963) and Clarke (1976). The models incorporate natural mortality, growth and recruitment and allow for a time delay between spawning and recruitment. The models are based on difference rather than differential equations so that time changes in a series of steps rather than continuously. They implicitly account for the age structure of the stock without having to know it. A summary of the development of the models is given in Quinn and Deriso (1999).

5.4 Including different age classes; age-structured models

Age-structured, or dynamic pool models consider the effects of growth, mortality, and recruitment on a stock composed of several different cohorts or year classes. Sections 5.4.1 and 5.4.2 include the classical assessments referred to as virtual population analysis and the yield per recruit analysis. Section 5.4.3 discusses a length-based Thompson and Bell model and this is the predictive equivalent of length-based cohort analysis that gives a simpler description of non-steady state situations.

5.4.1 Virtual population analysis

Virtual population analysis (VPA) and cohort analysis use the numbers of fish caught during commercial fishing operations to estimate historic fishing mortality and stock numbers in a cohort of fish. The catch curves discussed previously were constructed from the decrease in numbers in successive age groups (pseudo-cohorts) in the same length–frequency distribution (i.e. $N_{1.1}$, $N_{1.2}$ and $N_{1.3}$ in Fig. 4.16 in Chapter 4).

The numbers of individuals in a single age group, or cohort, can also be followed over successive time intervals (i.e. $N_{1.1}$, $N_{2.2}$ and $N_{3.3}$ as shown in Fig. 4.16, Chapter 4). This avoids the problems associated with variable recruitment, and is the basis of virtual population methods. From Equation 4.50, the number of fish surviving from one year (N_t) to the next year (N_{t+1}) is given by:

$$N_{t+1} = N_t \exp[-(F + M)] \qquad (5.20)$$

The number of fish dying is therefore, $N_t(1 - \exp[-Z])$. The catch (C_t) is the proportion dying due to fishing, and may be estimated from the catch equation:

$$C_t = [F_t/Z]N_t(1 - \exp[-(F + M)]) \qquad (5.21)$$

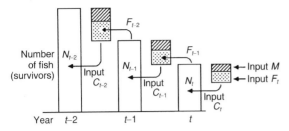

Fig. 5.9 Virtual population analysis (VPA): open bars represent the number of survivors (*N*), line-shaded bars represent natural mortality (*M*), and stippled bars represent fishing mortality (*F*).

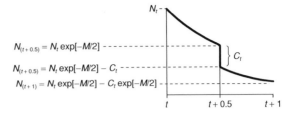

Fig. 5.10 The step-wise reduction in numbers which is the basis of cohort analysis. Numbers present over time are estimated by the formulae at the left of the diagram.

Rearranging Equation 5.21 allows the estimation of the number of fish, N_t, and dividing Equation 5.20 by Equation 5.21 produces Gulland's (1965) equation:

$$N_{t+1}/C_t = (F_t + M) \exp[-(F_t + M)]/F_t(1 - \exp[-(F_t + M)])$$ (5.22)

This equation cannot be solved for *F* algebraically, but may be solved by using iterative methods on a computer. VPA involves working backwards from the most recent year (*t* + 1). Inputs required are *M*, an initial value of *F*, and the catch, (*C*) in numbers for each year (Fig. 5.9).

Cohort analysis (Pope, 1972) is a procedure which provides an approximation to VPA results, and is one that can be completed on either a calculator, or preferably, a computer spreadsheet program. In cohort analysis the exponential decrease in fish numbers is replaced by a step-wise function (Fig. 5.10), in which the catch from an age group is assumed to be taken instantaneously in the middle of the year. If N_t fish are present at the start of the time period *t* (one year), and M is the natural mortality per year, there will be $N_t \exp[-M/2]$ fish halfway through the year, just before the catch is 'removed'. After the catch (C_t) is taken, there will be $N_t \exp[-M/2] - C_t$ fish left. The number of fish remaining at the end of the year (N_{t+1}) will be ($N_t \exp[-M/2] - C_t$) $\exp[-M/2]$; this may be rearranged to give:

$$N_t = (N_{t+1} \exp[M/2] + C_t) \exp[M/2]$$ (5.23)

Survival over the year N_{t+1}/N_t is $\exp[-M - F]$, and therefore:

$$F_t = -\ln[N_{t+1}/N_t] - M$$ (5.24)

The three steps in using the equations for both VPA and cohort analysis involve working backwards from the most recent year (year *t* in Fig. 5.9).

1 The catch (C_t), an initial value for fishing mortality (F_t), and the known value for natural mortality, *M*, are substituted in Equation 5.21 to estimate the number in the oldest group N_t as:

$$N_t = C_t / [F_t/Z] N_t (1 - \exp[-(F + M)])$$

Using virtual population analysis, steps two and three become:
2 Equation 5.22 is used to estimate F_{t-1} by iteration
3 Equation 5.20 is used to estimate N_{t-1}

Alternatively, using cohort analysis, steps two and three become:
2 Equation 5.23 is used to estimate N_t
3 Equation 5.24 is used to estimate F_t

A numerical example is provided in Box 5.5 'Cohort analysis example'.

Virtual population analysis and cohort analysis may provide reasonable estimates of fishing mortality on young, or the least recent, cohorts, as long as the stock is heavily fished (i.e. if *F* is greater than *M*). The resulting estimates of *F* for the younger groups are relatively insensitive to the choice of the initial value of *F*, but it is usually the older groups for which estimates are most urgently required for fisheries management.

5.4.2 The classical yield per recruit model

Yield per recruit models examine the trade-off between capturing a large number of smaller fish

---

Box 5.5

Cohort analysis example

An example of cohort analysis (from Jones, 1981) is shown in Table B5.5.1 and the calculations for ages 3 and 4 are as follows (beginning from age 4):

Step 1:

$N_4 = C_4 / [F/Z](1 - \exp[-Z])$

$= 250/([0.4/0.6] (1 - \exp[-0.6])) = 831$

Step 2:

$N_3 = (N_4 \exp[M/2] + C_3) \exp[M/2]$

$= (831 \exp[0.2/2] + 300) \exp[0.2/2] = 1347$

Step 3:

$F_3 = -\ln[N_4/N_3] - M$

$= -\ln[831/1347] - 0.2 = 0.283$

Steps 2 and 3 are continued to obtain estimates for N and F for previous years before starting on a new cohort.

Table B5.5.1 Cohort analysis based on the numbers of fish caught in a single cohort of fish over four years. Input parameters are $M = 0.2$, and $F_4 = 0.4$.

Age (t)	Numbers caught (C_t)	Numbers in each age (N_t)	$S_t = \exp(-Z)$	Z_t	F_t
1	50	2334	0.80	0.22	0.02
2	200	1866	0.72	0.33	0.13
3	300	1347	0.62	0.48	0.28
4	250	831			0.40

---

early in their lifespan and capturing a smaller number of larger fish later in their lifespan. Growth overfishing is said to occur when many small individuals are caught before they grow to a size at which the stock biomass is maximized. The classical Beverton and Holt (1957) yield per recruit model has been commonly used to estimate the level of fishing and the mean age of capture required to maximize yield per recruit. In the model the yields estimated are relative – that is, they are related to the number of recruits and this is usually unknown. Yield is usually expressed as grams per recruit – a yield of 3 g per recruit means a yield of 3 grams for each individual recruited, or 3 tonnes for each million recruits.

The model considers the effects of age at first capture, mortality and growth on yield (Fig. 5.11). It assumes a steady-state situation in which the age structure of the stock is the same from year to year and natural and fishing mortalities are constant from the mean age at first capture. Gear selectivity is modified to a 'knife-edge' selection curve, so it may be assumed that, after the

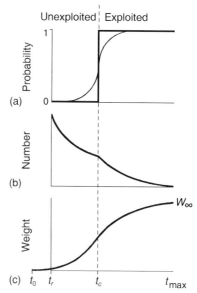

Fig. 5.11 Assumptions of the yield per recruit model are that: (a) the selection curve (light line) is modified to a knife edge curve (heavy line); (b) natural mortality, M, is constant after the age at recruitment, t_r, and fishing mortality, F, is constant after the age at first capture, t_c; and (c) weight is predicted by the von Bertalanffy equation.

mean age at first capture (t_c), all fish coming into contact with the fishing gear have an equal probability of capture (Fig. 5.11a). Natural mortality, M, is assumed to be constant after the age at recruitment, t_r, and fishing mortality, F, is constant after the age at first capture, t_c, so that stock numbers, N, decrease in a two-stage survival curve (Fig. 5.11b). Weight, W, is predicted by the von Bertalanffy equation (Fig. 5.11c).

Under the steady-state assumption, the total yield in any one year from all age classes (all pseudo-cohorts) is the same as that from a single cohort over its whole lifespan. The yield (catch in weight) of a single year class of fish from age at first capture (t_c) to some maximum age (t_{max}) is an integral of the fishing mortality (F), the number of fish present (N) and their mean weight (W). In summation terms the yield is given as:

$$Y = \Sigma F_t N_t W_t \ldots \qquad (5.25)$$

where Σ indicates the summation from t_c to t_{max}. Several functions can be substituted for the components of Equation 5.25. The number of fish surviving at different ages can be written as follows:

- at the age at recruitment, t_r: $N_{tr} = R$
- at the age at first capture, t_c: $N_{tc} = R \exp[-M(t_c - t_r)]$

- at any subsequent age, t: $N_t = N_{tc} \exp[-(M + F)(t - t_c)]$

$$= R \exp[-M(t_c - t_r) - (M + F)(t - t_c)]$$

The fraction of recruitment, R, surviving until age t is therefore $N_t/R = \exp[-M(t_c - t_r) - (M + F)(t - t_c)]$

In the same way that $(1 - \exp[x])^3$ can be expanded to give $(1 - 3\exp[x] + 3\exp[2x] - \exp[3x])$, the von Bertalanffy growth equation in terms of weight can be expanded and substituted, with the previous expressions, in Equation 5.25. By integration, the expression for yield per recruit, simplified for use on calculators by FAO (Sparre et al., 1989) becomes:

$F \exp[-M(t_c - t_r)] W_\infty$ multiplied by a summation of the following four terms:

$$\{1 \exp[-0K(t_c - t_0)]/(M + F + 0K)\} +$$
$$\{-3 \exp[-1K(t_c - t_0)]/(M + F + 1K)\} +$$
$$\{3 \exp[-2K(t_c - t_0)]/(M + F + 2K)\} +$$
$$\{-1 \exp[-3K(t_c - t_0)]/(M + F + 3K)\} \qquad (5.26)$$

An example of using the yield per recruit equation is provided in Box 5.6.

Yield per recruit curves are commonly used to suggest the changes in yield which would result from altering fishing effort (and therefore fishing

---

Box 5.6

Yield per recruit example

The example below is based (loosely) on the spotted whiting, *Sillaginoides punctatus*, using parameters within the range of those estimated by Jones (1979). The natural mortality rate is $M = 0.5$, the age of recruitment is $t_r = 1$ year, and the mean age at capture $t_c = 1.5$ years. The von Bertalanffy growth parameters are $K = 0.3 \; y^{-1}$ and $W_\infty = 900$ g with $t_0 = 0$. The yield per recruit for a value of $F = 1$, for example, may be calculated by using Equation 5.26 broken up into five parts as below.

1) $FW_\infty \exp[-M(t_c - t_r)] = 1 \times \exp[-0.5 \times (1.5 - 1.0)]$
$= 700.921$

which is multiplied by a summation of the following four terms:

2) $+1 \exp[-0K(t_c - t_0)]/(M + F + 0K) = +1/(0.5 + 1.0) =$
$+0.667$

3) $-3 \exp[-1K(t_c - t_0)]/(M + F + 1K) = -3 \times \exp[-1 \times$
$0.3 \times 1.5]/(0.5 + 1.0 + 1 \times 0.3) = -1.063$

4) $+3 \exp[-2K(t_c - t_0)]/(M + F + 2K) = +3 \times \exp[-2 \times$
$0.3 \times 1.5]/(0.5 + 1.0 + 2 \times 0.3) = +0.581$

5) $-1 \exp[-3K(t_c - t_0)]/(M + F + 3K) = 1 \times \exp[-3 \times$
$0.3 \times 1.5]/(0.5 + 1.0 + 3 \times 0.3) = -0.108$

The yield per recruit is therefore equal to $700.921 \times$ $(0.667 - 1.063 + 0.581 - 0.108) = 53.97$ g per recruit for a value of $F = 1$. Repeating these calculations for a range of fishing mortalities, say from $F = 0$ to $F = 1.5$ in steps of 0.1, allows the construction of a yield per recruit curve. These tedious calculations can be done more readily on the computer spreadsheet described in Appendix A4.11 'Yield per recruit model'.

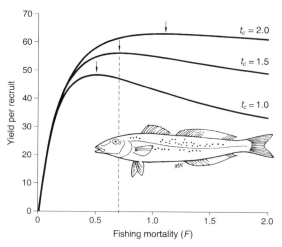

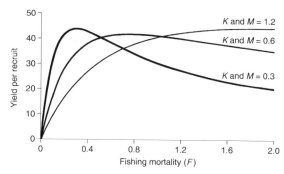

Fig. 5.13 Yield per recruit curves for different values of growth and mortality.

Fig. 5.12 Yield per recruit curves based on the spotted whiting, *Sillaginoides punctatus*, for ages at first capture of 1.0, 1.5 and 2.0 years. The vertical dashed line represents a present fishing mortality of $F = 0.7$ yr^{-1}. Arrows indicate maximum yields (Y_{max}) for each age at first capture.

mortality), and by delaying the age at first capture. The curves for the spotted whiting, *Sillaginoides punctatus*, shown in Fig. 5.12 illustrate this. If the fishing mortality at present is 0.7 yr^{-1}, and the age of first capture is 1 year, the curves suggest that yield could be increased in two ways. The lowest curve suggests that yield could be increased by reducing fishing mortality from 0.7 to 0.5 yr^{-1}. The middle curve suggests that if fishing mortality was kept at 0.7 yr^{-1}, delaying the age at first capture would also increase yield.

The main disadvantage of the classical model is its assumptions of a steady state, whereas, in a stock where exploitation is increasing, the numbers in the various age groups, and even the stock parameters are likely to be changing. The exclusion of considerations of recruitment and spawning stock size also means that the model provides no guide to the level of fishing at which the stock is affected by recruitment overfishing; this level of overfishing, more serious than growth overfishing, refers to the level at which recruitment is too low to replace the losses due to mortality.

From a practical viewpoint, the model works best when applied to species with low mortality

rates (say when natural mortality is less than 0.6) in which case yield per recruit curves reach maxima at relatively low levels of fishing mortality (Fig. 5.13). If mortality rates are high, as they often are in tropical species, a yield per recruit curve may not reach a maximum within a reasonable range of fishing mortality values. In these cases, the results of yield per recruit analyses may be quite misleading, often suggesting that an extremely high (or sometimes infinite!) fishing mortality is required to secure the maximum yield. At higher levels of fishing mortality, assumptions of constant recruitment may be violated and recruitment may decline in spite of marginal predicted increases in yield. Even if the mean age at first capture, t_c, is greater than the age at maturity, high fishing mortality may reduce the stock's spawning biomass to levels below that required to maintain recruitment. The critical spawning biomass is imprecisely known in most species, but may be between 20% and 50% of the unexploited levels.

In cases where a yield per recruit curve becomes broad and flat-topped, the choice of the fishing mortality that secures the maximum yield (F_{max}) is not only difficult but risky – the upper curve in Fig. 5.13 suggests that large increases in effort are required to gain a marginal increase in yield. At high levels of fishing mortality, recruitment may decline in spite of marginal predicted increases in yield. Because of the difficulty of estimating maximum yield and the risk of overexploitation associated with it, many fisheries scientists are recommending lower targets than

Box 5.7
Reference points based on yield per recruit

In fisheries management, a commonly used target or reference point is $F_{0.1}$ – the value of F at which the slope of the yield per recruit curve is 10% of its initial value. The calculation of $F_{0.1}$ is shown in Table B5.7.1 for the flat-topped curve in Fig. B5.7.1.

Table B5.7.1 Values of F, yield per recruit (YPR), and the slope of the yield curve. $F_{0.1}$ is marked by an asterisk.

F	YPR	Slope
0	0	—
0.1	0.015	0.150
0.2	0.0275	0.125
0.3	0.0364	0.089
0.4	0.0446	0.082
0.5	0.0494	0.048
0.6	0.0546	0.052
0.7	0.0584	0.038
0.8	0.0614	0.030
0.9	0.0636	0.022
1.0	0.0659	0.023
1.1*	0.0674	0.015
1.2	0.0688	0.014

The third column gives the difference in successive yield per recruit values divided by the difference in F as an approximation of the slope of the curve. The value of F at which the slope is 0.1 of the initial slope is $F_{0.1} = 1.1$. There is no biological reason for selecting $F_{0.1}$ as a reference point but, as this level of fishing mortality is lower than that required to obtain the maximum yield it is believed to provide greater profitability and a buffer against recruitment overfishing (Gulland & Boerema, 1973; Gulland 1983). Other reference points used in fisheries management are discussed in Chapter 6.4.3.

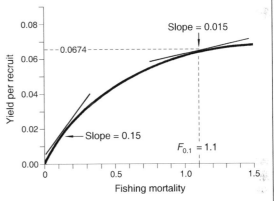

Fig. B5.7.1 A yield curve showing $F_{0.1}$, the point at which the slope of the curve is one-tenth of the value of the initial slope.

the maxima suggested by yield per recruit curves (see Box 5.7 'Reference points based on yield per recruit').

Developments of the model have increased its flexibility. These include the incorporation of different patterns of growth and mortality in the model. Growth can be allowed to vary seasonally or empirical values can be used. The use of cohort analysis allows not only the estimation of age-related mortality, but values of recruitment strength, which can be used in the model to give direct estimates of yield rather than yield per recruit. The following section describes a more direct model, which escapes the knife-edge selectivity assumption as well as the steady-state assumption of the yield per recruit model.

5.4.3 The Thompson and Bell model

The basis for the so-called Thompson and Bell model for a single cohort was presented in Section 5.3.2. In this present section the model is extended to include the contribution of several age classes. As an example, a small fishery in which the target stock is made up of several year classes is considered. Each year, 5 million fish are recruited into the stock at age one, and are vulnerable to the fishing gear at age two.

The growth parameters are $K = 0.25$ yr^{-1}, $L_{\infty} = 50$ cm, and $t_0 = 0$. The instantaneous rate of natural mortality is $M = 0.3$ m^{-1}, and weight (g) is related to length (cm) by the equation $W = 0.01L^3$. A worksheet can be arranged as in

Table 5.6 Yield, including catches from seven year classes, based on a recruitment of 5 million fish, a fishing mortality of $F = 0.2$ yr^{-1}, and an age of first capture of 2 years.

1 Age (years)	2 M	3 F	4 Length (cm)	5 Weight (g)	6 Number surviving	7 Catch numbers	8 Catch (t)
1	0.3	0	11.06	13.53	5 000 000	0	0.00
2	0.3	0.2	19.67	76.10	3 704 091	582 978	44.36
3	0.3	0.2	26.38	183.58	2 246 645	353 594	64.91
4	0.3	0.2	31.61	315.84	1 362 659	214 466	67.74
5	0.3	0.2	35.67	453.85	826 494	130 080	59.04
6	0.3	0.2	38.84	585.92	501 294	78 898	46.23
7	0.3	0.2	41.31	704.96	304 050	47 854	33.74
						Total =	316.02

Table 5.6, with mortalities listed in columns 2 and 3; a more realistic situation would be where F gradually increases with increasing age and vulnerability to the gear, but for the sake of simplicity, all fish of two years of age and older are assumed to be equally vulnerable.

The calculations are best completed on a spreadsheet model for which the model described in Appendix A4.10 'Biomass and biovalue model including fishing mortality – prawn example' can be adapted. As suggested in Table 5.6, length, weight, numbers surviving, numbers caught, and catch weight (yield) can be estimated as before, but in this case, as recruitment is known, yield is absolute rather than relative. The first column contains separate age classes in the stock so that the total at the bottom of the final column is the total catch from all exploited age classes.

The example shown in Table 5.6 is for a fishing mortality of $F = 0.2$ (for ages of two years and over) and the total yield of all exploited age classes combined is 316 tonnes. The calculations can be repeated for a range of F values, say from $F = 0.1$ to $F = 1.0$ in steps of 0.1. Results of 211 tonnes for $F = 0.1$, 316 tonnes for $F = 0.2$, 364 tonnes for $F = 0.3$ etc. are graphed in the curve at the top of Fig. 5.14. The curve at the bottom of Fig. 5.14 shows the results of repeated runs of the program for different ages at first capture with fishing mortality held at $F = 0.2$. This suggests that, at this level of fishing mortality, delaying the

age of first capture until age three (by mesh size regulations, for example) would maximize yield.

This model, adapted to include the effects of mesh selectivity, is similar to the one used in the International Council for the Exploration of the Sea (ICES) (northeast Atlantic) area, and is also known as the age-based Thompson and Bell method (Sparre et al., 1989). The model can incorporate changes in fishing patterns, and different mortalities can be applied to different age classes, thus avoiding the yield per recruit model's assumption of a steady state. Because the model need not be based on knife-edge selectivity (fishing mortality often increases gradually with increasing size), an often more realistic yield curve is produced, which usually does not approach an asymptote as the yield per recruit model does in certain cases. However, such curves, in common with yield per recruit curves, provide little indication of the level of fishing mortality at which recruitment overfishing occurs. This is more related to the shape of the stock–recruitment curve, and the degree of environmentally-induced fluctuations about that curve.

5.5 Simulation and ecosystem models

A fisheries simulation model consists of a simple, and often pictorial, representation of how a

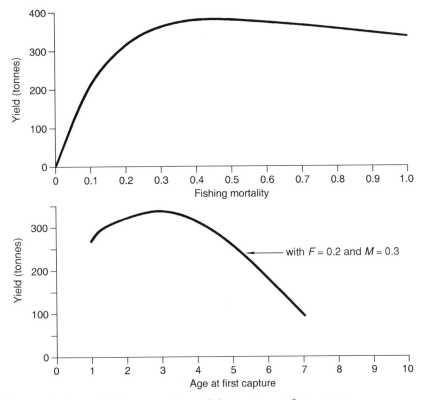

Fig. 5.14 Yield versus: (top) annual fishing mortality; and (bottom) age at first capture.

complex system behaves in various circumstances. Such models are often used to predict the possible outcomes of various management strategies. Predictions of simple biomass dynamic models, based on stock growth as in surplus production models, can be checked against the past history of a fishery before being used to predict future behaviour. The models can be elaborated to include stock age structure, economic considerations and ecosystems. And they can be made more realistic by including the possibility of chance variation or random catastrophic events in their equations. In all cases, however, the models are always an approximation to, and a simplification of, reality.

Models can be either deterministic or stochastic. In a deterministic model the parameters and relationship between them are constant; in this case the model will always give the same results for the same set of input parameters. In a stochastic model, one or more of the input parameters is allowed to vary in a random way over the run of

the model. The stochastic parameters could vary within a range above or below the predicted mean or could be taken randomly from a probability density function, say a normal distribution with the predicted value being the mean with a particular variance. For example, the catchability coefficient, q, can be randomized by adding to its mean value a random variable, $N(0,\sigma^2)$, with a mean of zero and a variance of σ^2.

The least satisfying model, at least to the person assessing the stock, is one that is merely descriptive. That is, although the model may simulate observed data well, its components have no biological meaning. A fisheries manager, however, may be less concerned about the lack of realism in a 'black-box' descriptive model as long as it appears to describe and mimic reality. A more satisfying model is one that seeks to explain the observed data with components and inputs that have some biological meaning. Such a model would include parameters, variables and

Box 5.8
Sensitivity analyses to set research priorities

The output, or results, from a model is influenced to a greater or lesser degree by each of its input parameters. The output may be sensitive (exhibit large changes) in relation to variations in one parameter, but may be relatively insensitive (exhibit small changes) in relation to others. The degree of sensitivity of a particular model to its various input parameters may be used to suggest which parameters are important, and hence to set priorities in research. Other things being equal, it makes sense to expend a greater research effort on obtaining more precise estimates of parameters to which a model is most sensitive. A simple analysis of model sensitivity involves changing one parameter at a time by varying amounts, and measuring whether this produces a weak or a strong perturbing effect on the model's output.

In the context of the present section, input parameters of growth, mortality, and critical age, produce an output in terms of yield per recruit (YPR). The sensitivity analysis below is based on the example given in Box 5.6 in which the unperturbed output for YPR, for a fishing mortality of $F = 1.0 \, \text{yr}^{-1}$, is 53.8 g. The changes induced in YPR values by perturbations in three of the input parameters, the growth coefficient, K, natural mortality, M, and the age at recruitment, t_r, are shown in Table B5.8.1. The 'correct' estimate of each parameter represents the unperturbed state (zero perturbation) shown in the middle column of the table, where the value of yield per recruit is 53.8 g per recruit.

Example changes, which are best given as percentages, are that an underestimate of K by 10% will result in the value of YPR decreasing from 53.8 to 43.4, a change of 19%. An underestimate of M by 10%, on the other hand, will result in the value of YPR increasing from 53.8 to 58.4, a change of 9%. This suggests that the YPR model is more sensitive to variations in (or incorrect estimates of) the value of the growth parameter K than the value of the natural mortality rate M. In other words, research should be directed towards obtaining more precise estimates of growth, rather than estimates of mortality.

Table B5.8.1 Values of yield per recruit for perturbations in three input parameters.

| | Percentage change in input parameters | | | | | | | | |
	−40%	−20%	−10%	−5%	0%	+5%	+10%	+20%	+40%
Parameter values of yield per recruit									
K	17.8	33.9	43.4	48.5	53.8	59.2	64.8	76.4	100.5
M	75.8	63.5	58.4	56.0	53.8	51.7	49.7	46.0	39.6
t_r	44.0	48.7	51.2	52.5	53.8	55.2	56.6	59.5	65.7

relationships that are believed to represent real events and interactions in nature.

Because of the widespread use of computers, models can be based on simple, or at least more meaningful, difference equations. The simplest approach to modelling is to consider that changes to a fish stock occur in a series of discrete time steps, usually one year. This obviates the need for differential equations and results in a more transparent model made up of a sequence of difference equations representing various processes. Even in more complex models, the basic processes affecting a fish stock in each time step include biomass additions by recruitment and individual growth in weight as well as decreases through natural and fishing mortality.

The value of simulation models in fisheries management rests with their ability to simulate the fishery's behaviour under various scenarios. The model should react to perturbations, say changes in fishing effort, in the same way as the fish stock being modelled and return to its original or equilibrium state when the perturbation is removed. Most systems are stable but only locally so. That is the system is only stable within certain limits and, if pushed beyond these limits, will not return to equilibrium; note the situation illustrated in Fig. 5.4 which suggests that if a stock is driven below some critical value, stock growth becomes negative and extinction is inevitable.

It should be noted that simulation models are more conveniently developed by writing computer programs in one of the common languages. These programs are more suited to complete large numbers of calculations for many different year classes in a single time period, usually a year, before looping back to do the same for the following time period. With Monte Carlo simulations, a stochastic model draws random values from some probability distribution in a similar manner to games of chance, from which the name is derived; the simulation is run many thousands of times to determine the probabilities of particular outcomes resulting from certain management actions.

In the present chapter, spreadsheet programs are used to develop simulation models and these have the advantage of being more transparent – that is, intermediate values and meaningful steps can be seen in cells, columns and rows. Transparency is not an inconsequential matter, particularly when demonstrating the working and assumptions behind models to colleagues, fisheries managers and the fishing industry.

5.5.1 A biomass dynamic simulation model

In this section, the Schaefer model parameters estimated from the barramundi data (Section 5.2) are used to develop a model that simulates longer-term fisheries behaviour. A spreadsheet model is included in Appendix A4.12 'Biomass dynamic simulation model'. This surplus

production simulation model, like previous biomass dynamic models, does not explicitly include growth and mortality.

The Schaefer parameters of stock biomass, catch and catch rates from the non-equilibrium model shown in Fig. 5.5 are $B_\infty = 4126$, $r = 0.5324$ and $q = 0.00000568$. The values of fishing effort can be altered to simulate different scenarios. Each year's biomass (B_{t+1}) can be expressed in terms of the present year's biomass (B_t) plus the rate of change in biomass minus the catch, C_t:

$$B_{t+1} = B_t + rB_t(1 - B_t/B_\infty) - C_t \tag{5.27}$$

The model described in Appendix A4.12 allows the maximum biomass to vary randomly by plus or minus 40% by multiplying B_∞ in the above equation by $[0.6 + 0.8(\text{RAND}())]$ – see Box 5.9 'Incorporating randomness'. Equation 5.27 now becomes:

$$B_{t+1} = B_t + rB_t(1 - B_t/(B_\infty[0.6 + 0.8(\text{RAND}())]))$$
$$- C_t \tag{5.28}$$

Catches are predicted as $C_t = qf_tB_t$, where f is fishing effort and q is the catchability coefficient.

Simulation models can be used to provide advice on the likely outcome of alternative management policies such as how stock biomass will respond to various levels of fishing effort. Where the aim of fisheries management is to conserve a minimum stock biomass to maintain recruitment levels, for example, the simulation model can be run many times to find the appropriate level of fishing effort. Fig. 5.15 shows stock biomass trends with constant rates of fishing effort of 25 000, 50 000 and 100 000 units of hundreds of metres of net per day (hmnd).

Because of the stochastic nature of the model, each run of the model will be different. However, the runs suggest that effort level of 25 000 hmnd will maintain stock biomass at about 3000 tonnes. An increase in fishing effort to 100 000 hmnd will cause the biomass to get dangerously close to zero and possible stock extinction. An appropriate management question may be to ask what level of fishing effort would maintain stock biomass at a minimum of 2500 tonnes for 90% of all years.

The model in Appendix A4.12 used to produce

Box 5.9
Incorporating randomness

Each run of a deterministic simulation model will produce identical results. However, such unchanging outcomes do not match natural fluctuations in reproduction, recruitment, growth, mortality and abundance. A stochastic model includes one or more factors that vary randomly, ideally with a frequency and a magnitude that matches actual fluctuations observed in the history of the fishery. Annual fluctuations are common but events with different frequencies can be allowed for. If an inshore stock, for example, is affected by floods that are known to occur randomly but with a frequency of one every 5 years on average, this frequency can be built into the model.

A model is made stochastic with variations away from the deterministic predictions of submodels – for example, recruitment can be allowed to vary around the value predicted by the stock–recruitment relationship equation in the model. Variation can be included by selecting random values from the residuals in a time series of observations or at random from a probability density function (PDF) that describes the residuals (Monte Carlo modelling). Realism is enhanced by incorporating randomness in the simulations.

Using stock biomass as an example of incorporating randomness, next year's biomass (B_{t+1}) can be expressed in terms of the present year's biomass (B_t) as $B_{t+1} = B_t + rB_t(1 - B_t/B_\infty)$ where r is the stock growth rate and B_∞ is the stock

biomass at the carrying capacity of the environment. As B_∞ is affected by density-independent factors, its value can be allowed to vary randomly with equal probability between limits by using the random number function built into computer programs (for example, RND in BASIC and RAND() in the spreadsheet program Microsoft Excel). These return a value equal to or greater than 0 and less than 1. Thus, variations with equal probability between plus or minus 25% could be included by multiplying B_∞ by [0.75 + 0.5 (RAND())]. Fig. B5.9.1 shows stock growth using the following stochastic equation:

$$B_{t+1} = B_t + rB_t(1 - B_t/(B_\infty[0.75 + 0.5 \text{ RAND()}]))$$

An alternative that recognizes the chances of a random value being closer to the mean than further away, is to assume that there is a normally distributed error around the mean with a particular standard deviation, σ.

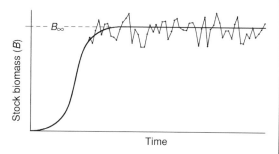

Fig. B5.9.1 Growth in stock biomass with randomly generated fluctuations, or white noise, about the mean, B_∞.

the graphs in Fig. 5.15 can be adapted to simulate values other than biomass. From a commercial viewpoint, a processor may be interested in changes in total catch whereas fishers may be more interested in how catch rates would respond to reductions in fishing effort.

5.5.2 An age-structured simulation model

The age-structured methods introduced in Section 5.4.3 can also be adapted as simulation

models. The inclusion of different age classes results in an increase in complexity, at least in the number of equations required, and the model could be more conveniently built using a regular programming language. However, for the purposes of demonstration, an age-structured model is developed in Appendix A4.13 'Age-structured simulation model' in which the number of age classes is restricted to four.

The spreadsheet program described in Appendix A4.13 completes the required calculations in

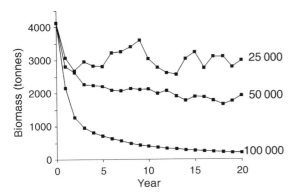

Fig. 5.15 Barramundi biomass simulated stochastically over 20 years with fishing efforts of 25 000, 50 000 and 100 000 hmnd.

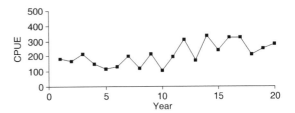

Fig. 5.16 A simulation of CPUE over 20 years. CPUE values have generally increased following a reduction in fishing effort after year 10. Graph produced using the 'Age-structured simulation model' in Appendix A4.13.

annual steps for the 20 years in column 1 (Table 5.7). Column 2 contains the number of annual recruits; the initial recruitment (of one million in the example) appears at the top of the second column and after this, is predicted from the Ricker stock–recruitment equation with random fluctuations within plus or minus 25%. The spawning stock in column 6 consists of the sum of fish of ages 2 to 4 from the previous year.

The numbers in age classes 2 to 4 (which are susceptible to the fishing gear) are shown in columns 3, 4, and 5. Fishing mortality is calculated as $F = fq$ where f is the fishing effort and q is the catchability coefficient, which in this case is allowed to vary by plus or minus 40%. The annual fishing effort in boat days is entered in column 7 and may be altered for different scenarios. In the example, fishing effort is 6000 boat days for years 1 to 10 and then reduced drastically to 2000 boat days for years 11 to 20 to illustrate just one particular scenario.

Important outputs of the model for management purposes are spawning stock size (column 6), total catch weight (column 12) and the catch rate (column 13). The program in Appendix A4.13 can be designed to graph any one of these columns depending on the type of advice required. As an example, Table 5.7 and Fig. 5.16 simulates CPUE over 20 years under the levels of fishing effort entered in column 7. Following the drastic reduction in fishing effort (from 6000 to

2000 units) after year 10, the CPUE values (column 13 in Table 5.7) have generally increased and the spawning stock size (column 6) appears to have recovered to its initial values. Questions that a fishery manager may ask include 'what minimum reduction in effort (from 6000) would be required to allow spawning stock sizes to remain above 700 000 in future years'?

The above age-structured model uses a series of difference equations in time steps of one year. The result of each step is then used as the start value for the next time period. It can be extended to include economic and commercial aspects as in the instructional simulation model FISHSIM (King, 1996) which is interactive and allows users to enter the biological and economic parameters of a fishery. In the above spreadsheet model (Table 5.7), fish prices, running costs and fixed costs would have to be added to the seven cells at the top left of the spreadsheet. An additional column could include fishing costs calculated as [(RC × fishing days) + (FC × vessels)] where RC is the running costs per day, and FC is the annual fixed cost per fishing unit or vessel. A second additional column could contain the annual profit calculated as [fishing costs – (catch weight × fish price)].

5.5.3 Ecosystem models

The necessity of an ecosystem approach to fisheries management has encouraged the development of ecosystem models that include interactions between species and ecosystems. Four multispecies models, including production

Table 5.7 Spreadsheet produced as part of the program 'Age-structured simulation model' (in Appendix 4.13).

1 Yr	2 Recruit Age 1	3 Age 2	4 Numbers Age 3	5 Age 4	6 Spawning stock	7 Fishing effort	8 Fishing mortality	9 Catch weight Age 2	10 Age 3	11 Age 4	12 Total	13 Catch rate
1	1 000 000	548 812	301 194	165 299	1 015 305	6000	0.461	425 911	398 351	282 613	1 106 876	184
2	1 011 526	548 812	189 888	104 213	842 912	6000	0.559	495 844	292 376	207 428	995 648	166
3	834 041	555 137	172 222	59 588	786 948	6000	0.966	739 729	391 096	174 928	1 305 753	218
4	901 839	457 731	115 922	35 963	609 617	6000	0.822	547 903	236 474	94 836	879 213	147
5	957 948	494 940	110 464	27 976	633 380	6000	0.579	459 386	174 732	57 205	691 322	115
6	1 021 520	525 733	152 254	33 981	711 968	6000	0.566	479 775	236 790	68 318	784 883	131
7	1 005 996	560 622	163 780	47 431	771 833	6000	0.903	715 045	355 997	133 276	1 204 318	201
8	1 171 274	552 102	124 696	36 429	713 227	6000	0.514	467 383	179 899	67 939	715 222	119
9	798 563	642 809	181 169	40 918	864 895	6000	0.871	800 256	384 373	112 225	1 296 855	216
10	1 079 005	438 261	147 686	41 624	627 570	6000	0.476	348 718	200 265	72 964	621 946	104
11	992 250	592 170	149 454	50 363	791 987	2000	0.220	242 757	104 413	45 485	392 655	196
12	1 131 116	544 558	260 795	65 820	871 174	2000	0.302	296 179	241 732	78 867	616 778	308
13	1 319 319	620 769	220 855	105 770	947 395	2000	0.144	172 654	104 683	64 809	342 147	171
14	919 589	724 057	294 858	104 904	1 123 819	2000	0.250	333 274	231 295	106 376	670 946	335
15	907 124	504 681	309 371	125 985	940 038	2000	0.195	185 597	193 890	102 070	481 557	241
16	1 292 483	497 840	227 836	139 664	865 341	2000	0.303	271 369	211 649	167 719	650 736	325
17	1 086 851	709 330	201 753	92 332	1 003 414	2000	0.285	366 451	177 628	105 086	649 165	325
18	1 178 337	596 476	292 694	83 250	972 421	2000	0.176	199 613	166 929	61 377	427 919	214
19	902 661	646 685	274 454	134 676	1 055 815	2000	0.188	229 891	166 273	105 474	501 637	251
20	850 709	495 391	294 021	124 783	914 194	2000	0.239	218 697	221 206	121 360	561 263	281

models, virtual population analysis, bioenergetics models, and Ecopath with Ecosim have been reviewed by Latour et al. (2003). Not the least because of its accessibility and the support available for its use, Ecopath with Ecosim has been widely used to examine the trophic dynamics in a fisheries context in many parts of the world. The software package, available free of charge from www.ecopath.org, is based on the original Ecopath model of a coral reef ecosystem in Hawaii (Polovina, 1984) and the software has been extended and developed by many others (Christensen & Pauly, 1992; Pauly et al., 2000) from whose work the following is précised.

The Ecopath model creates a static mass-balanced snapshot of the resources in an ecosystem and their interactions, represented by trophically-linked biomass groups including phytoplankton, zooplankton, fish etc. If the system is in equilibrium, gains in each group will equal losses due to respiration, flow to detritus, and other exports. Thus production moves from one trophic level to the next to reach top predators and fisheries (Fig. 5.17).

Inputs to Ecopath include some parameters that will be available from stock assessment including biomass, total mortality estimates and catches. Other information needed relates to diet and consumption. The basic equations of Ecopath first divide production into components:

Production = catch + predation + net migration + biomass accumulation + other mortality

And, within each group:

Consumption = production + respiration + unassimilated food

The model requires input of three of the following four parameters: biomass, production/biomass ratio (or total mortality), consumption/biomass ratio, and ecotrophic efficiency for each of the functional groups in the model. Here, the ecotrophic efficiency expresses the proportion of the production that is used in the system (it

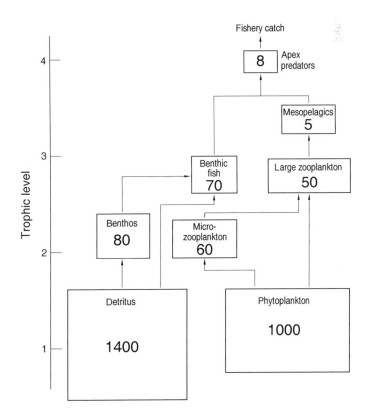

Fig. 5.17 Flow of production through a hypothetical and simplified ecosystem. The numbers represent the biomass of each group (in kg/km^2). Losses are due to respiration, other exports, and flow to detritus.

incorporates all production terms apart from 'other mortality'). If all four basic parameters are available for a group, the program can instead estimate either biomass accumulation or net migration. Ecopath sets up a series of linear equations to solve for unknown values establishing mass-balance in the same operation. The approach, its methods, capabilities and pitfalls are described in detail by Christensen and Walters (2004).

As the name suggests, Ecosim provides simulations into the future based on Ecopath's static description of ecosystem structure. By doing repeated simulations, Ecosim allows for the fitting of predicted biomasses to time-series data to evaluate fisheries and environmental effects. This allows managers to predict the effects of alternate management actions on fish stocks and ecosystems. Ecospace, also part of the software suite, can be used for spatial analysis based on the Ecopath model.

The elements of the package allow biological and policy analysis so far available only for areas where detailed multispecies models have been constructed over years by teams of experts. However, the Ecopath package relies on relatively few inputs that can be provided by a multidisciplinary group of scientists knowledgeable about a specific ecosystem. An example is a study of Prince William Sound, Alaska, where a workshop of specialists defined the ecosystem and its components for the model. Participants contributed estimates for input parameters and provided information on spatial and temporal changes to construct a model showing the trophic levels and the relative biomasses of biotic components. Since the 1989 Exxon Valdez oil spill in Prince William Sound, research teams had focused on the status of particular biological components of the ecosystem but an overall analysis was not available until the Ecopath model was constructed (Okey & Pauly, 1999).

5.5.4 Risk assessment

Forward projections of models into the future provide an opportunity for fisheries managers to review the effects of proposed management actions. Using such simulations, managers have access to assessments that enable them to review the likely consequences of alternative management strategies and consider the risks associated with each. The wide adoption of the precautionary approach to fisheries management is increasing the need to consider risk and, in the face of overexploitation in so many fisheries, there is increasing public pressure on fisheries managers to adopt low-risk policies.

The consideration of risk in fisheries, reviewed by Francis and Shotton (1997), involves both assessment and management stages. Risk assessment involves formulating advice for fisheries managers in a manner that makes clear the uncertainties and consequences associated with each alternative management option. Uncertainties are due to the lack of knowledge about complex systems, including the natural variations and fluctuations in many biological processes, particularly recruitment. Risk management involves taking uncertainty into account and allows decisions on options to achieve objectives within an acceptable level of risk.

The following example uses the non-equilibrium surplus production model presented in Section 5.2.2. The model is based on a stock of barramundi with a virgin biomass of $B_\infty = 4126$ tonnes, an intrinsic rate of increase of $r = 0.5234$ and a catchability coefficient of $q = 0.0000057$. Fishing effort is measured in hundreds of metres of net used per day (hmnd). For the sake of the example, the objective is to keep stock biomass above half the original unexploited level, say 2000 tonnes. This is the limit or threshold reference point, below which stock biomass should not be allowed to fall. The concept of reference points and indicators are detailed in Section 6.4.3.

The model design is described in Appendix A4.14 'Risk assessment model' and this is used to simulate biomass and catch over a 20-year period (Fig. 5.18). At simulated levels of fishing effort equal to or below 40 000 hmnd, the stock biomass remains above the threshold level of 2000 tonnes. At a fishing effort of 55 000, however, stock biomass drops below this threshold

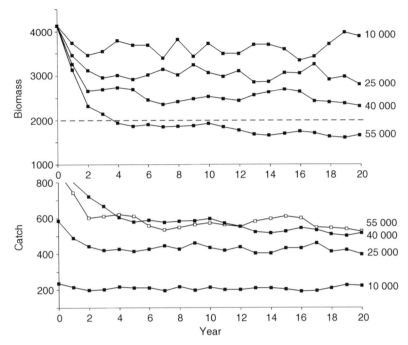

Fig. 5.18 Top: Simulated changes in biomass in a barramundi stock over a 20-year period for four levels of fishing effort (hmnd). The target reference level of biomass (2000 tonnes) is indicated by the broken line. Bottom: Simulated catches over 20 years for the same four levels of effort.

value after four years. Of the trialled values, the 'safe' fishing effort of 40 000 hmnd secures an annual catch that varies between 500 and 650 tonnes per year (the lower graph in Fig. 5.18). What is likely to be particularly convincing for those in the industry that want to increase effort, is that catches do not appear to change even at the higher (and more risky) fishing effort level of 55 000 hmnd.

As illustrated in this example, management often involves attempting to maximize one performance measure subject to a constraint on a second measure. That is, an objective seeks to maximize some desirable condition subject to there being some low probability of an undesirable condition occurring. Here the objective is to maximize the catch subject to the condition that there is less than 20% chance that the stock biomass will decrease below 50% of the virgin biomass, B_∞; that is, subject to the condition that $P(B < 0.5B_\infty) < 0.2$.

In Monte Carlo risk assessment, random values are drawn from some probability distribution,

and a large number of simulations (between 1000 and 10 000) is completed to arrive at a statistically viable result; that is, to generate the probability of some event, usually a negative one, occurring. For demonstration purposes, Fig. 5.19 is based on a much smaller number of replications for each of the 20 years obtained from the spreadsheet program in Appendix A.14. Each of the points in the graph represents the probability that biomass will drop below the reference level of 2000 tonnes for the particular level of fishing effort shown at the right of each line. At a fishing effort of 40 000 hmnd, the probabilities of stock dropping below 2000 tonnes are always below the threshold and acceptable probability of 0.2. At all higher levels of effort trialled, the probability of the stock crashing is higher than 0.2 for all years after the first three.

Risk assessment allows managers and stakeholders to evaluate a range of options with some knowledge of the likely consequences of each. Uncertainty is incorporated by providing advice in terms of probabilities of targets being met and

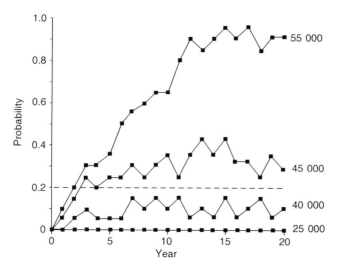

Fig. 5.19 The probability in each of the following 20 years of the barramundi stock biomass falling below 2000 tonnes. The reference probability level of 0.2 or 20% is shown as a broken line. Lack of smoothness in the lines is due to the small number of replicates for each point.

things going wrong. Often managers are under pressure to increase fishing effort in a fishery, say by either allowing a longer fishing season or by permitting some new gear development that makes fishing more efficient. Risk analyses can be used to show the industry and stakeholders the consequences of doing this. If, for example, increasing fishing effort by 20% is predicted to increase catches by only 5% but greatly increase the risk of the stock collapsing, those involved may have second thoughts.

Exercises

Answers to selected exercises are available at www.blackwellpublishing.com/king

Exercise 5.1

Deep-water snappers are caught by artisanal fishers using hand reels and droplines in depths of over 200 m off many tropical coasts. The approximate fishing effort (in trips per year) and the combined catch of two deep-water snapper species, *Etelis* and *Pristipomoides*, are shown below. Estimate fishing effort to secure the maximum sustainable yield, f_{MSY}, using both the Schaefer and Fox models.

Trips per year	Catch (tonnes)
106	60
215	86
321	110
491	136
540	176
608	141
650	155
680	141
885	180
920	208
1100	178
1410	160

Exercise 5.2

Molluscs are collected from four different
sandbanks (numbers 1 to 4) by subsistence fishers.
Each sandbank has a different area, and
approximately 60 subsistence fishers traditionally
fish on each of the sandbanks. A biologist
estimates that the areas of the sandbanks and
the annual catch from each are as shown in the
table below.

Sandbank number	Area of sandbank (km²)	Total catch (tonnes) per year
1	20	9.0
2	10	7.2
3	7	5.4
4	5	3.6

Use a surplus yield model to estimate the
maximum sustainable yield per km² of sandbank,
and the number of subsistence fishers per km² of
sandbank required to obtain it. In what way does
the data used differ from the usual data used in
surplus yield models? What are the assumptions
in using these data in a surplus yield model?

Exercise 5.3

Deep-water caridean shrimps are caught in traps
which have an unknown area of influence. As traps
spaced at more than about 50 m apart appear not
to compete with each other, it may be assumed
(in the absence of currents) that a single trap
attracts shrimps in from a circle with a radius of
half this distance. The shrimps are estimated to
have a natural mortality of $M = 0.66 \text{ yr}^{-1}$, and a
habitat area of 200 km². During a survey in this
area, mean catch rates were 2 kg per trap per
night. Assuming that the traps are 50% efficient,
use Gulland's approximate yield equation to
estimate the maximum sustainable yield for
the stock. On what assumptions does this
estimate depend?

Exercise 5.4

The von Bertalanffy growth parameters for a
species of penaeid prawn are $K = 0.2 \text{ m}^{-1}$, $L_\infty = 55$
mm carapace length, and $t_0 = 0$. The instantaneous
rate of natural mortality, $M = 0.25 \text{ m}^{-1}$. Weight (g) is
related to carapace length (mm) by the equation
$W = 0.0007L^3$, and the value of the catch is US$5
per kg at recruitment (in month 4), increasing by
10% each month. In the fishery, prawns are
recruited from adjacent mangrove areas
in April at 4 months of age.

(a) When would the unexploited stock reach
a maximum biomass?
(b) When would the unexploited stock reach
a maximum biovalue?

Approach these exercises by analysing the
changes in length, weight, relative numbers,
biomass, and biovalue of a single recruitment of
prawns in the form of Table 5.3. Then draw graphs
of both the relative biomass and relative biovalue
against months since recruitment.

Exercise 5.5

Consider the prawn species in Exercise 5.4 as an
exploited stock. Fishing is equivalent to imposing
a fishing mortality rate of $F = 0.20 \text{ month}^{-1}$ during
every month fished. The fishery may extend over a
9 month 'season' from April to December inclusive,
but some of the initial months could be closed
to fishing.

(a) If the first month (April) is closed to fishing, i.e.
the season is reduced to eight months, are there
any gains in the total weight and value of the catch
over the season? Are there any gains in the number
of prawns left in the stock at the close of the
season (at the end of December)?
(b) The fishermen wish to increase their gear
efficiency (which would result in increasing the
catchability coefficient, q, and therefore F by 25%).
However, the Fisheries Agency is concerned that
the stocks are already overfished, and the breeding
stock left after the end of the fishing season is too

small. A compromise is to allow the gear changes but to delay opening the fishery for two months. What changes in catch value would result from these changes? Would the aims of the Fisheries Agency be accomplished – i.e. to reduce overfishing and allow a greater breeding stock to be present after the end of the fishing season (after December)?

Exercise 5.6

A stock of herring, *Clupea harengus*, has the following population parameters:
- Growth: $K = 0.48$ yr^{-1}, $W_\infty = 180$ g, and $t_0 = 0$;
- Mortality: $M = 0.36$ yr^{-1};
- Critical ages: $t_r = 0.25$ yr, and $t_c = 0.75$ yr.

Construct a yield per recruit curve for the herring for values of $F = 0$ to $F = 1.0$ yr^{-1}, in steps of 0.1. If $F = 0.6$ at present, by what percentage would F have to be reduced in order to maximize yield and by what percentage would yield change as a result?

Exercise 5.7

Based on the length–frequency data for the lobster, *Panulirus penicillatus*, the population parameters are:
- Growth: $K = 0.39$ yr^{-1}, $W_\infty = 1.5$ kg, and $t_0 = 0$;
- Mortality: $M = 0.5$ yr^{-1};
- Critical ages: $t_r = 0.25$ yr, and $t_c = 1.0$ yr.

If fishing mortality is presently $F = 0.4$, use the computer spreadsheet program described in Appendix A4.11 to estimate the percentage change in yield that would result from delaying the lobster's age at first capture to 1.5 years.

Exercise 5.8

If fishing mortality on the lobster (Exercise 5.7) is presently $F = 0.4$, complete a sensitivity analysis to determine if the estimation of yield per recruit is most sensitive to errors in either K or M.

Exercise 5.9

A temperate-water species of herring has the following population parameters:
- Growth in weight: $K = 0.25$ yr^{-1}, $W_\infty = 350$ g, and $t_0 = 0$;
- Mortality: $M = 0.38$ yr^{-1}.

In a small fishery, an average of 1.2 million young herring are recruited at two years of age, when they are fully vulnerable to the fishing gear. The catch is essentially made up of six age classes (years 2 to 7 inclusive). Construct a computer spreadsheet table similar to Table 5.6, and record the yield for runs of the model for $F = 0$ to $F = 1$ in steps of 0.1. Draw the yield curve.

Exercise 5.10

Build the simulation model illustrated in Fig. A4.19 in Appendix 4.13 on a computer spreadsheet and enter a constant fishing effort of 2000 hours in the 'fishing effort' column. Advise managers on likely average catch rates if the level of fishing effort is: (a) increased by 25%; and (b) decreased by 25%. (Hint: run the model ten times for each of the three situations.)

Exercise 5.11

Change the graph produced by the simulation model shown in Fig. A4.19 to show spawning stock size over the 20 years. Use it to recommend a level of fishing that would maintain the spawning stock numbers at levels above 800 000 for 90% of all simulation runs (that is, there will be only a 10% risk that spawning stock numbers will drop below this level).

6 Fisheries management

6.1 Introduction

The purpose of fisheries management is to ensure that catches from a fish stock are ecologically sustainable in the long-term and benefits to fishers and communities are maximized. Although stock sustainability is a universal requirement, the benefits that management should deliver will vary from case to case depending on the type of fishery being managed. In many parts of the world, seafood is a vital source of protein and the benefits of fisheries management relate to food security rather than profits.

As discussed in Chapter 3, fisheries can be classified as either subsistence, artisanal or industrial. A subsistence fishery is one in which a major part of the catch is used by the fishers and their families for food and a lesser part is often sold. An artisanal fishery is a small-scale, low-cost, and labour intensive fishery in which the catch is sold and consumed locally, and an industrial fishery involves large vessels making catches that are marketed around the world. The distinction between these categories, however, is often blurred; increased consumerism in many communities relying on subsistence fisheries, for example, is resulting in a higher proportion of catches being sold to purchase other items regarded as essential.

Much of the methodology described in Chapter 5 has been developed for the assessment and management of large, industrial fisheries based on unit stocks in cooler climates. These large-scale fisheries usually involve highly efficient vessels equipped with gear including trawls, seines and longlines to target open sea, continental shelf or upwelling species. The target species are often spatially homogeneous, more or less evenly distributed over fishing areas, and catch rates may be a reasonable index of abundance; that is, continually falling catch rates will reflect a stock that is in decline. Overexploitation may be less common because fishing is likely to cease once catch rates decrease below some break-even point. The stock assessment methods presented in Chapter 5 may provide useful guidance for the management of such fisheries.

Subsistence and artisanal fisheries, however, are frequently based on species that are spatially heterogeneous – that is, sedentary or less mobile species that are widely but unevenly distributed around coastlines. The separation of substocks, differing spatial densities and the concentration of fishing effort near towns and villages all contribute to making catch rates a poor index of overall stock abundance; falling catch rates in a particular fishing area, however, may indicate local depletion. Some of the assessment methods discussed in Chapter 5 have been developed or adapted for use on artisanal and particularly tropical fisheries; such is the case with the use of length-based methods in place of the age-based ones used on cooler-water species (on the assumption that tropical species are more difficult to age than temperate species). Some characteristics related to the assessment and management of these extremes of large-scale, commercial fisheries and small-scale artisanal fisheries are summarized in Table 6.1.

The usual collection of data through logbooks or port sampling in large-scale fisheries is not possible in most small-scale fisheries, in which fishers operate from small communities spread out over large areas. This prompts the question of

Table 6.1 Characteristics of large-scale commercial fisheries and small-scale artisanal or community fisheries.

	Large-scale, industrial	Small-scale, artisanal, community
Resource	Often spatially homogeneous. Unit stocks or substocks of pelagic or demersal species.	Often spatially heterogeneous. Often less mobile or sedentary but dispersed along coasts.
Fishery	Commercial. Few units concentrated at landing sites and ports.	Artisanal; often subsistence. Many units, often spread out over coastlines.
Data collection	From logbooks, sampling etc.	Anecdotal information only.
Stock assessment	Estimates of stock size. CPUE useful as abundance index. Targets can be set.	Often not practicable. CPUE not useful as abundance index. Fisher information can be used.
Management	Often state-based or 'top-down'.	Community-based.
Regulations	Control of catch and effort. Quotas, licences etc.	Traditional measures and tabus. Closed seasons or areas.
Compliance	Strong enforcement required.	Self-regulation (by communities).

how useful catch and effort data would be in any case. For example, if fluctuations in catch per unit effort merely reflect the movement of fishers from one area of high stock density to another, such as between separated reefs, very little can be learned about overall stock abundance from CPUE data.

Large commercial fisheries are often managed, sometimes with varying degrees of industry involvement, by centralized government agencies. However, it is usually not feasible to manage a large number of community or artisanal fisheries without those involved in fishing playing a key role. Cooperative management (or co-management) and community-based fisheries management are discussed in Section 6.3. The fisheries management process, which is presented in Section 6.4, involves setting policies, objectives and strategies, all of which should be clearly documented in a fisheries management plan as discussed in Section 6.4.4.

Although management of large-scale fisheries often involves placing restrictions on the amount of fishing or the quantity of fish caught, such controls are unsuitable for many small-scale fisheries, particularly in subsistence fisheries where most of the daily catch is required as food. Alternative

management controls, particularly the closure of areas and seasons to fishing, are more appropriate in many community fisheries. In addition, nationally-imposed fisheries regulations are all but impossible to enforce in artisanal fisheries in which many thousands of fishers operate along extensive coastlines or in many different islands. Fisheries regulations and their enforcement are discussed in Section 6.5.

6.2 The need for fisheries management

The idea that natural resources open to the use of all will inevitably become overexploited was described in Hardin's (1968) tragedy of the commons. An individual fisher exploiting a resource that is regarded as common property will not be motivated to take any action to conserve the stock as any sensible management measure taken is likely to benefit someone else. Immature fish deliberately avoided by one fisher, for example, may be caught by another fisher. As fishing increases, and catch rates continue to decline, each fisher is motivated to fish harder in order to maintain profits. But any voluntary reduction in fishing by an individual will benefit others by allowing them to catch more rather than benefit the stock by allowing it to increase. This gloomy picture suggests that the stock in any fishery with unrestricted entry will always become depleted.

6.2.1 Biological overfishing

Even until quite recently the seas have been regarded as an inexhaustible larder from which seafood could be taken perpetually to feed the world's growing populations (see Box 6.1 'The inexhaustible larder?'). Today, many fish stocks are overexploited and the chances of developing fisheries on new species are slim. One of the main reasons for the poor state of many of the world's fisheries is the lack of controls on access, which has resulted in too many fishers hunting too few fish.

In spite of the concerns about surplus production models and their unfortunate history of overestimating sustainable yields, they are useful in emphasizing that fisheries resources are finite. As Schnute and Richards (2002) claim, the surplus production paradigm lies at the heart of any rational theory of fishing. In the early stages of development in a fishery, each increase in fishing effort is rewarded by a corresponding increase in the annual catch or yield (Fig. 6.1). At this stage, catch rates will be high, encouraging the entry of more fishing units into the fishery. As fishing effort (say the number of boats) continues to grow, the resulting increases in yield will not be as great, and mean catch rates (catch per unit effort or CPUE) will continue to decline. Eventually, the catch from the stock will reach the maximum sustainable yield (MSY) at some level of fishing effort, f_{MSY}. This overall scenario holds true even though environmentally-induced fluctuations in stock abundance may make the existence of a steady MSY doubtful and f_{MSY} may be considerably lower than that predicted by the symmetrical parabolic model shown. If fishing

---

Box 6.1
The inexhaustible larder?

'I believe then that the cod fishery, the herring fishery, pilchard fishery, the mackerel fishery, and probably all the great sea fisheries are inexhaustible: that is to say that nothing we do seriously affects the numbers of fish. And any attempt to regulate these fisheries seems consequently from the nature of the case to be useless.'

T.H. Huxley (1884) Inaugural Address, Fisheries Exhibition Lit., 4, 1–22.

The words of T.H. Huxley in his address at the London Fisheries Exhibition in 1883 may appear naive to us today when many of the world's fish stocks are overexploited. But in Huxley's time, fishing boats still used sails, there were no large international fishing fleets, and the world's

population was less than one-third of what it is today. Perhaps Huxley could have heeded the words of Malthus who, over 50 years before, had contended that the world's population was increasing faster than its means of subsistence. However, considering the lack of hunting skills in humans, and the assumed productivity of the oceans which cover over 70% of the world's surface, Huxley's statement may not have appeared overly optimistic to an audience in 1883. The view that the sea represents an inexhaustible food larder has persisted until quite recently, even though many of the major fish stocks of the northern Atlantic had shown signs of reduction by the end of the nineteenth century. By the time the International Council for the Exploration of the Sea (ICES) was formed in 1902, steam trawlers had replaced sail, and it had been established conclusively that catch rates decline as fishing effort increases. A history of the science and management of fisheries is given in Smith (2002).

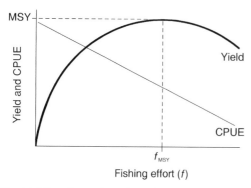

Fig. 6.1 The relationship of yield and catch per unit effort (CPUE) to fishing effort showing the maximum sustainable yield (MSY) and the fishing effort required to take it (f_{MSY}).

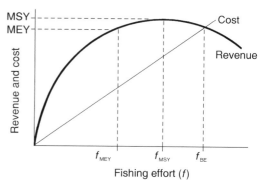

Fig. 6.2 The relationship of fishing revenue and fishing costs to fishing effort showing the maximum sustainable yield (MSY), maximum economic yield (MEY) and the economic break-even point (where the two lines cross). The points f_{MEY}, f_{MSY}, and f_{BE}, represent the levels of fishing effort that correspond to these points.

effort is further increased, the adult stock will be reduced to the extent that recruitment overfishing occurs and insufficient young fish are produced to maintain the stock. If the number of recruits is less than the number lost, both catch rates and catches will continue to decrease, on the average, year after year.

6.2.2 Economic overfishing

The surplus production model was used in one of the first attempts at an economic analysis of a fishery (Gordon, 1954). The relationship between revenue, fishing costs, and fishing effort (Fig. 6.2) is used to show how a fishery with no limits on the number of participating fishers (an open-access fishery) will become overexploited in the economic as well as the biological sense. In the face of large and increasing world populations, fisheries economics involves the use of scarce resources to satisfy practically unlimited demand (Hannesson, 2002).

In an economic sense, excess fishing effort occurs when the revenue generated by a marginal increase in effort is less than the cost of that increase. Assuming that the cost of fishing is directly proportional to the amount of fishing, total fishing costs will increase linearly with fishing effort. The total revenue curve, which is the product of yield and fish price per unit weight, will have the same shape as the yield curve shown

previously in Fig. 6.1. Total profit (or economic rent) from the fishery is maximized where the distance between the revenue curve and the cost line is the greatest. This point, referred to as the maximum economic yield (MEY), is obtained at a level of fishing effort (f_{MEY}) that is lower than that required to maximize yield in weight (f_{MSY}). Economic rent refers to 'excess' profits; i.e. the profits earned in excess of the level required to cover all fishing costs, produce a return to fishers and provide an incentive to go fishing. In a fishery where there is no control on fishing effort (an open-access fishery) more people will be attracted into the fishery by high profit levels until, at the economic break-even point (f_{BE}), total costs will equal total catch value and economic rents will have been dissipated. An example of the calculations involved is given in Box 6.2 'Economic yield example'.

If a fish stock is regarded as a common property resource over which no individual has exclusive property rights, fish may be regarded as belonging to the community. A farmer who manages his stock (a private property resource) will get the benefits of good management. However, if a fisher decides not to catch small fish, for instance, to allow them to grow to a more marketable size, others may catch them before he does. In other words, a sensible management decision taken by an individual fisher may not

Box 6.2
Economic yield example

Calculations may be completed in the form of a worksheet such as that based on the barramundi example in Table 5.1 in Chapter 5. For the sake of the example only, assume that fishers received a price of US$3 per kg for their catch. Yield in weight (column 2) may be multiplied by the price to give yield in revenue (column 3). A graph of revenue (column 3) versus fishing effort (column 1) would be similar to that shown in Fig. 6.2. If all fixed and running costs were US$22 per 100 m of net used, the total fishing costs would be as shown in column 4 of Table B6.2.1.

The fishing cost line cuts the yield (revenue) curve at the break-even, or open-access point, where earnings from the catch just balance fishing costs. Profit from the fishery is at its greatest at the maximum economic yield (MEY), where the distance between the cost line and the revenue curve is greatest. Total revenue is calculated as the yield (kg) multiplied by the unit price (US$ per kg):

$$\text{Revenue} = (af + bf^2) \times \text{price} \qquad (1)$$

where α and b are the constants in Schaefer's model (Equation 5.6).

Total fishing costs are estimated as fishing effort multiplied by the cost per unit effort, and at the open-access point, fishing costs equal fishing revenue so that:

$$f_{BE} \times \text{costs} = (af_{BE} + bf_{BE}^2) \times \text{price}$$

This equation simplifies to give the open-access fishing effort, f_{BE}, as:

$$f_{BE} = [(\text{costs/price}) - a]/b \qquad (2)$$

In the example for the barramundi, $f_{BE} = 89\,106$ hmnd and the level of fishing effort that maximizes profit occurs at one half this value at $f_{MEY} = 44\,553$ hmnd. The same caveats for MSY hold for MEY: environmentally-induced fluctuations in stock abundance from year to year and variations in catchability make setting a constant level of fishing effort to maximize profits a risky undertaking. However, from a biological viewpoint, setting f_{BE} as a target fishing effort is conservative as it is always lower than f_{MSY}.

Table B6.2.1 Hypothetical revenue and costs for the barramundi example in Chapter 5 (Table 5.1).

1 Effort (hmnd)	2 Yield (kg)	3 Total revenue ($1000s)	4 Total costs ($1000s)
0	0	0	0
20 000	430 120	1290	440
40 000	696 480	2089	880
60 000	799 080	2397	1320
80 000	737 920	2214	1760
100 000	513 000	1539	2200

result in any personal benefit. There is no incentive in open-access fisheries for individual fishers to restrict their effort as long as costs remain less than revenue. Each individual in the fishery is motivated to compete for a maximum share of the resource, and has little incentive to practise conservation. If demand remains high in relation to fishing costs in an unmanaged fishery, it is almost inevitable that fishing effort will continue to increase, and the stock will become overexploited in both an economic and biological sense.

The above is based on the view of a fish stock as a common property resource. However, areas of coastal water in some countries are held and managed under traditional ownership by adjacent community groups. Fisheries resources in these areas are managed by mechanisms such as tabu and custom, in effect, as a private property resource (see Section 6.3 and Chapter 3, Box 3.3 'Customary marine tenure'). Private ownership of fisheries resources, in the sense of land and stock held by a farmer, has been suggested as a way of overcoming some of the problems of common property resource management in large commercial fisheries (Keen, 1991). At the 1983

FAO World Fisheries Conference in Rome, it was recognized that open access to unmanaged fisheries had resulted in competition for limited resources, overcapitalization, and the depletion of stocks. To prevent such consequences, one of the recommendations of the conference was that governments should seek to ensure that fishers have clearly defined fishing rights, and that the allowable catches do not exceed the productivity of the resources. Methods of addressing the rights of fishers include either allocating rights to the capacity to fish (through input controls such as licence allocation) and allocating rights to a specified share of the resource (output controls in the form of catch quotas). Individual transferable quotas (ITQs) usually provide an individual with the right to harvest one or more species and the right to lease or transfer quota to others. Property rights in the form of fishing licences and quotas are discussed in Section 6.5.

6.3 Managers and stakeholders

Although fisheries management cannot be undertaken by an individual, it can be implemented by cooperating participants in a fishery. If all participants in a fishery agree to take some management action, say by not fishing on spawning aggregations of the target species, there may be no need for an external management body, such as a government fisheries agency, to be involved. Although abiding by such informal agreements between participants may be possible in small local fisheries it is usually not possible in larger fisheries operating from different fishing centres or ports. Usually a controlling group or organization has to be involved in fisheries management and possibilities include a community, collective or cooperative to which all fishers belong and a disinterested party such as a government fisheries authority.

The problem with fisheries being managed solely by fishers and fishers' organizations is that there are many other individuals and groups that can claim interests in either the use of the resource or the fishing area. Obvious stakeholders are the fishers, boat-builders, processors, and suppliers of gear. Other stakeholders with interests in the resource and the environment could include recreational fishers, nature groups and conservationists. In inshore fisheries in particular, other water-users can include recreational divers and boat operators. In addition, the general public has an interest in benefiting from any excess profits gained from the exclusive exploitation of a resource that belongs to all.

6.3.1 Fisheries management authorities

In theory, fisheries can be managed either by the state, usually by a government agency, or by communities or groups of resource users. In many parts of the world, coastal communities, far from centres of government, operate under traditional systems of resource stewardship referred to as customary marine tenure or CMT (see Box 3.3 in Chapter 3). Under CMT, the community or its leaders constitute the managing authority. In most countries, however, a government ministry, department or agency is assigned the overall task of fisheries management and does so for the benefit of all citizens. Even in cases in which the control of fisheries by fishing communities is encouraged, the government itself and its appointed agency represent the ultimate authority and are the arbiters in matters to do with fisheries development and management.

Once policies have been set by government, it is up to fisheries managers in the authority or agency to determine and implement the actions necessary to achieve the policy goals. Fisheries managers require a much wider range of skills, in politics, economics, and sociology, than those involved in stock assessment. The differences in skill requirements have been used to emphasize the fact that stock assessment and fisheries management should be the responsibilities of different people (Hilborn & Walters, 1992). However, the reality is that in many parts of the world, because of lack of staff numbers and expertise in government agencies, the same people are involved in both tasks.

Fisheries management involves managing people and their behaviour rather than managing fish. Negotiation and social skills rate highly in the attributes required by fisheries managers as key tasks involve securing agreement with participants to restrict either fishing effort or catches and to take actions to protect key habitats. In the case of valuable commercial fisheries in which access has been limited there is often considerable pressure from those outside the fishery to gain entry. Fisheries managers have to be able to make, and stand by, difficult decisions in order to ensure the sustainability of stocks. On the other hand, firmness must be tempered by an empathy with those in the catching sector of the fishery; many government fisheries agencies have developed an unfortunate adversarial position in relation to fishers. Although fisheries managers are required to be guardians of publicly-owned resources, they should not view their role as valiant defenders of fish stocks against the depredation of voracious fishers.

One of the effects of total control of fisheries by a government agency is that fishers feel disenfranchised from the management process and are less likely to respect restrictions and regulations that they have had no hand in devising. Fishers feel no ownership of state-developed management plans, which may suffer in practical details from not having any user input. Management is imposed on a fishery and any fisheries regulations have to be strongly enforced, often at considerable expense.

In the case of total control by fishers or fishing communities, on the other hand, rules and restrictions are often arrived at by common agreement and are more likely to be respected by participants. If all fishers support communally-made decisions, no enforcement is necessary and the lack of legal status of the fisheries management plan becomes irrelevant. However, the lack of legal status may be a problem if community rules cannot be applied to those outside the community. Communal or fisher management plans may also suffer from a lack of scientific and technical input, including advice on the most effective management measures. The characteristics of government management and communal management are compared in Table 6.2.

Perhaps because of its disinterested status, a government agency will find it is marginally easier to make hard management decisions such as reducing fishing effort in the case of an overexploited fishery. A fishing community, on the other hand, will often find it more difficult to take necessary actions if they have a negative effect on the livelihoods of local fishers. A government agency will also find it easier to balance the demands of a wide range of interest groups, including non-fishers, in a dispassionate manner. A group of

Table 6.2 Comparisons of state and fisher management of fisheries.

	State management	Fisher or communal management
Fisheries management plans	No stakeholder ownership of management plans	Plan may suffer from lack of technical/scientific input
Controls and regulations	Controls may be impractical without stakeholder input	Controls imposed by communities may lack legal status
Enforcement	Strong enforcement required	Self-regulated or enforced by communities
Conservation	Fishing effort relatively easy to reduce	Fishing effort difficult to reduce due to self interest of managers
Stakeholders	Needs of all interest groups balanced	Non-fisheries interest groups may be disadvantaged

fishers is less likely to take into account the views of non-fishers, particularly if these run counter to fishers' interests.

For large commercial fisheries, particularly those that straddle country boundaries, a centralized management authority allows the use of sophisticated tools such as satellite-based vessel monitoring systems (VMS – see Section 6.5.4 'Compliance and enforcement') to track and control fishing vessels.

In most circumstances, a middle course between the extremes of state and communal control offers some of the advantages of each without many of the disadvantages. Under this arrangement, referred to as co-management, government agencies and user groups cooperate to take responsibility for, and manage, a fishery resource. There is, in fact, a continuum from total state control to total communal control that involves increasing levels of stakeholder involvement that can be illustrated as follows:

Total state management > some fisher input > equal control > some state input > total fisher management

The following two sections, 6.3.2 and 6.3.3, cover the involvement of stakeholders in commercial fisheries and in community fisheries respectively.

6.3.2 Co-management in commercial fisheries

In co-management arrangements in commercial fisheries, there are various levels of fisher participation ranging from government authorities holding meaningful consultations with fishers to the acceptance of major management responsibilities by fishers – this corresponds to movement from left to right along the continuum shown above. It should be emphasized here that the case of a government authority making a presentation of a predetermined management plan to a meeting of stakeholders who have had no input into the plan's preparation is not co-management.

Although many fisheries authorities will engage in consultations with stakeholders, higher levels of fisher involvement are less common. One

example of a higher level of fisher involvement in co-management is the case of Japanese fishers' cooperatives in which only members of the cooperative are permitted to fish. The cooperative has the authority to regulate fishing and those who do not abide by cooperative rules risk being expelled. One of the earliest examples of fisheries co-management is in the case of a small-scale fishery in the Lofoten Islands, Norway. Here, a seasonal cod fishery was affected by conflicts between large numbers of fishers using different fishing gear. Early government regulations were unsuccessful until the 1980s when co-management was implemented. Representatives from the different fishing gear groups formed committees, made rules on fishing times and areas and elected fishers to act as inspectors. The co-management system has changed little over time and still continues to operate (Jentoft et al., 1998).

As the cost of centralized government rises and users are expected to pay for fisheries research and management, it will become increasingly important to involve fishers in the management of fisheries resources. The sharing of responsibilities results in management aims being widely agreed to, and necessary regulations being more readily accepted and supported.

One means of establishing co-management is through the formation of groups of stakeholders with the purpose of providing management advice on particular fisheries. These groups, sometimes called management advisory committees, or MACs, are usually made up of representatives of government agencies and elected representatives of key stakeholder groups. Stakeholder groups may include fishers and processors as well as non-fisheries groups such as recreational fishers and other users of the marine environment.

One difficulty is that particular stakeholder groups often have different objectives. Seafood processors, for example, may want to maximize catch to provide a greater factory throughput whereas fishing vessel operators will want to maximize the economic yield from the fishery. The stakeholders will therefore need to

Box 6.3

Development stages in an unmanaged fishery

Development stages in an unmanaged fishery as summarized in Fig. B6.3.1. After the resource is 'discovered', usually by fishers, a growth phase follows in which more fishers, encouraged by high catch rates, enter the fishery. By the time the fully exploited phase is reached, abundance, and often catch rates, have decreased to about half their original values. In the overexploited phase, catches are high and catch rates are low as fishers compete for fish that are low in abundance. In the collapse phase, total catches fall and this is followed by effort decreasing as fishers leave the fishery. If fishing stops or continues at a low level, abundance may increase slowly. However, some populations require a minimum number of individuals to ensure reproductive success (the Allee effect – see Box 3.5 in Chapter 3) and may become locally extinct.

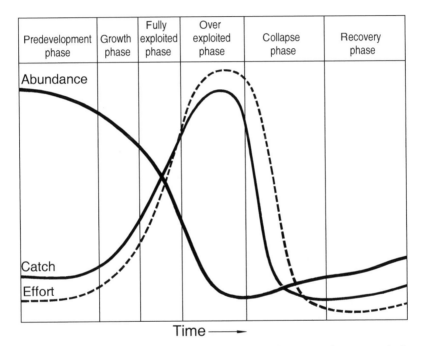

Fig. B6.3.1 Development stages of an unmanaged fishery. Adapted from several sources including Gulland (1983) and Csirke and Sharp (1984).

appreciate the tradeoffs between various aims and agree on the management objectives for the fishery.

Co-management and MACs are more straightforward to establish in fisheries that are currently managed on the basis of providing rights to fishers, either those that hold a licence in a restricted entry fishery or have an individual quota to harvest a certain proportion of a total allowable catch. Many organizations are actively promoting co-management in commercial fisheries, and an example is the manual on the topic published by the Secretariat of the Pacific Community (Watt, 2001).

6.3.3 Community-based fisheries management

In many parts of the world, particularly in tropical regions, coastal communities are dependent on catches of seafood to provide essential protein. These communities, which are spread out on long coastlines and on small islands, are often located a considerable distance from the centres of population where government fisheries agencies are located. Typically, subsistence fisheries are characterized by a large number of fishers using many different fishing methods to make small individual catches of a great variety of species.

In spite of the obvious importance of subsistence fishing in providing food, government fisheries agencies usually concentrate their management efforts on commercial fisheries that are often based in city and town ports. Although most countries have enacted national laws to protect fish stocks, enforcing such laws in widespread rural areas is difficult, particularly in communities that operate under their own traditional governing structure. Many isolated communities resent what they see as outside 'interference' by central governments.

Fishing communities are often repositories of valuable traditional knowledge concerning fish stocks and have a high level of awareness of the marine environment (Johannes, 1982). In addition, many communities have some degree of control, either legal or traditionally assumed, of adjacent waters. Collectively, these factors provide an ideal basis on which communities can be encouraged and motivated to manage their own marine resources. If the community is encouraged to set its own conservation rules, as many have done in the past, they are more likely to be respected and under community ownership: management measures are enforced by the communities themselves. Because communities play the key role, this type of management is referred to as community-based fisheries management. A case study from the Pacific island of Samoa is given in Box 6.4 and the process is summarized in Fig. 6.3.

Community-based fisheries management is being promoted for subsistence fisheries in many parts of the world (Pinkerton, 1989; Amos, 1993; Johannes, 1998; Pomeroy, 1998; King & Fa'asili, 1999). A guide to the literature on traditional

community-based fishery management in the Asia–Pacific region is given in Ruddle (1994). A manual for use in South Pacific islands with methods that can be adapted to suit local situations, cultures and customs has been published by the Secretariat of the Pacific Community in New Caledonia (King & Lambeth, 2000).

Often the most difficult task in establishing community-based fisheries management is in overcoming the initial reluctance of government agencies to encourage rural communities to take actions for which they see themselves responsible. The agency may feel a loss of power in placing the initiative for marine conservation in the hands of fishing communities. However, experience suggests that a government agency promoting community-based management gains in many ways; the agency is seen to be active in rural areas, gains public support and, because communities assume responsibility for enforcing their own fisheries regulations, gets credit for reducing enforcement costs. However, the main reason for encouraging community-based fisheries management is that it may represent the only chance for subsistence fisheries to be exploited on a sustainable basis. Regardless of national legislation and enforcement, the responsible management of subsistence fisheries resources will only be achieved when fishing communities themselves see it as their responsibility rather than that of the government (King & Fa'asili, 1999).

With all the current enthusiasm for community-based fisheries management, it must be accepted that there are many problems that local community management cannot address. Some environmental problems that affect local fisheries involve activities and areas some distance from the affected area and therefore beyond the control of a local community. For example, fish catches may be falling in a community fishing area because silt from a nearby river is killing corals in the lagoon. The silt may be the result of poor farming techniques or the logging of timber in hills many kilometres away from the village. Such problems can only be addressed by an integrated effort by government agencies and community groups working together. Integrated

Box 6.4
Community-based fisheries management in Samoa

A community-based fisheries extension project, supported under Australian aid funds, began in Samoa in 1995 (King & Fa'asili, 1999). A culturally acceptable extension process was developed which recognized the village *fono* (council) as the prime instigator of change, while still allowing ample opportunities for the wider community to participate (Fig. 6.3). Participating villages were required to have an awareness of problems with the marine environment and fisheries resources, a concern for these problems and a willingness to take actions to solve them.

Following an indication of interest, a village *fono* meeting was arranged to provide the community with information to allow either acceptance or refusal of the extension programme. If the *fono* accepted, it was then asked to arrange for meetings of several village groups, including separate meetings for women and untitled men (*aumaga*). Over a series of meetings, each group held separate meetings to discuss their marine environment and fish stocks, decide on key problems, determine causes, propose solutions, and plan remedial actions. Problem/solution trees were recorded on portable white boards by trained facilitators. Finally, a village fisheries advisory committee was formed, with three people nominated from each group, to prepare a draft village fisheries management plan (assisted by fisheries agency staff) for discussion and approval by the village *fono*.

Each fisheries management plan listed the resource management and conservation undertakings of the community, and the servicing and technical support required from the fisheries agency. If the plan was accepted, the *fono* then appointed a fisheries management committee to oversee the working of the plan. The time taken from initial contact to approval of the plan by each village community averaged 13.4 weeks and, so far, over 60 coastal villages have developed fisheries management plans. A joint agency/community assessment of how each community plan was working was conducted about six months after the plan's inception.

In their plans, communities included undertakings to support and enforce government laws banning the use of plant-derived poisons, bleaches and explosives to kill fish. Traditional destructive fishing methods such as smashing of coral to catch sheltering fish were also banned. Most villages made their own rules to enforce national laws banning the capture of fish less than a minimum size, and some set their own (larger) minimum size limits. Some villages placed controls on the use of nets and the use of underwater torches for spear fishing at night. Community conservation measures included collecting crown-of-thorns starfish, *Acanthaster planci*, banning the removal of beach sand and dumping of rubbish in lagoon waters. An unexpectedly large number of villages, over 30, chose to establish their own small areas closed to fishing. Fisheries agency actions to support community undertakings included providing technical advice and assistance with the restocking of giant clams in village fish reserves.

coastal management (ICM) takes into account the interdependence of ecosystems, and the involvement of many different agencies (for example, those responsible for agriculture, forestry, fisheries, public works and water supply) and other stakeholders.

If a community is encouraged to impose its own fisheries rules and regulations, these cannot be permitted to replace (or even compromise) national fisheries regulations. For example, if a minimum size limit is imposed on a particular species under national regulations, communities may be allowed to locally set and enforce a higher but not a lower size limit. Any fisheries regulations imposed, whether by governments or communities, to allow fish stocks to recover will result in decreased catches, at least in the short term. This means that communities that rely on

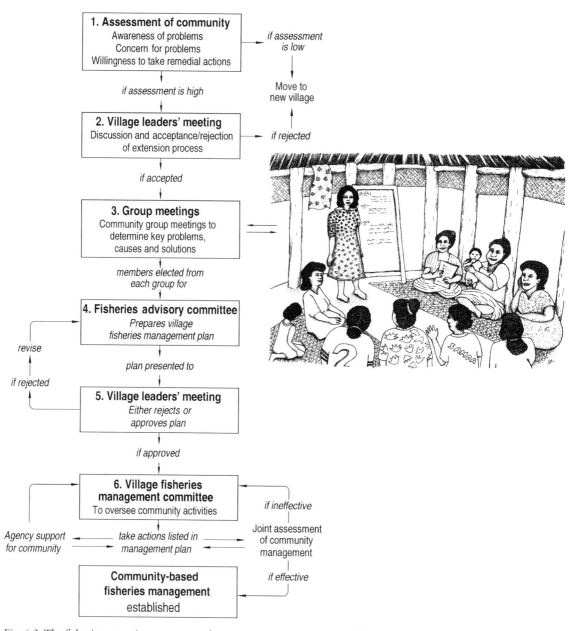

Fig. 6.3 The fisheries extension process used to promote community-based fisheries management in Samoa. The process includes several village group meetings to identify problems, discuss effects and propose solutions. Adapted from King and Lambeth (2000). Group meeting drawing by Jeremy King.

seafood may need some short-term support when taking stringent management measures.

6.4 The management process

The highest level of the fisheries management process (Table 6.3) involves the setting of policies and this is usually the prerogative of governments. Policy goals can be based on a range of social, economic and political objectives, all resulting in catches that are within the biological limits of the resource. For example, a policy may

Table 6.3 The step-wise fisheries management process from policies to actions and evaluation.

Policy goals
Broad economic, social and ecological aims to benefit society
▼
Objectives with reference points (targets)
Practical level goals relating to policies Targets set as reference points
▼
Strategies
Plan of action to achieve the specific objectives
▼
Actions (tactics)
Specific actions and regulations required under each strategy
▼
Indicators and performance measures
Indicators of performance that will be used to measure performance in relation to reference points
▼
Review and evaluation
Review of performance in relation to objectives and policies

aim for economic efficiency by allowing a small number of efficient fishing units to harvest a total allowable catch. On the other hand, a policy may aim to maximize participation and employment by allowing a larger number of less efficient fishing units to harvest the same quantity of fish. The collecting of biological, ecological, economic, sociological and cultural data for an information system to assist policy-makers and managers is discussed in Evans and Grainger (2002).

Ideally, policy-makers take account of the results of stock assessment, particularly with respect to upper limits on exploitation, and the predicted outcomes of alternative management choices. In some instances, shallower political motives, say to maximize votes, may be realized by allowing excessive or uncontrolled access to a fishery and this will inevitably lead to overexploitation. And eventually, the unenviable task of reducing the number of fishers in the overexploited fishery falls to fisheries managers in the designated authority.

Fisheries managers, usually employed by a government ministry, department or agency, are responsible for translating the policies into specific objectives. An objective is some target that is aimed for, or some goal that is set to be achieved. In fisheries management, objectives can include ecological, biological, economic, social and political goals that address set policies. A strategy is a plan of action designed to achieve a particular objective. If a fisheries management objective is to maintain production from a fishery at a particular level, one particular strategy would be to maintain the spawning stock size at a particular biomass, say at 50% of the virgin spawning biomass. Reference points and indicators should be devised that allow an evaluation of the degree to which objectives have been met. Data collection is then aimed at quantifying the variables used as indicators. This section describes the steps in the management process summarized in Table 6.3, excluding management actions which are described later in Section 6.5.

6.4.1 Management policies and objectives

Historically, the main aim of fisheries management has been to maximize sustainable yield (MSY) as it was believed that the level of fishing required to take it would not overexploit the target stock. However, bitter experience has shown that fishing at this level is too high and MSY is more appropriate as an extreme upper limit rather than a target.

More recently, biological targets have been set at lower and less risky levels of fishing effort and fisheries policies have been extended to address economic, social and environmental objectives. The broad objectives of fisheries management may, therefore, include the conservation of fisheries resources and the marine environment, the maximization of economic returns from the fishery, and payment of fees to the community from profits made by the exploitation of a public resource. Subsuming all these objectives is the need to ensure that fisheries are exploited on an ecologically sustainable basis. The development of fisheries management policy is usually the responsibility of government and takes into account wide-ranging considerations relating to the well-being of the country's citizens. Ideally, policies are formulated in consultation with the stakeholders involved, particularly communities in the case of subsistence fisheries and commercial fishers and processors in the case of commercial fisheries.

In some cases, it will be government policy to develop a small-scale artisanal fishery that maximizes participation and local food security. Following this policy, fisheries managers set objectives to ensure as many individuals as possible have an opportunity to share the resource. Thus the strategy could be to allow many smaller fishing units into the fishery and limit their efficiency by either restricting the type of fishing gear used or by placing quotas to limit individual catches. At the other extreme, for a commercial fishery in which most of the catch is exported, it may be government policy to maximize profitability and export income. A strategy for achieving this objective may involve having a small number of highly efficient fishing units in the fishery.

The difficulties involved in managing a fisheries resource are related to the number and types of user groups, and the distribution and mobility of the fish stock. Conflicts arise in many fisheries where the same stock is exploited by different user groups, particularly commercial and non-commercial fishers. An example is where juvenile fish are caught in inshore areas by recreational or artisanal fishers, and adults of the same species are caught offshore by commercial fishers. In this case, inshore catches will reduce the numbers of fish growing and migrating out to join the offshore stock, and commercial catches offshore may reduce the breeding stock which produces new crops of juveniles in inshore areas. Allocating resources between artisanal and commercial fishers is a difficult task, and in many cases, involves weighing the advantages of employment and food security from subsistence and artisanal fisheries against the advantages of export income from commercial fisheries.

The value of recreational fisheries is often high in terms of both sociological and economic benefits. In some areas, participation in recreational fishing is particularly high, and the amount spent by anglers is more than the value of the commercial catch. In amateur fisheries the management objective may be to maximize participation by restricting catches to a small bag limit. However, some recreational fisheries may be managed to maximize expenditure, as in the case of requiring game fishers to pay a high licence fee to catch fish such as marlin. The choices in this case are between maximizing participation and maximizing recreational expenditure. From a political viewpoint, large numbers of recreational fishers and amateur fishing clubs represent a powerful lobbying force which has a strong influence on fisheries management policy.

Resource allocation problems also exist where exploited species are distributed over the coastline of several adjacent countries or states. In such cases, allocating a share of the total resource to fishers from different countries is often the subject of difficult negotiations. The problems are even more severe in cases where different parts of the same unit stock, e.g. juveniles, adults and spawners, are unequally distributed in different countries. Take the example of a migrating fish species which grows as it moves along the coastlines of different countries. Scientific advice may suggest that younger fish should be protected by management measures to prevent growth overfishing. However, fishers in a country with access only to young fish are unlikely to support regulations that prevent them catching small individuals, which are going to grow and be caught later by fishers of another country. On a larger scale, many highly migratory stocks, particularly some species of tuna, are fished off the coasts of many countries, and in the open ocean by many distant-water fleets.

Under the Law of the Sea Treaty, a coastal country has control of offshore areas, an exclusive economic zone (EEZ), out to 200 nautical miles from its coastline or outer reefs (see Box 6.5 'Exclusive economic zones'). In the case of stocks shared by two or more states (straddling stocks),

Box 6.5
Exclusive economic zones

Under European law, coastal countries have had jurisdiction over a narrow band adjacent to their coastlines since the seventeenth century; the width of this band was set at three nautical miles, the range over which a cannon could be successfully used. The ownership of the open sea, however, was in dispute until the late 1950s, when a series of United Nations conferences produced the Law of the Sea Treaty. The treaty allowed countries to take control of offshore areas out to 200 nautical miles from their coastlines or outer reefs. Within this area of open sea, known as the exclusive economic zone (EEZ), a country has the right to control economic activity which includes exploiting and managing fisheries resources. Under the treaty, many coastal countries gained control over large areas of ocean. Even in the largest of the world's oceans, the Pacific, most of the open sea is controlled by individual small countries, which in some cases became the custodians of valuable pelagic fisheries.

The declaration of EEZs has resulted in a fairer distribution of economic power. The developed nations, no longer assured of a cheap source of fish from the seas of other countries, were forced to negotiate with the owners of the resources – often poorer countries. Many coastal countries allow foreign nations, including Japan, USA, USSR, Korea and Taiwan, to fish for open-water species such as tuna in their EEZs under either joint ventures or bilateral (or multilateral) agreements.

Joint ventures are partnerships between foreign and local fishers, and provide foreign investment and expertise, as well as a progressively increasing involvement of the local partners. Bilateral and multilateral agreements require foreign fishermen to pay fees for access to fish stocks not fully utilized by national fishers. By allowing foreign vessels into their EEZs for access fees, small nations can avoid the foreign debt involved in building their own fishing fleets. The choice between the two types of arrangement involves comparing the short-term gains from access fees under bilateral agreements with the possible long-term gains from increasing local involvement under joint ventures.

Whichever country harvests the resource, the responsibility for resource management rests with the country controlling the EEZ. It is in that country's interest (and not necessarily in the foreign fishing country's interest) to ensure that catches taken are sustainable.

the Law of the Sea Convention states that the countries shall seek to 'agree upon the measures necessary to coordinate and ensure the conservation and development of such stocks'.

6.4.2 Management objectives and strategies

As fisheries management must often address social, political, legal, economic and biological factors, the overall objectives of fisheries management will almost always involve compromise. This reflects the impossibility of optimizing several individual, and seemingly worthwhile, outcomes at once.

If, for instance, the objective is to maximize profits from a resource without biologically overfishing it, it may be considered that it is more efficient economically to take a large catch every few years than to take a smaller catch every year. However, it could be that this strategy, although economically efficient, is socially unacceptable, causing an irregular income for local fishers, and a discontinuous supply of product to local processors. However, irrespective of economic and social objectives, the overriding considerations are to ensure the long-term sustainability of fisheries resources and minimize disruption to marine ecosystems.

Objectives have been summarized by Clark (1985), Hilborn and Walters (1992) and Jennings

Table 6.4 Example objectives and strategies.

Example objective	Example strategies
Biological/ecological	
To maintain spawning stock biomass at 300 000 tonnes	Spawning areas or seasons closed to fishing Constant escapement harvesting strategy
To reduce catches of key bycatch species by 15%	Test exclusion devices and require installation on all commercial vessels
Economic	
To maximize profits by maintaining fishing effort at f_{MEY}	Remove (inefficiency) controls on fishing gear and reduce number of fishing licences
To increase export earnings by 12%	Conduct seafood handling training for fishers and implement seafood safety regulations
Social/political	
To increase participation (in an artisanal fishery) by 20%	Place bag limits (individual quotas) on fishing units Ban use of overly-efficient gear
To limit conflicts between recreational and commercial fishers by 35%	Ban commercial fishing at weekends and in depths of less than 10 m

et al. (2001) and some examples with possible strategies to achieve them are presented in Table 6.4. Objectives are categorized as biological, economic, social or political objectives although it will be noted that many address more than one of these categories. Strategies are put into effect by the application of the fisheries regulations described in Section 6.5.

Objectives should include reference points and indicators that allow an assessment of the degree to which they have been achieved; these are discussed in Section 6.4.3. Once the objectives of fisheries management are set, strategies must be developed to achieve them. Ideally, a strategy should be designed to cover a period of several years in a stable fishery but may be for a shorter period in either a developing or overexploited fishery. For example, in a stable fishery that is not under threat, an appropriate long-term strategy might be to maintain fishing mortality at its present level. However, in a fishery that is believed to be overexploited, an appropriate strategy could be to reduce fishing mortality by a certain

percentage per year; this strategy is obviously not a long-term one. In any case, strategies must address specific objectives referring to quantities and time periods wherever possible.

Strategies may be based on stock sizes and the amount harvested. These strategies, which are usually long-term, include a constant exploitation rate strategy, a constant stock size strategy and constant catch strategy (Table 6.5). Under a constant harvest rate strategy, a constant proportion of a stock is taken each year. An example is the trochus, *Trochus niloticus*, harvest in the Cook Islands in which fisheries agency staff assess trochus stocks and permit a total allowable catch (TAC) of one third of the total population size. In a larger fishery, with a fluctuating stock size, a highly variable TAC may have consequences on processing plants that hire staff and have export contracts to fulfil.

Under a constant stock size strategy, catches are set at a level that allows a constant number of mature adults to spawn each year. This may involve monitoring and truncating fishing

Table 6.5 Three harvesting strategies.

Strategy	Catch . . .	Assessments	Typical tactics/results
Constant harvest rate	. . . is constant proportion of fishable stock	Annual assessment of fishable stock size	Total allowable catch adjusted each year
Constant escapement or constant stock size	. . . allows escapement of a constant number of fish to spawn	Annual assessment and monitoring	Gear restrictions (mesh sizes) and fishing season adjustments
Constant catch (fixed quota)	. . . does not vary from year to year	No annual assessment required	Stock underexploited when stock numbers are high and vice-versa

seasons or imposing gear restrictions (say on mesh sizes in nets) to allow the set number of mature individuals to survive until the following spawning season.

A constant catch strategy involves taking a fixed quota or constant catch each year. The danger is that natural fluctuations in recruitment and subsequent stock abundance are not allowed for and the fishery will be underexploited when stock numbers are high and overexploited when stock numbers are low. New Zealand has had a set quota of 200 000 tonnes for the hoki (whiptail, blue hake or blue grenadier) fishery for many years although this has been adjusted upwards when strong year classes have boosted the stock.

Constant stock size strategies reveal nothing about recruitment levels at other stock sizes, and it has been proposed that strategies should be varied in order to allow comparisons between alternative management strategies (see Box 6.6 'Adaptive management strategies').

6.4.3 Reference points and indicators

Part of the management process involves translating an objective into a target for the exploitation of a fishery. A target or reference point can be defined in terms of stock size (say, a particular number of spawners) or fishing (say, a particular fishing mortality) or any other biological, ecological, economic, or social objective. A target reference point (TRP) suggests a desirable level of effort and production at which management

actions, including controls on fishing and on catch, should be aimed (FAO, 1999).

However, many TRPs, particularly those designed to maximize yield, have resulted in depleted fish stocks and there appears to be a mismatch between the precision of assessment and the precision of management (Caddy & Mahon, 1995). In spite of considerable investment in stock assessment methodology, fisheries around the world have been overexploited through over-running the target reference points proposed by fishery scientists.

For a particular objective, a target reference point (TRP) represents an optimum position. In the past a common objective has been to maximize yield from a fishery and, in this case, the value of fishing mortality, F_{MSY}, required to secure the maximum sustainable yield, MSY, has been set as a TRP. However, TRPs that maximize yield have resulted in the overexploitation of fisheries and it is now considered more appropriate to prescribe lower and less risky rates of exploitation. TRPs in terms of the fishing mortality intended to maximize yield, such as F_{max} and F_{MSY}, are being replaced by lower levels of F such as $F_{0.1}$ and $2/3F_{MSY}$. In these cases, $F_{0.1}$ is the value of F at which the slope of the yield per recruit graph is one-tenth of its value near the origin (see Fig. B5.7.1 in Chapter 5) and $2/3F_{MSY}$ is two-thirds of the value of the fishing mortality needed to obtain the MSY. There are no scientific reasons for the choice of these lower values of F other than that they are believed to increase

Box 6.6
Adaptive management strategies

'You never know what is enough
unless you know what is more than enough.'

William Blake, 1757–1827.

A conservative strategy of gradually increasing fishing effort has been recommended for new and developing fisheries (Gulland, 1984). One or two additional vessels or fishing units can be allowed into the fishery each year over a period of many years, during which the responses of the stock to increasing fishing effort are monitored. However, this gradual acquisition of fishery-dependent data on the stock will often be at the expense of forgone revenue from the fishery if the stock is underexploited over many years of development. From an assessment point of view, the effects of small increments in effort on the stock (considering its inherent variability) are hard to detect.

In an established fishery that has been exposed to steady levels of fishing effort over a long period there will be a paucity of information on how the fishery will respond to changes in management strategy. In such a fishery, stock size may not vary greatly (there is said to be little contrast) and it is impossible to know how it would respond to increases in fishing effort. In particular, there is no information on how recruitment would change over a range of stock sizes.

One of the criticisms of applying a single management strategy to a fishery is that there are no results from alternative strategies with which to make comparisons. The scientific method is to compare results from experimental and control situations, or to make observations over a range of different experimental conditions. An alternative is to gain some empirical evidence by experimentation, by observing how the stock behaves under different strategies including

different levels of exploitation. The deliberate manipulation of a fishery in order to learn more about it is called adaptive management (Walters & Hilborn, 1976; Walters, 1986; Hilborn & Walters, 1992).

Adaptive management is often difficult to put into practice without affecting the livelihoods of those fishing. It may be easier to design and apply such experiments in a fishery based on a species that exists in widely-separated substocks between which exchange rates are low; examples include exploited species that inhabit separated seamounts, reefs, bays and estuaries. Each substock is regarded as an experimental unit, in which alternative strategies (different levels of fishing effort, for example) are applied and the responses to each experimental condition recorded. In an ideal experimental situation, only one factor (such as the level of fishing effort) should be altered at a time, and the experimental design should include control areas that are left unfished, replicates in which each treatment is applied to more than one area, and the random or systematic selection of areas in which different levels of exploitation are applied (Hilborn & Walters, 1992). The analysis of catch rates and yield from a number of different village fishing areas, each with different levels of fishing effort, in American Samoa presented in Chapter 5 (Box 5.3) is an example of an 'unplanned' adaptive management experiment brought about by a fortuitous situation.

As suggested in the quote at the top of this box, the only reliable way to know the sustainable catch level in a fishery is by exceeding it. But the deliberate overfishing of a single spatially homogenous stock in order to learn more about it would be unacceptable on social grounds. The precautionary approach (FAO, 1995b) encourages fisheries managers to control access from the start of a fishery and place conservative limits on fishing mortality and ecosystem impacts.

profitability and provide a buffer against over-fishing (Gulland & Boerema, 1973).

The concept of reference points is illustrated in Fig. 6.4. After an objective has been translated into a target reference point (the upper horizontal line in Fig. 6.4) there is a need to evaluate and monitor how well, or otherwise, the management strategies are performing in relation to the objective. This involves the use of an appropriate indicator. If, for example, the single objective for a particular fishery was to maintain a certain level of spawning stock biomass, some estimate, proxy or index of this could be used as an indicator. The mean catch rate of sexually mature individuals, for example, would be one practical way of measuring the relative size of the spawning stock (note, however, that the use of CPUE as an index of stock size assumes that the catchability coefficient, q, remains constant – see Section 4.2.4). An alternative would be to aim for an exploitation rate such that the average size of fish caught is equal to, or greater than, the average size at maturity; at this rate, at least 50% of individuals would have an opportunity to reproduce. For the sake of the example, say the reference point target is set at 110 kg of sexually mature fish per hour trawled. The fishing strategy and fishing regulations are then applied with the aim of keeping the spawning stock biomass indicator at or above the target reference level.

To avoid overexploitation there is a need to define a threshold or limit reference point (LRP) that indicates the fishery is in an undesirable situation and immediate management action is needed. In the case of the example given in Fig. 6.4, an LRP of 60 kg of mature fish per hour has been considered unacceptable because the low number of spawners that this suggests may cause recruitment to fail. If the indicator drops below the LRP it triggers urgent, pre-planned actions that have been designed to allow the spawning stock to recover. In general, an LRP can be some minimum level (e.g. a dangerously low spawning biomass) or some maximum level (e.g. an unreasonably high mortality rate).

Management performance is measured as the vertical distance between the indicator and the target reference point – e.g. the indicator is about 45% above target in year 4 and 20% below target in year 9 of Fig. 6.4. The value of plotting an indicator as in Fig. 6.4 is that changes in the indicator and performance can be followed over time. Irrespective of fishing levels, the indicator will fluctuate from year to year, but a continuing decrease in performance over time would be a cause for concern. The decreasing trend in the indicator after year 2 would make fisheries managers uneasy, particularly when it dropped below the target reference point after year 7 and continued to fall. At this stage some pre-planned actions

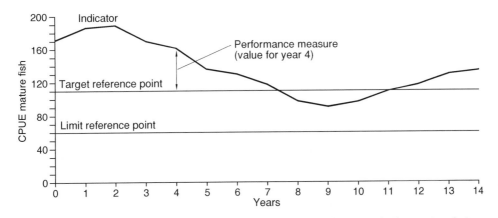

Fig. 6.4 Values of an indicator (catch rate of sexually mature individuals in this example) by year in relation to reference levels (shown as horizontal lines). Performance is measured as the vertical distance between the indicator and the target reference point. Adapted from FAO (2003).

would have to be taken, such as increasing mesh sizes in trawl nets and reducing the length of the fishing season.

Indicators can be devised to track ecological, economic and social objectives as well as biological (stock-related) ones. For many clearly stated biological and economic objectives, the indicators are self-evident; if an economic objective, for example, was to maintain fishing effort at f_{MEY}, the indicator would be a measure of effective fishing effort in the fishery. Many social and political objectives, however, are often more difficult to quantify; for example, if the objective is to increase the well-being of members of a fishing community, an indicator that measures well-being may be more difficult to quantify. Gathering data for assessment and management is reviewed in Evans and Grainger (2002).

In a fishery that is overexploited, an appropriate objective might be to increase the stock biomass over a number of years and the indicator would be expected to increase by a certain percentage per year for a certain number of years. In this case, the target reference will be in the form of a line with a positive slope rather than the horizontal one shown in Fig. 6.4.

In spite of criticisms of the maximum sustainable yield (MSY) as an objective (Larkin, 1977) it remains useful in many fisheries as an extreme upper limit or limit reference point (LRP). The fishing effort required to obtain MSY is f_{MSY} and this is often a dangerously high level of exploitation. The use of a particular level of fishing effort, f, as a target reference point is also dangerous in that it does not always have a constant relationship with fishing mortality. Small improvements in fishing technology, for example, will result in increases in the catchability coefficient and therefore increases in fishing mortality, even though fishing effort remains the same. That is, *effective* effort may increase even if *apparent* effort does not (see Box 6.7 'Technology creep' in Section 6.5). For these reasons, the regulation of fishing mortality, F, is a more direct way of controlling exploitation and it is more usually used in biological reference points.

Maximizing profits from a fishery is an appropriate objective in a commercial fishery, particularly if the major part of the catch is exported. In this case, the corresponding target reference point would be F_{MEY}. The advantage of maintaining fishing mortality at the level required to achieve maximum economic yield, MEY, is that it is lower than that required to maximize yield in weight and the possibility of recruitment overfishing is reduced. Social problems associated with fishing at MEY is that large excess profits may be made by a relatively few fishers, usually those that are licensed to participate in the fishery. The view may be held that the community, which has had to forgo its right to fish in favour of those participating in the fishery, is entitled to share in these profits. In such cases, excess profits from the exploitation of a public resource may be shared with the community by mechanisms such as resource taxes, often in the form of licence fees.

In many cases, management objectives are based on biological and economic reference points, in terms of fishing mortality and spawning stock biomass and a range of these are shown in Table 6.6. Indicators can also be used to reflect broader social and ecological objectives such as those addressing the impacts of fishing on the ecosystem. An example of an ecological objective is 'to reduce catches of key bycatch species by 15% within 3 years'. The obvious indicator for this objective is the CPUE of each of the key bycatch species and a target reference value would be set at 15% less than the current CPUE.

With increasing concerns relating to marine ecosystems, it is desirable to set limit reference points using a precautionary approach (FAO, 1995b; Garcia, 1996; Caddy, 1998). The precautionary principle is commonsense and simple – its elements are that managers must take into account uncertainty and anticipate the possibility of ecological damage, rather than react to it as it occurs. The precautionary principle states that the absence of full scientific certainty should not preclude action where irreversible or major detrimental effects are possible. Before proceeding with an activity, the degree of caution should be proportional to the degree of potential impacts

Table 6.6 Examples of reference points or values based on spawning stock biomass, B, and fishing mortality, F. From FAO (1999).

B_{lim}	lowest acceptable spawning biomass
B_{LOSS}	lowest biomass ever observed
B_{MSY}	biomass at the maximum sustainable yield
B_{pa}	biomass set as a target under the precautionary approach
$B_{x\%R}$	biomass at which the average recruitment is x percent of the maximum predicted by the spawner–recruit relationship
$F_{0.1}$	F at which slope of yield per recruit graph is one-tenth of its value near the origin
F_{MSY}	F corresponding to the maximum sustainable yield
F_{MCY}	F corresponding to the maximum consistent, or long-term, yield without reducing stock below a predetermined level
F_{MEY}	F corresponding to the maximum economic yield
F_{pa}	F set as a target under the precautionary approach
F_{crash}	F that would cause recruitment to fail
F_{lim}	F set at the highest that is acceptable based on some specified criterion

and the ability to take remedial action in the case of deleterious effects. One of the precautionary measures involves setting reasonable limit reference levels, and taking firm actions when these are approached. Under a particular management strategy, the risks of an indicator dropping below a reference point may be estimated by modelling as described in Chapter 5 (Section 5.5). Repeated runs of a stochastic simulation model can be used, for example, to count the number of times that the indicator drops below a reference point in 10 000 simulation runs.

Most fisheries have more than one objective, including ones that address social, economic, ecological and biological goals and some of these may be conditional. For example, an objective could be to maximize employment (for which the indicator would be the number of people employed in the fishery) subject to the condition that stock size remains above 50% of the virgin biomass (for which the indicator would be from survey results or catch rate data). There may also

be several indicators used in combination; for example, a minimum mean size of fish in the catch and a minimum spawning stock biomass could both be used as limit reference points either in conjunction or separately to trigger urgent management action.

Quantitative ecosystem indicators for fisheries management have been the subject of a symposium (Cury & Christensen, 2005) at which it was concluded, among other things, that a suite of indicators is required (relating to different data, groups, and processes) and that aggregated indicators can provide a quick evaluation of the state of marine ecosystems.

The removal of several species from an ecosystem is likely to cause significant changes, particularly if keystone species are affected (see 1.4.1 'Coastal waters'). A reduction in species diversity is believed to result in a higher rate of ecosystem collapse and longer recovery times (Worm et al., 2006; Section 6.5.3. 'Controls to protect ecosystems'). In multispecies fisheries, exploitation results in a shift towards more smaller and often less valuable species in the catches and this could be used as a diversity-based reference point. That is, a limit reference point could be set at a particular level of species diversity (say, the number of different species per unit effort) or at a mean size of species in the catches. The indicators dropping below these limit reference points would trigger some pre-planned management actions.

Maintaining a minimum or buffer stock size may improve the stability of catches from year to year and this may make better use of fishing fleet capacity and processing facilities. In stocks reduced by heavy fishing, catches consist of relatively few year classes, and may therefore fluctuate according to the strength of each newly-recruited year class. Although a buffer stock size may minimize fluctuations in catches due to variations in recruitment, it could be at the expense of higher long-term catches.

Objectives may include ensuring (at least within some level of probability) that the spawning stock biomass does not decrease below some minimum value, B_{lim}. The decline of many fisheries has been due to overexploitation and recruitment failure at low levels of the spawning stock. Many of the world's great fisheries, particularly those based on clupeid species such as sardines and herrings, have collapsed due to recruitment failure (Cushing, 1971). Although many species show great resilience to reductions in stock levels, recruitment failure is the ultimate fate of all stocks reduced below some critical level. The minimum level of spawning stock required to keep recruitment at a sufficient level is often not known, although previous low spawning stock levels that have caused no reductions in recruitment may be used as a reference limit. Whether or not a species can maintain recruitment levels in the face of a low spawning stock biomass depends on the shape of the stock–recruitment relationship. As described in Chapter 4, the level of recruitment is often highly variable, even in the absence of fishing. Theoretical stock–recruitment relationships with different consequences for exploitation are shown in Fig. 6.5.

Species with low fecundity and delayed maturity often have a Beverton and Holt-type stock recruitment curve, which has a low gradient (curve A in Fig. 6.5). In this case, reductions in the spawning stock size will cause decreases in recruitment from the start of exploitation (at S_A). And in the case of overexploited stocks, a reduction in fishing effort will allow only a slower rate of recovery. Species such as cod, however, with a Ricker dome-shaped stock–recruitment curve

(curve B), are more resilient to exploitation. Moderate reductions in the spawning stock size will cause compensatory increases in the level of recruitment. Curve B suggests that, for this example, a minimum spawning stock (S_B) of 50% of the virgin stock is required to maintain high levels of recruitment. Species such as penaeid prawns or shrimps may have a stock–recruitment relationship, as shown in curve C, which is flat over a large range of spawning stock sizes.

It is the steepness of the initial part of the curve which determines whether recruitment either decreases gradually, or collapses suddenly at low stock levels. If the initial slope is low (curve A), recruitment will decrease steadily with reductions in stock size, and managers often have many years of warnings in the form of reduced catch rates of mature individuals. However, if the initial slope is high (curve C), reducing the stock to very low levels will cause recruitment to collapse without warning. Curve C suggests that, for this particular species, recruitment would collapse at a spawning stock level (S_C) of about 35% of the unexploited stock level.

6.4.4 Management plans

Many fisheries are managed in a *laissez-faire* manner, in which a management authority abstains from involving itself with a fishery unless a crisis occurs. Under this type of management, the authority will need to take actions (and apply regulations) in a reactive way by responding to crises and problems as they occur. In the worst case, the management agency will be faced with the unenviable task of reducing fishing effort in a fishery that has been overexploited. Having to reduce effort in an overexploited fishery is a difficult and unpalatable undertaking. The remedial measures required may include such draconian steps as reducing fleet size, or closing a fishery for a number of years, with severe social, economic and political repercussions. This style of management is unacceptable, particularly with the present poor state of fisheries resources.

Objectives, strategies and necessary actions should be clearly documented and well-

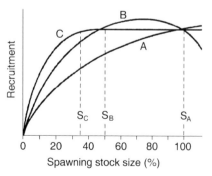

Fig. 6.5 Theoretical stock–recruitment relationships illustrating different consequences of spawning stock size reduction. The virgin (unexploited) biomass is indicated by the vertical line at 100%.

publicized in a fisheries management plan. Having a plan allows management authorities, groups or communities to carry out actions in a proactive way and thereby reduce the possibility of problems occurring. As suggested in the widely adopted precautionary approach, preventing overfishing in the first place is preferred to reacting to it after it happens. Ideally, the plan will have been produced with at least some degree of stakeholder input and a written plan therefore will be an important record of what was agreed to and why. This is important as different stakeholder groups may have different aims and must reach agreement on the management objectives for the fishery (see Section 6.3.2 'Co-management in commercial fisheries').

Management plans may address a particular number of resource species or a particular fishing area (or ecosystem). The former is justified if the target species are few and the ecosystem is not complex or threatened (e.g. a management plan for a local tuna longline fishery). The latter is necessary if the fishery is a multispecies one and the ecosystem is complex or threatened (e.g. a management plan for a large bay or an estuary). Whichever of the two is appropriate, environmental issues involved with a particular fishery must always be addressed in the plan.

Fisheries management plans follow many different formats according to the requirements of the various authorities. Those produced in the USA, for example, are required to include a measurable definition of overfishing for each managed stock. Nevertheless, the following list of headings suggests the general elements of a good plan.

1 **Background, history and status of the fishery.** A history of the fishery, with tables of annual catches and fishing effort, if available. Developments that have affected the fishery and the marine environment should be included.

2 **Present status of, and threats to, the fishery.** Current and foreseen threats from overexploitation and environmental degradation should be clearly documented with as much supporting evidence as possible.

3 **Policy goals.** Broad economic, social and ecological aims (usually set by the government)

should be clearly stated with a history of how policies have changed.

4 **Objectives and reference points.** This section translates the policy into practical-level goals or objectives and the expected results of implementing the plan should be clearly presented. Reference points, the desired positions in relation to the objectives, must be defined (e.g. to increase the size of the spawning stock by 15% over five years).

5 **Strategies.** The plan of action needed to achieve specific objectives.

6 **Actions.** The specific actions (including the imposition of fisheries regulations) must be detailed – some standardized format such as the logical framework matrix, or logframe, can be used to summarize expected outputs (Table 6.7). Reference to the timing of planned actions should be included preferably in the form of a time line, an example of which is shown in Fig. 6.6.

7 **Risk analyses.** Document what will be done when things go wrong, as they undoubtedly will. Bad weather may delay surveys, equipment may not function, agreements with stakeholders may take longer to achieve than planned; the risks to achieving results will be many. Risks and remedial actions can be included in columns of the logframe matrix shown in Table 6.7. The risks of an indicator dropping below a reference point may be estimated by repeated runs of a simulation model as described in Chapter 5 (Section 5.5).

8 **Monitoring – indicators and performance measures.** A programme for regular monitoring and evaluating results against the reference point. The indicator must be defined and methods of obtaining indicator values detailed.

9 **Evaluating and reporting.** An allowance for an annual review of performance – the vertical distance between the reference point and the indicator (see Fig. 6.4).

10 **Financing the plan.** The cost of implementing the plan, including the costs of surveys, research and enforcement, must be presented. Methods of having the users and beneficiaries of the resource pay for management costs must be considered.

Whether a fisheries management plan is developed under co-management arrangements or

Table 6.7 A variation of a logical framework matrix, or logframe, which includes objectives, actions, reference points, assessments of performance, possible risks and remedial actions to be taken if necessary. Definitions are included in the first row and an example is given in the second row.

Objective	Strategy/action	Reference point (target)	Assessment of performance	Risks	Remedial actions
Definition The expected result or goal	Tasks required to achieve the objective	Desired position in relation to the objective. Includes quantity and timing	Methods used to quantify indicator and measure performance	Risks to achieving the objective	Actions to be taken if output is not achieved
Example To increase the size of the spawning stock	Extend season closed to fishing by 3 weeks	15% increase in number of mature individuals (over 20 cm) within 3 years	Underwater transects at a minimum of 6 sites	Extension of closed season ignored by many fishers	1 Seek more stakeholder support 2 Increase publicity 3 Increase enforcement

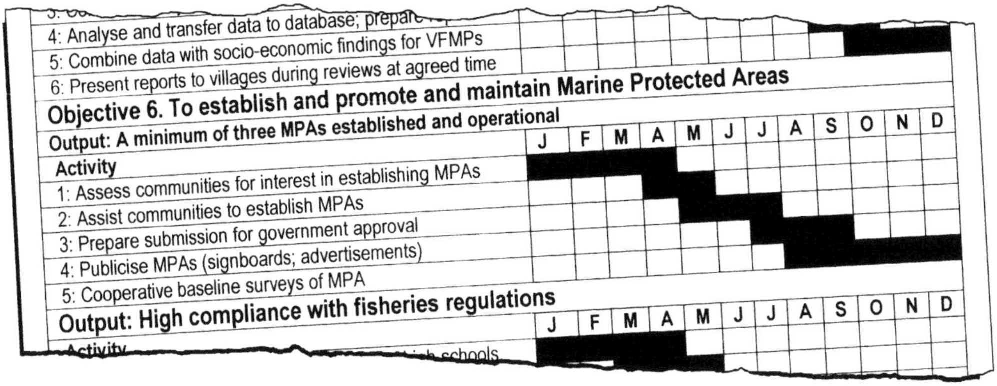

Fig. 6.6 Portion of a management plan showing time lines for activities leading to the establishment of community-owned marine protected areas.

not, the plan must be made available to all stakeholders. Objectives, strategies and management actions must be clearly presented with justifications. The plan should be copied or printed and contained within a cover so that its appearance enhances its status as an important document. An executive summary should be included for those not wanting or needing to read the detailed document.

Fisheries management plans must conform to the United Nations 1982 Convention on the Law of the Sea, the FAO Code of Conduct for Responsible Fisheries (1995a) and the precautionary approach to management. Plans, for example, are required to include the management of bycatch species, impacts on the ecosystem and the effect of fishing gear on marine habitats. In cases where the management of the coastal environment, including estuaries, salt marshes, mangrove forests, and seagrass beds is not the responsibility of the fisheries management authority, it is necessary to link the fisheries management plan to other relevant government authorities. Fisheries objectives must be integrated into the general multiple-use framework of coastal area management.

6.5 Management actions

Excessive fishing effort is the major cause of depleted fish stocks and the situation is aggravated by damage to marine ecosystems. Thus in many cases, the most important management actions involve restricting or even decreasing fishing effort, reducing the incidental catch of non-target species and setting aside marine reserves to protect ecosystems.

Management actions include the conventional fisheries regulations or technical measures that are imposed on a fishery to support a strategy designed to achieve predefined objectives. It is unlikely that any single management measure will produce the desired results, and a combination of several regulations is often applied. The regulations discussed in this section are used either to restrict or control fishing effort (input controls such as restricting the number of fishing licences), restrict the catch (output controls such as quotas) or protect marine ecosystems.

Regulations such as controls on the type of fishing gear allowed are included under input controls, and those such as size limits to protect small individuals are included under output controls. It is convenient to separate controls that have a broader effect in protecting the marine environment – these measures include the closing of areas to fishing and are included under the heading of environmental controls.

The controls summarized in Fig. 6.7 can be imposed by a management authority which may be a government agency, a stakeholder group, a community or a combination of these under a co-management system. Some controls are designed to protect particular parts of the stock. Small individuals can be protected by regulations such as minimum mesh sizes, legal minimum lengths, and closures, and the breeding stock can be protected by closed seasons. Besides applying regulations to the fishing operation itself, controls can be applied at any convenient point in the post-harvest chain. It may be easier, for example, to prevent small fish (below a legal minimum size) being purchased by a few processors than it is to inspect and regulate the catches of a large number of fishers.

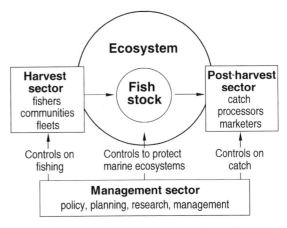

Fig. 6.7 A fishery system showing the relationship between the management sector, the harvest sector, the post-harvest sector and the stock as part of an ecosystem. Management controls are used to contain fishing (input controls), restrict the catch (output controls) and protect the marine environment.

6.5.1 Input controls (on fishing and fishing effort)

Limiting the number of fishing units

Licence limitation, as opposed to an open-access system, has been introduced in many fisheries but not enough to prevent the widespread depletion of resources. In licence-limited fisheries, a number of vessels, or vessel owners, are given licences to fish. The number of licences issued is set at a level at which it is believed capable of imposing some predetermined level of fishing mortality, or taking a total allowable catch.

The problems of a licensing system are the difficulties in devising fair methods of selecting licence-holders and, in the case of highly profitable fisheries, the jealousy and potential lobbying by fishers excluded from the fishery. A proportion of the excess profits from restricted-access fisheries may be collected by mechanisms such as taxes or licence fees. In most restricted-access fisheries, licences may be transferred from one fisher to another, and the value of a licence will reflect its future earning ability. If fishers retain all, or a high proportion of the economic rent from a fishery, licences can acquire a very high value. In terms of loan repayments, this places a high cost burden on new entrants, and makes it difficult for younger fishers entering the fishery. Purchasers of licences, faced with large loan repayments, may work harder by fishing longer hours, and in rougher weather, than the original licence-holders. The result of this is that effective fishing effort increases even though the number of licences remains the same. Unless other regulations prevent it, licence-holders may try to maximize effective fishing effort by improving equipment, or by purchasing larger boats or engines, which can lead to overexploitation and overcapitalization.

An issue connected with licence limitation is a vessel replacement policy. In a fishery in which there are no replacement restrictions, fishers may replace their vessels with larger and more efficient ones, resulting in an increase in effective fishing effort. At the other extreme, not allowing vessel replacement will result in an increase in the age and inefficiency of vessels in the fishery, and force fishers to operate unsafe boats. Compromises such as allowing replacement within certain limits of boat length, carrying capacity, or engine power are often used.

One of the problems associated with licence-based fisheries is that the efficiency of fishing may increase even though the number of fishing units does not. Examples include the use of the global positioning system (GPS) to set nets and traps in areas of known high stock density more quickly, resulting in more fishing time per fishing trip. Tuna longline fishers are able to use satellite sea surface temperature information to set longlines on oceanic fronts or 'hotspots' and this has increased the efficiency of fishing effort (see Box 6.7 'Technology creep').

As an example of the effect of increasing effective effort, consider the case in which 100 licences are issued on the basis that this represents the fishing effort required to secure the maximum economic yield (MEY) from a particular developing fishery. If the licensed vessels subsequently make collective improvements in fishing methods that increase catches by 50% for the same effort, this is equivalent to having 150 licensed vessels at the original level of efficiency.

Limiting the efficiency and types of fishing gear

Another way of controlling effective fishing effort is by limiting the efficiency and types of fishing gear. Fishing gear can be limited in size, type and number, and some types of gear and fishing methods can be banned. Trawl nets can be limited to a maximum headline length, hook and line gear can be restricted to a maximum number of hooks per line, and gill nets can be restricted to a maximum length and hanging ratio. Some highly efficient or destructive gear may be banned in particular fisheries. For example, the use of scuba gear may be banned in dive fisheries for abalone and tropical lobsters, tangle nets and mono-filament gill nets may be illegal in some fisheries, and purse seine nets may be disallowed in some pelagic fisheries.

Gear regulations are often imposed to allow resources to be shared by a larger number of fishers. These regulations can be appropriate in the case of artisanal fisheries, for instance, where the resource provides employment or food for a large number of fishers; in this context employment and the provision of food may be chosen in favour of efficiency. In some cases, the survival of a resource species depends on inefficient exploitation!

In a strictly commercial fishery, where profits are the objective, there is a good case for allowing fishers to become as efficient as possible. From an economic point of view, gear regulations usually lead to economic inefficiency, and raise the cost of catching fish. This places commercial fishers at a disadvantage with respect to competition from other regions, or from other sources of protein. From an economic viewpoint, it is preferable to have a smaller number of efficient fishing units than a larger number of inefficient ones. Hence, in the case of a commercial fishery threatened by overexploitation, many fisheries management authorities would choose to reduce the number of fishers in preference to imposing gear restrictions (see Box 6.8 'Licence buy-back schemes').

Total fishing effort each year can also be limited by restricting fishing to a set number of fishing days. In some fisheries in which both commercial and recreational fishers are involved,

Box 6.7
Technology creep

Management objectives based on controlling fishing effort, rather than catches, suffer from the fact that increases in efficiency will cause increases in *effective* effort even though *apparent* effort remains the same. The gradual increase in the efficiency of fishing methods and gear is sometimes referred to as technology creep. Often the acquisition of new technology is not gradual at all. The competitive nature of fishers, as well as the fact that gear purchases are tax deductible in many countries, often results in new developments spreading rapidly through a fishing community.

The result of any improvement in efficiency is that one unit of fishing effort, say one hour of fishing, will take a greater proportion of the stock than before. Fig. B6.7.1 shows catch per unit effort (CPUE) under three different conditions. Curve A shows the reduction in CPUE over time under conditions in which fishing efficiency is unchanged. Curve B shows a situation in which there is some development of the fishing gear that is adopted by the fleet in year 7. Curve C shows a gradual improvement in efficiency that keeps CPUEs at a relatively high level even though the stock is being depleted.

If fishing effort had an unchanging efficiency (curve A) the reduction in CPUE would suggest that the stock size had been reduced to half its virgin biomass by year 8. Technology creep, therefore, poses particular problems in stock assessment when the assessors are unaware that effective effort is increasing. Increases in efficiency result in increased profitability under quota-based management, as less fishing time is required to secure the same catch. On the other hand, increases in efficiency under licence-based management may result in overexploitation unless fishing time is reduced.

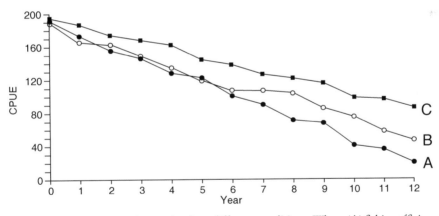

Fig. B6.7.1 Changes in CPUE over time under three different conditions. When: (A) fishing efficiency remains constant; (B) fishing efficiency increases in years 7 and 8; (C) fishing efficiency increases steadily over time.

commercial fishing is banned on weekends to avoid conflict between the two groups. This, and other forms of closures have the secondary effect of reducing annual fishing effort. Closures, which have an important role in protecting marine ecosystems, are discussed in Section 6.5.3.

Minimum mesh sizes and escape gaps
Minimum mesh sizes in nets, and escape gaps in traps are applied in many fisheries to allow small individuals to escape and grow to a more valuable market size (i.e. to prevent growth overfishing). A further aim may be to allow

---

Box 6.8
Licence buy-back schemes

Many licence-managed fisheries suffer from excessive fishing effort, and any increase in economic viability will involve taking management measures to decrease fishing effort. This would, most reasonably, mean a reduction in the number of vessels in the fishery. Various strategies to achieve this end have been considered including vessel 'buy-back' schemes. In a programme to reduce excessive fishing effort in inlets, bays and lakes in Victoria, Australia, the government has instituted a voluntary buy-back programme that allows licence-holders to sell their licence to the programme and agree not to re-enter the fishery for a period of five years.

In large commercial fisheries, a buy-back scheme can be financed by fishers contributing to a fund, or being charged a levy, to purchase a number of licences which are then removed from the fishery. The problem is that if a fishery is overexploited to the extent that a buy-back scheme is required, profits will usually be so low that fishers are unlikely to be able to afford to finance it. An alternative is for the government to provide a loan to finance a buy-back scheme, but repayments will place a large financial burden on fishers. As reduction in effort in an overexploited fishery is unlikely to result in an *immediate* improvement in catch rates, fishers may not be in a position to repay the loan in the short term; in this case, an incrementally-increasing tax system following a licence buy-back programme is an option.

---

individuals to reach a size at which they can reproduce at least once before capture.

Sensibly, the regulation should be applied only when there is some information on the selectivity of the fishing gear in relation to target species (see Section 3.4.1). Ideally, although this is not often the case, there should be estimates of the survival prospects of small individuals which pass through the escape gaps, or the meshes of the net. The selectivity of a species fished by a specific fishing gear is affected by the species' shape. In respect to trawl nets, for example, selection in species with an irregular shape, or long lateral appendages (such as shrimps and prawns), occurs over a wide size range of individuals (curve A in Fig. 6.8). Selection in species with a regular or smooth shape, however, occurs over a narrow size range (curve B in Fig. 6.8). This suggests that minimum mesh size regulations applied to trawls used on species such as prawns are less efficient in conserving small individuals than those applied to fusiform-shaped fish. The effect of increasing the minimum mesh size in trawl nets is shown in the right-hand curve C in the figure.

A problem with controlling mesh sizes in trawl nets and purse seines is that the target species, bycatch and other material may be caught in such

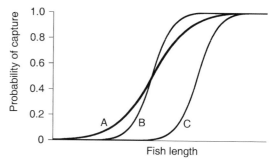

Fig. 6.8 Mesh selectivity curves for species with (A) an irregular shape, and (B) a regular shape. Curve C suggests the effects of an increase in mesh size in nets used to capture species B.

quantities that they block the meshes, making selection ineffective. Also, conventional diamond-shaped meshes tend to close up in heavily loaded trawl nets, making the escape of small individuals difficult. In trawl nets it is also possible for the fisher to distort the meshes deliberately, by adjusting the angle of attack of the otterboards. However, regulations requiring the use of square mesh netting, which is less likely to distort under pressure than conventional diamond-shaped netting, appears to improve the selectivity of nets under most conditions.

In gear such as gill nets which catch a restricted range of fish sizes, a case can be made that increasing mesh sizes exacerbates recruitment overfishing. In gill nets, small fish swim through the meshes, and larger fish bounce off the net as their heads cannot penetrate the meshes. The selection curve is therefore in the shape of a bell-shaped or normal curve (see Section 3.4.1). Gill nets with a small mesh size will therefore catch fish with a small mean length, and not catch larger fish. As large fish produce a much larger number of eggs than smaller fish (because of the usual power curve relationship between fecundity and length), increasing the minimum mesh size will remove the protection afforded to larger fish, and can result in a reduction in the reproductive output of the stock.

6.5.2 Output controls (on the catch)

Quotas

One of the key output controls on fisheries involves the estimation and application of a total allowable catch (TAC). In a global catch quota system, the TAC becomes the quota and fishers compete to secure a large personal catch before the quota is reached, and the fishery is closed. The consequences of this are that fishing seasons become shorter, and costs increase as fishers overcapitalize to gain a competitive edge. In addition, safety is compromised as fishers feel compelled to go to sea irrespective of the weather, and boat maintenance during the fishing season is minimal. Other disadvantages are similar to those for seasonal closures, in which fishing is discontinuous, and, after the season, boats and factories sit idle. An alternative to global catch quotas is to apply catch quotas to individuals, or to individual fishing units.

Individual transferable quotas (ITQs) are the best-known form of property rights in fisheries management. ITQs are allocated to individual fishers, who then have a guaranteed share of the total allowable catch of a particular resource species. The proportion of the TAC allocated to a fishing unit may be decided on the basis of either

equal quotas to all fishers, or be based on past history in the fishery, vessel size, or according to a point score system based on a combination of these. It is also possible to offer ITQ units by auction or tender. An ITQ allocation can be either harvested or traded by the fisher. A guaranteed share of the resource allows a fisher to go fishing when market demand is high, and to spend non-fishing time on boat and gear maintenance. A result of this is that catches are often harvested over a longer period. The capacity to trade an ITQ means that less efficient fishers can sell part or all of their quota to more efficient operators for the market price of the quota.

Problems with ITQs include the difficulties of estimating the total allowable catch and enforcing the quotas. If the TAC was overestimated by scientists, for example, the benefits listed above would be negated, and fishers would be compelled to compete for as large a share of the catch as possible, as in global quota management. On the other hand, if the TAC was underestimated, fishers would lose confidence in the management system. If quotas are based on the landed or marketed catch weight, there will be a temptation for fishers to 'high-grade' catches by dumping small individuals and less valuable species at sea. If this occurs, catch rates will be underestimated, and their use in the assessment of fish stocks will be suspect. Scientists require information on actual catch per unit effort, and this necessitates the validation of landings against the catch at sea. This could involve the random placement of observers on fishing vessels, and a subsequent comparison of catches made by boats with observers on board against those without observers.

It is difficult to apply quotas in a multispecies fishery, unless quotas are estimated for each of the component species and the fishery is closed when the quota for the most susceptible species is reached. Such conservative management would encourage discarding of the most susceptible species once their quotas are reached or result in lower than optimum yields from the other less susceptible species.

It is difficult for authorities to ensure that individual quotas are not exceeded. If quotas are based on catches recorded in logbooks, there is a strong motivation for fishers to under-report catches, and if quotas are based on the catch sold to registered processors, there is a temptation to sell to 'black market' buyers and restaurants. Enforcement which involves the inspection of vessels, or numerous landing ports and black markets for the catch, is expensive.

Quotas, or bag limits, are often applied to recreational fishers, if only to allow a limited resource to be shared by a large number of individuals. Bag limits are usually given in terms of a maximum number or weight of fish allowed to be taken by one person per day. The enforcement costs of applying bag limits to a large recreational fishery, where fishing and landing occurs over a large area, is high. Costs may be reduced by adapting bag limit rules to make it illegal for non-professional fishers to have more than a certain weight of fish (or fish fillets) in their possession at any one time. Some amateur fishers, sometimes referred to as 'shamateurs', make large catches which are sold on the black market, resulting in resentment from other recreational and, particularly, commercial fishers.

Size limits

Limiting the size of individuals retained is one of the oldest of regulations applied to fisheries. The regulation involves returning captured individuals smaller than a prescribed minimum size to the sea. Minimum legal sizes have been applied to many species of molluscs, crustaceans and fish, and are particularly applicable to recreational fishers. In commercial fisheries where catches are large, measuring each individual in the catch and rejecting undersize ones often represents an unreasonable amount of additional labour for fishers. Restrictions on gear sizes (such as mesh size in nets) theoretically achieve the same ends as size limits, but prevent the fish being caught in the first place.

Traditionally, the reasons for the application of a size limit have been to prevent the marketing of fish considered too small, and to allow individual fish to spawn at least once before capture. The former reason is related to growth overfishing,

and whether or not size limits are effective in this respect depends on the growth and mortality rates of the target species. If growth rates are high relative to mortality, then the economic value of the catch may be increased by delaying the size at first capture (see biomass models in Chapter 5).

Allowing individual fish to spawn at least once before capture is related to the prevention of recruitment overfishing. However, minimum legal lengths applied to species with high fecundity may not result in an increased recruitment unless stock sizes are very low (curve C in Fig. 6.5). On the other hand, in some heavily exploited fish stocks, increases in minimum legal lengths have been shown to result in small, but significant benefits. Although it may be difficult to justify the imposition of size limits in many species (particularly in those with high mortality rates, and high fecundity), it remains popular with fisheries managers, perhaps because there is general public sympathy for such regulations. From a social viewpoint, well-publicized and enforced size limits are a constant public reminder of the need for conservation.

The regulation is only applicable in fisheries where the catch is not harmed by the catching method, such as molluscs gathered by hand, or crustaceans caught in traps. Although some shallow-water fish caught on hooks may survive well if returned to the water immediately, this type of regulation has little application to trawl-caught and deep-water fish species. The probabilities of fish surviving after being hauled to the surface from deep water and released are likely to be small. Ideally, the survival probabilities of animals which have been caught, measured, and released should be assessed before the imposition of minimum legal size regulations. These probabilities will vary greatly with the method of capture, the post-capture handling, and the delay before individuals are returned to the sea.

Minimum legal sizes pose problems when applied to species which are sexually dimorphic (in which each sex has a different ultimate size), and to those which are sequentially hermaphroditic (which change sex). In such species, the application of minimum sizes may concentrate fishing pressure on one particular sex. In a sexually dimorphic species where, for example, females grow to a larger size than males, most of the legally-sized fish in the catch will be female. This can also be the effect in the case of protandrous hermaphrodites, where males change sex to become females at a particular age or size.

The application of minimum legal sizes assumes that fishers are able to identify the young of the regulated species, and are familiar with the prescribed method of measurement. In some species, the young have quite a different appearance from adults. Minimum legal sizes are prescribed in many alternative ways. In fish, measurements include caudal fork length and total length, and in crustaceans, carapace length, carapace width, total length, and tail (abdomen) length can be used. In some cases, one particular method of measurement may be more practical, or more easily enforced, than others. If lobsters are processed at sea, for instance, and only tails are kept for freezing, minimum legal sizes in terms of tail length are the most sensible option.

If the aim of applying minimum sizes is to protect breeding individuals, it should be realized that species reach sexual maturity at particular ages, rather than at particular lengths or weights. Areas at the limits of the stock's range, or where environmental conditions are less than optimal for the species, may contain stunted individuals. These individuals may grow more slowly and mature at a smaller size (although perhaps at the same age) as individuals elsewhere, and the application of a lower legal minimum length may be justified in these areas. However, different minimum sizes applied to the same species in adjacent areas often create enforcement problems. Some exploited species are distributed over the coastline of several adjacent countries or states, in which the management authorities have applied different minimum legal sizes on the grounds of variations in growth rates between the different areas. Nevertheless, fishers can circumvent the regulations by catching small individuals in one area where they are legally undersize, and landing the catch in another area where the same individuals are of legal size. There are also problems in

the case of migrating fish which grow as they move along the coastlines of different countries. A coastal community with access to young fish may be opposed to the conservation of small individuals when they are going to be caught later, when above legal size, by other fishers.

An upper size limit has also been applied to some species, in which it is usually justified on the grounds that larger individuals produce a disproportionately greater number of eggs than smaller individuals. However, upper size limits are most acceptable to the fishing industry in species where larger individuals are less marketable than smaller individuals. Large trochus shells, for example, which are often pitted due to a lifetime of parasitic attack, are of less use in the manufacture of pearl-shell buttons, and larger lobsters are less acceptable to the restaurant trade than smaller plate-size ones. Maximum legal sizes have also been applied to species that accumulate toxins with age. Large sharks, for instance, may accumulate unacceptably high levels of mercury over their long lifespan. In a few cases, large individuals have been protected for aesthetic and emotional reasons: large groupers, for example, have been protected on coral reefs frequented by tourists and recreational divers.

Rejection of females, or gravid females

Making it illegal to retain females, or females bearing eggs, is a method of conservation that is widely favoured by the fishing industry. Yet the biological benefits of protecting females in fast growing species with high fecundity are not always obvious from a consideration of stock–recruitment relationships. Curve C in Fig. 6.5 suggests that the breeding stock can be reduced to low levels without affecting recruitment (although at very low stock levels, recruitment and the fishery are likely to collapse suddenly). Modelling of many species has indicated threshold stock levels, below which population cannot be depleted without affecting egg production and resource sustainability.

Regulations protecting females can only be applied sensibly to species in which the sexes can be distinguished easily by fishers, and where the catching method does not harm the individuals caught. The sex of most fish cannot be determined by external examination, and other regulations, such as the application of closed seasons during the spawning period, may have to be considered as alternatives. In most crustaceans the sexes are readily distinguished, and regulations making it illegal to retain ovigerous, or 'berried', crustaceans such as lobsters and crabs are commonly used.

6.5.3 Controls to protect ecosystems

Although excessive exploitation is undoubtedly the cause of depleted fish stocks, the effects of overexploitation are indistinguishable from those caused by pollution and habitat loss. It is pointless to address the problem of a depleted fish stock by reducing fishing effort or imposing a lower catch quota if the key threats to its recovery are environmental. Marine ecosystems are threatened by many non-fisheries activities as well as by fishing. The negative effects of non-fisheries activities on the marine environment have been discussed in Chapter 1 on ecosystems and include reclamation, siltation, eutrophication, and pollution. All of these are capable of affecting food webs and key habitats including nursery and spawning areas.

Non-fisheries threats to fish stocks in many parts of the world are exacerbated by rapidly increasing local human populations. One of the most common causes of environmental degradation on the coasts of many populous countries is the inadequate treatment of human sewage. In many Pacific islands, for example, the eutrophication of lagoons has resulted in algal growths covering and killing coral with a resulting increase in harmful algal blooms and a resulting decrease in the reef species on which communities depend for food (King, 2004).

The loss of marine biodiversity is increasingly impairing the ocean's capacity to provide seafood and recover from perturbations. In an analysis of global fisheries data, researchers found that the collapse of ecosystems is faster and recovery is slower in areas where species diversity is low

(Worm et al., 2006). Their projections, which are somewhat contentious, suggest that all major fisheries will be based on stock sizes of less than 10% of their original size by the middle of the present century. However, these trends are not irreversible and the restoration of biodiversity has the potential to increase productivity fourfold and decrease variability by 21% on average.

In the face of decreasing catches, even in strictly-managed fisheries, many scientists are advocating that the narrowly-based management of a single resource species be replaced by the more broadly-based management of ecosystems that support all marine species (Kessler et al., 1992; FAO, 2003). In this section, ecosystem-based management is discussed with additional approaches including the use of 'ecolabelling' on seafood products from sustainable fisheries. The use of marine protected areas, which many claim to be the most effective of all management measures, is also discussed.

Ecosystem-based fisheries management

Ecologically sustainable development (ESD) is based on the sustainable use both of species and ecosystems, the maintenance of essential ecological processes, and the preservation of biological diversity. Also, marine ecosystems are valuable for reasons other than commercial fishing, including the prevention of coastal erosion, climate regulation, nutrient storage, the maintenance of biodiversity and recreation. The marine environment is in demand, and under pressure, from many different sources, including commercial fishing, subsistence fishing, recreational fishing, aquaculture, tourism, water sports, shipping, coastal development and industry.

Under ESD guidelines, the management of a fishery is regarded as a subset of managing the ecosystem of which the resource species is a component. The management of a resource species in isolation from its environment, including predator and prey species, can mean that catch rates decline, not due to fishing pressure, but because its environment is deteriorating, or because the tenuous links between it and other species are unbalanced.

Perhaps the best way to have sustainable fisheries and protected ecosystems is by the holistic management of large marine ecosystems or LMEs (see Section 1.4 'Marine ecosystems'). In spite of the difficulty in considering the impacts of exploitation on the entire ecosystem, the large areas involved do provide some attractive management options. Perhaps the most useful of these is the setting aside of sub-areas in which fishing is either banned or restricted. An example of an LME in which issues of exploitation and conservation are balanced is the Australian Great Barrier Reef, which is controlled by a single management authority. However, a disadvantage of managing an LME is the need to patrol such a large area and control access. In spite of the creditable attempts to manage Antarctic waters under the Convention on the Conservation of Antarctic Marine Living Resources, for example, finfishes continue to be exploited by pirate fishers.

Many studies have been conducted to assess ecosystems and particularly the effects of human activities. A multidisciplinary review of ecosystems in Australia used a report card system which assigned a score to various ecosystems from excellent (no effects of human activities) to poor (very serious effects) in order to prioritize areas where actions were necessary (Zann, 1995). The fact that such a study was multidisciplinary emphasizes that the assessment of ecosystems is usually beyond the capacity of a single government agency, such as one involved in stock assessment and fisheries management.

More recently, fisheries management that has a broader ecosystem focus has been referred to as an ecosystem approach to fisheries, or EAF (FAO, 2003) or ecosystem-based management, or EBM, by the Word Wildlife Fund and others. One of the constraints to the implementation of ecosystem-based fisheries management is the fact that fisheries and the marine environment are often the responsibilities of two or more different government agencies. In many countries, the different agencies compete rather than cooperate with each other, making joint management difficult.

Implementing ecosystem-based management involves a similar process to that suggested in Section 6.4.4 for fisheries management plans. Additional work is involved in identifying and involving stakeholders as well as in determining environmental threats, and setting limit reference levels. An ecosystem-based limit reference level, for example, could be based on the mean size of species in the catch or on the diversity of species in the catch; both of these tend to decrease with increasing exploitation (see Section 6.4.3 'Reference points and indicators'). If the mean size or the number of species dropped below the relevant limit reference levels, drastic management actions would be initiated. The extent of such a plan can vary in scope from an ecosystem as small as an individual estuary for some inshore species to an entire ocean in the case of some pelagic species such as tuna.

Actions of the public, consumers and conservation groups

Alternative approaches to improving the environmental sustainability of fisheries include the use of 'ecolabelling' on seafood products from fisheries that have been assessed by an impartial third party as being sustainable (FAO, 2005). The Marine Stewardship Council (MSC), founded in 1997 by the World Wide Fund for Nature and Unilever, has established a widely-recognized ecolabelling system. The MSC blue ecolabel may be displayed on seafood from a fishery that has been certified as operating in an environmentally responsible way and that does not contribute to overfishing. In awarding MSC certification, the three key considerations are the condition of the fish stock, the impact of the fishery on the marine ecosystem, and the fishery management system.

Ecolabelling depends on public awareness and sympathy for conservation as well as consumer willingness to seek out and perhaps pay a premium for ecolabelled products. It has been suggested that although ecolabelling, as currently practised, may have sporadic success in some eco-conscious markets it is unlikely to stimulate global improvement of fisheries management (Gardiner & Viswanathan, 2004).

The increasing consumer awareness of the need for the conservation of marine species is growing. Public reactions are particularly strong when the incidental catching of endangered or charismatic species is involved. A public campaign persuaded consumers to boycott canned tuna from a purse seine fishery in the eastern tropical Pacific, where dolphins were accidentally caught when they accompanied schools of tuna. This resulted in canneries sourcing tuna from fisheries in which dolphins were not killed and the canned product could be advertised and labelled as 'dolphin-safe' or 'dolphin-free'. Similarly, catches of turtles in prawn (shrimp) fisheries have been avoided by fitting turtle exclusion devices (TEDs) to trawl nets, and the prawns can thus be preferentially marketed as 'turtle-free'.

In the 1990s, there was growing public concern about accidental catches of leatherback turtles in a longline fishery for swordfish in Hawaii. In 1999, conservation groups initiated legal proceedings against fishing interests and the US National Marine Fisheries Service. Subsequently, a court ordered the closure of the fishery and about 500 people working in positions associated with the industry became unemployed. The fishery is now operating under new requirements and regulations.

Marine protected areas
The goals of an ecosystem-based approach to fisheries management include the sustainable use both of species and ecosystems, the maintenance of essential ecological processes, and the preservation of biological diversity at all levels, from the ecosystem to the gene. These goals relate to a strategic approach to protecting large-scale ecosystems from pollution and abuse. It is hard to disagree with the aims of ecosystem-based fisheries management and the concept is in great favour with the public and conservation groups. However, the concept is often vague and its aims and required actions are ill-defined. One tactical approach that meets general approval is the setting aside of marine protected areas to protect particular habitats, and to provide refuges for marine flora and fauna, as well as allowing for the often competing demands of particular user groups.

The term marine protected area (MPA) refers to a wide range of marine areas to which various types and levels of restrictions are applied to protect living and non-living things. The IUCN definition of a marine protected area (MPA) is any area of intertidal or subtidal terrain, together with its overlying waters and associated flora, fauna, historical and cultural features, which has been reserved by legislation or other effective means to protect part or all of the enclosed environment (Kelleher & Recchia, 1998). The IUCN categories of MPAs are given in Table 6.8.

MPAs range in size from very small community-managed coastal areas to large offshore areas. In 2006, the USA created the world's largest MPA (or ocean sanctuary) that encloses the Northwestern Hawaiian Islands. The islands are scattered over 360 000 km² of ocean, an area slightly larger than Australia's Great Barrier Reef Marine Park at 350 000 km². Although the Great Barrier Reef has a greater diversity of species, the Northwestern Hawaiian Islands has a large number of endemic species: out of more than 7000

Table 6.8 The six IUCN protected area management categories.

No.	Category	Managed mainly for ...
Ia	Strict Nature Reserve	science
Ib	Wilderness Area	wilderness protection
II	National Park	ecosystem protection and recreation
III	Natural Monument	conservation of specific natural features
IV	Habitat/Species Management Area	conservation through management intervention
V	Protected Landscape/Seascape	landscape/seascape conservation and recreation
VI	Managed Resource Protected Area	the sustainable use of natural ecosystems

marine species about a quarter of them, including the endangered Hawaiian monk seal, are found nowhere else on earth. Prior to the declaration of this large MPA only 1% of the world's oceans were protected in reserves.

Large MPAs are generally under the control of a single country, but the Ligurian Sea Cetacean Sanctuary is an international one established for the protection of marine mammals (cetaceans). It covers an area of approximately 84 000 km² in waters off the coasts of France, Monaco, and Italy in the Mediterranean basin, and is a major feeding ground for fin whales.

MPAs can be established for a multitude of reasons including protecting such historical sites as shipwrecks and important cultural sites. Not all MPAs exclude extractive processes such as fishing but from a fisheries management viewpoint, interest is in areas in which fishing is prohibited. These areas, sometimes called fish reserves, fish sanctuaries or no-take areas, are believed to protect exploited fish stocks as well as marine diversity and ecosystems. Reviews of MPAs and their roles in fisheries are given in Roberts and Polunin (1991), Fairweather and McNeil (1992), Polunin (2000), Anon (2001), Halpern and Warner (2002), Reynolds et al. (2002), Russ and Alcala (2004) and Sale et al. (2005).

From a fisheries point of view, the productivity of an exploited stock is generally greater than that of an unexploited one, and is maximized when a stock is at a level well below its maximum biomass (see Chapter 5). Marine protected areas, therefore, are likely to contain stabilized populations in which biomass production is relatively low. Fishery benefits of protected areas are usually stated as being related to the maintenance of the natural age structure of the stock, and protection of the spawning biomass.

In less mobile and sessile species, MPAs allow the aggregation of parental stocks, and these are believed to result in increased recruitment and migration to surrounding unprotected areas. The role of MPAs in enhancing fisheries in adjacent areas also depends on whether they act as sources or sinks for post-settlement fish. Post-settlement fish either migrate from the protected areas, or aggregate in them; the latter may be because of habitat disturbance in fished areas, or habitat enhancement within protected areas.

If a newly-protected site has been heavily fished in the past, there is obviously a greater potential for positive changes. In reef habitats, growths of coral and algae often increase as do the size and abundance of some other invertebrate and vertebrate species present. Other species will move into the newly-established MPA, presumably attracted by the growth of algae and other food species. Territorial and less mobile fish are more likely to benefit from an MPA as are site-attached species including gastropods, sea urchins, sea cucumbers and lobsters. Many species (but not all) will reproduce in an MPA with larger areas being more effective in this respect than smaller ones. For many species, large closed areas appear to be necessary: 40% of the North Sea, for example, was closed to protect cod from fishing during the spawning season in the spring of 2001.

The increase in size of individuals in MPAs potentially results in a much greater egg production. Egg production is related to parental fish volume which has a cubic relationship to fish length; this suggests that, for example, if a fish is allowed to double in size, potential egg production will increase by 2^3, or eight times. However, as individuals grow and can feed on larger prey, they may move up the food chain with a cascading effect on the ecosystem (Juanes et al., 2002).

The possible movements of individuals out of an MPA are summarized in Fig. 6.9. Larvae produced in the MPA (shown by a heavy circle) will drift in the water body and may either settle within the boundaries of the MPA or drift over the boundaries and become distributed over an area referred to as the dispersal envelope. A large MPA in particular is likely to be self-recruiting with many of the larvae produced within its boundaries settling within the MPA itself. Larvae drifting out of the MPA will be affected by net current flow across the area, and the dispersal envelope may take up an elliptical shape as suggested in Fig. 6.9. This emphasizes the importance of positioning an MPA such that prevailing

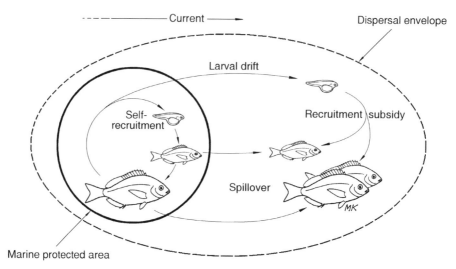

Fig. 6.9 Larvae produced in an MPA (heavy circle) may either settle within its boundaries (resulting in self-recruitment) or drift within a dispersal envelope stretching away from the MPA. Areas outside the MPA are enhanced by spillover, the net movement of juveniles and adults from the MPA.

currents will maximize larval drift and settlement in depleted or targeted areas. However, fish larvae may be able to detect the presence of, and to swim towards, reefs several kilometres away (Wolanski et al., 1997) and this suggests that MPA-derived larvae may actively move to, and repopulate, nearby reefs.

In MPAs designed to increase fish production, the expectation is that larvae will drift and settle, in what is called a recruitment subsidy, within a fished area outside the source location. Juvenile or adult fish may also move out of MPAs as spillover, perhaps in response to increased crowding and competition. Tagging studies in South Africa suggest that excess stocks of fish in reserves move to adjacent exploited areas (Attwood & Bennett, 1994).

Besides protecting biodiversity, an MPA established in part of a fish stock's range will provide a bank of individuals and a hedge against stock depletion by overfishing. A network of MPAs may also provide a hedge against localized and large environmental fluctuations. For fisheries enhancement, some smaller and linked reserves enable a significant proportion of larvae to disperse to surrounding fished areas (Sale et al., 2005). Subsistence fishers in Samoa, for example,

appear to have benefited from a network of MPAs managed by individual communities (see Box 6.9 'Community-owned MPAs in Samoa'). An expectation of improved catches of seafood in nearby areas is the key motivation for most communities wanting to establish MPAs. However, overly-optimistic expectations should not be encouraged by authorities, as benefits may lie some distance in the future, and a prolonged level of community commitment is required (King & Fa'asili, 1998). However, community involvement in establishing MPAs is crucial. The imposition of large MPAs by government authorities with little stakeholder consultation is likely to generate resentment in local communities that have lost access to fishing areas. The result of this resentment is that the level of poaching is likely to be high as will be government enforcement costs.

The effects of environmental fluctuations on a population may be mitigated if some part of it is protected by a marine reserve (Pauly et al., 2002). Fig. 6.10 illustrates the sizes of three populations in which numbers vary with environmental fluctuations over time. The exploited population (middle curve), which is afforded some protection by a marine reserve, is able to survive less than favourable environmental conditions.

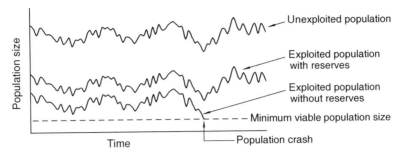

Fig. 6.10 The effects of environmental fluctuations on population size may be mitigated in one that is protected by a marine reserve. The bottom (exploited and unprotected) population crashes when it drops below the minimum number of individuals required to successfully reproduce. Adapted from Jackson in Pauly et al. (2002).

Although MPAs are likely to sustain and even enhance fishery yields in areas surrounding them, there are significant gaps in scientific knowledge and these are being ignored in some uncritical advocacy; the science is reviewed and gaps identified by Sale et al. (2005). The collaboration of scientists, managers and communities is advocated to design and run adaptive management experiments to answer crucial questions on connectivity (the linking of local populations through dispersal) and recruitment subsidies in areas outside the MPA. Such knowledge would allow scientists to offer better advice on the size, placement and spacing of MPAs.

Although the results of establishing MPAs are likely to be overwhelmingly positive, studies are also necessary to identify negative effects. One likely effect is trophic cascading in which protecting large piscivores in MPAs leads to a decline in the numbers of their prey species, and the subsequent increase in abundance of still lower trophic groups. Also, although MPAs may act as banks that protect fish stocks they will not protect marine ecosystems from widespread or more pervasive environmental threats such as coral bleaching. An eight-year study in Papua New Guinea (Jones et al., 2004) revealed a decline in coral cover and a resulting decline in fish biodiversity both in MPAs and in areas open to fishing.

A side benefit of MPAs, but one of considerable importance to fisheries managers, is that the task of enforcement and compliance can be simplified, and management costs reduced. The broad controls applied to MPAs often replace a large number of different fisheries regulations, including bag limits, legal minimum sizes, and gear controls, which can be more costly to enforce.

Closures as fisheries management tools

The permanent closures discussed previously have a beneficial effect on fisheries but they have much broader uses in conservation, many of which are unrelated to fisheries. However, some non-permanent or seasonal closures as well as closures in key habitat areas are common tools applied by fisheries managers to address management objectives. Fishing can be banned either during particular times or seasons (temporal closures), or in particular areas (spatial closures), or during a combination of both; for example, a particular spawning area can be closed on a seasonal basis.

If the period of recruitment in a particular species is short and well defined, a closed season at the time of recruitment can allow small individuals to grow to a more marketable size, and, depending on natural mortality rate, the total catch may be increased (see Section 5.3). In some fisheries on short-lived species, where catches are mainly from a single cohort, fishing grounds can be closed at the time of recruitment and re-opened when young fish reach an optimum size.

Closed seasons during a well-defined spawning period can allow adults to breed without interference. Seasonal closures are important in the management of aggregating species such as the Nassau grouper, *Epinephelus striatus*, which forms dense spawning aggregations at known

Box 6.9
Community-owned MPAs in Samoa

Under a community fisheries extension programme in the Pacific island of Samoa, coastal villages developed their own community-based fisheries management plans (see Box 6.4). Each plan set out the resource management and conservation undertakings of the community and over 60 villages have chosen to establish small marine protected areas (MPAs) in part of their traditional fishing areas (King & Fa'asili, 1998; A. Mulipola, pers. com., 2006). Each community's expectation is that, by banning fishing in part of its traditional fishing area, fish catches in adjacent areas would eventually improve.

In some cases, village communities proposed to establish MPAs in unsuitable positions (e.g. areas of bare sand or coral rubble) and technical information was provided to encourage the selection of a more appropriate site. Some villages initially elected to have very large MPAs and a few wanted to ban fishing in their entire lagoon area. In such cases, communities were asked to balance the perceived fish production advantages of a large reserve against the sociological disadvantages of banning fishing in a large proportion of the village's fishing area. In the latter case, although young men would still be able to go fishing beyond the reef, women (who traditionally collect echinoderms and molluscs in subtidal areas) would be particularly disadvantaged in losing access to shallow-water fishing areas.

Although by social necessity many of the Samoan community-owned reserves are small, their large number, often with small separating distances, forms a network of fish refuges. Such a network may maximize linking of larval sources and suitable settlement areas and provide the means by which adjacent fishing areas are eventually replenished with marine species through reproduction and migration. As the Samoan MPAs are being managed by communities with a direct interest in their success, compliance with bans on fishing is high and there are none of the enforcement costs associated with national reserves. Most villages with reserves actively enforce their own rules, and apply often severe penalties, including traditional fines of pigs or canned goods, for infringements.

A subsequent household survey revealed that fishers in villages with community-owned MPAs made average catch rates of 2.8 kg per person per hour whereas fishers in villages without MPAs made average catch rates of 1.8 kg per person per hour. Although this difference is highly significant, care must be taken in drawing conclusions as there is the possibility that people in villages joining the community-based extension programme were already better and more aware fishers.

---

sites in the Caribbean. Closed areas can also be used to protect the key habitats of exploited species. Shallow-water mangrove habitats, for instance, are known to be nursery areas for many species and are closed to commercial fishing in many coastal areas.

In stocks of molluscs and crustaceans, closures are sometimes imposed during periods when individuals are in poor condition, i.e. when the meat recovery rate is low. Sometimes particular areas are closed to harvesting when seafood species become toxic to human consumers. A common seafood safety reason for imposing peri-odic closures is related to the appearance of harmful algal blooms, or HABs (see Section 1.6.3). Canadian authorities, for example, close shellfish beds when concentrations of domoic acid, derived from toxic diatoms, reach levels in seafood flesh that would result in amnesic shellfish poisoning in consumers.

Closing and opening times for seasonal clo-sures to protect newly-recruited individuals and spawning stocks are best not tied to fixed dates. Due to environmental variations, the timing of recruitment and spawning is usually different from year to year and the ability to vary the timing of the closed period is important. Under a floating closure, the extent of the closure may be included in legislation as a fixed number of weeks, but its start date can be based on monitoring information and varied from year to year. Usually, a senior government official is given the authority to declare the date of closing with sufficient advance warning to fishers and processors.

Rotational closures, (periodic harvesting or pulse fishing) where different areas are closed and opened on a rotational basis, may be a useful strategy if it is considered more efficient to take a large catch say every few years than to take a smaller catch every year. A case can be made that it is also less environmentally damaging to allow exploitation to be broken up by periods of recovery in marine ecosystems. Periodic closures may also result in an increased reproductive output; in abalone stocks, for example, where areas are pulse fished by divers, periodic closures may allow greater egg production. Unlike total seasonal closures, rotational closures allow a continuous supply of product to processors; they are, however, likely to complicate and therefore increase the costs of enforcement.

Whether or not temporal closures are effective in protecting juveniles and spawning females, a useful side effect is that overall annual fishing effort in the fishery is reduced (unless fishing effort merely becomes more concentrated in the open season). Temporal closures may also be imposed for social reasons. A ban on commercial fishing during the weekend, for example, can be used to decrease conflict between professional and recreational fishers.

6.5.4 Compliance and enforcement

The aspirations of genuine fishers and the aims of fisheries managers are under constant threat from those that ignore regulations designed to promote the sustainability of stocks. Managers are custodians of what can be regarded as a public resource which, ideally, is exploited by a few licensed individuals operating under a strict set of rules.

Fishers that break the rules, for example by fishing in closed areas or by catching small individuals, threaten the sustainability of the resource. Unauthorized or unlicensed fishers place additional pressure on the stock. As this additional unlicensed fishing effort has not been allowed for, the stock is likely to become overexploited. In effect, those that break the rules are guilty of stealing from the community that has interest in the common property resource.

Monitoring, control and surveillance (MCS)

The purpose of monitoring, control and surveillance (MCS) systems is to ensure that rules put in place to address conservation and management issues are implemented. The design of an MCS system differs from fishery to fishery but activities range from the relatively simple collection of catch data at landing points to costly surveillance cruises by patrol vessels.

The rise in illegal fishing activities emphasized the need for cooperative law enforcement between countries. At an international conference sponsored by Chile in 2000, the foundations were laid for the formation of the monitoring, control and surveillance (MCS) network. The network was established to improve the efficiency and effectiveness of MCS through enhanced cooperation, coordination, information collection and exchange among national governmental organizations and institutions responsible for fisheries.

Illegal, unreported and unregulated (IUU) fishing

On a world scale, the issue of illegal, unreported and unregulated (IUU) fishing is of serious and increasing concern (FAO, 2002). Illegal fishing refers to fishing carried out by unauthorized vessels whereas unreported fishing refers to fishing in which catches have not been reported to management authorities. Unregulated fishing refers to fishing activities carried out in the absence of management measures. All illegal fishing, particularly by unauthorized vessels or pirate fishing vessels, frustrates management efforts to ensure the sustainability of fish stocks. Illegal fishing usually benefits the wrongdoers at the expense of legal fishers; this would be the result, for example, if only licensed fishers adhere to conservation

Fisheries conservation regulations are often ignored by the public unless there is an appreciation of the need for such regulations. Here, crabs are caught by people setting traps off a pier in San Francisco in spite of the presence of many notices banning fishing for crabs.

regulations (such as respecting closed areas and seasons) and contribute to management costs. To counteract illegal, unreported and unregulated fishing, countries are being encouraged to restrict access to fisheries and to develop rigorous monitoring, control and surveillance measures in the interests of sustainability. If only vessels licensed to fish are provided with access to port facilities and markets, pirate vessels would find it difficult to operate.

A powerful weapon in the fight against illegal fishing is a vessel monitoring system (VMS) which is being increasingly used by some management authorities. A VMS unit, coupled to a global positioning system (GPS), is installed on each fishing vessel and this beams information on the vessel's position, course and speed via satellite to a computer at the office of the management authority. VMS provides a method of detecting vessels moving in areas and seasons that are have been closed to fishing by management authorities. As VMS can accurately determine the position of a vessel at all times it has the potential to allow more complex, and hopefully more efficient, management systems. It gives managers, at least in theory, the ability to ensure that fish are harvested at optimum market size, yield is maximized, spawning stocks are protected and fishing is sustainable. However, in the case of a vessel illegally fishing in a closed area, most systems are not able to determine with sufficient certainty for prosecution purposes whether or not the vessel is actually fishing. Consequently, in the case of illegal fishing, evidence needs to be gathered from other sources to corroborate evidence obtained from the VMS.

Compliance

Although there is a need to reduce the amount of fishing on many stocks, most fisheries management strategies usually include several different regulations, which, to be effective, must be enforced. From a more local or single country viewpoint, it is usually government staff, often fisheries agency staff or police officers that have the task of enforcing national fisheries regulations. The use of fisheries agency staff in an enforcement capacity causes problems relating to the dual roles of the agency. Research staff are often required to work closely with fishers and seek their cooperation in obtaining information and data; these tasks are made more difficult if staff from the same agency have an enforcement role.

Compliance with fisheries regulations is usually higher in a fishery in which stakeholders have been involved in developing the management plan. It is even higher in small-scale fisheries being managed by communities under community-based fisheries management arrangements. Overall, any involvement of stakeholders in devising sensible regulations is likely to reduce the need for enforcement.

The first and most important aspect of enforcement is education, and prosecution should be regarded as a measure of last resort. Users of a resource, or managed area, should be made familiar with any regulations, and the reason for their imposition. Public meetings, radio talks, press articles, and poster displays may be all used to publicize regulations, and to provide the public with an appreciation of the need to have regulations. If the majority of users support the aims of the regulations, peer pressure becomes a strong deterrent to those disregarding the law.

Indeed, in some cases, education is the only practical way to change attitudes towards overexploitation and environmentally damaging practices. An extreme example is where explosives and commercial poisons are used by members of isolated coastal communities. The use of commercially available poisons such as bleaches (sodium hydrochlorite) and insecticides, as well as explosives represents a serious threat to coral reef ecosystems and the long-term viability of fisheries in many part of the world. Fishers using such destructive fishing methods are often tolerated, and sometimes highly regarded, in the community, as the catches are usually shared. Because of the isolated fishing locations, as well as the lack of public cooperation, enforcement staff have difficulty in apprehending offenders. Public education appears to be the only method of ensuring that the use of such methods is seen as contrary to the long-term interests of the community. If public attitudes are turned against illegal fishing, the practice will be self-policing at the community level. A public education campaign could include short-term measures, such as a series of talks given to community groups, and the distribution of posters emphasizing the antisocial nature of using explosives and industrial chemicals for fishing. Regulations could enforce the inclusion of warning labels on certain chemicals sold. All bleaching agents, for instance, could include an adhesive label with a message warning against its use in fishing and emphasizing the long-term damaging effects to the environment and fish production. Longer-term methods include teachers introducing aspects of the marine environment and conservation into high school curricula.

Enforcement

Although prosecution should be regarded as a measure of last resort, necessary regulations must be rigorously enforced. Regulations which are imposed but unenforced, either due to insufficient enforcement staff, or to overly complex and impractical rules, will fall into disrepute. If regulations are unenforced, benefits will accrue to those who ignore the regulations at the expense of those who fish according to the rules. Penalties applied should be significant to the offender, and relevant to the offence. Although a small fine may be appropriate in the case of an amateur taking undersize fish, the commercial poaching of high-value species, such as lobster or abalone, should attract a large fine and gear confiscation to act as an effective deterrent. Enforcement staff should receive training in public relations, fisheries management, evidence collecting, and court procedures. If an infringement cannot be dealt with by education, cases taken to court should have a high probability that the prosecution will be successful.

Enforcement costs often account for a substantial proportion of the total costs of managing a fishery or marine protected area. The cost of transport for enforcement staff using vehicles and boats is high. This is particularly so in the case of open sea fisheries (involving fisheries patrol vessels), and in coastal fisheries where there are a large number of fish landing sites. Enforcement staff usually work in pairs, for safety reasons, and in order that corroborative evidence is available in the case of prosecutions. In addition, the preparation of documents for prosecution is expensive in terms of non-field time. In the worst case, the cost of policing regulations that are intended to maximize profits in a fishery could be greater than the benefits gained.

The cost of enforcing regulations should be considered when any alternative management strategies are proposed. In some cases, it may be preferable to apply a less direct regulation which is cheaper to police, than a more direct one that is expensive. Enforcing legal minimum length regulations, for example, is likely to be expensive in a large artisanal fishery, in which there are a large number of fishers working over an extensive geographic area. However, if catches were marketed at a limited number of outlets, it would be a simpler task to inspect fish at the point of sale instead of at the point of capture. In this case, a regulation making it illegal to sell rather than to catch undersize fish would be easier to enforce. Although some undersize fish may still be caught, fishers would soon avoid taking smaller individuals which are legally unmarketable.

The main problem is not so much in enforcing fisheries controls, but in convincing the community that they are necessary. In the past, when populations were small, and fishing methods were less efficient, catches made by one person had very little effect on catches made by others. But the human population and its demand for seafood have grown beyond that which can be supported by finite resources. In addition, there are other claims on the marine environment from development and industry. The dilemma is, that as demand for fisheries resources is increasing, the ability of the marine environment to sustain them may be decreasing. The freedom to catch fish, or to use the marine environment, without control, is now more than ever likely to be at the expense of someone else's freedom to do the same thing. Some of these freedoms must be sacrificed to allow the continuing use of the marine environment and its resources by present and future generations. The renewability of fisheries resources depends on accepting controls which not only protect fish stocks, but ensure that the environment in which they live does not deteriorate.

Exercises

Answers to selected exercises are available at www.blackwellpublishing.com/king

Exercise 6.1

This question relates to the deep-water snapper fishery in Exercise 5.1 in Chapter 5. The present cost of fishing is approximately US$600 per trip and the price that fishers receive for the catch is US$3.50 per kg. Graph yield (as revenue) and fishing costs against effort, and estimate f_{MEY} based on the Schaefer model.

Exercise 6.2

The figure below represents the life cycle of a snapper that is caught inshore by artisanal fishers from villages A and B using beach seine nets, and caught offshore by licensed commercial fishers using hook and dropline gear. The artisanal fishers make large catches from January to March when there are large numbers of small individuals in inshore areas. The commercial fishers, from a town 15 km south of the bay, fish offshore all year round, but make the highest catch rates in April when newly-recruited individuals appear offshore. Fishers from village B also catch larger mature fish

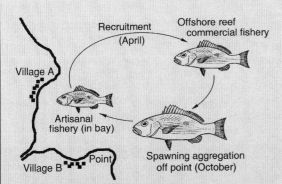

Artisanal and commercial fisheries on a species of snapper.

using hooks and lines from small boats during a spawning aggregation off the point in October.

Catches have been falling in both the inshore and offshore fishery. Each group (artisanal and commercial fishers) blames the other for falling catch rates. It is politically undesirable to reduce the number of offshore boats (which are licensed by the government fisheries authority).

Propose alternative fisheries controls that could be applied to the inshore and offshore fisheries. Consider the advantages and disadvantages of each and compare the relative costs and difficulties of enforcing each proposed regulation. Discuss the possible sociological implications of imposing the regulations.

Exercise 6.3

(a) Discuss three methods of delaying the age of first capture of exploited species – that is, of protecting small individuals of exploited marine species in general.
(b) Which of the three methods is likely to be most suitable for protecting sub-adult penaeid prawns or shrimps that recruit from nursery areas over a short period each year?

Exercise 6.4

A species with growth parameters $K = 0.3$ yr^{-1} and $W_\infty = 900$ g is subject to a natural mortality of $M = 0.5$ yr^{-1}. The age at recruitment is $t_r = 1.0$ years and the age at first capture is $t_c = 2.5$ years. Use a yield per recruit model (as in Fig. B5.7.1 in Chapter 5) to estimate $F_{0.1}$ for use as a target reference point.

Exercise 6.5

Use the stochastic simulation model described in Chapter 5 and Appendix A4.14 ('Risk assessment model'), to determine the maximum fishing effort that would allow the stock biomass to remain above 2500 tonnes for 90% of all future years.

References

Agnew, D.J., Beddington, J.R. and Hill, S.L. (2002) The potential use of environmental information to manage squid stocks. *Canadian Journal of Fisheries and Aquatic Sciences* **59**, 1851–7.

Alcala, A.C. and Gomez, E.D. (1979) Recolonization and growth of hermatypic corals in dynamite-blasted coral reefs in the central Visayas, Philippines. *Proceedings of the International Symposium on Marine Biogeography and Evolution in the Southern Hemisphere. NZDSIR Information Series* **2**, 645–61.

Allen, K.R. (1963) Analysis of stock-recruitment relations in Antarctic fin whales. *Rapports et Procès-verbaux des Réunions, Commission Internationale pour l'Exploration Scientifique de la Mer Méditerranée* **164**, 132–7.

Amos, A.M. (1993) Traditionally based marine management systems in Vanuatu. *Traditional Marine Resource Management and Knowledge Information Bulletin* **2**, 14–17.

Anon (1987) Tagwatch 'whiting tagging programme' SAFISH. Department of Fisheries, South Australia **11.1**, 25–6.

Anon (1991) *Northern Territory Barramundi Fishery Management Plan.* Fisheries Division, Department of Primary Industry and Fisheries. Northern Territory, Australia.

Anon (2001) Scientific consensus statement on marine reserves and marine protected areas. *Annual Meeting of the American Association for the Advancement of the Sciences.* National Center for Ecological Analysis and Synthesis, University of California.

Anon (2005) On introducing super tilapia (GIFT) farming in India. *Fishing Chimes* **25** (5), India.

Attwood, C.A. and Bennett, B.A. (1994) Variation in dispersal of galjjoen (*Coracinus capensis*) (Teleostei: Coracinidae) from a marine reserve. *Canadian Journal of Fisheries and Aquatic Sciences* **51**, 1247–57.

Azam, F., Fenchel, T., Field, J.G., Gray, J.S., Meyer-Reil, L.A. and Thingstad, F. (1983) The ecological role of water-column microbes in the sea. *Marine Ecology Progress Series* **10**, 257–63.

Bagenal, T. (ed.) (1974) *Ageing of Fish. Proceedings of an International Symposium on the Ageing of Fish held at the University of Reading, England, 19–20 July 1973.* Unwin Brothers Limited, Old Woking, Surrey.

Bagenal, T. (ed.) (1978) *Methods for Assessment of Fish Production in Fresh Waters.* Blackwell Scientific Publications, Oxford.

Baillon, B. (1988) Aging of tropical reef fish by density of daily otolith increments. Workshop on Pacific Inshore Fishery Resources, South Pacific Commission, Noumea. Background Paper 4.

Barber, I. and Poulin, R. (2002) Interactions between fish, parasites and disease. In: P. Hart and J. Reynolds (eds) *Handbook of Fish Biology and Fisheries*, Vol. I, pp. 59–388. Blackwell Science, Oxford.

Basson, M., Rosenberg, A.A. and Beddington, J.R. (1988) The accuracy and reliability of two new methods for estimating growth parameters from length-frequency data. *Journal du Conseil, Conseil International pour l'Exploration de la Mer* **44**, 277–85.

Beamish, R.J. and McFarlane, G.A. (1983) The forgotten requirement for age validation in fisheries biology. *Transactions of the American Fisheries Society* **112**, 735–43.

Beamish, R.J. and McFarlane, G.A. (1987) Current trends in age determination and methodology. In: R.C. Summerfelt and G.E. Hall (eds) *Age and Growth of Fish.* pp. 15–42. Iowa State University Press.

Beddington, J.R. and Cooke, J.G. (1983) The potential yield of fish stocks. *FAO Fisheries Technical Paper* **242**.

Bell, R.S., Channells, P.W., MacFarlane, J.W., Moore, R. and Phillips, B.F. (1987) Movements and breeding of the ornate rock lobster, *Panulirus ornatus*, in Torres Strait and on the north-east coast of Queensland. *Australian Journal of Marine and Freshwater Research* **38**, 197–210.

Beverton, R.J.H. and Holt, S.J. (1956) A review of methods for estimating mortality rates in fish populations, with special reference to sources of bias in catch sampling. *Rapports et Procès-verbaux des Réunions, Commission Internationale pour l'Exploration Scientifique de la Mer Méditerranée* **17A**, 1–153.

Beverton, R.J.H. and Holt, S.J. (1957) On the dynamics of exploited fish populations. *Fisheries Investigations*, Series 2, 19, Ministry of Agriculture, Fisheries and Food, United Kingdom.

Bhattacharya, C.G. (1967) A simple method of resolution of a distribution into Gaussian components. *Biometrics* **23**, 115–35.

Birkeland, C. (1997) Symbiosis, fisheries and economic development on coral reefs. *Trends in Ecology and Evolution* **12**, 364–7.

Boudouresque, C.F. (1999) The Red Sea–Mediterranean link: unwanted effects of canals. In: T. Sandlund, P.J. Schei and A. Viken (eds) *Invasive Species and Biodiversity Management*, pp. 213–28. Kluwer, Dordrecht.

Bowen, W.D. (2000) Reconstruction of pinniped diets: accounting for complete digestion of otoliths and cephalopod beaks. *Canadian Journal of Fisheries and Aquatic Sciences* **57**, 898–905.

Boy, R.L. and Smith, B.R. (1984) *Design Improvements to Fish Aggregation Device (FAD) Mooring Systems in General Use in Pacific Island Countries*. Handbook No. 24, South Pacific Commission, New Caledonia.

Boyle, P.R. and Rodhouse, P.G. (2005) *Cephalopods: Ecology and Fisheries*. Blackwell Science, Oxford.

Brander, K. (1981) Disappearance of common skate, *Raja batis*, from Irish Sea. *Nature* **290**, 48–9.

Brouwer, S. and Griffiths, M.H. (2004) Age and growth of Argyrozona (Pisces: Sparidae) in a marine protected area: an evaluation of methods based on whole otoliths, sectioned otoliths and mark-recapture. *Fisheries Research* **67** (1), 1–12.

Brown, B.E. (1997) Coral bleaching: causes and consequences. *Coral Reefs* **16**, 129–38.

Brown, I.W. and Fielder, D.R. (eds) (1991) *The Coconut Crab: Aspects of the Biology and Ecology of Birgus latro in the Republic of Vanuatu*. Australian Centre for International Agricultural Research (ACIAR) Monograph 8.

Buchanan-Wollaston, H.J. and Hodgson, W.C. (1929) A new method of treating frequency curves in fisheries statistics, with some results. *Journal du Conseil, Commission Internationale pour l'Exploration Scientifique de la Mer Méditerranée* **4**, 207–25.

Butler, T.H. (1980) Shrimps of the Pacific coast of Canada. *Canadian Bulletin of Fisheries and Aquatic Sciences* **202**.

Caddy, J.F. (1998) A short review of precautionary reference points and some proposals for their use in data poor situations. *FAO Fisheries Technical Paper* **379**.

Caddy, J.F. and Mahon, R. (1995) Reference points for fisheries management. *FAO Fisheries Technical Paper* **347**.

Campana, S.E. (2005) Otolith science entering the 21st century. *Marine and Freshwater Research* **56**, 485–95.

Campana, S.E. and Thorrold, S.R. (2001) Otoliths, increments and elements: keys to a comprehensive understanding of fish populations. *Canadian Journal of Fisheries and Aquatic Sciences* **58**, 30–8.

Cassie, R.M. (1954) Some uses of probability paper in the analysis of size frequency distributions. *Australian Journal of Marine and Freshwater Research* **5**, 513–22.

Castro, J.I., Woodley, C.M. and Brudek, R.L. (1999) A preliminary evaluation of the status of shark species. *FAO Fisheries Technical Paper* **380**.

Castro, M. and Erzini, K. (1988) Comparisons of two length-frequency based packages for estimating growth and mortality parameters using samples with varying recruitment patterns. *Fishery Bulletin*, U.S. **86** (4), 645–53.

Christensen, V. and Pauly, D. (1992) ECOPATH II – a software for balancing steady-state ecosystem models and calculating network characteristics. *Ecological Modelling* **61**, 169–85.

Christensen, V. and Walters, C.J. (2004) Ecopath with Ecosim: methods, capabilities and limitations. *Ecological Modelling* **172**, 109–39.

Cinner, J. (2005) Socioeconomic factors influencing customary marine tenure in the Indo-Pacific. *Ecology and Society* **10** (1), 36.

Clark, C.W. (1985) *Bioeconomic Modelling and Fisheries Management*. John Wiley & Sons, New York.

Clark, R.B. (2001) *Marine Pollution*, 5th edn. Oxford University Press.

Clarke, C.W. (1976) A delayed recruitment model of population dynamics, with an application to baleen whale populations. *Journal of Mathematical Biology* **3**, 381–91.

Collette, B.B. and Nauen, C.E. (1983) Scombrids of the world. An annotated and illustrated catalogue of tunas, mackerels, bonitos and related species known to date. *FAO Fisheries Synopsis* **125**, Vol. 2.

Conand, C. (1989) The fishery resources of Pacific

island countries. Part 2. Holothurians. *FAO Fisheries Technical Paper* 272.2.

Copeland, J.W. and Lucas, J.S. (eds) (1988) *Giant Clams in Asia and the Pacific*. Australian Centre for International Agricultural Research (ACIAR) Monograph 9.

Cowx, I.G. (1983) Review of the methods for estimating fish population size from survey removal data. *Fisheries Management* 14, 67–82.

Cowx, I.G. (1996) (ed.) *Stock Assessment in Inland Fisheries*. Fishing News Books, Oxford.

Cowx, I.G. (2002) Recreational fishing. In: P. Hart and J. Reynolds (eds) *Handbook of Fish Biology and Fisheries*, Vol. II, pp. 367–85. Blackwell Science, Oxford.

Cruz, A. and Lombarte, A. (2004) Otolith size and its relationship with colour patterns and sound production. *Journal of Fish Biology* 65, 1512–25.

Csirke, J. and Sharp, G.D. (1984) Reports of the expert consultation to examine changes in abundance and species composition of neritic fish resources. *FAO Fisheries Report* 291.

Csirke, J. and Vasconcellos, M. (2005) Fisheries and long-term climate variability. *Review of the State of World Marine Fisheries Resources, FAO Fisheries Technical Paper* 457.

Cury, P.M. and Christensen, V. (2005) Quantitative ecosystem indicators for fisheries management: introduction. *ICES Journal of Marine Science* 62, 307–10.

Cushing, D.H. (1971) The dependence of recruitment on parent stock in different groups of fishes. *Journal du Conseil International pour l'Exploration de la Mer* 33, 340–62.

Cushing, D.H. (1990) Plankton production and year-class strength in fish populations: an update of the match/mismatch hypothesis. *Advances in Marine Biology* 26, 249–93.

Dalzell, P., Adams, T.J.H. and Polunin, N.V.C. (1996) Coastal fisheries in the Pacific islands. *Oceanography and Marine Biology: an Annual Review* 34, 395–531.

Depczynski, M. and Bellwood, D.R. (2005) Shortest recorded vertebrate lifespan found in a coral reef fish. *Current Biology* 15 (8), 288–9.

Deriso, R.B. (1980) Harvesting strategies and parameter estimation for an age-structured model. *Canadian Journal of Fisheries and Aquatic Sciences* 37, 268–82.

Dittmar, T., Hertkorn, N., Kattner, G. and Lara, R.J. (2006) Mangroves, a major source of dissolved organic carbon to the oceans. *Global Biogeochemical Cycles* 20.

Doherty, P.J. and Williams, D.McB. (1988) The replenishment of coral reef fish populations. *Oceanography and Marine Biology, an Annual Review* 26, 487–551.

English, E., Wilkinson, C. and Baker, V. (eds) (1997) *Survey Manual for Tropical Marine Resources*, 2nd edn. Australian Institute of Marine Science.

Evans, D. and Grainger, R. (2002) Gathering data for resource monitoring and fisheries management. In: P. Hart and J. Reynolds (eds) *Handbook of Fish Biology and Fisheries*, Vol. II, pp. 84–102. Blackwell Science, Oxford.

Fabens, A.J. (1965) Properties and fitting of the von Bertalanffy growth curve. *Growth* 29, 265–89.

Fairweather, P.G. and McNeil, S. (1992) Ecological and other scientific imperatives for marine and estuarine conservation. *Proceedings of the Fourth Fenner Environment Conference*, Sydney, Australia.

FAO (1995a) *Code of Conduct for Responsible Fisheries*. FAO, Rome.

FAO (1995b) Precautionary approach to fisheries. Part 1: Guidelines on the precautionary approach to capture fisheries and species introductions. *FAO Fisheries Technical Paper* 350, Part 1.

FAO (1999) Indicators for sustainable development of marine capture fisheries. *FAO Technical Guidelines for Responsible Fisheries*, No. 8.

FAO (2002) Implementation of the international plan of action to deter, prevent and eliminate illegal, unreported and unregulated fishing. *FAO Technical Guidelines for Responsible Fisheries*, No. 9.

FAO (2003) The ecosystem approach to fisheries. *FAO Technical Guidelines for Responsible Fisheries*, No. 4, Suppl. 2.

FAO (2004) *The State of World Fisheries and Aquaculture*. FAO, Rome.

FAO (2005) *Guidelines for the Ecolabelling of Fish and Fishery Products from Marine Capture Fisheries*. FAO, Rome.

Fenchel, T. (1988) Marine planktonic food chains. *Annual Review of Ecological Systems* 19, 19–38.

Fletcher, W.J. (1993) Coconut crabs. In: A. Wright and L. Hill (eds) *Nearshore Marine Resources of the South Pacific*, pp. 643–82. Forum Fisheries Agency, Honiara; Institute of Pacific Studies, Suva.

Fletcher, W.J. and Amos, M. (1994) *Stock Assessment of Coconut Crabs*. Australian Centre for International Agricultural Research (ACIAR) Monograph MN29.

Forsgren, E., Reynolds, J.D. and Berglund, A. (2002) Behavioural ecology of reproduction in fish. In: P. Hart and J. Reynolds (eds) *Handbook of Fish*

Biology and Fisheries, Vol. I, pp. 225–48. Blackwell Science, Oxford.

Fournier, D.A., Sibert, J.R., Majkowski, J. and Hampton, J. (1990) MULTIFAN a likelihood-based method for estimating growth parameters and age composition from multiple length frequency data sets illustrated using data for southern bluefin tuna (*Thunnus maccoyii*). *Canadian Journal of Fisheries and Aquatic Sciences* 47, 301–17.

Fournier, D.A., Hampton, J. and Sibert, J.R. (1998) MULTIFAN-CL a length-based, age-structured model for fisheries stock assessment, with application to South Pacific albacore, *Thunnus alalunga*. *Canadian Journal of Fisheries and Aquatic Sciences* 55, 2105–16.

Fox, W.W. (1970) An exponential surplus yield model for optimising exploited fish populations. *Transactions of the American Fisheries Society* 99 (1), 80–8.

Francis, R.I.C.C. and Shotton, R. (1997) 'Risk' in fisheries management: a review. *Canadian Journal of Fisheries and Aquatic Sciences* 54, 1699–715.

Frimodt, C. (1995) *Illustrated Multilingual Guide to the World's Commercial Warm Water Fish*. Fishing News Books, Oxford.

Garcia, B. (2005) Seaweed processing teaching to boost industry. *Minda News* IV (187).

Garcia, S.M. (1996) The precautionary approach to fisheries and its implications for fishery research, technology, and management: an updated review. *FAO Fisheries Technical Paper* 350.2, 1–75.

Garcia, S., Sparre, P. and Csirke, J. (1987) A note on rough estimates of fisheries resources potential. *Fishbyte* 5 (2), 11–16.

Garcia, S., Sparre, P. and Csirke, J. (1989) Estimating surplus production and maximum sustainable yield from biomass data when catch and effort time series are not available. *Fisheries Research* 8, 13–23.

Gardiner, P.R. and Viswanathan, K.K. (2004) Ecolabelling and fisheries management. *WorldFish Center Studies and Reviews* 27, 44.

Gill, A.C. and Mooi, R.D. (2002) Phylogeny and systematics of fishes. In: P. Hart and J. Reynolds (eds) *Handbook of Fish Biology and Fisheries*, Vol. I, pp. 15–42. Blackwell Science, Oxford.

Gillett, R. (1988) Pacific islands trochus introductions. South Pacific Commission Workshop on Pacific Inshore Fishery Resources, New Caledonia. Background Paper 61.

Gjosaeter, P., Dayaratne, P., Bergstad, O.A., Gjosaeter, H., Sousa, M.I. and Beck, I.M. (1984) Ageing tropical fish by growth rings in the otoliths. *FAO Fisheries Circular* 776.

Glynn, P.W. (1993) Coral bleaching: ecological perspectives. *Coral Reefs* 12, 1–17.

Glynn, P.W. (1996) Coral reef bleaching: facts, hypotheses and implications. *Global Change Biology* 2, 495–509.

Glynn, P.W. (1997) Bioerosion and coral reef growth: a dynamic balance. In: C. Birkeland (ed.) *Life and Death on Coral Reefs*, pp. 68–95. Academic Press, New York.

Gooding, R.M., Polovina, J.J. and Dailey, M.D. (1988) Observations of deepwater shrimp, *Heterocarpus ensifer*, from a submersible off the island of Hawaii. *Marine Fisheries Review*, U.S. 50 (1), 32–9.

Gordon, H.S. (1954) The economic theory of a common property resource: the fishery. *Journal of Political Economy* 62, 124–42.

Grey, D.L., Dall, W. and Baker, A. (1983) *A Guide to the Australian Penaeid Prawns*. Northern Territory Government Printing Office, Darwin.

Gulland, J.A. (1965) Estimation of mortality rates. *International Council for the Exploration of the Sea*, CM 3, 1.

Gulland, J.A. (1970) Food chain studies and some problems in world fisheries. In: J.H. Steele (ed.) *Marine Food Chains*, pp. 296–315. Oliver and Boyd, UK.

Gulland, J.A. (ed.) (1977) *Fish Population Dynamics*. John Wiley, London.

Gulland, J.A. (1983) *Fish Stock Assessment: a Manual of Basic Methods*. FAO/Wiley Series on Food and Agriculture. John Wiley & Sons, Chichester.

Gulland, J.A. (1984) Advice on target fishery rates. *ICLARM Fishbyte* 2 (1), 8–11.

Gulland, J.A. and Boerema, L.K. (1973) Scientific advice on catch levels. *National Oceanographic and Atmospheric Administration Fishery Bulletin*, U.S. 71 (2), 325–35.

Gulland, J.A. and Holt, S.J. (1959) Estimation of growth parameters for data at unequal time intervals. *Journal du Conseil, Conseil International pour l'Exploration de la Mer* 25 (1), 47–9.

Haddon, M. (2001) *Modelling and Quantitative Methods in Fisheries*. Chapman & Hall/CRC, New York.

Hale, H.M. (1976) *The Crustaceans of South Australia*. Government Printer, South Australia, Adelaide.

Hallegraeff, G.M., Grundle, D., Marshall, J.A., Dowdney, J. and Holmes, A. (2005) Range extension of the red-tide dinoflagellate Noctiluca into Tasmanian waters. 5th Workshop of the Australian Research Network on Algal Toxins. Moreton Bay Research Station, Brisbane, Queensland.

Halpern, B. and Warner, R. (2002) Marine reserves have rapid and lasting effects. *Ecology Letters* 5, 361–6.

Hampton, J. and Majkowski, J. (1987) An examination of the accuracy of the ELEFAN computer programs for length-based stock assessment. In: D. Pauly and G.R. Morgan (eds) *Length-based Methods in Fisheries Research*, pp. 193–202. International Center for Living Aquatic Resources Management (ICLARM), Manila, Philippines, and KISR, Kuwait.

Hannesson, R. (2002) The economics of fisheries. In: P. Hart and J. Reynolds (eds) *Handbook of Fish Biology and Fisheries*, Vol. II, pp. 249–69. Blackwell Science, Oxford.

Hardin, G. (1968) The tragedy of the commons. *Science* 131, 1243–7.

Harvell, C.D., Mitchell, C.E., Ward, J.R., Altizer, S., Dobson, A.P., Ostfeld, R.S. and Samuel, M.D. (2002) Climate warming and disease risks for terrestrial and marine biota. *Science* 296 (5576), 2158–62.

Health Canada (1990) *Action Towards Healthy Eating – Canada's Guidelines for Healthy Eating and Recommended Strategies for Implementation*. Public Works and Government Services Canada, Ottawa.

Hilborn, R. and Walters, C.J. (1992) *Quantitative Fisheries Stock Assessment: Choice, Dynamics and Uncertainty*. Chapman & Hall, New York.

Hjort, J. (1914) Fluctuations in the great fisheries of northern Europe viewed in the light of biological research. *Rapports et Procès-verbaux des Réunions, Commission Internationale pour l'Exploration Scientifique de la Mer Méditerranée* 20, 1–228.

Hoegh-Guldberg, O. (1999) Climate change, coral bleaching and the future of the world's coral reefs. *Marine and Freshwater Research* 50, 839–66.

Holthuis, L.B. (1980) FAO species catalogue. Vol. 1. Shrimps and prawns of the world. An annotated catalogue of species of interest to fisheries. *FAO Fisheries Synopsis* 125.

Hughes, T.P., Baird, A.H., Bellwood, D.R., Card, M., Connolly, S.R., Folke, C., Grosberg, R., Hoegh-Guldberg, O., Jackson, J.B., Kleypas, J., Lough, J.M., Marshall, P., Nyström, M., Palumbi, S.R., Pandolfi, J.M., Rosen, B. and Roughgarden, J. (2003) Climate change, human impacts, and the resilience of coral reefs. *Science* 301 (5635), 929–33.

Hurst, R. and Bagley, N. (1992) Trawl gear performance – a spreading issue. New Zealand. *Professional Fisherman*, October, 34–40.

Hutchings, J.A. (2002) Life histories of fish. In: P. Hart and J. Reynolds (eds) *Handbook of Fish Biology and Fisheries*, Vol. I, pp. 149–74. Blackwell Science, Oxford.

Isaac, V.J. (1990) The accuracy of some length-based methods for fish population studies. *ICLARM Technical Report* 27.

Isaksen, B., Valdemarsen, J.W., Larsen, R.B. and Karlsen, L. (1992) Reduction of fish by-catch in shrimp trawl using a rigid separator grid in the aft belly. *Fisheries Research* 13, 335–52.

Jennings, S., Kaiser, M.J. and Reynolds, J.D. (2001) *Marine Fisheries Ecology*. Blackwell Science, Oxford.

Jentoft, S., McCay, B.J. and Wilson, D.C. (1998) Social theory and fisheries co-management. *Marine Policy* 22 (4), 423–36.

Jobling, M. (2002) Environmental factors and rates of development and growth. In: P. Hart and J. Reynolds (eds) *Handbook of Fish Biology and Fisheries*, Vol. I, pp. 97–122. Blackwell Science, Oxford.

Johannes, R.E. (1982) Traditional conservation methods and protected marine areas in Oceania. *Ambio* 11 (5), 258–61.

Johannes, R.E. (1998) Government-supported, village-based management of marine resources in Vanuatu. *Ocean and Coastal Management* 40, 165–86.

Johannesson, K.A. and Mitson, R.B. (1983) Fisheries acoustics. A practical manual for aquatic biomass estimation. *FAO Fisheries Technical Paper* 240.

Jones, G.K. (1979) *Biological Investigations on the Marine Scale Fishery in Spencer Gulf*. Green Paper, South Australian Department of Fisheries.

Jones, G.P., McCormick, M.I., Srinivasan, M. and Eagle, J.V. (2004) Coral decline threatens fish biodiversity in marine reserves. *Proceedings of the National Academy of Sciences of the USA* 101, 8251–3.

Jones, R. (1976) Growth of fishes. In: D.H. Cushing and J.J. Walsh (eds) *The Ecology of the Seas*, pp. 251–79. Blackwell Science, Oxford.

Jones, R. (1981) The use of length composition in fish stock assessments (with notes on VPA and cohort analysis). *FAO Fisheries Circular* 734.

Jones, R.J., Hoegh-Guldberg, O., Larkum, A.W.D. and Schreiber, U. (1998) Temperature-induced bleaching of corals begins with impairment of the CO_2 fixation mechanism in zooxanthellae. *Plant, Cell and Environment* 21, 1219–30.

Juanes, F., Buckel, J.A. and Scharf, F.S. (2002) Feeding ecology of piscivorous fishes. In: P. Hart and J. Reynolds (eds) *Handbook of Fish Biology and Fisheries*, Vol. II, pp. 267–83. Blackwell Science, Oxford.

Kaiser, M.J. and de Groot, S.J. (eds) (2000) *Effects of Fishing on Non-target Species and Habitats*. Blackwell Science, Oxford.

Kaiser, M.J. and Jennings, S. (2002) Ecosystem effects of fishing. In: P. Hart and J. Reynolds (eds) *Handbook of Fish Biology and Fisheries*, Vol. II, pp. 342–66. Blackwell Science, Oxford.

Kamler, E. (1992) *Early Life History of Fish: an Energetics Approach*. Chapman & Hall. London.

Kantha, S.S. (1987) Ichthyotoxins and their implications to human health. *Asian Medical Journal* **30** (8), 458–70.

Kato, S. and Schroeter, S.C. (1985) Biology of the red sea urchin, *Strongylocentrotus franciscanus*, and its fishery in California. *Marine Fisheries Review* **47** (3), 1–20.

Keen, E. (1991) Ownership and productivity of marine fishery resources. *Fisheries* **16** (4), 18–22.

Keenan, C.P. and Blackshaw, A. (eds) (1999) *Mud Crab Aquaculture and Biology. ACIAR Proceedings* **78**.

Kelleher, G. and Recchia, C. (1998) Lessons from protected areas around the world. *PARKS* **8** (2), 1–4.

Kenchington, R.A. and Hudson, B.E.T. (1988) *Coral Reef Management Handbook*. UNESCO.

Keough, M.J. and Downes, B.J. (1982) Recruitment of marine invertebrates: the role of active larval choices and early mortality. *Oecologia* (Berlin) **54** (3), 348–52.

Kessler, W.B., Salwasser, H., Cartwright, C.W. and Caplan, J.A. (1992) New perspectives for sustainable natural resource development. *Ecological Applications* **2**, 221–5.

King, M.G. (1977) Cultivation of the Pacific oyster, *Crassostrea gigas*, in a non-tidal pond. *Aquaculture* **11** (2), 123–36.

King, M.G. (1984) The species and depth distribution of deep-water caridean shrimps (Decapoda: Caridea) near some southwest Pacific islands. *Crustaceana* **47** (2), 174–91.

King, M.G. (1986) The fishery resources of Pacific island countries. Part 1. *FAO Fisheries Technical Paper* **272.1**, 192–203.

King, M.G. (1987) The distribution and ecology of deep-water caridean shrimps (Crustacea: Natantia) near tropical Pacific islands. *Bulletin of Marine Science* **42** (2).

King, M.G. (1990) *The Development of Deep-water Snapper and Shrimp Fisheries in Mauritius*. FAO Terminal Report, Project MAR/88/004.

King, M.G. (1996) FISHSIM: a generalized fishery simulation model. *Naga, ICLARM Quarterly* **18** (4), 34–7.

King, M.G. (2004) *From Mangroves to Coral Reefs; Sea Life and Marine Environments in Pacific Islands*. South Pacific Regional Environment Programme, Apia, Samoa.

King, M.G. and Butler, A.J. (1985) The relationships of life-history patterns to depth distribution in deep-water caridean shrimp (Crustacea: Caridea). *Marine Biology* **86**, 129–38.

King, M.G. and Fa'asili, U. (1998) A network of small, community-owned fish reserves in Samoa. *PARKS* **8** (2), 11–16.

King, M.G. and Fa'asili, U. (1999) Community-based management of subsistence fisheries in Samoa. *Fisheries Management and Ecology*, UK **6**, 133–44.

King, M.G. and Lambeth, L. (2000) *Fisheries Management by Communities: a Manual on Promoting the Management of Subsistence Fisheries by Pacific Island Communities*. Secretariat of the Pacific Community, New Caledonia.

Labrosse, P., Kulbicki, M. and Ferraris, J. (2002) *Underwater Visual Fish Census Surveys*. Secretariat of the Pacific Community, New Caledonia.

Lalumera, G.M., Calamari, D., Galli, P., Castiglioni, S., Crosa, G. and Fanelli, R. (2004) Preliminary investigation on the environmental occurrence and effects of antibiotics used in aquaculture in Italy. *Chemosphere* **54** (5), 661–8.

Larkin, P.A. (1977) An epitaph for the concept of maximum sustainable yield. *Transactions of the American Fisheries Society* **106** (1), 1–11.

Latour, R.J., Brush, M.J. and Bonzek, C.F. (2003) Towards ecosystem-based fisheries management: strategies for multispecies modelling and associated data requirements. *Fisheries* **28** (9), 10–22.

Lea, E. (1910) On the methods used in herring investigations. *Publications de Circonstance, International Council for the Exploration of the Sea* **53**.

Lee, Chan L. and Lynch, P.W. (eds) (1997) *Trochus: Status, Hatchery Practice and Nutrition. ACIAR Proceedings* **PR079**.

Lee, R.M. (1920) A review of the methods of age and growth determination by means of scales. *Fishery Investigations*, London, Series 2, **4** (2).

Lery, J.M., Prado, J. and Tietze, U. (1999) Economic viability of marine capture fisheries. Findings of a global study and an interregional workshop. *FAO Fisheries Technical Paper* **377**.

Lockwood, S.J. (1988) *The Mackerel: its Biology, Assessment and the Management of a Fishery*. Fishing News Books, Oxford.

Ludwig, D., Hilborn, R. and Walters, C. (1993) Uncertainty, resource exploitation, and conservation: lessons from history. *Science* **260**, 17–36.

MacArthur, R.H. and Wilson, E.O. (1967) *The Theory of Island Biogeography*. Princeton University Press, USA.

MacDonald, P.D.M. and Green, P.E.J. (1985) *Mix 2.2: an Interactive Program for Fitting Mixtures of Distributions*. Icthus Data Systems, Hamilton, Ontario.

Marten, G.G. and Polovina, J.J. (1982) A comparative study of fish yields from various tropical ecosystems. In: D. Pauly and G.I. Murphy (eds) *Theory and Management of Tropical Fisheries*, ICLARM Conference Proceedings 9, pp. 255–85. International Center for Living Aquatic Resources Management (ICLARM), Manila, Philippines.

Martha, K., Jones, M., Fitzgerald, D.G. and Sale, P.F. (2002) Comparative ecology of marine fish communities. In: P. Hart and J. Reynolds (eds) *Handbook of Fish Biology and Fisheries*, Vol. I, pp. 341–58. Blackwell Science, Oxford.

Martin, J.W. and Davis, G.E. (2001) An updated classification of the recent Crustacea. Natural History Museum of Los Angeles County, *Science* Series 39.

Matsumoto, W.M. (1961) *Identification of Larvae of Four Species of Tuna from the Indo-Pacific Region. 1. The Carlsberg Foundation's Oceanic Expedition Round the World 1928–30 and Previous Dana Expeditions*. Dana Report 5.

McClanahan, T., Kakamura, A., Muthiga, N., Yebio, M. and Obura, D. (1996) Effects of sea urchin reductions on algae, coral, and fish populations. *Conservation Biology* 10, 136–54.

McCormick, M.I. and Choat, J.H. (1987) Estimating total abundance of a large temperate-reef fish using visual strip-transects. *Marine Biology* 96, 469–78.

McLachlan, A. (1983) Sandy beach ecology. In: A. McLachlan and T. Erasmus (eds) *Sandy Beaches as Ecosystems*, pp. 321–80. Dr W. Junk, The Hague.

Megrey, B.A. (1989) Review and comparison of age-structured stock assessment models from a theoretical and applied points of view. *American Fisheries Society Symposium* 6, 8–48.

Metcalfe, J., Arnold, G. and McDowall, R. (2002) Migration. In: P. Hart and J. Reynolds (eds) *Handbook of Fish Biology and Fisheries*, Vol. I, pp. 175–99. Blackwell Science, Oxford.

Misund, O.A., Kolding, J. and Freon, P. (2002) Fish capture devices in industrial and artisanal fisheries and their influence on management. In: P. Hart and J. Reynolds (eds) *Handbook of Fish Biology and Fisheries*, Vol. II, pp. 12–36. Blackwell Science, Oxford.

Mooi, R.D. and Gill, A.C. (2002) Historical biogeography of fishes. In: P. Hart and J. Reynolds (eds) *Handbook of Fish Biology and Fisheries*, Vol. I, pp. 43–68. Blackwell Science, Oxford.

Moyle, P.B. and Cech, J.I. (2004) *Fishes: an Introduction to Ichthyology*, 5th edn. Prentice Hall, New Jersey.

Munro, J.L. (1982) Estimation of the parameters of the von Bertalanffy growth equation from recapture data at variable time intervals. *Journal du Conseil, Conseil International pour l'Exploration de la Mer* 40, 199–200.

Munro, J.L. (1984) Yields from coral reef fisheries. *Fishbyte* 2 (3), 13–15.

Murphy, G.I. (1968) Pattern in life history and the environment. *American Naturalist* 102, 391–403.

Myers, R.A. (2002) Recruitment: understanding density-dependence in fish populations. In: P. Hart and J. Reynolds (eds) *Handbook of Fish Biology and Fisheries*, Vol. I, pp. 123–48. Blackwell Science, Oxford.

National Marine Fisheries Service (NMFS) (2002) Workshop on the Effects of Fishing Gear on Marine Habitats off the Northeastern United States. October 23–25, 2001, Boston, Massachusetts. Northeast Region Essential Fish Habitat Steering Committee. Woods Hole, Massachusetts, USA.

Nybakken, J.W. and Bertness, M.D. (2005) *Marine Biology: an Ecological Approach*, 6th edn. Pearson Education Inc., San Francisco.

O'Brien, C.M., Fox, C.J., Planque, B. and Casey, J. (2000) Fisheries: climate variability and North Sea cod. *Nature* 404, 142.

Oda, D.K. and Parrish, J.D. (1981) Ecology of commercial snappers and groupers introduced to Hawaiian reefs. *Proceedings of the Fourth International Coral Reef Congress*, Manila, Philippines, 1, 59–67.

Okey, T.A. and Pauly, D. (eds) (1999) *A Trophic Mass-balance Model of Alaska's Prince William Sound Ecosystem, for the Post-spill Period 1994–1996*, 2nd edn. Fisheries Centre Research Report 7 (4), University of British Columbia, Vancouver.

Olson, R.E. (1986) Marine fish parasites of public health importance. In: D.E. Kramer and J. Liston (eds) *Seafood Quality Determination*, pp. 339–355. Elsevier Science Publishers, Amsterdam.

Pace, M.L., Cole, J.J., Carpenter, S.R. and Kitchell, J.F. (1999) Trophic cascades revealed in diverse ecosystems. *Trends in Ecology and Evolution* 14, 483–8.

Panella, G. (1974) Otolith growth patterns: an aid in age determination in temperate and tropical fishes. In: T.B. Bagenal (ed.) *Ageing of Fish, Proceedings of an International Symposium on the Ageing of Fish held at the University of Reading, England, 19–20*

July 1973, pp. 28–39. Unwin Brothers Limited, Old Woking, Surrey.

Parker, K. (1980) A direct method for estimating northern anchovy, *Engraulis mordax*, spawning biomass. *Fisheries Bulletin*, U.S. 78, 5541–4.

Parkinson, B. (1982) *The Specimen Shell Resources of Fiji*. South Pacific Commission, New Caledonia.

Pauly, D. (1980) On the interrelationships between natural mortality, growth parameters, and mean environmental temperature in 175 fish stocks. *Journal du Conseil, Conseil International pour l'Exploration de la Mer* 39, 175–92.

Pauly, D. (1983) Some simple methods for the assessment of tropical fish stocks. *FAO Fisheries Technical Paper* 234.

Pauly, D. (1986) On improving operation and use of the ELEFAN programs. Part II. Improving the estimation of L infinity. *Fishbyte* 4 (1), 18–20.

Pauly, D. and Caddy, J.F. (1985) A modification of Bhattacharya's method for the analysis of mixtures of normal distributions. *FAO Fisheries Circular* 781.

Pauly, D. and Christensen, V. (2002) Ecosystem models. In: P. Hart and J. Reynolds (eds) *Handbook of Fish Biology and Fisheries*, Vol. II, pp. 211–27. Blackwell Science, Oxford.

Pauly, D. and David, N. (1980) An objective method for determining growth from length-frequency data. ELEFAN 1: User's instructions and program listing. *International Center for Living Aquatic Resources Management (ICLARM) Newsletter* 3 (3), 13–15.

Pauly, D. and Palomares, M-L. (2005) Fishing down marine food webs: it is far more pervasive than we thought. *Bulletin of Marine Science* 76 (2), 197–211.

Pauly, D., Christensen, V., Dalsgaard, J., Froese, R. and Torres, F.C. (1998) Fishing down marine food webs. *Science* 279, 860–3.

Pauly, D., Christensen, V. and Walters, C.J. (2000) Ecopath, Ecosim, and Ecospace as tools for evaluating ecosystem impact of fisheries. *ICES Journal of Marine Science* 57, 697–706.

Pauly, D., Christensen, V., Guénette, S., Pitcher, T.J., Sumaila, U.R., Walters, C.J., Watson, R. and Zeller, D. (2002) Towards sustainability in world fisheries. *Nature* 418, 689–95.

Payne, A.G., Agnew, D.J. and Pierce, G.J. (2006) Trends and assessment of cephalopod fisheries. *Fisheries Research* 78, 1–3.

Pella, J.J. and Tomlinson, P.K. (1969) A generalized stock production model. *Bulletin Inter-American Tropical Tuna Commission* 13, 419–96.

Penn, J.W. (1975) The influence of tidal cycles on the distribution pathways of *Penaeus latisulcatus* Kishinouye in Shark Bay, Western Australia. *Australian Journal of Marine and Freshwater Research* 26, 93–102.

Pérez-Farfante, I. and Kensley, B.F. (1997) Penaeoid and sergestoid shrimps and prawns of the world (keys and diagnoses for the families and genera). *Editions du Museum national d'Histoire naturelle* 175, Paris.

Peterson, D.L., Bain, M.B. and Haley, N. (2000) Evidence of declining recruitment of Atlantic sturgeon in the Hudson River. *North American Journal of Fisheries Management* 20, 231–8.

Phillips, B.F., Morgan, G.R. and Austin, C.M. (1980) Synopsis of biological data on the western rock lobster *Panuliris cygnus* (George, 1962). *FAO Fisheries Synopsis* 128.

Pianka, E.R. (1970) On r and K selection. *American Naturalist* 104, 592–7.

Pierce, G.J. and Guerra, A. (1994) Stock assessment methods used for cephalopod fisheries. *Fisheries Research* 21, 255–86.

Pinkerton, E. (ed.) (1989) *Co-operative Management of Local Fisheries*. University of British Columbia Press.

Pitcher, C.R. (1993) Spiny lobster. In: A. Wright and L. Hill (eds) *Nearshore Marine Resources of the South Pacific*, pp. 539–608. Forum Fisheries Agency, Honiara; Institute of Pacific Studies, Suva.

Pitcher, C.R., Skewes, T.D., Dennis, D.M. and Prescott, J.H. (1992) Estimation of the abundance of the tropical lobster *Panulirus ornatus* in Torres Strait, using visual transect-survey methods. *Marine Biology* 113, 57–64.

Pitcher, T.J. (2002) A bumpy old road: size-based methods in fisheries assessment. In: P. Hart and J. Reynolds (eds) *Handbook of Fish Biology and Fisheries*, Vol. II, pp. 189–210. Blackwell Science, Oxford.

Pitcher, T.J. and Hart, P.J.B. (1982) *Fisheries Ecology*. Croom Helm, London.

Pitcher, T.J. and MacDonald, P.D.M. (1973) An integration method for fish population fecundity. *Journal of Fish Biology* 4, 549–53.

Polacheck, T., Hilborn, R. and Punt, A.E. (1993) Fitting surplus production models: comparing methods and measuring uncertainty. *Canadian Journal of Fisheries and Aquatic Sciences* 50, 2597–607.

Polovina, J.J. (1984) Model of a coral reef ecosystem I. The ECOPATH model and its application to French frigate shoals. *Coral Reefs* 3, 1–11.

Polovina, J.J. (1986a) A variable catchability version of the Leslie model with application to an intensive fishing experiment on a multispecies stock. *Fishery Bulletin*, U.S. 84 (2), 423–8.

Polovina, J.J. (1986b) Assessment and management of deepwater bottom fishes in Hawaii and the Marianas. In: J.J. Polovina and S. Ralston (eds) *Tropical Snappers and Groupers: Biology and Fisheries Management*. Westview Press Inc., Boulder and London.

Polunin, N.V.C. (2000) Marine protected areas, fish and fisheries. In: P. Hart and J. Reynolds (eds) *Handbook of Fish Biology and Fisheries*, Vol. II, pp. 291–318. Blackwell Science, Oxford.

Polunin, N.V.C. and Pinnegar, J.K. (2002) Trophic ecology and the structure of marine food webs. In: P. Hart and J. Reynolds (eds) *Handbook of Fish Biology and Fisheries*, Vol. I, pp. 301–320. Blackwell Science, Oxford.

Pomeroy, R.S. (1998) A process for community-based fisheries co-management. *Naga, ICLARM Quarterly* **21** (1), 71–5.

Pope, J.G. (1972) An investigation of the accuracy of virtual population analysis. *ICNAF Research Bulletin* **9**, 65–74.

Potter, E.C.E. and Russell, I.C. (1994) Comparison of the distribution and homing of hatchery-reared and wild Atlantic salmon, *Salmo salar* L., from north-east England. *Aquaculture and Fisheries Management* **25**, Suppl. 2, 31–44.

Prentice, E.F. and Rensel, J.E. (1977) Tag retention of the spot prawn, *Pandalus platyceros*, injected with coded wire tags. *Journal of the Fisheries Research Board of Canada* **34**, 2199–203.

Prescott, J. (1988) *Tropical Spiny Lobsters: an Overview of their Biology, the Fisheries and the Economics with Particular Reference to the Double Spined Rock Lobster P. penicillatus*. South Pacific Commission Workshop on Pacific Inshore Fishery Resources, New Caledonia. Working Paper 18.

Preston, G.L. (1993) Beche-de-mer. In: A. Wright and L. Hill (eds) *Nearshore Marine Resources of the South Pacific*, pp. 371–408. Forum Fisheries Agency, Honiara; Institute of Pacific Studies, Suva.

Punt, A.E. and Hilborn, R. (1997) Fisheries stock assessment and decision analysis: the Bayesian approach. *Reviews in Fish Biology and Fisheries* **7**, 35–63.

Punt, A.E. and Smith, A.D.M. (2001) The gospel of maximum sustainable yield in fisheries management: birth, crucifixion and reincarnation. In: J.D. Reynolds, G.M. Mace, K.H. Redford and J.G. Robinson (eds) *Conservation of Exploited Species*, pp. 41–66. Cambridge University Press.

Quinn, T.J., II and Deriso, R.B. (1999) *Quantitative Fish Dynamics*. Oxford University Press.

Randall, J.E., Allen, G.R. and Steene, R.C. (1998) *Fishes of the Great Barrier Reef and Coral Sea*. University of Hawaii Press, Honolulu.

Rasero, M. and Pereira, J.M.F. (2004) Application of depletion methods to estimate stock size in the squid *Loligo forbesi* in Scottish waters (UK). *Fisheries Research* **69** (2), 211–27.

Reynolds, J.D., Dulvy, N.K. and Roberts, C.M. (2002) Exploitation and other threats to fish conservation. In: P. Hart and J. Reynolds (eds) *Handbook of Fish Biology and Fisheries*, Vol. II, pp. 319–341. Blackwell Science, Oxford.

Richards, L.J. and Schnute, J.T. (1986) An experimental and statistical approach to the question: Is CPUE an index of abundance? *Canadian Journal of Fisheries and Aquatic Sciences* **43** (6), 1214–27.

Ricker, W.E. (1954) Stock and recruitment. *Journal of the Fisheries Research Board of Canada* **11**, 555–623.

Ricker, W.E. (1975) Computation and interpretation of biological statistics of fish populations. *Fisheries Research Board of Canada Bulletin* **191**.

Roberts, C.M. and Polunin, N.V.C. (1991) Are marine reserves effective in management of reef fisheries? *Reviews in Fish Biology and Fisheries* **1**, 65–91.

Roel, B.A. and Butterworth, D.S. (2000) Assessment of the South African chokka squid *Loligo vulgaris reynaudii*. Is disturbance of aggregations by the recent jig fishery having a negative impact on recruitment? *Fisheries Research* **48**, 213–28.

Roper, C.F.E., Sweeney, M.J. and Nauan, C.E. (1984) FAO species catalogue. Vol. 3. Cephalopods of the world. An annotated and illustrated catalogue of species of interest to fisheries. *FAO Fisheries Synopsis* **125**.

Ruddle, K. (1994) A guide to the literature on traditional community-based fishery management in the Asia-Pacific tropics. *FAO Fisheries Circular* **869**.

Ruiz, G.M. and Dobbs, F.C. (2003) Biological invasions: consequences for parasites, pathogens, emerging diseases, and fisheries in the marine environment. ASFB Annual Conference, New Zealand.

Russ, G.R. and Alcala, A.C. (2004) Marine reserves: long-term protection is required for full recovery of predatory fish species. *Oecologia* **138**, 622–7.

Russell, E.S. (1931) Some theoretical considerations on the 'overfishing' problem. *Journal du Conseil International pour l'Exploration de la Mer* **6**, 3–20.

Ryther, J.H. (1969) Photosynthesis and fish production in the sea. *Science* **166**, 72–6.

Saenger, P.E. (1994) Mangroves and salt marshes. In: L.S. Hammond and R.N. Synnot (eds) *Marine Biology*, pp. 238–256. Longman Cheshire, Australia.

Sale, P.F., Cowen, R.K., Danilowicz, B.S., Jones, G.P., Kritzer, J.P., Lindeman, K.C., Planes, S., Polunin, N.V.C., Russ, G.R., Sadovy, Y.J. and Steneck, R.S. (2005) Critical science gaps impede use of no-take fishery reserves. *Trends in Ecology and Evolution* **20** (2), 74–80.

Salini, J., Brewer, D., Farmer, M. and Rawlinson, N. (2000) Assessment and benefits of damage reduction in prawns due to use of different bycatch reduction devices in the Gulf of Carpentaria. *Australian Fisheries Research* **45**, 1–8.

Santilan, N. and Williams, R.J. (1999) Mangrove transgression into salt marsh environments in southeast Australia. *Global Ecology and Biogeography* **8**, 117–24.

Saunders, W.B. and Landman, N.H. (1987) *Nautilus*. Plenum Publishing Corporation, New York and London.

Saville, A. (1977) Survey methods of appraising fisheries resources. *FAO Fisheries Technical Paper* **171**.

Schaefer, M.B. (1954) Some aspects of the dynamics of populations important to the management of commercial marine fisheries. *Bulletin, Inter-American Tropical Tuna Commission* **1**, 25–56.

Schaefer, M.B. (1957) A study of the dynamics of the fishery for yellowfin tuna in the Eastern Tropical Pacific Ocean. *Bulletin, Inter-American Tropical Tuna Commission* **2**, 247–85.

Schnute, J.T. (1985) A general theory for the analysis of catch and effort data. *Canadian Journal of Fisheries and Aquatic Sciences* **42**, 414–29.

Schnute, J.T. and Richards, L. (2002) Surplus production models. In: P. Hart and J. Reynolds (eds) *Handbook of Fish Biology and Fisheries*, Vol. II, pp. 103–26. Blackwell Science, Oxford.

Shelley, C.C. and Southgate, P.C. (1988) Reproductive periodicity and morphometry of *Hippopus hippopus* and *Tridacna crocea*. In: J.A. Copland and J.S. Lucas (eds) *Giant Clams in Asia and the Pacific*. Australian Centre for International Agricultural Research (ACIAR) Monograph 9.

Shepherd, J.G. (1982) A versatile new stock-recruitment relationship for fisheries, and the construction of sustainable yield curves. *Journal du Conseil, Conseil International pour l'Exploration de la Mer* **40** (1), 67–75.

Shepherd, S.A. and Laws, H.M. (1974) Studies on southern Australian abalone (genus *Haliotis*). II. Reproduction of five species. *Australian Journal of Marine and Freshwater Research* **25**, 49–62.

Shepherd, S.A., Tegner, M.J. and Guzman del Proo, S.A. (1992) *Abalone of the World*. Fishing News Books, Oxford.

Sinderman, C.L. (1990) *Principal Diseases of Marine Fish and Shellfish*, 2nd edn. Academic Press, San Diego.

Sloan, N.A. (1985) Echinoderm fisheries of the world: a review. In: B. Keegan and B. O'Connor (eds) *Echinodermata*, pp. 109–24. Balkema, Rotterdam.

Smith, C. and Reay, P. (1991) Cannibalism in teleost fish. *Reviews in Fish Biology and Fisheries* **1**, 41–64.

Smith, D.C. (1982) Age and growth of jackass morwong (*Nemadactylus macropterus* Bloch & Schneider) in eastern Australian waters. *Australian Journal of Marine and Freshwater Research* **33**, 245–53.

Smith, D.C. (1983) Annual total mortality and population structure of jackass morwong (*Nemadactylus macropterus* Bloch & Schneider) in eastern Australian waters. *Australian Journal of Marine and Freshwater Research* **34**, 253–60.

Smith, G.B. and van Nierop, M. (1986) Abundance and potential yield of spiny lobster *(Panulirus argus)* on the Little and Great Bahama Banks. *Bulletin Marine Science* **39**, 646–56.

Smith, T.D. (2002) A history of fisheries and their science and management. In: P. Hart and J. Reynolds (eds) *Handbook of Fish Biology and Fisheries*, Vol. II, pp. 61–83. Blackwell Science, Oxford.

Sotka, E.E. and Thacker, R.W. (2005) Do some corals like it hot? *Trends in Ecology and Evolution* **20** (2), 59–62.

Sparre, P. and Hart, P.J.B. (2002) Choosing the best model for fisheries assessment. In: P. Hart and J. Reynolds (eds) *Handbook of Fish Biology and Fisheries*, Vol. II, pp. 270–90. Blackwell Science, Oxford.

Sparre, P., Ursin, E. and Venema, S.C. (1989) Introduction to tropical fish stock assessment. Part I – Manual. *FAO Fisheries Technical Paper* **306.1**.

Stearns, S.C. (1976) Life history tactics: a review of the ideas. *Quarterly Review of Biology* **51**, 3–47.

Stephenson, T.A. and Stephenson, A. (1972) *Life Between Tide Marks on Rocky Shores*. Freeman & Co., San Francisco.

Summerfelt, R.C. and Hall, G.E. (1987) *Age and Growth of Fish*. Iowa State University Press.

Sutton, V. (2005) Custom, tradition and science in the South Pacific: Fiji's new environmental management act and vanua. *Journal of South Pacific Law* **9** (2).

Thresher, R.E. and Gunn, J.S. (1986) Comparative analysis of visual census techniques for highly mobile, reef-associated piscivores (Carangidae). *Environmental Biology of Fishes* **17** (2), 93–111.

Thurow, F. (1971) Changes of population parameters of cod in the Baltic. International Council for the Exploration of the Sea, Special Meeting on Cod and Herring in the Baltic, Cod 5.

Trendall, J.T. (1988) Growth of the spiny lobster *Panulirus ornatus*, in the Torres Strait. Workshop on the Pacific Inshore Fishery Resources, South Pacific Commission, Noumea. Background Paper 90.

Tsukamoto, K., Nakai, I. and Tesch, W. (1998) Do all freshwater eels migrate? *Nature* **396**, 635.

van Marlen, B. (2000) Technical modifications to reduce the by-catches and impacts of bottom fishing gears. In: M.J. Kaiser and S.J. de Groot (eds) *Effects of Fishing on Non-target Species and Habitats*, pp. 253–68. Blackwell Science, Oxford.

Veron, J.E.N. (1986) Distribution of reef-building corals. *Oceanus* **29**, 27–31.

Walkenback, J. (1999) *Microsoft Excel 2000 Power Programming with VBA*. IDG Books Worldwide Inc., Foster City.

Walters, C.J. (1986) *Adaptive Management of Renewable Resources*. Macmillan, New York.

Walters, C.J. and Hilborn, R. (1976) Adaptive control of fishing systems. *Journal of the Fisheries Research Board of Canada* **33**, 145–59.

Wass, R.C. (1982) The shoreline fishery of American Samoa – past and present. In: J.L. Munro (ed.) *Marine and Coastal Processes in the Pacific*, pp. 51–83. UNESCO-ROSTSEA, Jakarta, Indonesia.

Watson, R. and Pauly, D. (2001) Systematic distortions in world fisheries catch trends. *Nature* **414**, 534–6.

Watt, P. (2001) *A Manual for the Co-management of Commercial Fisheries in the Pacific*. Secretariat of the Pacific Community, New Caledonia.

Wetherall, J.A. (1986) A new method for estimating growth and mortality parameters from length-frequency data. *Fishbyte* **4** (1), 12–14.

Wetherall, J.A., Polovina, J.J. and Ralston, S. (1987) Estimating growth and mortality in steady-state fish stocks from length-frequency data. In: D. Pauly and G.R. Morgan (eds) *Length-based Methods in Fisheries Research*. International Center for Living Aquatic Resources Management (ICLARM), Manila, Philippines and KISR, Kuwait.

Whitehead, P.J.P. (1985) FAO species catalogue. Vol. 7. Clupeoid fishes of the world (Suborder Clupeoidae). An annotated and illustrated catalogue of the herrings, sardines, pilchards, sprats, shads, anchovies and wolf-herrings. Part 1. Chirocentridae, Clupeidae and Pristigasteridae. *FAO Fisheries Synopsis* **125**.7, Part 1.

Whitehead, P.J.P., Nelson, G.J. and Wongratana, T. (1988) FAO species catalogue. Vol. 7. Clupeoid fishes of the world (Suborder Clupeoidae). An annotated and illustrated catalogue of the herrings, sardines, pilchards, sprats, shads, anchovies and wolf-herrings. Part 2. Engraulidae. *FAO Fisheries Synopsis* **125**.7, Part 2, 305–579.

Wolanski, E., Doherty, P. and Carleton, J. (1997) Directional swimming of fish larvae determines connectivity of fish populations on the Great Barrier Reef. *Naturwissenschaften* **84**, 262–8.

Worm, B., Barbier, E.B., Beaumont, N., Duffy, E., Folke, C., Halpern, B.S., Jackson, J.B.C., Lotze, H.K., Micheli, F., Palumbi, S.R., Sala, E., Selkoe, K.A., Stachowicz, J.J. and Watson, R. (2006) Impacts of biodiversity loss on ocean ecosystem services. *Science* **314**, 787–90.

Wright, P.J., Woodroffe, D.A., Gibb, F.M. and Gordon, J.D.M. (2002) Verification of first annulus formation in the illicia and otoliths of white anglerfish, *Lophius piscatorius*, using otolith microstructure. *ICES Journal of Marine Science* **59** (3), 587–93.

Young, J.A. and Muir, J.F. (2002) Marketing fish. In: P. Hart and J. Reynolds (eds) *Handbook of Fish Biology and Fisheries*, Vol. II, pp. 37–60. Blackwell Science, Oxford.

Youngs, W.D. and Robson, D.S. (1978) Estimation of population number and mortality rates. In: T.B. Bagenal (ed.) *Methods for the Assessment of Fish Production in Freshwaters*. IBP Handbook 3, pp. 137–64. Blackwell Science, Oxford.

Zann, L.P. (1995) *Our Sea, Our Future: Major Findings of the State of the Marine Environment Report for Australia*. Great Barrier Reef Marine Park Authority, Townsville.

Further reading

Alcala, A.C. and Russ, G.R. (1990) A direct test of the effects of protective management on abundance and yield of tropical marine resources. *Journal du Conseil International pour l'Exploration de la Mer* **46**, 40–7.

Allen, G.R. (1985) Snappers of the world. An annotated and illustrated catalogue of lutjanid species known to date. *FAO Fisheries Synopsis* **125**, Vol. 6.

Allen, K.R. (1966) Some methods for estimating exploited populations. *Journal of the Fisheries Research Board of Canada* **23**, 1553–74.

Allen, K.R. (1971) Relation between production and biomass. *Journal of the Fisheries Research Board of Canada* **28**, 1573–81.

Beddington, J.R. and May, R.M. (1977) Harvesting natural populations in a randomly fluctuating environment. *Science* **197**, 463–5.

Brey, T. and Pauly, D. (1986) *Electronic Length-frequency Analysis. A Revised and Expanded User's Guide to ELEFAN 0, 1 and 2*. Berichte aus dem

Institut fur Meereskunde un der Christian Albrechts Universitut, Kiel, **149**.

Brown, I.W. (1993) Mangrove crabs. In: A. Wright and L. Hill (eds) *Nearshore Marine Resources of the South Pacific*, pp. 609–42. Forum Fisheries Agency, Honiara; Institute of Pacific Studies, Suva.

Caddy, J.F. (1983) The cephalopods: factors relevant to their population dynamics and to the assessment and management of stocks. *FAO Fisheries Technical Paper* **231**.

Caddy, J.F. (ed.) (1989) *Marine Invertebrate Fisheries: their Assessment and Management*. John Wiley & Sons, New York.

Caddy, J.F. (1998) A short review of precautionary reference points and some proposals for their use in data-poor situations. *FAO Fisheries Technical Paper* **379**.

Caddy, J.F. and Csirke, J. (1983) Approximations to sustainable yield for exploited and unexploited stocks. *Oceanographie Tropicale* **18** (1), 3–15.

Caddy, J.F. and Gulland, J.A. (1983) Historical patterns of fish stocks. *Marine Policy* **7** (4), 267–78.

Cochrane, K.L. (2002) Management measures and their application. FAO *Fisheries Technical Paper* **424**.

Copeland, J.W. and Grey, D.L. (eds) (1990) *Management of Wild and Cultured Sea Bass/Barramundi (Lates calcarifer). Australian Centre for International Agricultural Research (ACIAR) Proceedings* **20**.

Csirke, J. and Caddy, J.F. (1983) Production modeling using mortality estimates. *Canadian Journal of Fisheries and Aquatic Sciences* **40**, 43–51.

Cunningham, S. and Gréboval, D. (2001) Management of fishing capacity: a review of policy and technical issues. *FAO Fisheries Technical Paper* **409**.

Cushing, D.H. (1975) *Marine Ecology and Fisheries*. Cambridge University Press.

Drew, H.C., Mitchell, C.E., Ward, J.R., Altizer, S., Dobson, A.P., Ostfeld, R.S. and Samuel, M.D. (2002) Climate warming and disease risks for terrestrial and marine biota. *Science* **296** (5576), 2158–62.

FAO (1996) Precautionary approach to capture fisheries and species introductions elaborated by the technical consultation on the Precautionary Approach to capture fisheries (including species introductions), Lysekil, Sweden, 6–13 June 1995. *FAO Technical Guidelines for Responsible Fisheries*, No. 2.

FAO (2001) *An Evaluation of the Suitability of the CITES Criteria for Listing Species Exploited by Fisheries in Marine and Large Freshwater Bodies (Revised)* FI:SLC/2000/2.

Fox, W.W. (1975) Fitting the generalized stock production model by least squares and equilibrium approximation. *Fisheries Bulletin* **73**, 23–7.

Gayanilo, F.C., Soriano, M. and Pauly, D. (1989) *A Draft Guide to the Compleat ELEFAN*. ICLARM Software 2.70, Contribution 435. International Center for Living Aquatic Resources Management, Manila, Philippines.

Hart, P.J.B. and Reynolds, J.D. (2002) Banishing ignorance: underpinning fisheries with basic biology. In: P. Hart and J. Reynolds (eds) *Handbook of Fish Biology and Fisheries*, Vol. I, pp. 1–12. Blackwell Science, Oxford.

Hart, P.J.B. and Reynolds, J.D. (2002) The human dimension of fisheries science. In: P. Hart and J. Reynolds (eds) *Handbook of Fish Biology and Fisheries*, Vol. II, pp. 1–10. Blackwell Science, Oxford.

Hilborn, R. and Sibert, J. (1988) Adaptive management of developing fisheries. *Marine Policy*, April, 112–21.

Hilge, V. (1977) On the determination of stages of gonad ripenesss in female bony fishes. *Meeresforsch* **25**, 149–55.

Huse, G., Giske, J. and Salvanes, A.G.V. (2002) Individual-based models. In: P. Hart and J. Reynolds (eds) *Handbook of Fish Biology and Fisheries*, Vol. II, pp. 228–48. Blackwell Science, Oxford.

Jackson, J.B.C., Kirby, M.X. and Berger, W.H. (2001) Historical overfishing and the recent collapse of coastal ecosystems. *Science* **293**, 629–38.

Johannes, R.E. (1978) Reproductive strategies of coastal marine fishes in the tropics. *Environmental Biology of Fishes* **3**, 65–84.

Johannes, R.E. (1981) *Words of the Lagoon: Fishing and Marine Lore in the Palau District of Micronesia*. University of California Press.

King, M.G. (1985) *The Life-history of the Goolwa Cockle, Donax (Plebidonax) deltoides (Bivalvia: Donacidae), on an Ocean Beach, South Australia*. South Australian Department of Fisheries Report 85.

King, M.G. (1989) *Fisheries Research and Stock Assessment in Western Samoa*. FAO Terminal Report, Project SAM/8852.

Krause, J., Hensor, E.M.A. and Ruxton, G.D. (2002) Fish as prey. In: P. Hart and J. Reynolds (eds) *Handbook of Fish Biology and Fisheries*, Vol. I, pp. 284–98. Blackwell Science, Oxford.

Leslie, P.H. and Davis, D.H.S. (1939) An attempt to determine the absolute number of rats in a given area. *Journal of Animal Ecology* **8**, 94–113.

Lucas, J.S. (1988) Giant clams: description, distribution and life history. In: J.W. Copeland and J.S. Lucas (eds) *Giant Clams in Asia and the Pacific*. Australian Centre for International Agricultural Research (ACIAR) Monograph 9, pp. 21–32.

MacDonald, P.D.M. and Pitcher, T.J. (1979) Age-groups from size-frequency data: a versatile and efficient method of analyzing distribution mixtures. *Canadian Journal of Fisheries and Aquatic Sciences* **36**, 987–1001.

Magnusson, K.G. (1995) An overview of multispecies VPA – theory and applications. *Review in Fish Biology and Fisheries* **5**, 195–212.

Nash, W.J. (1993) Trochus. In: A. Wright and L. Hill (eds) *Nearshore Marine Resources of the South Pacific*. Forum Fisheries Agency, Honiara; Institute of Pacific Studies, Suva.

Pauly, D. (1984) Length-converted catch curves: a powerful tool for fisheries research in the tropics (Part 2). *Fishbyte* **2** (1), 17–19.

Pauly, D. and Murphy, G.I. (eds) (1982) *Theory and Management of Tropical Fisheries. ICLARM Conference Proceedings* **9**, International Center for Living Aquatic Resources Management (ICLARM), Manila, Philippines.

Pauly, D., Christensen, V. and Walters, C.J. (2000) Ecopath, Ecosim, and Ecospace as tools for evaluating ecosystem impact of fisheries. *ICES Journal of Marine Science* **57**, 697–706.

Pauly, D., Preikshot, D. and Pitcher, T. (eds) (1998) *Back to the Future: Reconstructing the Strait of Georgia Ecosystem*. Fisheries Centre Research Reports **6** (5).

Popper, A.N. (2003) Effects of anthropogenic sounds on fishes. *Fisheries* **28**, 24–31.

Ralston, S. and Polovina, J.J. (1982) A multispecies analysis of the commercial deep-sea handline fishery in Hawaii. *Fisheries Bulletin*, U.S. **80**, 435–48.

Roberts, C.M. (1995) Rapid build-up of fish biomass in a Caribbean marine reserve. *Conservation Biology* **91** (4), 815–26.

Sale, P.F. (ed.) (1991) *The Ecology of Fishes on Coral Reefs*. Academic Press, New York.

Saville, A. (1964) Estimation of the abundance of a fish stock from egg and larval surveys. In: J.A. Gulland (ed.) *On the Measurement of Abundance of Fish Stocks, Rapports et Procès-verbaux des Réunions, Commission Internationale pour l'Eexploration Scientifique de la Mer Méditerranée* **155**, 164–70.

Schnute, J.T. (1977) Improved estimates from the Schaefer production model: theoretical considerations. *Journal of the Fisheries Research Board of Canada* **34**, 583–603.

Schwartz, C.S. and Seber, G. (1999) Estimating animal abundance: Review III. *Statistical Science* **14**, 427–56.

Seber, G.A.F. (1982) *The Estimation of Animal Abundance and Related Parameters*. Charles Griffin and Co., London.

Shepherd, S.A. and Hearn, W.S. (1983) Studies on southern Australian abalone (Genus *Haliotis*). IV. Growth of *H. laevigata* and *H. ruber*. *Australian Journal of Marine and Freshwater Research* **34**, 461–75.

Shotton, R. (ed.) (2000) *Use of Property Rights in Fisheries Management. Proceedings of the FishRights99 Conference*, Fremantle, Western Australia 11–19 November 1999. Workshop Papers. *FAO Fisheries Technical Paper* **404/2**.

Sibert, J. (1986) Tuna stocks in the Southwest Pacific. Infofish Tuna Trade Conference, Bangkok, Thailand, 25–27 February 1986.

Sims, N.A. (1993) Pearl oysters. In: A. Wright and L. Hill (eds) *Nearshore Marine Resources of the South Pacific*, pp. 409–30. Forum Fisheries Agency, Honiara; Institute of Pacific Studies, Suva.

South, G.R. (1993) Seaweeds. In: A. Wright and L. Hill (eds) *Nearshore Marine Resources of the South Pacific*, pp. 683–705. Forum Fisheries Agency, Honiara; Institute of Pacific Studies, Suva.

Thompson, W.F. and Bell, F.H. (1934) Biological statistics of the Pacific halibut fishery. 2. Effect of changes in intensity upon total yield and yield per unit of gear. *Report of the International Fisheries (Pacific Halibut) Commission* 8.

Vermeij, G.J. (1993) Biogeography of recently extinct marine species: implications for conservation. *Conservation Biology* **7**, 391–7.

Walford, L.A. (1946) A new graphic method of describing the growth of animals. *Biological Bulletin*, Marine Biology Laboratory, Woods Hole, USA, **90**, 141–7.

Walter, G.G. (1973) Delay-differential equation models for fisheries. *Journal of the Fisheries Research Board of Canada* **30**, 939–45.

Walters, C.J. (1980) Comment on Deriso's delay-difference population model. *Canadian Journal of Fisheries and Aquatic Sciences* **37**, 2365.

Walters, C.J. and Hilborn, R. (1978) Ecological optimization and adaptive management. *Annual Review of Ecology and Systematics* **9**, 157–88.

Williams, T. and Bedford, B.C. (1974) The use of otoliths for age determination. In: T.B. Bagenal (ed.) *Ageing of Fish. Proceedings of an International Symposium on the Ageing of Fish held at the University of Reading, England, 19–20 July 1973*. Unwin Brothers Limited, Old Woking, Surrey.

Wright, A. (1993) Shallow water reef-associated finfish. In: A. Wright and L. Hill (eds) *Nearshore Marine Resources of the South Pacific*, pp. 203–84. Forum Fisheries Agency, Honiara; Institute of Pacific Studies, Suva.

Appendix 1: Fisheries symbols and formulae

Mesh selectivity

P probability of capture
L length
L_c mean length at capture

Logistic curve (for trawl net selectivity):
   probability of capture $(P) =$
   $1/(1 + \exp[-r(L - L_c)])$

Normal curve (gill nets):
   $P = [1/(s\sqrt{2\pi})] \times \exp[-(L - \bar{L})^2/2s^2]$

Catch and fishing effort

N numbers of fish in a stock
C catch
f fishing effort
q catchability coefficient
v vulnerability
CPUE
or C/f catch per unit of effort

Length and age

L_r and t_r mean length and age at recruitment
L_c and t_c mean length and age at first capture
L_m and t_m mean length and age at reaching sexual maturity
L_{max} and t_{max} maximum length and maximum age (longevity)

Growth

L_t length at age t
K von Bertalanffy growth coefficient
L_∞ asymptotic length
t_0 age at zero length

von Bertalanffy equation:
   $L_t = L_\infty(1 - \exp[-K(t - t_0)])$

Ford–Walford plot (L_{t+1} versus L_t):
   $L_\infty = a/(1 - b); K = -\ln[b]$

Gulland–Holt plot (growth rate versus mean length):
   $L_\infty = -a/b; K = -b$

von Bertalanffy plot ($-\ln[1 - L_t/L_\infty]$ versus t):
   $K = b; t_0 = -a/b$

Back-calculations for hard part analysis:
   $L_x = L_p(S_x/S_p)$ in which . . .
L_p present fish length,
L_x fish length at ring formation x
S_p present hard part length
S_x hard part length at ring x

Stock/recruitment

R number of recruits
S number in spawning stock (spawners)

Beverton and Holt: $R = aS/(b + S)$
Ricker: $R = aS \exp[-bS]$

Mortality

F instantaneous rate of fishing mortality
M instantaneous rate of natural mortality
Z instantaneous rate of total mortality
E exploitation rate ($= F/Z$)
N_t number at time t
N_0 initial number
S number of survivors

Exponential decay equation:
   $N_t = N_0 \exp[-Zt]$

Survivors per year:

$S = N_{t+1}/N_t = \exp[-Z]$

As percentages:

Survival is $100 \exp[-Z]$ and mortality is $100 (1 - \exp[-Z])$

Catch equation:

$C_t = (F/Z) N_t (1 - \exp[-Zt])$

Linear catch curve:

$\ln[N_t] = \ln[N_0] - Zt$

Yield

C catch (qfB)
MSY   maximum sustainable yield
f_{MSY} fishing effort at which MSY is achieved
Y/R   yield per recruit

Change in biomass from year t to $t + 1$

$B_{t+1} = B_t + rB_t (1 - B_t/B_\infty) - C_t$

Schaefer: $C = af + bf^2$

$f_{MSY} = -a/2b \ldots$ or $r/2q$
$MSY = -a^2/4b \ldots$ or $rB_\infty/4$

Fox: $C = f \exp[a + bf]$

$f_{MSY} = -1/b$
$MSY = (-1/b) \exp[a - 1]$

Yield per recruit:

$Y/R = F \exp[-M(t_c - t_r)] \; W_\infty$ multiplied by a summation of the following four terms:

$\{1 \exp[-0K(t_c - t_0)]/(M + F + 0K)\}+$
$\{-3 \exp[-1K(t_c - t_0)]/(M + F + 1K)\}+$
$\{3 \exp[-2K(t_c - t_0)]/(M + F + 2K)\}+$
$\{-1 \exp[-3K(t_c - t_0)]/(M + F + 3K)\}$

Appendix 2: Standard deviation and confidence limits

The frequencies of measurements from a large sample from a population can be graphed as the common bell-shaped curve, or a normal curve, shown in Fig. A2.1. Plotted heights from a large sample of adult human males, for example (shown at the left of Fig. A2.1), suggest that most individuals are around the mean height of 1.7 m and that a few very short and very tall individuals are at the lower and upper tails of the graph respectively. From this graph, the probability of a single individual (chosen at random) having a height between 1.8 and 1.9 m is indicated by the shaded area under the curve.

The spread (or dispersal) of values around the mean (the 'width' of the normal curve) can be measured by the standard deviation. In the normal curve on the right-hand side of Fig. A2.1, the shaded areas under the curve show that 68.26% of all values fall between one standard deviation (σ) of the mean (μ). Similarly, 95% of all values lie between the limits of $\mu \pm 1.96\ \sigma$.

The standard deviation of the mean of n measurements from a population with a standard deviation of σ is $\sigma/\sqrt{n}$. In the same way that we used the sample mean as an estimate of the population mean, we can use the sample standard deviation (s) to estimate the population standard deviation (σ). Conventionally, the standard deviation of a mean is referred to as its standard error (SE) and is computed as $s/\sqrt{n}$. It is possible to attach confidence limits at a chosen level of probability to the estimate of the true mean. At the 0.95 level of probability (95% confidence limits) we would compute $\bar{x} \pm 1.96 \times$ SE and be able to say that the true mean has a probability of 0.95 of falling within these limits and conversely a probability of 0.05 of falling outside. That is, we are 95% confident that the true mean lies between these limits.

However, the use of the value 1.96 for 95% confidence limits applies to large samples (>30). For smaller samples, it is necessary to use tables to find a value to substitute for 1.96. The confidence limits are then calculated as $\bar{x} \pm t \times$ SE where the value of t is taken from the table given in Appendix 7. The values in the table include an

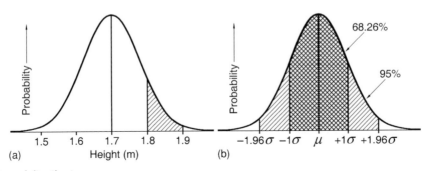

Fig. A2.1 Normal distributions.

allowance for the uncertainty involved in estimating the population standard deviation from a small sample.

The sea cucumber data summarized in Chapter 4, Table 4.1 are used here as an example. First, the variance (s^2) is calculated from the data in the table:

$$\text{variance } (s^2) = (\Sigma x^2 - \tfrac{1}{n}(\Sigma x)^2)/(n-1) =$$
$$(601 - 3481/7)/6 = 17.286 \qquad (A2.1)$$

The standard deviation (s) is calculated as $\sqrt{(s^2)} = \sqrt{17.286} = 4.158$. The standard error of the mean (SE) is then calculated as:

$$SE = s/\sqrt{n} = 4.158/2.646 = 1.571 \qquad (A2.2)$$

The confidence limits are calculated as $\bar{x} \pm t_{0.05} \times$ SE where $t_{0.05}$ is the value read from the table given in Appendix 7. The value of t for 6 degrees of freedom is 2.447 at the 95% level and the confidence limits are $8.43 \pm (2.447 \times 1.571)$ or 8.43 ± 3.844. In terms of percentages the value of 3.844 is 46% of the mean value of 8.43. Thus there is 95% confidence (or 0.95 probability) that the true stock size lies somewhere between (1315 − 46%) and (1315 + 46%), or between 710 and 1920 sea cucumbers.

Appendix 3: Correlation and regression

If we have quantitative data on two attributes, say x and y, for a sample of individuals and wish to examine the relationship between the two, the first step is to produce a scatter plot. The two attributes used here as an example are the carapace lengths (mm) and total lengths (mm) in a small sample of the penaeid prawn, *Penaeus latisulcatus*. The raw data, the paired measurements of carapace length and total lengths, are given in the first two columns of Table A3.1. The other columns and summations will be used in subsequent calculations.

A scatter plot based on the raw length data in the first two columns of Table A3.1 is shown in the left hand side of Fig. A3.1. The plot suggests that total length and carapace length are positively related and that the relationship appears linear. It is now appropriate to examine the

degree of correlation between the two measurements. For this it is necessary to calculate the correlation coefficient, r.

Correlation

The correlation coefficient (r) is usually calculated to express the degree of association (or correlation) between the two variables, and takes values between zero (no correlation) and one (perfect correlation). The sign of the value of r relates to either a negative (downhill) or positive (uphill) slope.

$$r = [\Sigma xy - (\Sigma x \Sigma y/n)]/\sqrt{[(\Sigma x^2 - (\Sigma x)^2/n)}$$
$$\times (\Sigma y^2 - (\Sigma y)^2/n)] \tag{A3.1}$$
$$= [55689 - (323 \times 1266/8)]/\sqrt{[(14335 -}$$
$$(104329/8)) \times (1266 - (1602756/8))]}$$
$$= 4574.25/4577.94 = 0.9992$$

Table A3.1 Worksheet based on carapace length (mm) and total length (mm) data for the penaeid prawn, *Penaeus latisulcatus*. Summations (indicated by Σ) are shown at the bottom of the columns. The sample size (n), the mean of the x values ($\bar{x}$) and the mean of the y values ($\bar{y}$) are shown below the table.

Carapace length (mm) x	Total length (mm) y	xy	x^2	y^2
20	87	1740	400	7569
26	105	2730	676	11 025
33	135	4455	1089	18 225
38	148	5624	1444	21 904
44	173	7612	1936	29 929
49	187	9163	2401	34 969
55	211	11 605	3025	44 521
58	220	12 760	3364	48 400
$\Sigma x = 323$	$\Sigma y = 1266$			
$(\Sigma x)^2 = 104\ 329$	$(\Sigma y)^2 = 1\ 602\ 756$	$\Sigma xy = 55\ 689$	$\Sigma x^2 = 14\ 335$	$\Sigma y^2 = 216\ 542$

Sample size (n) = 8
Mean of x's ($\bar{x}$) = $\Sigma x/n$ = 40.38
Mean of y's ($\bar{y}$) = $\Sigma y/n$ = 158.25

The correlation coefficient of $r = 0.999$ is very close to one, indicating that there is a high degree of correlation between total length and carapace length in the sample; its positive value reflects the fact that the slope is positive.

Although the relationship between the two variables is convincing in this example, there will be times when the value of the correlation coefficient is lower and some uncertainty exists. In these cases, statistical tables can be consulted to determine if the value of the correlation coefficient, r, is significant; that is, to find the probability of the correlation arising by chance alone, or whether the relationship found in the sample is applicable to the population as a whole.

For this we can test the significance of the deviation of r from zero (no relationship) by consulting tables of critical values of r for the relevant number of degrees of freedom (Appendix 7). As two parameters have been used in the preparation of the table of critical values there are $(n - 2)$ degrees of freedom. Using 6 degrees of freedom, the critical value of r from the table is 0.707 at the 95% confidence level. As the value of r in the prawn example (0.999) is greater than the critical value, we can be 95% certain that the relationship is significant.

Linear regression

In the case of this example, it is relatively easy to draw a line by eye through the data points; that is, to use a ruler to draw a line that best fits the data points. A better alternative is to mathematically construct a 'line of best fit' through the data by linear regression; such a line minimizes the sum of vertical differences between the data points and the fitted line.

In the previous examination of the relationship between variables using correlation the assumption was that the data are bivariate-normal (that is, each of two measurements were normally distributed). In regression analysis, a distinction is made between the independent variable (x) and the dependent variable (y). Two of the requirements of linear regression are that for each value of x there is a corresponding true value of y

and that a linear relationship exists between the two variables. The general equation of a straight line is:

$$y = a + bx \qquad (A3.2)$$

where y and x are the two variables (different length measurements in this case), a is the intercept on the y-axis, and b is the slope. In Equation A3.2, the slope, b, is estimated by:

$$b = [\Sigma(x_i - \bar{x})(y_i - \bar{y})]/\Sigma(x_i - \bar{x})^2 \ldots \qquad (A3.3)$$

or, in a form that is more easily calculated using data from the worksheet given in Table A3.1:

$$b = [n\Sigma xy - \Sigma x\Sigma y]/[n\Sigma x^2 - (\Sigma x)^2] \qquad (A3.4)$$
$$= [(8 \times 55689) - (323 \times 1266)]/$$
$$[(8 \times 14335) - 104329] = 3.535$$

The intercept, a, is then estimated by:

$$a = \bar{y} - (b\bar{x}) \qquad (A3.5)$$
$$= 158.25 - (3.535 \times 40.38) = 15.51$$

A regression line using the calculated values of a and b is shown on the right-hand side of Fig. A3.1. The regression line may be drawn through the predicted values of y (total length) for any two chosen values for x (carapace length) by substituting the value of the intercept, a, and the value of slope, b, in the linear equation $y = a + bx$. For example, for carapace lengths of 20 mm and 40 mm, the regression line would pass through the predicted values of 86.2 mm and 156.9 mm respectively. The slope, or gradient, of the regression line, which is positive in this case ($b = 3.535$) reflects the fact that total length increases 3.535 times with every unit increase in carapace length. As the value of the intercept ($a = 15.51$) is not equal to zero, the two length measurements are not proportional over the whole size range of prawns (at least not for very small prawns).

Although it is instructive to go through the above calculations (at least once), many calculators and computer programs include built in procedures to complete linear regressions. Fig. A3.2 shows a spreadsheet from Microsoft Excel and a guide to the use of the software is given in Walkenback (1999). The program has worksheet functions that are entered in the three cells in the

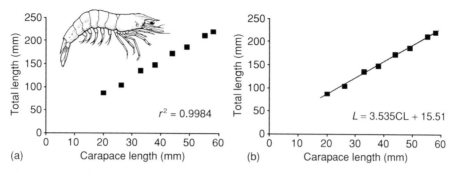

(a)   Carapace length (mm)   (b)   Carapace length (mm)

Fig. A3.1 (a) A scatter plot of carapace lengths and total lengths for the penaeid prawn, *Penaeus latisulcatus*. The coefficient of determination r^2 is 0.9984 and the correlation coefficient r is therefore 0.999. (b) Regression line is TL = 15.51 + 3.535CL for the same data.

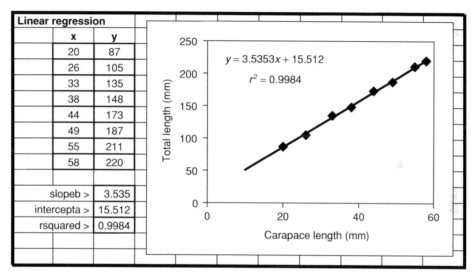

Fig. A3.2 A spreadsheet for linear regression. A scatter plot chart has been added with a trendline. Spreadsheet and graph produced using Microsoft Excel.

lower part of the spreadsheet to estimate the slope, b, as SLOPE(C3:C10,B3:B10), the intercept, a, as INTERCEPT(C3:C10,B3:B10) and the coefficient of determination, r^2, as RSQ(C3:C10,B3:B10). A scatter plot graph has been added at the right of the spreadsheet. In Excel, under the chart menu 'Add trendline' item, an available option is to print the equation of the line as well as the value of r^2 in the chart area as shown in Fig. A3.2.

Confidence limits for the intercept and slope

In many cases of linear relationships, the confidence limits of the intercept and slope can be

estimated. Using the penaeid prawn data as an example, the variance of x and y is estimated from Equation A2.1 (with values from Table A3.1) as:

$$s_{(x)}^2 = [\Sigma x^2 - 1/_n(\Sigma x)^2]/(n-1) = (14335 - [(1/8) \times 104329])/(8-1) = 184.84$$

$$s_{(y)}^2 = [\Sigma y^2 - 1/_n(\Sigma y)^2]/(n-1) = (216542 - [(1/8) \times 1602756])/(8-1) = 2313.93$$

The variance (s^2) about the regression line is estimated as:

$$s^2 = [(n-1)/(n-2)] \times [s_{(y)}^2 - (b^2 \times s_{(x)}^2)] \quad (A3.6)$$
$$= [7/6] \times [2313.93 - (3.535^2 \times 184.84)]$$
$$= 1.167 \times [2313.93 - 2310.50] = 4.00$$

The variances of the slope (b) and the intercept (a) are estimated as:

$$s_{(b)}^2 = [^1/_{(n-2)}] \times [(s_{(y)}^2/s_{(x)}^2) - b^2] \qquad (A3.7)$$
$$= 0.167 \times [12.52 - 12.5] = 0.003$$

$$s_{(a)}^2 = s_{(b)}^2 \times [\, (^{(n-1)}/_n \times s_{(x)}^2) + \bar{x}^2] \qquad (A3.8)$$
$$= 0.003 \times [\, (0.875 \times 184.84)$$
$$+ 40.38^2] = 5.38$$

Standard deviations for the slope and intercept are calculated as the square roots of the variances as $s_{(b)} = 0.055$ and $s_{(a)} = 2.32$.

Using a value of t from Appendix 7 with $(n-2)$ degrees of freedom (as there are two parameters, the slope and the intercept), 95% confidence limits can be estimated for the intercept and the slope.

For the intercept, a, confidence limits are:

$$a \pm (s_{(a)} \times t)] \qquad (A3.9)$$
$$= 15.51 \pm (2.32 \times 2.45) = 15.51 \pm 5.68$$
$$= \text{confidence interval of } 9.83 \text{ to } 21.19$$

For the slope, b, confidence limits are:

$$b \pm (s_{(b)} \times t)] \qquad (A3.10)$$
$$= 3.535 \pm (0.055 \times 2.45) = 3.535 \pm 0.135$$
$$= \text{confidence interval of } 3.40 \text{ to } 3.67$$

Interestingly, the confidence interval for the intercept does not include zero, and any hypothesis of direct proportionality between total length and carapace length over the entire length range (in which case the intercept would be zero) must be rejected at the 95% level. Perhaps total length has a different relationship to carapace length in juvenile prawns.

Appendix 4: Spreadsheet models

An alternative to estimating a model's parameters by natural logarithmic transformation and graphical methods is the use of a computer to seek values of the parameters that result in a best fit between the predictions of the model and a set of 'real' or observed data. A commonly used measure of best fit is least squares, which involves a search for the parameters that minimize the differences between the model's predicted (or expected) values and the observed data. Although the least squares method is frequently used (including widely in this book) there are alternatives, including maximum likelihood methods (a description of these is given in Haddon, 2001).

In Chapter 4, the von Bertalanffy growth parameters were estimated by converting the growth equation to a linear form, and using linear regression analysis. Here the non-linear least squares method is used to fit a growth curve to a set of mean length and age data for a large sample of fish (Table A4.1). This involves a direct search for the von Bertalanffy parameters, the values of K, L_∞ and t_0, that best fit a curve through the length at age data in Table A4.1. That is, a curve of the form:

$$L_t = L_\infty(1 - \exp[-K(t - t_0)]) \qquad \text{(A4.1)}$$

The least squares method minimizes the differences between the observed data and a curve in the same way that linear regression results in a value of the intercept and slope that minimizes the differences between the observed data and a straight line.

A computer spreadsheet design is shown in Table A4.2. The age and the observed length data (from Table A4.1) are entered in columns A and B in rows seven and below. Values of the three growth parameters are entered in the input cells C2, C3 and C4. Use fixed values of $L_\infty = 41$ in C3 and $t_0 = 0$ in C4 and enter a trial or initial 'guessed' value of K in C2. It is the value of K that will be optimized in the least squares process. The lengths predicted by the model are now calculated in column C by entering the von Bertalanffy equation (Equation A4.1).

The differences between observed values (column B) and expected or predicted values (column C) are called residual errors or 'noise' about a model's line of best fit. As observations can lie both above and below the line of best fit, (the errors can be either positive or negative) the differences are squared to remove negative values. The squared differences (squared residuals or SR) between the observed lengths and the predicted lengths are entered in column D. The sum of all the squared residual errors (SSR) between all observed data and expected values is now calculated in a cell at the bottom of column D as SSR = Σ(observed − expected)^2. The objective is now to search for the parameter values that minimize SSR and thus result in the model that best fits the observed data.

Table A4.1 Mean lengths of fish in nine age classes.

Age (years)	1	2	3	4	5	6	7	8	9
Mean length (cm)	10.7	15.8	26.6	28.0	33.0	34.6	36.3	36.6	38.0

Table A4.2 Upper, left-hand corner of a spreadsheet designed to estimate growth parameters from length at age data by non-linear least squares. The completed spreadsheet is shown in Fig. A4.1.

	A	B	C	D	E	F
1	Growth – fitting a curve to length at age					
2		K >	0.25			
3		Linf >	41		Age	Length
4		tzero >	0		0	=$C$3*(1–EXP(–$C$2*(E4–$C$4)))
5					=E4+1	=$C$3*(1–EXP(–$C$2*(E5–$C$4)))
6	Age	L obs	L exp	SR	=E5+1	=$C$3*(1–EXP(–$C$2*(E6–$C$4)))
7	1	10.7	=$C$3*(1–EXP(–$C$2*(A7–$C$4)))	=(B7–C7)^2	=E6+1	=$C$3*(1–EXP(–$C$2*(E7–$C$4)))
8	=A7+1	15.8	=$C$3*(1–EXP(–$C$2*(A8–$C$4)))	=(B8–C8)^2	=E7+1	=$C$3*(1–EXP(–$C$2*(E8–$C$4)))

Although not necessary for the analysis, an age–length key is included in columns E and F. Values from these two columns (as well as the observed data in column B) are graphed on the spreadsheet shown in Fig. A4.1. The initial or seed value of $K = 0.25$ yr^{-1} has produced expected lengths (in column C) and a curve that does not match the observed lengths very well. The seed value of K can now be repeatedly changed – do this manually, say in steps of 0.01 cm, until the value of SSR reaches a minimum. This will occur when the value of K provides the growth curve of

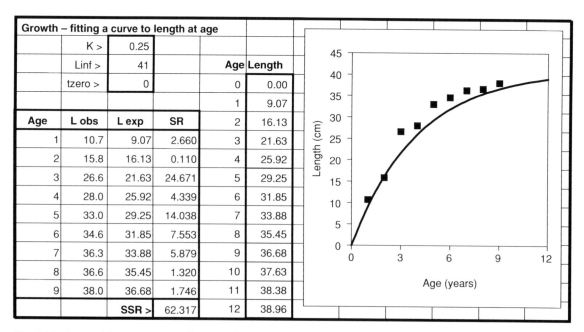

Fig. A4.1 A spreadsheet program to fit a growth curve to length at age data. A trial value of $K = 0.25$ has been entered with other parameters fixed at $L_\infty = 41$ and $t_0 = 0$. Spreadsheet and graphs produced using Microsoft Excel.

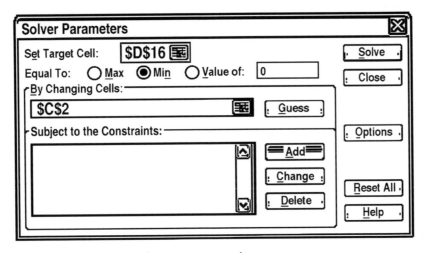

Fig. A4.2 The Excel Solver box for the growth parameter example.

best fit (when the expected lengths are closest to the observed lengths).

A manual search for the value of K that minimizes the SSR is relatively easy in this case because the initial seed value was reasonably close to the 'real' value and there was only a single parameter to estimate. More usually, trial and error procedures involve many repeated runs and a computer is best suited for such a repetitive task. The Solver function in Microsoft Excel 2003 is used here to complete a search for the minimum SSR and to identify the optimum values of the parameters.

The Microsoft Excel add-in program Solver is located under the Tools menu. The Solver parameter box (Fig. A4.2) requires a target cell, $D$16 in this example, which contains the sum of squared residuals (SSR) that will be minimized. The option 'Min' should be selected in the 'Equal To' row. The cell $C$2 is entered in the 'By Changing Cells' box; this cell contains the seed value of K that the program will change in an attempt to minimize the SSR. No constraints are required in the border provided. Once the target cell and the cells to change have been entered, click the 'Solve' button.

Solver works by trying different values for the parameter (by default 100 iterations are allowed) until the one that results in a minimum value for the sum of squared residuals (SSR) is found. An optimum value of $K = 0.301$ in cell $C$2 is found at a minimum SSR value of 14.9 and the resultant von Bertalanffy curve appears to fit the data quite well (Fig. A4.3). If the residuals are distributed normally, the average growth of the sampled stock will be represented by the expected growth curve.

The computer-based search for an optimum value for a single parameter was accomplished easily and Solver located an overall or global minimum value of SSR above the best estimate of a parameter. However, in some cases, particularly if more than one parameter had been involved and the initial seed values had been far from the 'real' or optimum values, there may have been local minima which confound the search for a global minimum (Fig. A4.4). If more than one minimum exists, an automatic search may stop at a local minimum (and thus suggest a poor estimate of the parameter) because search movements in either direction will increase the value of the SSR.

Now use Excel Solver to automatically try different combinations of the two parameters, K and L_∞, while leaving t_0 fixed at zero. Given the increased flexibility, it would be reasonable to expect that a better fitting growth curve could be found than in the single-parameter case. Remembering that a value of SSR of 14.9 was obtained when two of the three parameters were fixed,

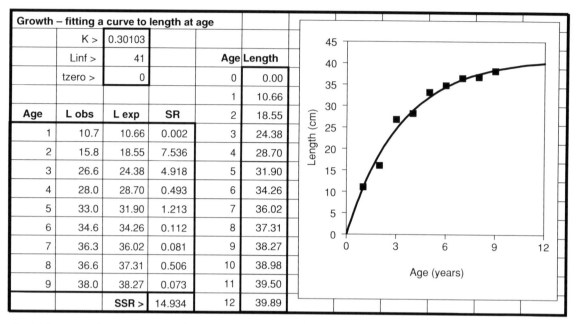

Growth – fitting a curve to length at age					
	K >	0.30103			
	Linf >	41		Age	Length
	tzero >	0		0	0.00
				1	10.66
Age	L obs	L exp	SR	2	18.55
1	10.7	10.66	0.002	3	24.38
2	15.8	18.55	7.536	4	28.70
3	26.6	24.38	4.918	5	31.90
4	28.0	28.70	0.493	6	34.26
5	33.0	31.90	1.213	7	36.02
6	34.6	34.26	0.112	8	37.31
7	36.3	36.02	0.081	9	38.27
8	36.6	37.31	0.506	10	38.98
9	38.0	38.27	0.073	11	39.50
		SSR >	14.934	12	39.89

Fig. A4.3 A spreadsheet program to fit a growth curve to length at age data. The Solver function has been used to minimize the SSR and located an optimum value of $K = 0.30$. Other parameters were fixed at $L_\infty = 41$ and $t_0 = 0$. Spreadsheet and graphs produced using Microsoft Excel and the Solver function.

a lower value of SSR would suggest a better fit with only one parameter fixed. Finally, use Solver to estimate all three parameters: experiment by using different initial values, increasing precision, and decreasing convergence limits. Hold each parameter constant in turn and allow Solver to adjust the other two.

When Solver is used to estimate all three parameters, using reasonable guesses of the three parameters as seed values, estimates of

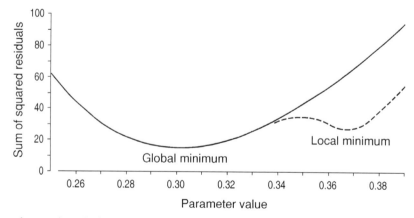

Fig. A4.4 Sum of squared residuals (goodness of fit) for different seed values of a single parameter (the growth coefficient, K). A global minimum of SSR = 14.9 is found at $K = 0.30$. If another minimum had occurred (say as suggested by the broken line) a computer-based minimization routine may locate the local minimum rather than the global one.

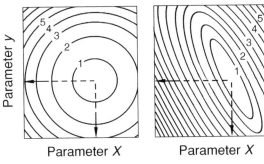

Fig. A4.5 Contour diagrams of sum of squared residuals for a two-parameters model. The broken lines suggest the optimum combination of parameter values in each case. The diagram on the right indicates some correlation between parameters.

$K = 0.33$ yr^{-1}, $L_\infty = 40.2$ cm and $t_0 = 0.16$ years are found with an SSR of 13.875. The task of estimating three unknowns from only nine data pairs is a difficult one, and if reasonable initial values of the three parameters are not used, Solver may locate local minima rather than the global minimum. This is because there may be different combinations of parameters that result in a growth curve that fits the data nearly as well as the 'real' parameters indicated by the global SSR minimum.

Contour diagrams of the sum of squared residuals are shown in Fig. A4.5 for a hypothetical case in which just two parameters are to be estimated. Different values of the SSRs are shown as contours and the parameter values are on the x and y axes. The graph at the left shows an idealized example in which an optimal combination of the two parameters is located in the central low area of the circle. The graph at the right shows an example in which the contours are diagonal ellipses, indicating that the two parameters are not completely independent of one another – that is, that the parameters are correlated in some way.

A4.1 Growth from a single length–frequency sample – mackerel example

The growth parameters in the mackerel example (Chapter 4, Fig. 4.17) were estimated using a graphical method. The parameters can be estimated more directly by non-linear least squares and a computer spreadsheet design is shown in Table A4.3. The age and the observed length data are entered in columns A and B in rows seven and below (17, 24, 29, 33 and 35 cm). The expected lengths are estimated by the von Bertalanffy equation, $L_t = L_\infty(1 - \exp[-K(t - t_0)])$, in column C. The equation uses initial guesses of the parameters that are entered in the input cells C2, C3 and C4. The squared differences (squared residuals, SR) between the observed lengths and the predicted lengths are entered in column D; at the bottom of this column the SRs are added as the sum of the squared residuals (SSR). It is the SSR that the Solver function will attempt to minimize to find the best values of the growth parameters.

Initial but reasonable guesses of the growth parameters have to be used in cells C2, C3 and C4 as Solver has the difficult task of estimating three unknowns from only five data pairs. There may be different combinations of the parameters that result in a growth curve fitting the data equally well.

In Fig. A4.6, Solver found a minimum residual sum of squares of SSR = 0.163 with parameter values of $K = 0.35$ yr^{-1}, $L_\infty = 41.2$ cm and $t_0 = -0.52$ years. An age–length key can be included on the spreadsheet in columns E and F by inserting the estimated parameter values in the growth equation (Equation 4.22). The age–length key columns are then used to produce the growth curve, which includes the observed lengths from column B, shown in Fig. A4.6. If the residuals are distributed normally, the average growth of the sampled stock will be represented by the expected growth curve.

A4.2 Growth from multiple length–frequency samples – banana prawn example

The growth parameters estimated using the von Bertalanffy plot for the banana prawn in Chapter 4, Fig. 4.21, can also be estimated by non-linear least squares on a computer spreadsheet. The

Table A4.3 Upper, left-hand corner of a spreadsheet designed to estimate growth parameters. The completed spreadsheet is shown in Fig. A4.6.

	A	B	C	D	E	F
1		Growth – fitting a curve to length at age data				
2		K >	0.2			
3		Linf >	39		Age	Length
4		tzero >	−0.3		0	=$C$3*(1−EXP(−$C$2*(E4−$C$4)))
5					=E4+1	=$C$3*(1−EXP(−$C$2*(E5−$C$4)))
6	Age	L obs	L exp	SR	=E5+1	=$C$3*(1−EXP(−$C$2*(E6−$C$4)))
7	1	17	=$C$3*(1−EXP(−$C$2*(A7−$C$4)))	=(B7−C7)^2	=E6+1	=$C$3*(1−EXP(−$C$2*(E7−$C$4)))
8	=A7+1	24	=$C$3*(1−EXP(−$C$2*(A8−$C$4)))	=(B8−C8)^2	=E7+1	=$C$3*(1−EXP(−$C$2*(E8−$C$4)))

Growth – fitting a curve to length at age

Age	L obs	L exp	SR	Age	Length
	K >	0.34897			
	Linf >	41.1657			
	tzero >	−0.5209			
				0	6.84
				1	16.95
				2	24.09
1	17.0	16.95	0.002	3	29.12
2	24.0	24.09	0.007	4	32.67
3	29.0	29.12	0.014	5	35.17
4	33.0	32.67	0.111	6	36.94
5	35.0	35.17	0.029	7	38.18
		SSR >	0.163	8	39.06

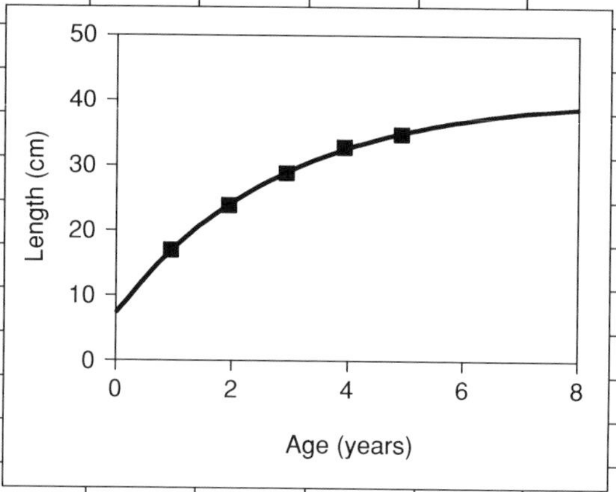

Fig. A4.6 A spreadsheet program to fit a growth curve to length at age data. Spreadsheet and graphs produced using Microsoft Excel and the Solver function.

spreadsheet shown in Fig. A4.7 is similar in design to that shown in Fig. A4.6. The age and mean length data from Table 4.8 in Chapter 4 are entered in the first two columns. The expected lengths are estimated by the von Bertalanffy equation, $L_t = L_\infty(1 - \exp[-K(t - t_0)])$. As before, the equation uses initial guesses of K, L_∞, and t_0 entered into cells at the top of the spreadsheet.

Solver is used to locate the minimum of the squared residuals (SSR) – note that zero has to be entered in the top cell of the squared residuals

column (for age 8.1 weeks) as this point may have been affected by the selectivity of the trawl nets and is excluded from the analysis. In the spreadsheet shown in Fig. A4.7, Solver found a minimum residual sum of squares of SSR = 20.74 with parameter values of $K = 0.029$ wk^{-1}, $L_\infty = 46.7$ mm and $t_0 = -6$ weeks. Under the assumption of normally-distributed residuals, the average growth of the sampled stock will be represented by the expected curve.

The results differ from those obtained using the

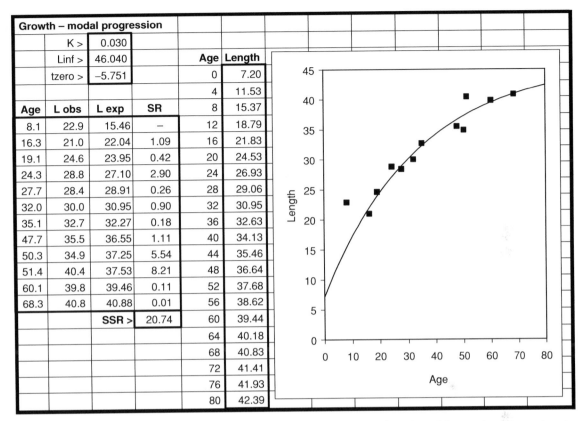

Growth – modal progression					
	K >	0.030			
	Linf >	46.040		Age	Length
	tzero >	–5.751		0	7.20
				4	11.53
Age	L obs	L exp	SR	8	15.37
8.1	22.9	15.46	–	12	18.79
16.3	21.0	22.04	1.09	16	21.83
19.1	24.6	23.95	0.42	20	24.53
24.3	28.8	27.10	2.90	24	26.93
27.7	28.4	28.91	0.26	28	29.06
32.0	30.0	30.95	0.90	32	30.95
35.1	32.7	32.27	0.18	36	32.63
47.7	35.5	36.55	1.11	40	34.13
50.3	34.9	37.25	5.54	44	35.46
51.4	40.4	37.53	8.21	48	36.64
60.1	39.8	39.46	0.11	52	37.68
68.3	40.8	40.88	0.01	56	38.62
		SSR >	20.74	60	39.44
				64	40.18
				68	40.83
				72	41.41
				76	41.93
				80	42.39

Fig. A4.7 A spreadsheet program to fit a growth curve to modal progression data. Spreadsheet and graphs produced using Microsoft Excel and the Solver function.

von Bertalanffy plot and this reflects the fact that there are often different combinations of growth parameters that result in a curve that appears to fit the data equally well. Also, the least squares analysis had the difficult task of estimating three unknowns from the data set; different runs could be tried with L_∞ locked to the previously estimated value and leaving Solver to estimate the remaining two parameters.

A4.3 Growth on a seasonal basis – surf clam example

Following the cohorts in the length–frequency histograms for the surf clam, *Donax deltoides*, (Chapter 4, Fig. 4.23) suggests seasonal growth. Here the von Bertalanffy curve is modified by the inclusion of a sine wave (Pitcher & MacDonald, 1973; Haddon, 2001) as:

$$L_t = L_\infty(1 - \exp-[C\sin(2\pi(t-s)/12) + K(t-t_0)])$$
(A4.2)

which has two parameters in addition to the usual von Bertalanffy growth model: C reflects the amplitude of the oscillation above and below the regular (non-seasonal) growth curve and s is the starting time for the sine wave.

Using the *Donax* data in Fig. 4.23 of Chapter 4, the major cohort that appears on 1 January at a modal length of 12.5 mm shell length (at which stage it is given the arbitrary age of 8 weeks) can be followed over time. Data for this cohort are entered on a computer spreadsheet with relative ages in column A and modal lengths in column B. As in previous spreadsheet models, trial input parameters are entered in the cells at the top of the spreadsheet. The equation for the seasonal growth curve (Equation A4.2) is entered in column C. Column D contains the squared residuals and their sum is entered at the bottom of the column.

Solver may have difficulty locating a global minimum value when trying to adjust five parameters and some experimentation is necessary. Adjust the parameters in the five input cells manually until predicted lengths in column C become close in value to the observed lengths in column B. In the spreadsheet shown in Fig. A4.8, Solver found a minimum SSR value of 35.58, which estimated an estimated L_∞ of 67.8 mm; as the largest animal found was 61 mm, a value of $L_\infty = 64$ mm could be expected. Try fixing L_∞ at 64 mm and allowing Solver to estimate the remaining four parameters. Also try fixing L_∞, K and t_0 and allowing Solver to adjust the remaining two parameters C and s. Note that setting C to zero reduces the equation to the unseasonalized growth equation. The graph shown at the right of the spreadsheet plots observed lengths (as points) and predicted lengths (as a curve) against age.

A4.4 Growth from hard part analysis – morwong example

Age–length data presented in Chapter 4, Table 4.11, for the morwong, *Nemodactylus macropterus*, are summarized in Table A4.4.

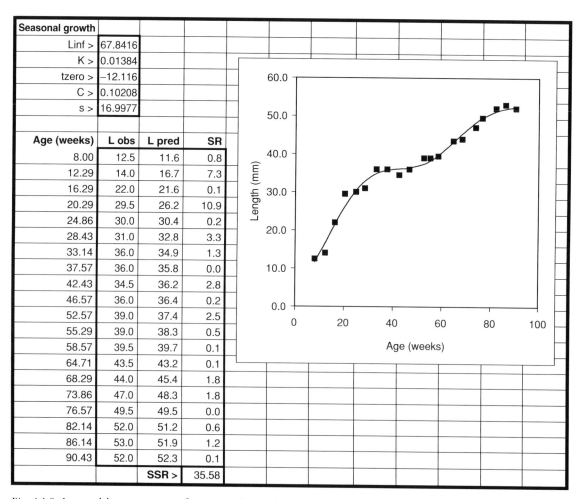

Seasonal growth			
Linf >	67.8416		
K >	0.01384		
tzero >	−12.116		
C >	0.10208		
s >	16.9977		
Age (weeks)	**L obs**	**L pred**	**SR**
8.00	12.5	11.6	0.8
12.29	14.0	16.7	7.3
16.29	22.0	21.6	0.1
20.29	29.5	26.2	10.9
24.86	30.0	30.4	0.2
28.43	31.0	32.8	3.3
33.14	36.0	34.9	1.3
37.57	36.0	35.8	0.0
42.43	34.5	36.2	2.8
46.57	36.0	36.4	0.2
52.57	39.0	37.4	2.5
55.29	39.0	38.3	0.5
58.57	39.5	39.7	0.1
64.71	43.5	43.2	0.1
68.29	44.0	45.4	1.8
73.86	47.0	48.3	1.8
76.57	49.5	49.5	0.0
82.14	52.0	51.2	0.6
86.14	53.0	51.9	1.2
90.43	52.0	52.3	0.1
		SSR >	35.58

Fig. A4.8 A spreadsheet program to fit a seasonal growth curve to modal progression data. Spreadsheet and graphs produced using Microsoft Excel and the Solver function.

Table A4.4 Sample size (N), mean length (L) in cm for male morwong, *Nemodactylus macropterus*, by age group. Data from Smith (1982).

	III	IV	V	VI	VII	VIII	IX	X	XI
N	108	134	110	69	49	29	20	7	4
L	27.2	30.3	32.6	34.6	35.9	37.3	38.6	40.6	42.5

The data from Table A4.4 can best be analysed using non-linear least squares on a computer spreadsheet similar in design to that shown previously in Fig. A4.7 for modal progression. The age and mean length data from Table A4.4 are entered in the first two columns. The expected lengths are estimated by the von Bertalanffy equation as $L_t = L_\infty (1 - \exp[-K(t - t_0)])$. As before, the equation uses initial guesses of K, L_∞, and t_0 entered into the appropriate cells at the top of the spreadsheet shown in Fig. A4.9. Solver is then used to locate the minimum of the sum of squared residuals (SSR). As the sample sizes for the last two age groups (X and XI) are small, try using Solver again with the last two cells in the SR column (for ages 10 and 11) left blank to exclude

them from the least square analysis. In the spreadsheet shown in Fig. A4.9, Solver found a sum of squared residuals (SSR) of 0.097 with parameter values of $K = 0.20$ yr^{-1}, $L_\infty = 43.2$ cm and $t_0 = -1.93$ years. An age–length key has been included on the spreadsheet to enable a growth curve to be constructed; the graph shown is prepared from the two age–length key columns and the raw data from the observed length (L obs) column.

A4.5 Mean length at sexual maturity – tiger prawn example

The number of female tiger prawns (*Penaeus semisulcatus*) with ripe ovaries by length class is given in Chapter 4, Table 4.14. A logistic curve

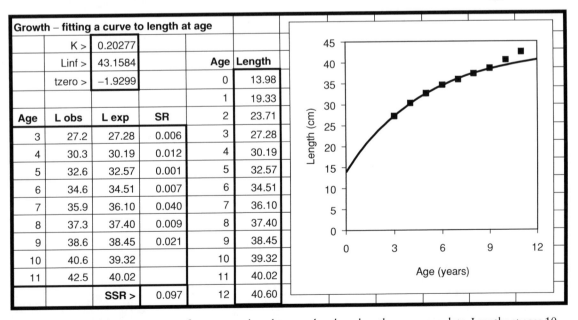

Fig. A4.9 A spreadsheet program to fit a curve to length at age data based on the morwong data. Lengths at ages 10 and 11 are excluded from the analysis. Spreadsheet and graphs produced using Microsoft Excel and the Solver function.

may be fitted to the proportion (P) of sexually mature individuals by length (L) using:

$$P = 1/(1 + \exp[-r(L - L_m)]) \qquad (A4.3)$$

where r is the slope of the curve, and L_m is the mean length at sexual maturity, or the length which corresponds to a proportion of 0.5 (or 50%) in reproductive condition. The computer spreadsheet shown in Fig. A4.10 uses Solver to locate a minimum SSR and thereby find optimal values of L_m and r. The program produces a logistic curve about a mean length at sexual maturity (L_m) of 15.17 cm.

A4.6 Stock–recruitment models

Two models are commonly used to suggest the relationship between recruitment and stock size (see Section 4.3.6). The Beverton and Holt (1957) equation suggests that recruitment approaches an asymptote at high stock densities:

$$R = aS/(b + S) \qquad (A4.4)$$

where R is the number of recruits, S is the number of individuals in the spawning stock and a and b are the parameters of the curve. The Ricker (1954, 1975) model describes the situation where recruitment reaches a maximum before decreasing at higher levels of stock abundance:

$$R = aS \exp(-bS) \qquad (A4.5)$$

The constants in both the Ricker model and the Beverton and Holt model can be estimated by non-linear least squares in the model design

suggested in Table A4.5. As in previous spreadsheets, trial parameter input cells are placed at the top of the spreadsheet. Trial values of a and b in the Ricker model are placed in C3 and C4 and the values of a and b in the Beverton and Holt model are placed in cells E3 and E4; these values can be initial but reasonable guesses as the program will use Solver to minimize the sum of squared residuals and optimize the parameters (note that a and b in each model are unrelated). The observed stock and recruitment data from Table 4.15 in Chapter 4 are entered in columns A and B below row 7.

Variations in data around stock–recruitment curves often have a log-normal distribution that reflects occasional large recruitment values (Hilborn & Walters, 1992; Haddon, 2001). The stock–recruitment equations are therefore log transformed to normalize the error structure and to allow the use of ordinary least squares. The log-transformed Ricker equation, $\ln(a) - bS$, is entered in cell C8 and below. The observed values (the log values of R/S) minus the expected values (in column C) are entered into column D; at the bottom of this column the sum of squared residuals is calculated. Finally, the untransformed Ricker equation, $aS \exp(-bS)$, is entered in column E. The process is now repeated in columns F, G and H for the Beverton and Holt model in which the log-transformed equation is $\ln(a) - \ln(b + S)$ and the untransformed equation is $aS/(b + S)$.

Solver is used to find values for a and b in the Ricker model and then, separately, to find the values of a and b in the Beverton and Holt model

Table A4.5 Upper, left-hand corner of a spreadsheet to estimate stock–recruitment curves.

	A	B	C	D	E	F	G	H
1	Stock–recruitment curves							
2			Ricker		B&H			
3		a >	5.05859	a >	7946.635			
4		b >	0.0002464	b >	624.695			
5								
6	Stock	Recruit	Rick.log	SR	R curve	B&H log	SR	B&H curve
7	0				0			0
8	900	4000	=LN($C$3)–($C$4*A8)	=(LN(B8/A8)–C8)^2	=$C$3*A8*EXP(–$C$4*A8)	=LN($E$3)–LN($E$4+A8)	=(LN(B8/A8)–F8)^2	=$E$3*A8/($E$4+A8)

Length at maturity								Length	Prop.
	r >	0.65						7.5	0.01
	Lm >	15.17						8.5	0.01
	Pmax >	0.62						9.5	0.02
								10.5	0.05
Length	**N sample**	**N mature**	**P mature**	**P adjust.**	**P predict**	**SR**		11.5	0.08
12.5	109	5	0.05	0.07	0.15	0.006		12.5	0.15
13.5	73	7	0.10	0.15	0.25	0.010		13.5	0.25
14.5	42	10	0.24	0.38	0.39	0.000		14.5	0.39
15.5	48	21	0.44	0.71	0.55	0.023		15.5	0.55
16.5	321	158	0.49	0.79	0.70	0.008		16.5	0.70
17.5	458	215	0.47	0.76	0.82	0.004		17.5	0.82
18.5	771	396	0.51	0.83	0.90	0.005		18.5	0.90
19.5	535	280	0.52	0.84	0.94	0.010		19.5	0.94
20.5	180	85	0.47	0.76	0.97	0.043		20.5	0.97
21.5	29	18	0.62	1.00	0.98	0.000		21.5	0.98
22.5	1	1	*					22.5	0.99
					SSR >	0.109		23.5	1.00

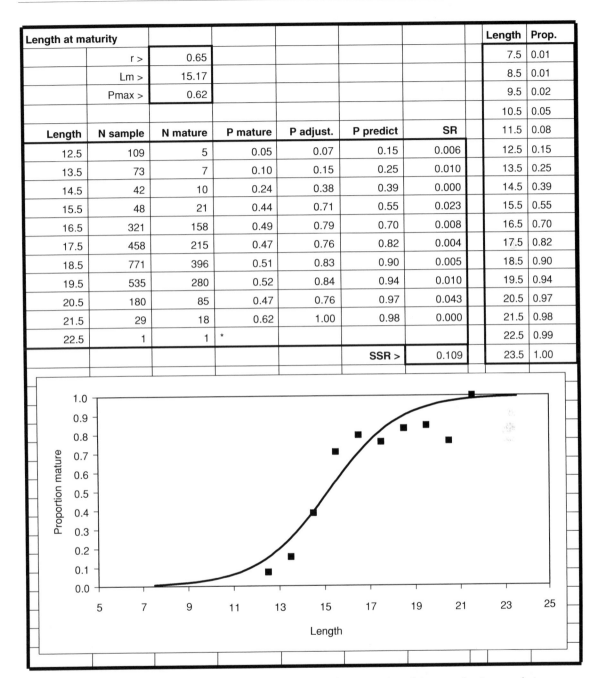

Fig. A4.10 A spreadsheet program to fit a logistic curve relating the proportion of the reproductive population to length for *Penaeus semisulcatus* in the Persian Gulf. Spreadsheet and graphs produced using Microsoft Excel and the Solver function.

(remembering that *a* and *b* in each model are unrelated). The untransformed values in columns E and H are then graphed (with the raw recruitment values in column B) against stock values in column A as shown in Fig. A4.11; a zero has been added to the initial cells in columns A, E and H only to allow the graphs to begin at the origin. Judging by the closeness of the values of the sum

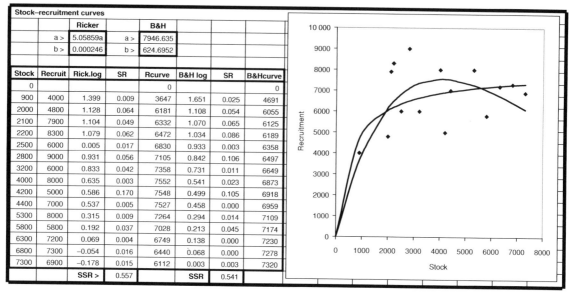

Stock–recruitment curves							
		Ricker		B&H			
	a >	5.05859a	a >	7946.635			
	b >	0.000246	b >	624.6952			
Stock	Recruit	Rick.log	SR	Rcurve	B&H log	SR	B&Hcurve
0				0			0
900	4000	1.399	0.009	3647	1.651	0.025	4691
2000	4800	1.128	0.064	6181	1.108	0.054	6055
2100	7900	1.104	0.049	6332	1.070	0.065	6125
2200	8300	1.079	0.062	6472	1.034	0.086	6189
2500	6000	0.005	0.017	6830	0.933	0.003	6358
2800	9000	0.931	0.056	7105	0.842	0.106	6497
3200	6000	0.833	0.042	7358	0.731	0.011	6649
4000	8000	0.635	0.003	7552	0.541	0.023	6873
4200	5000	0.586	0.170	7548	0.499	0.105	6918
4400	7000	0.537	0.005	7527	0.458	0.000	6959
5300	8000	0.315	0.009	7264	0.294	0.014	7109
5800	5800	0.192	0.037	7028	0.213	0.045	7174
6300	7200	0.069	0.004	6749	0.138	0.000	7230
6800	7300	−0.054	0.016	6440	0.068	0.000	7278
7300	6900	−0.178	0.015	6112	0.003	0.003	7320
		SSR >	0.557			SSR	0.541

Fig. A4.11 A spreadsheet program to fit stock–recruitment curves. The dome-shaped line is a Ricker curve and the continually rising line is a Beverton and Holt curve. Spreadsheet and graphs produced using Microsoft Excel and the Solver function.

of squared residuals for each model it appears that either could be reasonably used to describe the data.

A4.7 Length-converted catch curve – lobster example

A length-converted catch curve is a plot of $\ln[F_{(L1-L2)}/\Delta t]$ against $[(t_{L1} + t_{L2})/2]$ and the length–frequency data for the spiny lobster (from Fig. 4.46 in Chapter 4) is used here as an example. Table A4.6 shows the design of a spreadsheet program in which the von Bertalanffy growth parameters are entered in cells $C2 and $C3 as $K = 0.39$ yr^{-1} and $L_\infty = 121$ mm respectively. The length classes and the corresponding frequencies (from Table 4.19 in Chapter 4) are entered in columns A to D below row six. In column E, the relative age, t, at the lower limit of the length class is calculated using the inverse of the von Bertalanffy equation with t_0 set at zero:

$$t = -(1/K)\ln[1 - L_t/L_\infty] \tag{A4.6}$$

In column F, the value of Δt, the time taken for the species to grow through a particular length class is entered. The value of Δt between L_1 and L_2 (the upper and lower limits of the length class) is estimated as $\ln[1 - L_1/L_\infty]/K - \ln[1 - L_2/L_\infty]/K$ which gives:

$$\Delta t = \ln[(L_\infty - L_1)/(L_\infty - L_2)]/K \tag{A4.7}$$

In column G, mean ages, used as the independent (x) values in the length-converted catch curve, are approximated as the ages at the midpoint of each length class, $(L_1 + L_2)/2$ and this is also calculated using the inverse of the von Bertalanffy equation. In column H, values of $\ln[F_{(L1-L2)}/\Delta t]$ are entered using values in column D divided by values in column F.

In column I, a conditional formula is used to duplicate values in column H only for cells between the upper and lower age limits entered in cells E2 and E3. In the graph on the spreadsheet shown in Fig. A4.12, values in column H $(\ln[F_{(L1-L2)}/\Delta t])$ are plotted against values in column G (relative mean age). Values in column I are also plotted (using a heavier point symbol) as these are the points (between the age values in cells E2 and E3) through which the regression line is fitted. The value of total mortality estimated

Table A4.6 Top portion of the spreadsheet shown in Fig. A4.12 using length–frequency data for the female spiny lobster to construct a length-converted catch curve.

	A	B	C	D	E	F	G	H	I
1	Length–converted catch curve								
2		K >	0.39	First point >	2.5	Slope >	=SLOPE (I7: I44,G7:G44)		
3		Linf >	121	Last point >	4.8	Z >	=−G2		
4									
5					Age	Age	Age at	ln[F/dt]	ln[F/dt]
6	L1	L2	Mid L	F	at L1	change	mid L		regression
7	40	42	=(A7+B7)/2	2	=(−1/$C$2)*LN (1−(A7/$C$3))	=LN(($C$3−A7)/ ($C$3−B7))/$C$2	=(−1/$C$2)*LN (1−(C7/$C$3))	=LN(D7/F7)	=IF(G7<$E$2,", IF(G7>$E$3,",H7))
8	=A7+2	=B7+2	=(A8+B8)/2	2	=(−1/$C$2)*LN (1−(A8/$C$3))	=LN(($C$3 −A7)/ ($C$3−B7))/$C$2	=(−1/$C$2)*LN (1−(C8/$C$3))	=LN(D8/F8)	=IF(G8<$E$2,", IF(G8>$E$3,",H8))

from the length-converted catch curve shown in Fig. A4.12 is $Z = 0.99 \text{ yr}^{-1}$.

A4.8 Wetherall plot – scad example

Length–frequency data for the scad, *Decapterus macrosoma*, (from Chapter 4, Fig. 4.49) are used here to construct a Wetherall plot model. The plot consists of a straight line through the values of $(\bar{L} - L')$ plotted against a series of cut off points, L':

$$\bar{L} - L' = a + bL' \tag{A4.8}$$

$\bar{L}$ is the mean length of all fish equal to, or longer than, length L', which now becomes a series of

lower limits for the length intervals of fully vulnerable fish. Table A4.7 shows the design of a spreadsheet program in which the von Bertalanffy growth coefficient is entered in cell B2 as $K = 1.2 \text{ yr}^{-1}$. The length classes and the corresponding frequencies are entered in columns A to C below row five. In column D, the cells contain the midpoint of the length class multiplied by the frequency. In column E, the means of all fish longer than L' are calculated and in column F the values of mean lengths minus the cut off lengths L' are entered.

In column G, a conditional formula is used to duplicate the values in column F only for cells between the upper and lower length limits entered

Table A4.7 Top portion of the spreadsheet shown in Fig. A4.13 using length–frequency data for scad to construct a Wetherall plot.

	A	B	C	D	E	F	G
1	Wetherall plot						
2	K >	1.2	First point >	185	Slope >	=SLOPE (G6:G25, A6:A25)	
3	Linf >	=−INTERCEPT (G6:G5, A6:A5)/F2	Last point >	220	Z >	=−(1+F2)/ F2*B2	
4							
5	L1(L′)	L2		F	SumLF	MeanL	MeanL–L′	Regress.
6	140	145		6	=(A6+B6)/2*C6	=SUM(D6:$D$25)/ SUM(C6:$C$25)	=E6–A6	=IF(A6<$D$2,",IF (A6>$D$3,",F6))
7	145	150		9	=(A7+B7)/2*C7	=SUM(D7:$D$25)/ SUM(C7:$C$25)	=E7–A7	=IF(A7<$D$2,",IF (A7>$D$3,",F7))

Length-converted catch curve

K >	0.39		First point >	2.5		Slope >	−0.986	
Linf >	121		Last point >	4.8		Z >	0.986	

L1	L2	Mid L	F	Age at L1	Age change	Age at mid Lr	ln[F/dt]	ln[F/dt] regress.
40	42	41	2	1.029	0.064	1.061	3.44	
42	44	43	2	1.093	0.066	1.126	3.42	
44	46	45	1	1.159	0.067	1.192	2.70	
46	48	47	6	1.226	0.069	1.261	4.46	
48	50	49	13	1.296	0.071	1.331	5.21	
50	52	51	10	1.367	0.073	1.403	4.92	
52	54	53	24	1.440	0.075	1.478	5.76	
54	56	55	20	1.516	0.078	1.554	5.55	
56	58	57	27	1.593	0.080	1.633	5.82	
58	60	59	15	1.673	0.083	1.715	5.20	
60	62	61	22	1.756	0.085	1.799	5.55	
62	64	63	31	1.842	0.088	1.886	5.86	
64	66	65	25	1.930	0.092	1.975	5.61	
66	68	67	29	2.022	0.095	2.069	5.72	
68	70	69	30	2.117	0.099	2.166	5.72	
70	72	71	36	2.215	0.103	2.266	5.86	
72	74	73	44	2.318	0.107	2.371	6.02	
74	76	75	50	2.425	0.112	2.480	6.11	
76	78	77	41	2.536	0.117	2.594	5.86	5.86
78	80	79	37	2.653	0.122	2.713	5.71	5.71
80	82	81	30	2.775	0.128	2.838	5.46	5.46
82	84	83	30	2.903	0.135	2.970	5.40	5.40
84	86	85	28	3.038	0.142	3.108	5.28	5.28
86	88	87	22	3.181	0.151	3.255	4.98	4.98
88	90	89	30	3.331	0.160	3.410	5.23	5.23
90	92	91	34	3.492	0.171	3.576	5.29	5.29
92	94	93	20	3.663	0.183	3.753	4.69	4.69
94	96	95	12	3.846	0.197	3.943	4.11	4.11
96	98	97	13	4.043	0.214	4.148	4.11	4.11
98	100	99	13	4.257	0.233	4.371	4.02	4.02
100	102	101	14	4.490	0.257	4.616	4.00	4.00
102	104	103	7	4.747	0.285	4.886	3.20	
104	106	105	2	5.032	0.321	5.188	1.83	
106	108	107	6	5.353	0.367	5.530	2.79	
108	110	109	2	5.720	0.428	5.925	1.54	
110	112	111	3	6.148	0.515	6.393	1.76	
112	114	113	5	6.663	0.644	6.965	2.05	
114	116	115	3	7.307	0.863	7.703	1.25	
116	118	117	1	8.170	1.310	8.742	−0.27	

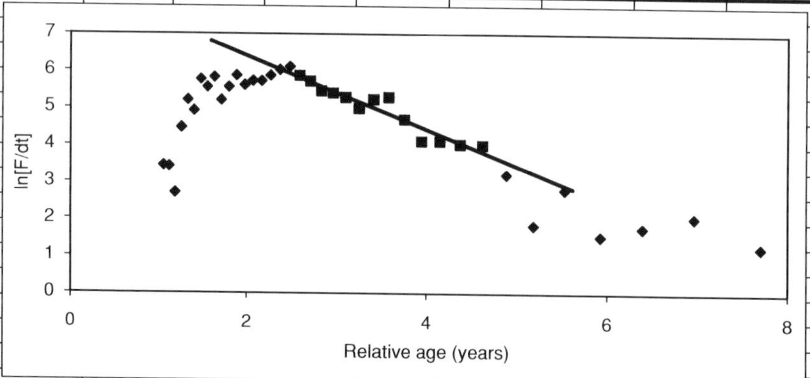

Fig. A4.12 A spreadsheet program to fit a length-converted catch curve. The regression line is fitted through the data points shown as large squares (see text). Spreadsheet and graphs produced using Microsoft Excel.

in cells D2 and D3. In the graph on the spreadsheet shown in Fig. A4.13, values in column F for $(\bar{L} - L')$ are plotted against values in column A for (L'). Values in column G are also plotted (using a heavier point symbol) as these are the points (between the length values in cells D2 and D3) through which the regression line is fitted.

The regression line in the Wetherall plot (Fig. A4.13) is fitted through all data representing the fully exploited part of the sample, often from one length interval to the right of the highest mode in the length–frequency data (185 mm in this case), but excluding length classes with small sample sizes (in this example, the final three length classes, which contain less than ten individuals). From the regression line, the value of Z/K is estimated from the slope as:

$$Z/K = -(1 + \text{slope})/\text{slope}$$
$$= -(1 - 0.265)/(-0.265) = 2.8 \qquad \text{(A4.9)}$$

As the value of the growth coefficient was estimated as $K = 1.2 \text{ yr}^{-1}$ from modal progression analysis, then $Z = 3.3 \text{ yr}^{-1}$. A useful additional result of the analysis is that L_∞ may be estimated from the intercept with the x-axis as:

$$L_\infty = -a/b = -63.515/(-0.2649) = 239.8 \text{ mm} \qquad \text{(A4.10)}$$

which has been entered in cell B3 of the spreadsheet (see Table A4.7).

A4.9 Non-equilibrium surplus yield

The following simple model is based on the time series of barramundi data (Table 5.1 in Chapter 5) to estimate the usual parameters of the Schaefer model, r, q and B_∞. From Equation 5.11 in Chapter 5, next year's biomass B_{t+1} will equal the present year's biomass (B_t) plus the surplus production minus the catch (C_t). As catch is equal to qfB where f is fishing effort:

$$B_{t+1} = B_t + rB_t(1 - B_t/B_\infty) - qf_t B_t \qquad \text{(A4.11)}$$

The model uses estimates of r, q and B_∞ and an initial or starting stock biomass, B_1, to calculate biomass, catch and CPUE for all subsequent years. In the barramundi data, fishing effort at the beginning of the data series was low enough to regard the initial biomass as close to its unexploited level; that is $B_1 \approx B_\infty$.

Calculations may be completed on a computer spreadsheet as suggested in Table A4.8 with three boxes (C2 to C4) at the top of the sheet to contain the values of B_∞, q and r. Below these input parameter cells, the spreadsheet is set up with seven columns as follows. Columns A, D and F contain the observed or actual data. Column A contains the effort values from the barramundi data in ascending order. Column D contains the corresponding catch values (in tonnes) and column F contains the observed CPUEs (in kg per hmnd) calculated as $1000 \times D8/A8$.

Column B contains the predicted biomass values. Cell B8 uses the trial value entered in C2. In the cells below this, biomass can be predicted by using Equation A4.11 for population growth, $B_{t+1} = B_t + rB_t(1 - B_t/B_\infty) - C_t$, with trial input values of B_∞ and r from boxes C2 and C3 respectively. Column C contains the predicted catch estimated as $C_t = qfB_t$ and column E contains the predicted CPUE estimated as C/f. Finally, column G is used to contain the values to be minimized, in this case the squared residual errors (SR). The cell at the bottom of column G contains the sum of the squared residuals (SSR) which will be used as a target cell by Solver.

Preliminary values of the parameters are now entered into the three cells at the top of the spreadsheet. As there are often many possible combinations of the three parameters that will provide a reasonable fit, it is important to use initial values that are realistic. A reasonable initial value of unexploited biomass B_∞ in an exploited fishery would be two to four times the annual catch and the value of q, which is the proportion of the stock taken by one unit of fishing effort, has a very small positive value. A realistic starting value for r, the intrinsic rate of stock growth, may be between 0.1 and 1. Starting values of $B_\infty = 4000$, $r = 0.5$ and $q = 0.000005$ have been entered in cells C2, C3 and C4 as seed values (Table A4.8).

Including graphs of catch rates and catch against fishing effort on the spreadsheet makes it

Wetherall plot						
K >	1.2	First point >	185	Slope >	−0.265	
Linf >	239.8	Last point >	220	Z >	3.33	

L1(L′)	L2	F	SumLF	MeanL	MeanL − L′	Regress.
140	145	6	855.0	188.40	48.40	
145	150	9	1327.5	188.67	43.67	
150	155	16	2440.0	189.04	39.04	
155	160	25	3937.5	189.63	34.63	
160	165	35	5687.5	190.46	30.46	
165	170	29	4857.5	191.50	26.50	
170	175	72	12420.0	192.27	22.27	
175	180	107	18992.5	193.98	18.98	
180	185	147	26827.5	196.41	16.41	
185	190	115	21562.5	199.94	14.94	14.94
190	195	122	23485.0	203.03	13.03	13.03
195	200	93	18367.5	206.78	11.78	11.78
200	205	83	16807.5	210.25	10.25	10.25
205	210	61	12657.5	214.13	9.13	9.13
210	215	37	7862.5	217.98	7.98	7.98
215	220	40	8700.0	220.96	5.96	5.96
220	225	17	3782.5	225.89	5.89	5.89
225	230	6	1365.0	231.14	6.14	
230	235	2	465.0	235.50	5.50	
235	240	3	712.5	237.50	2.50	

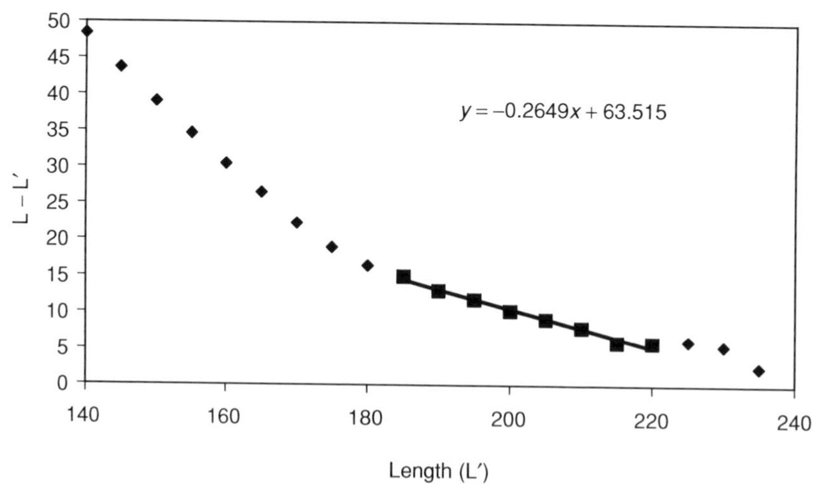

$$y = -0.2649x + 63.515$$

Fig. A4.13 A spreadsheet program to construct a Wetherall plot. The regression line is fitted through the data points shown as large squares (see text). Spreadsheet and graphs produced using Microsoft Excel.

Table A4.8 The top left-hand corner of an Excel spreadsheet designed to estimate non-equilibrium surplus yield.

	A	B	C	D	E	F	G
1	Non-equilibrium surplus production model (observation error)						
2		Binf >	4000		MSY >	=$C$3*$C$2/4	
3		r >	0.5		f(MSY) >	=$C$3/(2*$C$4)	
4		q >	0.000005				
5							
6		Predicted	Predicted	Observed	Predicted	Observed	
7	Effort	biomass	catch	catch	CPUE	CPUE	SR
8	15 700	=$C$2	=$C$4*A8*B8	432	=$C$4*B8*1000	=1000*D8/A8	=(E8-F8)^2
9	17 300	=B8+B8*$C$3*(1–B8/$C$2)–D8	=$C$4*A9*B9	382	=$C$4*B9*1000	=1000*D9/A9	=(E9-F9)^2

easier to observe how well or how badly the predicted data fits the observed data. The three parameters are then adjusted to improve the fit by minimizing the differences between all of the observed and all of the predicted data (i.e. by minimizing the value of SSR in the cell at the bottom of column G). Excel Solver can be used to automatically try different combinations of the three parameters to minimize the value of SSR. Experiment with Solver by using different initial values, increasing precision, and decreasing convergence limits; hold one parameter constant and allow Solver to adjust the other two. An automatic search involves small incremental changes to each of the parameters sequentially to locate an overall or global minimum value of SSR located above the best estimate of a parameter. If several minima exist, searches may stop at a local minimum (above a poor estimate of a parameter) because movements in either direction will increases the value of SSR. Fig. A.14 shows one run of the spreadsheet model in which Solver estimates of $B_\infty = 4126$, $r = 0.50318$ and $q = 0.0000057$ were obtained at a minimum SSR of 263.77.

Equations 5.7 and 5.8 in Chapter 5 can be used to enter f_{MSY} as $(r/2q)$ in cell F3 and MSY as $(rB_\infty/4)$ in cell F2. Yield can be estimated by recalling from Equation 5.5 that:

$$a = \text{CPUE}_\infty \text{ or } qB_\infty \qquad (A4.12)$$

and that:

$$b = -(\text{CPUE}_\infty q/r) \text{ or } -q^2B/r \qquad (A4.13)$$

Thus Equation 5.6 becomes:

$$C = qB_\infty f - (q^2B/r)f^2 \qquad (A4.14)$$

and this can be entered in an additional column (shown to the right of the graphs in Fig. A4.14). Note that the MSY is considerably lower, and obtained at a lower fishing effort, than was estimated by the equilibrium Schaefer model.

A4.10 Biomass and biovalue model including fishing mortality – prawn example

The biomass model suggested in Table 5.4 in Chapter 5 is extended to include exploitation by adding a column of values of fishing mortality, F. The value of $F = 0.3$ m^{-1} has been entered in one of the nine input boxes at the top of the spreadsheet (Fig. A4.15). If values of F varied with age, different individual values of F could have been added in each cell of the column.

Cohort numbers (column F) are calculated as in Tables 5.4 and 5.5 in Chapter 5 but with F included; i.e. $N_{t+1} = N_t \exp[-(M_t + F_t)]$ – note that M and F are the values from the previous month, t, not $t + 1$. The numbers dying from both natural and fishing mortality (column G) are not required for the calculations, but are added for interest; they may be estimated as equal to $N_t(1 - \exp[-(M_t + F_t)])$ or simply as $N_t - N_{t+1}$. Catch

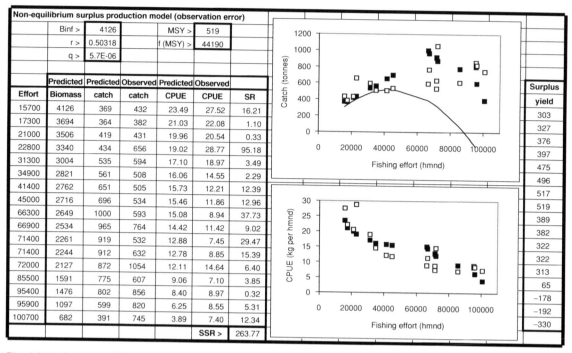

Non-equilibrium surplus production model (observation error)						
Binf >	4126		MSY >	519		
r >	0.50318		f (MSY) >	44190		
q >	5.7E-06					

Effort	Predicted Biomass	Predicted catch	Observed catch	Predicted CPUE	Observed CPUE	SR
15700	4126	369	432	23.49	27.52	16.21
17300	3694	364	382	21.03	22.08	1.10
21000	3506	419	431	19.96	20.54	0.33
22800	3340	434	656	19.02	28.77	95.18
31300	3004	535	594	17.10	18.97	3.49
34900	2821	561	508	16.06	14.55	2.29
41400	2762	651	505	15.73	12.21	12.39
45000	2716	696	534	15.46	11.86	12.96
66300	2649	1000	593	15.08	8.94	37.73
66900	2534	965	764	14.42	11.42	9.02
71400	2261	919	532	12.88	7.45	29.47
71400	2244	912	632	12.78	8.85	15.39
72000	2127	872	1054	12.11	14.64	6.40
85500	1591	775	607	9.06	7.10	3.85
95400	1476	802	856	8.40	8.97	0.32
95900	1097	599	820	6.25	8.55	5.31
100700	682	391	745	3.89	7.40	12.34
					SSR >	263.77

Surplus yield
303
327
376
397
475
496
517
519
389
382
322
322
313
65
−178
−192
−330

Fig. A4.14 A non-equilibrium surplus production model. Minimizing the sum of the squared residuals provides estimates of $B_\infty = 4126$, $r = 0.50318$ and $q = 0.0000057$. Graphs show observed data (open squares) and predicted data (solid squares) for catch and CPUE. A surplus yield curve is included in the upper graph. Graphs and spreadsheet produced using Microsoft Excel and the Solver function.

Biomass and biovalue model										
			Growth		**Length–weight**		**Mortality**		**Price**	
		K >	0.15	a >	0.0008	M >	0.3	Unit price >	3	
		Linf	45	b >	3	F >	0.3	% increase >	10	
		tzero	−0.44							
A	**B**	**C**	**D**	**E**	**F**	**G**	**H**	**I**	**J**	**K**
			Length	**Weight**	**Cohort**	**Numbers**	**Catch**	**Catch**	**Unit**	**Catch**
Age	**M**	**F**	**(mm)**	**(g)**	**numbers**	**dying**	**numbers**	**weight (kg)**	**price ($/kg)**	**value**
4	0.30	0.30	21.88	8.38	1000.00	451.19	225.59	1.89	3.00	5.67
5	0.30	0.30	25.10	12.65	548.81	247.62	123.81	1.57	3.30	5.17
6	0.30	0.30	27.87	17.32	301.19	135.90	67.95	1.18	3.63	4.27
7	0.30	0.30	30.26	22.16	165.30	74.58	37.29	0.83	3.99	3.30
8	0.30	0.30	32.31	26.99	90.72	40.93	20.47	0.55	4.39	2.43
9	0.30	0.30	34.08	31.66	49.79	22.46	11.23	0.36	4.83	1.72
10	0.30	0.30	35.60	36.10	27.32	12.33	6.16	0.22	5.31	1.18
11	0.30	0.30	36.91	40.23	15.00	6.77	3.38	0.14	5.85	0.80
12	0.30	0.30	38.04	44.02	8.23	3.71	1.86	0.08	6.43	0.53
			Numbers remaining =		4.52		Total weight =	6.81	Value =	25.06

Fig. A4.15 Biomass and biovalue model for an exploited banana prawn *Penaeus merguiensis* fishery. Unpublished data from Fisheries Research Centre, Bandar Abbas, Iran. Spreadsheet produced using Microsoft Excel.

numbers (column H) are estimated from the catch equation as $C_t = (F/Z)N_t(1 - \exp[-(M_t + F_t)])$, or by (F_t/Z_t) multiplied by the numbers dying (column G). Values in the remaining columns are calculated in the same way as in Table 5.4 and 5.5. The summations at the bottom of the table are the relative numbers remaining at the end of the fishing season, the total relative catch weight for the season, and the total relative catch value for the season respectively. The computer spreadsheet program shown in Fig. A4.15 can be run repeatedly for different levels of exploitation or other scenarios as suggested in Section 5.3.2.

A4.11 Yield per recruit model

In the classic yield per recruit model described in Section 5.4.2, the tedious calculations involved are best completed by constructing the spreadsheet model suggested in Table A4.9. Parameters used to illustrate the model are natural mortality rate, $M = 0.5$, age of recruitment $t_r = 1$ year and mean age at capture $t_c = 1.5$ years. The von Bertalanffy growth parameters used are $K = 0.3$ y^{-1} and $W_\infty = 900$ g with $t_0 = 0$. Table A4.9 shows a portion of the spreadsheet design in which the above input parameters are entered into cells B2 to B4 and D2 to D4; these parameters can then be changed for different species. Values of fishing mortality, F, from 0 to 1.5 in steps of 0.1 are entered in column A from cell A7 down.

The simplified expression for yield per recruit is:

$$F\exp[-M(t_c - t_r)]W_\infty \qquad \text{(term 1)}$$

multiplied by a summation of the following four terms:

$$\{1\exp[-0K(t_c - t_0)]/(M + F + 0K)\}+ \qquad \text{(term 2)}$$

$$\{-3\exp[-1K(t_c - t_0)]/(M + F + 1K)\}+ \qquad \text{(term 3)}$$

$$\{3\exp[-2K(t_c - t_0)]/(M + F + 2K)\}+ \qquad \text{(term 4)}$$

$$\{-1\exp[-3K(t_c - t_0)]/(M + F + 3K)\} \qquad \text{(term 5)}$$

The above five terms of the yield per recruit equation are entered in columns B to F respectively. The yield per recruit values in column G are obtained by multiplying the values in column A by the sum of the four values in columns C to F. The yield per recruit graph showing values of yield per recruit from Column G plotted against the values of fishing mortality in column A is shown on the completed spreadsheet in Fig. A4.16.

Experiment with the spreadsheet by increasing K and M to model a fast-growing short lifespan species and then by decreasing these parameters to model a slow-growing long lifespan species. With high values of K and M, such as found in many tropical species, the curve becomes flat-topped and does not reach a maximum yield within

Table A4.9 Top left-hand corner of the spreadsheet which is shown completed in Fig. A4.16.

	A	B	C	D	E	F	G
1	Yield per recruit						
2	K >	0.3	M>	0.5			
3	Winf >	900	tr>	1			
4	tzero >	0	tc>	1.5			
5							
6	F	Term 1	Term 2	Term 3	Term 4	Term 5	Y/R
7	0	=A7*$B$3* EXP(-$D$2* ($D$4-$D$3))	=1/(A7+$D$2)	=-3*EXP (-1*$B$2* ($D$4-$B$4))/ ($D$2+A7 +1*$B$2)	=3*EXP (-2*$B$2* ($D$4-$B$4))/ ($D$2+A7 +2*$B$2)	=-1*EXP (-3*$B$2* ($D$4-$B$4))/ ($D$2+A7 +3*$B$2)	=B7*SUM (C7:F7)

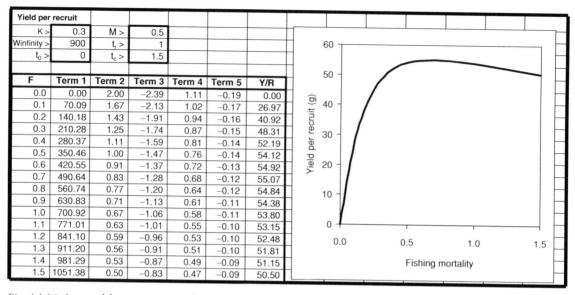

Yield per recruit						
K >	0.3	M >	0.5			
Winfinity >	900	t_r >	1			
t_0 >	0	t_c >	1.5			

F	Term 1	Term 2	Term 3	Term 4	Term 5	Y/R
0.0	0.00	2.00	−2.39	1.11	−0.19	0.00
0.1	70.09	1.67	−2.13	1.02	−0.17	26.97
0.2	140.18	1.43	−1.91	0.94	−0.16	40.92
0.3	210.28	1.25	−1.74	0.87	−0.15	48.31
0.4	280.37	1.11	−1.59	0.81	−0.14	52.19
0.5	350.46	1.00	−1.47	0.76	−0.14	54.12
0.6	420.55	0.91	−1.37	0.72	−0.13	54.92
0.7	490.64	0.83	−1.28	0.68	−0.12	55.07
0.8	560.74	0.77	−1.20	0.64	−0.12	54.84
0.9	630.83	0.71	−1.13	0.61	−0.11	54.38
1.0	700.92	0.67	−1.06	0.58	−0.11	53.80
1.1	771.01	0.63	−1.01	0.55	−0.10	53.15
1.2	841.10	0.59	−0.96	0.53	−0.10	52.48
1.3	911.20	0.56	−0.91	0.51	−0.10	51.81
1.4	981.29	0.53	−0.87	0.49	−0.09	51.15
1.5	1051.38	0.50	−0.83	0.47	−0.09	50.50

Fig. A4.16 A spreadsheet program used to calculate yield per recruit based on the whiting data in Box 5.6. Equations used are shown in Table A4.9. The effect of delaying the age at first capture may be seen by increasing the value of t_c. Discrepancies between the spreadsheet and the hand calculations in Box 5.6 are due to rounding errors. Spreadsheet and graph produced using Microsoft Excel.

reasonable values of fishing mortality. This effect is discussed in Section 5.4.2.

A4.12 Biomass dynamic simulation model

The simulation model described in Section 5.5.1 of Chapter 5 may be built on a computer spreadsheet. The example uses the Schaefer parameters $B_\infty = 4126$, $r = 0.5324$ and $q = 0.00000568$. The design of the model, used to predict changes in stock biomass, catch and catch rates, is suggested in Table A4.10 with the three boxes D2, D3 and D4 reserved to contain the input parameters of B_∞, r and q. Column A contains the years from 1 to 20 and column B contains the fishing effort data, which may be altered for different scenarios. Cell C8 contains the initial value of stock biomass (4126). In column C below cell C8, each year's biomass (B_{t+1}) can be expressed in terms of the present year's biomass (B_t) plus the rate of change in biomass minus the catch, C_t. That is:

$$B_{t+1} = B_t + rB_t(1 - B_t/B_\infty) - C_t \qquad (A4.15)$$

For the purpose of this exercise, the model is made stochastic by allowing the maximum biomass to vary randomly by ±40% by multiplying B_∞ in the above equation by [0.6 + 0.8(RAND())]:

$$B_{t+1} = B_t + rB_t(1 - B_t/(B_\infty[0.6 \\ + 0.8(RAND())])) - C_t \qquad (A4.16)$$

Column D contains catches predicted as $C_t = qf_tB_t$, where f is fishing effort and q is the catchability coefficient. Column E contains the catch per unit data calculated as $CPUE_t = C_t/f_t$ multiplied by 1000 to give the values in kg per unit effort.

Fig. A4.17 shows the spreadsheet model with simulated stochastic data over 20 years for a constant fishing effort of 25 000 units of hundreds of metres of net per day (hmnd). In this case biomass has been graphed at the right of the spreadsheet but either catch or CPUE data could have been graphed against fishing effort.

A4.13 Age-structured simulation model

The following age-structured model is developed

Table A4.10 Top left-hand corner of a spreadsheet model that simulates biomass, catch and CPUE over a 20-year period.

	A	B	C	D	E
1	Surplus production simulation model				
2				Binf > 4126	
3				r > 0.5324	
4				q > 0.00000568	
5					
6			Predicted	Predicted	Predicted
7	Time	Effort	biomass	catch	CPUE
8	0	25 000	=D2	=$D$4*B8*C8	=1000*D8/B8
9	=A8+1	25 000	=C8+C8*$D$3*(1–C8/$D$2* (0.6+0.8*RAND()))–D8	=$D$4*B9*C9	=1000*D9/B9

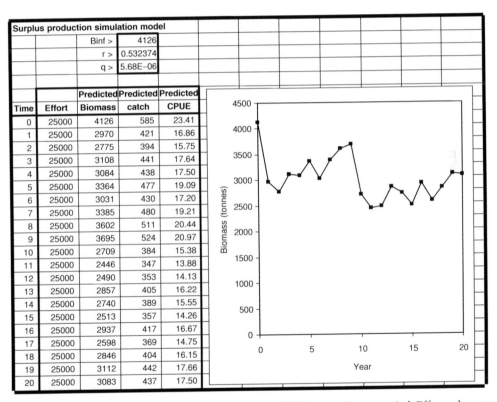

Fig. A4.17 A spreadsheet model to simulate biomass, catch and CPUE over a 20-year period. Effort values can be changed for different scenarios. Spreadsheet produced using Microsoft Excel.

using a spreadsheet program with the number of age classes restricted to four (to enable it to fit on a single spreadsheet page). The seven cells (D3 to D9) at the top left of the spreadsheet table shown in Table A4.11 are reserved for the input of the required parameters, which for the purpose of the example are:

- initial recruitment = 1 million
- catchability coefficient, $q = 0.00012$
- von Bertalanffy growth parameters, $K = 0.6$ yr

Table A4.11 The top left-hand corner of a spreadsheet for an age-structured simulation model.

	A	B	C	D	E	F	G	H	I	J	K	L	M
1	Age-structured simulation model												
2													
3	Initial recruitment> 1000000												
4	Catchability > 0.00012												
5	Growth K > 0.6				**Mean weight**								
6	Max weight Winf (kg) > 8				Age 1 > = $D$6*(1−EXP(−$D$5*1))^3								
7	Ricker stock recruit, a > 3				Age 2 > = $D$6*(1−EXP(−$D$5*2))^3								
8	Ricker stock recruit, b > 0.000001				Age 3 > = $D$6*(1−EXP(−$D$5*3))^3								
9	Natural mortality, M > 0.6				Age 4 > = $D$6*(1−EXP(−$D$5*4))^3								
10													
11	Year	Recruit	Numbers	Numbers	Numbers	Spawn	Fishing	Fishing	Catch w	Catch w	Catch w	Catch w	Catch
12		age 1	age 2	age 3	age 4	stock	effort	mort.	age 2	age 3	age 4	TOTAL	rate
13	1	=$D$3	B13*EXP(−$D$9) (copied right to E)			=SUM (C13: E13)	6000	=G13* $D$4* (0.6+ 0.8* RAND())	=mean weight at age *C13*($H13/ ($H13+$D$9)) *(1−EXP(−$H13−$D$9)) (copied right to J and K)			=SUM (I13: K13)	=L13/ G13
14	2	=$D$7* F13* EXP (−$D$8* F13)* (0.75+0.5 *RAND())	=B13* EXP (−$D$9)	C13*EXP (−$D$9−H13) (copied right to E)		=SUM (C14: E14)	6000	=G14* $D$4* (0.6+ 0.8* RAND())	=mean weight at age *C14*($H14/ ($H14+$D$9)) *(1−EXP(− $H14−$D$9)) (copied right to J and K)			=SUM (I14: K14)	=L14/ G14

- asymptotic weight = 8 kg
- constant in the Ricker stock–recruitment model, $a = 3$
- constant in the Ricker stock–recruitment model, $b = 0.000001$
- natural mortality, $M = 0.6$

The first step is to calculate the mean weights for all age classes (in this case, only four for simplicity) by inserting the von Bertalanffy equation $W_t = W_\infty (1 - \exp[-Kt])^3$ for $t = 1$ to 4 in cells F6 to F9 in Table A4.11.

The rest of the spreadsheet program is designed to complete the required calculations in annual steps for 20 years. The following points cover the column entries from left to right on the spreadsheet from row 13 and down.

- Column A contains the years from 1 to 20.
- Column B contains the number of new recruits (recruitment is assumed to occur at the beginning of the year). Cell B13 picks up the initial recruitment (of one million in the example) from cell $D$3 of the input panel.
- Column B, cells B14 and below, contain the Ricker stock–recruitment equation, $R = aS \exp[-bS]$, where a and b are constants from the input box and S is the spawning stock (in this case, the spawning stock consists of ages 2 to 4 from the previous year) – the value of S is the sum of numbers in ages 2, 3, and 4 from the previous year and is entered in column F. Recruitment is allowed to vary by ±25% by multiplying the stock recruitment equation by [0.75 + 0.5(RAND())].
- Columns C, D and E (row 13 only) contain the numbers in age classes 2 to 4 and these are estimated as $N_{t+1} = N_t \exp[-M]$ where N_t is the number in the column before it. Natural mortality is taken from cell $D$9 in the input panel.
- Columns C (row 14 and below) contain the numbers in age class 2 estimated as $N_{t+1} =$

$N_t\exp[-M]$ where N_t is the number in same cohort from the previous year reduced by natural mortality only (ages 2 and above are exploited).

- Columns D and E (row 14 and below) contain the numbers in age classes 3 and 4 estimated as $N_{t+1} = N_t\exp[-F-M]$ where N_t is the number in same cohort from the previous year reduced by both natural and fishing mortality (diagonal arrows in Fig. A4.18). Fishing mortality is taken from column H.
- Column F contains the total spawning stock; that is, the sum of numbers in age classes 2 to 4 from columns C, D and E.
- Column G contains the fishing effort in boat days. *These are the values that will be altered to simulate different scenarios.* For the time being, fishing effort is 6000 boat days for years 1 to 10 and is then reduced drastically to 2000 boat days for years 11 to 20 to illustrate just one particular scenario.
- Column H contains fishing mortality calculated as $F = fq$ where f is the effort taken from column G and q, the catchability coefficient, is taken from cell $D$4. Catchability is allowed to vary by ±40% by multiplying q by $[0.60 + 0.8(\text{RAND}())]$.
- Columns I, J and K contain the catch weights from years 2 to 4 (in this case, only fish from age classes 2 to 4 are vulnerable). The equation for catch is $C_t = N_tW_t(F/(F + M))(1 - \exp[-F-M])$ where N_t is the number of fish from

columns C, D and E and W_t is the weight at age from cells $F$6 to $F$9.
- Column L contains the total catch weights – the sum of cells in columns I, J and K.
- Column M shows the catch rates – the cells in column L divided by the cells in column G.

The simulation model can be either designed to graph CPUE as shown in Fig. A4.19 or used to graph other important characteristics of the fishery such as spawning stock size and total catch weight.

A4.14 Risk assessment model

The following example of risk assessment is based on a stock of barramundi (Section 5.2.2) with a virgin biomass of $B_\infty = 4126$ tonnes, an intrinsic rate of population increase of $r = 0.5234$ and a catchability coefficient of $q = 0.0000057$. Fishing effort is measured in hundreds of metres of net used per day (hmnd).

For the sake of the example, the management objective is to keep stock biomass above half the original unexploited level, say 2000 tonnes. The spreadsheet design (shown in Table A4.12) uses Equation A4.15 adapted so that the maximum biomass varies randomly by ±20%:

$$B_{t+1} = B_t + rB_t(1 - B_t/B_\infty[0.8 + 0.4(\text{RAND}())]) - qf_tB_t \quad (A4.17)$$

In the spreadsheet (Table A4.12) the parameters B_∞, r, and q are entered in cells B2 to B4 and the

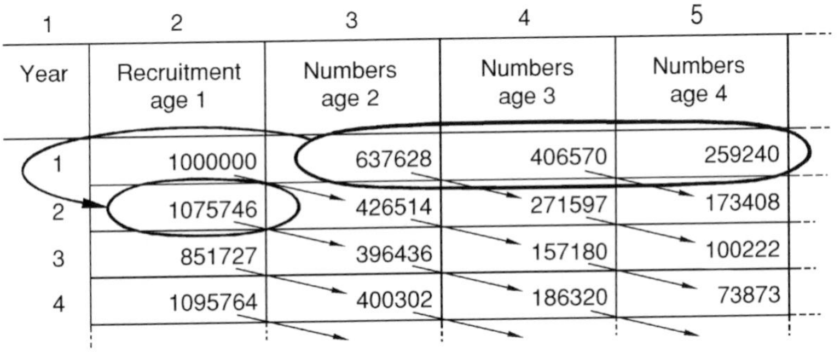

1	2	3	4	5	
Year	Recruitment age 1	Numbers age 2	Numbers age 3	Numbers age 4	
1	1000000	637628	406570	259240	
2	1075746	426514	271597	173408	
3	851727	396436	157180	100222	
4	1095764	400302	186320	73873	

Fig. A4.18 Portion of the spreadsheet shown in Table A4.11 showing how three age classes (large ellipse) contribute to the following year's recruitment (small ellipse). Diagonal arrows show the progress of cohorts with numbers reduced by natural and fishing mortality.

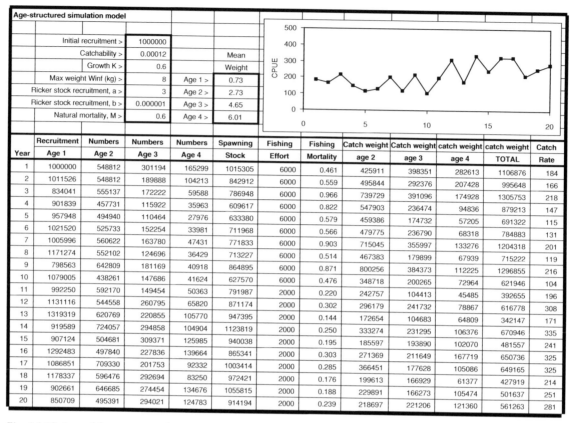

Age-structured simulation model												
	Initial recruitment >	1000000										
	Catchability >	0.00012		Mean								
	Growth K >	0.6		Weight								
	Max weight Winf (kg) >	8	Age 1 >	0.73								
	Ricker stock recruitment, a >	3	Age 2 >	2.73								
	Ricker stock recruitment, b >	0.000001	Age 3 >	4.65								
	Natural mortality, M >	0.6	Age 4 >	6.01								

Year	Recruitment Age 1	Numbers Age 2	Numbers Age 3	Numbers Age 4	Spawning Stock	Fishing Effort	Fishing Mortality	Catch weight age 2	Catch weight age 3	catch weight age 4	catch weight TOTAL	Catch Rate
1	1000000	548812	301194	165299	1015305	6000	0.461	425911	398351	282613	1106876	184
2	1011526	548812	189888	104213	842912	6000	0.559	495844	292376	207428	995648	166
3	834041	555137	172222	59588	786948	6000	0.966	739729	391096	174928	1305753	218
4	901839	457731	115922	35963	609617	6000	0.822	547903	236474	94836	879213	147
5	957948	494940	110464	27976	633380	6000	0.579	459386	174732	57205	691322	115
6	1021520	525733	152254	33981	711968	6000	0.566	479775	236790	68318	784883	131
7	1005996	560622	163780	47431	771833	6000	0.903	715045	355997	133276	1204318	201
8	1171274	552102	124696	36429	713227	6000	0.514	467383	179899	67939	715222	119
9	798563	642809	181169	40918	864895	6000	0.871	800256	384373	112225	1296855	216
10	1079005	438261	147686	41624	627570	6000	0.476	348718	200265	72964	621946	104
11	992250	592170	149454	50363	791987	2000	0.220	242757	104413	45485	392655	196
12	1131116	544558	260795	65820	871174	2000	0.302	296179	241732	78867	616778	308
13	1319319	620769	220855	105770	947395	2000	0.144	172654	104683	64809	342147	171
14	919589	724057	294858	104904	1123819	2000	0.250	333274	231295	106376	670946	335
15	907124	504681	309371	125985	940038	2000	0.195	185597	193890	102070	481557	241
16	1292483	497840	227836	139664	865341	2000	0.303	271369	211649	167719	650736	325
17	1086851	709330	201753	92332	1003414	2000	0.285	366451	177628	105086	649165	325
18	1178337	596476	292694	83250	972421	2000	0.176	199613	166929	61377	427919	214
19	902661	646685	274454	134676	1055815	2000	0.188	229891	166273	105474	501637	251
20	850709	495391	294021	124783	914194	2000	0.239	218697	221206	121360	561263	281

Fig. A4.19 Spreadsheet program for an age-structured simulation model. In this scenario, fishing effort is high for years 1 to 10 before being reduced for years 11 to 20. The graph shows CPUE against year. Spreadsheet and graphs produced using Microsoft Excel.

Table A4.12 Top portion of a spreadsheet to assess risk. Cell B9 and below are copied across 10 columns from B to K. Row 10 is copied down to cover years 0 to 20. Column L contains the proportion of biomass replicates below 2000.

	A	B . . . K	L
1			
2	Binf >	4126	
3	r >	0.5324	
4	q >	0.000005675	
5	f >	35 000	
6			
7		Predicted	Proportion of replicates
8	Year	biomass	less than 2000
9	0	=$C$2	=COUNTIF(B9:K9,"<2000")/10
10	1	=(B9+B9*$B$3*(1−B9/$B$2*(0.8+0.4*RAND())))−$B$4*$B$5*B9	=COUNTIF(B10:K10,"<2000")/10

first trial value of fishing effort, f, is entered in B5. The years 0 to 20 are entered in cells from A9 down. In column B, cell B9 picks up the virgin biomass from B2. Equation A4.17 is entered in cell B10 and copied down for the 20 years. Column L uses the COUNTIF function to check the 10 replicate values in columns B to K and print the number of times biomass is below the 2000 tonne threshold limit. Dividing this number by 10 gives the proportion, $P(B < 0.5B_\infty)$. The program can be run for different levels of fishing effort, f, to replicate the graphs in Fig. 5.18 in Chapter 5.

Appendix 5: Collection of length–frequency data

When a random sample of fish is measured, the idealized expectation is that this sample reflects the distribution of different sizes of fish in the total stock, or at least that part of the stock in the sampling vicinity. However, the distribution of lengths in the sample may be different from that in the population for several reasons, including:

- the selectivity of the sampling gear may result in particular sizes (often small fish) being under-represented in the sample;
- poor sampling techniques, e.g. when a sub-sample is taken in a biased way from a large catch;
- inaccurate measuring or recording of lengths by biologists;
- certain sizes of fish may predominate in the obtained sample just by chance.

Adjusting length–frequency data for gear selectivity

Length–frequency data, particularly those collected from commercial fishing, are often a poor representation of the actual relative abundance of fish of particular lengths in the stock. Because of the selectivity of fishing gear such as trawl nets, for example, smaller fish are less readily retained, and are under-represented in length–frequency distributions. Table A5.1 shows the data and Fig. A5.1 shows a length–frequency distribution for a sample of 420 fish from a commercial trawl catch.

In cases where the selectivity of the fishing gear is known, the probability of capture for each length class may be used to adjust the frequencies in the sample to reflect the actual abundance of smaller fish in the population. Adjusted frequencies may

Table A5.1 Length–frequency data from a sample collected from a trawl catch. The probability of capture is predicted from the logistic equation with $r = 0.4$, and $L_c = 14.5$ cm.

Length (cm)	Frequency (sample)	Probability of capture	Frequency (adjusted)
4	1	0.015	67.7
5	2	0.022	91.4
6	3	0.032	92.9
7	5	0.047	105.4
8	8	0.069	115.7
9	11	0.100	110.3
10	13	0.142	91.6
11	12	0.198	60.7
12	10	0.269	37.2
13	9	0.354	25.4
14	11	0.450	24.4
15	16	0.550	29.1
16	27	0.646	41.8
—	—	—	—
... (Table abbreviated) ...			
—	—	—	—
33	3	0.999	3.0
34	3	1.000	3.0
35	2	1.000	2.0
36	2	1.000	2.0
37	1	1.000	1.0

be calculated by dividing the number of fish in each length class by the probability of capture:

adjusted F = sampled F/probability of capture

(A5.1)

In the case of trawl data, the probability of capture, P, is predicted by the logistic curve (Chapter 3, Fig. 3.16), which has the equation $P = 1/(1 + \exp[-r(L - L_c)])$ in which L is fish length, r is the

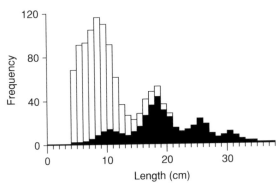

Fig. A5.1 A length–frequency distribution of 420 fish collected from a trawl catch; the original data are shown as the solid part of the histogram, and the data adjusted for mesh selectivity are shown as an outline.

slope parameter of the curve, and L_c is the mean length at first capture. If $r = 0.4$, and $L_c = 14.5$ cm in the example data, the values of probability of capture and adjusted frequencies are as shown in Table A5.1. An adjusted length–frequency distribution is shown in Fig. A5.1.

Sampling and measuring techniques

If the total catch made from the use of fishing or collecting gear is measured, a major bias in the sample is likely to be due to the selectivity of the gear. If the catch is too large to measure, an additional bias may result from the process of sub-sampling. In factory sampling, for example, the catch may have been sorted, or larger fish may be displayed at the top layer of a bin intended for auction; in this case it is important to measure fish from lower down in the bin as well as at the top. In sampling on board a fishing vessel, the catch may be unequally distributed when it is landed on the deck, or on the sorting table. In the case of a trawler, for example, smaller fish may collect towards the end of the net's codend.

One way of obtaining a sub-sample from a large catch and to avoid bias as much as possible, is by 'quartering'. This involves mixing and spreading the total catch across the deck, sorting table, or factory floor, and using a shovel to divide it into four visually equal quarters. Two diagon-

ally opposite quarters are retained for the sample, and the other two quarters are discarded; this reduces the sample to one-half of the total catch. If the sample is still too large to handle, the process may be repeated. The number of times this process is repeated will indicate whether the sample represents either one-half, one-quarter, one-eighth etc. of the total catch.

Fish should be measured as consistently and accurately as possible. If more than one person is involved in collecting length measurements, the standard way of measuring the particular species, and the accuracy of the measurements required should be agreed upon before the research programme starts. If lengths of a species are to be measured to the nearest cm, for example, the choice is whether to record the length as the closest whole cm below the actual measurement (e.g. a 32.6 cm fish is recorded as 32 cm), or to the nearest cm (e.g. 32.6 cm is recorded as 33 cm).

Often there is a predilection to favour particular numbers. Measurers may tend to round lengths to even rather than odd numbers, or to round numbers to end in five or zero (e.g. round a fish measuring 49.4 to 50 rather than to 49). An example is illustrated in the actual length–frequency data shown in Fig. A5.2, which shows a marked bias towards recording lengths as multiples of ten.

Smoothing the data – moving averages

A sample taken for length measurements will often contain a predominance of certain sizes of fish just by chance. The resulting length–frequency distribution will contain random fluctuations as well as more meaningful (or real) ones. The top length–frequency diagram in Fig. A5.3, for example, contains such fluctuations. Random fluctuations, and to a certain extent inaccuracies in measurements, may be allowed for by 'smoothing' the data. One way of doing this is to apply moving averages (sometimes called running means) to the data. For any set of frequencies (F_1, F_2, F_3 etc.), a moving average of order N has the following sequence of arithmetic means:

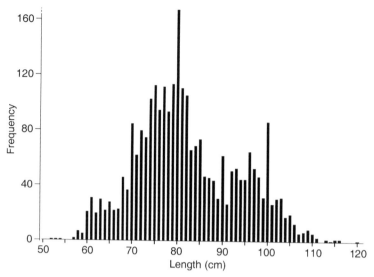

Fig. A5.2 Albacore length–frequency data (actual data from a government fisheries observer programme) which illustrates a marked bias towards recording lengths as multiples of ten.

Table A5.2 Lengths (cm) and frequencies (numbers) for a sample of fish. Columns three to five show frequencies as moving averages of orders N = 3, N = 5, and N = 7 respectively.

Length	Frequency	N = 3	N = 5	N = 7
4	0	0.00	0.00	0.00
5	0	0.00	0.00	0.14
6	0	0.00	0.20	1.00
7	0	0.33	1.40	2.14
8	1	2.33	3.00	3.57
9	6	5.00	5.00	6.14
10	8	8.00	8.60	8.00
11	10	12.00	11.00	10.29
12	18	13.67	13.00	12.00
13	13	15.67	14.00	13.57
14	16	14.00	15.40	13.86
15	13	15.33	13.80	13.43
16	17	13.33	12.60	12.14
17	10	11.33	11.20	11.43
18	7	8.67	10.20	11.14
—	—	—	—	—
... (Table abbreviated) ...				
—	—	—	—	—
35	1	4.00	3.00	4.00
36	2	1.33	2.60	2.14
37	1	1.00	0.80	1.86
38	0	0.33	0.60	0.57
39	0	0.00	0.20	0.43
40	0	0.00	0.00	0.14

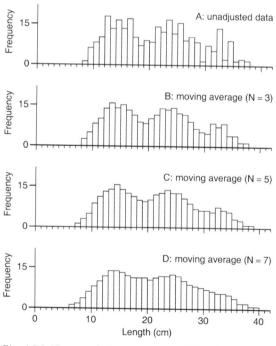

Fig. A5.3 Top graph shows unadjusted length–frequency data. Lower graphs show the same data smoothed by the application of moving averages of orders N = 3, N = 5, and N = 7 respectively.

$(F_1 + F_2 + \ldots + F_N)/N, (F_2 + F_3 + \ldots + F_{N+1})/N$, etc.

That is, applying a moving average of the order $N = 3$, a 'smoothed' frequency is calculated as the mean of three consecutive frequencies. In relation to the data in Table A5.2, the moving average frequency for a length of 15 cm is the mean of three frequencies, including the one on each side of 15 cm, that is:

$(F_{14} + F_{15} + F_{16})/3 = (16 + 13 + 17)/3 = 15.33$

In the length–frequency diagrams shown in Fig. A5.3, the unadjusted data in the top graph has so many (presumably random) fluctuations that it is difficult to see more meaningful peaks or modes in the graph. By applying a moving average, it may be possible to discern modes which may represent real properties of the sampled population, such as age classes. The smoothed data in graphs B and C suggest that there are three or more groups which may represent age classes; in graph B, if the modes at 14 cm, 23 cm, and 32 cm represent consecutive year classes, the length–frequency distribution may be used to analyse growth. In graph D, the data have been 'over-smoothed' (the order of $N = 7$ is too high), and meaningful components may have been lost in the process. The choice of smoothing factor (the value of N) is largely a matter of judgement.

Appendix 6: Bhattacharya plots

A method of separating a length–frequency distribution into its component normal distributions was developed by Buchanan-Wollaston and Hodgson (1929), and subsequently developed by Bhattacharya (1967). The Bhattacharya plot (Pauly & Caddy, 1985; Sparre et al., 1989) is a graphical method of separating a length–frequency distribution into a series of normal distributions or pseudo-cohorts. The input data consists of a length–frequency distribution which ideally has been corrected for the effects of gear selectivity.

The Bhattacharya method is based on approximating the assumed normal (bell-shaped) curve of a length–frequency distribution as a parabola, which is then converted to a straight line (a detailed description of this conversion is given in Sparre et al., 1989). The Bhattacharya plot, the straight line in Fig. A6.1, has the form:

$$dt(\ln[N]) = a + b(L) \qquad (A6.1)$$

where $dt(\ln[N])$ is the difference between the natural logarithms of the number in one length class and the number in the preceding length class, and L is the upper limit of the preceding length class. The straight line crosses the length axis at a point which is the mean (and the mode) of the length–frequency distribution; that is, the normal distribution has a mean of $(-a/b)$, and a standard deviation of $\sqrt{(-dL/\text{slope})}$, where dL is the length class interval.

If a length–frequency distribution consisted of a number of overlapping pseudo-cohorts (instead of just the one shown in Fig. A6.1, the Bhattacharya method may be used to separate the distribution into normal components as in graphs (a) to (d) of Fig. A6.2.

(a) The left-hand side of the length–frequency

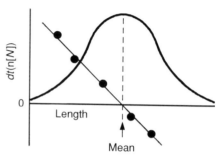

Fig. A6.1 The conversion of a parabola to a straight line.

distribution is assumed to consist solely of individuals in the first pseudo-cohort, and is uncontaminated by individuals in other overlapping pseudo-cohorts to the right.

(b) A Bhattacharya plot of $dt(\ln[N])$ against the upper limit of the preceding length class is used to estimate the mean of the first cohort from the point where the straight line crosses the length axis.

(c) By working 'backwards', the linear regression results are used to produce a normal curve which is assumed to contain the total number of individuals in the first pseudo-cohort.

(d) The number of individuals in the first cohort is subtracted from the total length–frequency distribution. The remaining distribution has a new 'clean' left hand side, from which a second cohort may be separated.

The Bhattacharya method of separating out normal components is demonstrated using the length–frequency data shown in Table A6.1. The frequencies in column 3 of Table A6.1 are referred to as N_{1+} to suggest that they include frequencies in all of the component pseudo-cohorts. Column 4 contains the natural logarithms of the frequencies. The values of $dt(\ln[N_{1+}])$ in column 5 are the differences between the natural logarithms

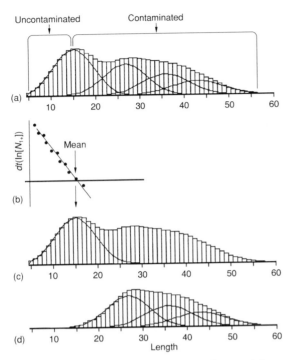

Fig. A6.2 The sequences of operation in the use of the Bhattacharya method of separating a length–frequency distribution into normal components.

of the frequency in the corresponding length class and the frequency in the preceding length class.

The values of $dt(\ln[N_{1+}])$ are graphed against L, the upper limit of the preceding length class, in Fig. A6.3. The straight line is fitted through the initial (uncontaminated) data points which appear linear; its slope is -0.10083, and the y-axis

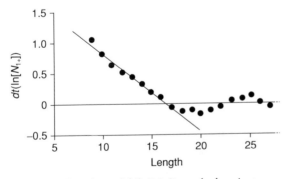

Fig. A6.3 The values of $dt(\ln[N_{1+}])$ graphed against L, the upper limit of the preceding length class. The straight line is fitted through the initial (uncontaminated) data points which appear linear; its slope is -0.10083, and the y-axis intercept is 1.74.

intercept is 1.74. The normal distribution of the first pseudo-cohort, therefore, has the following properties:

$$\text{mean} = (-a/b) = 17.3$$

$$\text{standard deviation} = \sqrt{(-dL/b)} = 3.15$$

The next step involves adding more columns to Table A6.1 and working 'backwards' using linear

Table A6.1 Length–frequency data and values of $dt(\ln[N_{1+}])$. The symbols * and ** denote the first point and last point included in the regression shown in Fig. A6.3.

1 L_1	2 L_2	3 N_{1+}	4 $\ln[N_{1+}]$	5 $dt(\ln[N_{1+}])$
8	9	1	0.000	—
9	10	3	1.099	*1.099
10	11	7	1.946	0.847
11	12	13	2.565	0.619
12	13	21	3.045	0.480
13	14	30	3.401	0.357
14	15	41	3.714	0.312
15	16	50	3.912	0.198
16	17	58	4.060	0.148
17	18	61	4.111	0.050
18	19	59	4.078	**−0.033
19	20	55	4.007	−0.070
20	21	50	3.912	−0.095
21	22	44	3.784	−0.128
22	23	41	3.714	−0.071
23	24	41	3.714	0.000
24	25	42	3.738	0.024
25	26	44	3.784	0.047
26	27	44	3.784	0.000
27	28	44	3.784	0.000
28	29	42	3.738	−0.047
29	30	41	3.714	−0.024
30	31	39	3.664	−0.050
31	32	37	3.611	−0.053
32	33	36	3.584	−0.027
33	34	36	3.584	0.000
34	35	34	3.526	−0.057
—	—	—	—	—
... (Table abbreviated) ...				
—	—	—	—	—
45	46	5	1.609	−0.336
46	47	3	1.099	−0.511
47	48	2	0.693	−0.405
48	49	1	0.000	−0.693
49	50	1	0.000	0.000

Table A6.2 Length–frequency data from Table A6.1, showing the estimated number of individuals in cohort one (column 8), and those in the remaining length–frequency distribution (column 9). The table is abbreviated in the lower part, where calculations are no longer necessary.

1 L_1	2 L_2	3 N_{1+}	4 $\ln[N_{1+}]$	5 $dt(\ln[N_{1+}])$	6 $dt(\ln[N_c])$	7 $\ln[N_1]$	8 N_1	9 N_{2+}
8	9	1	0.000	—			1.0	0
9	10	3	1.099	1.099	0.833		3.0	0
10	11	7	1.946	0.847	0.732		7.0	0
11	12	13	2.565	0.619	0.631		13.0	0
12	13	21	3.045	0.480	0.530		21.0	0
13	14	30	3.401	0.357	0.430		30.0	0
14	15	41	3.714	0.312	0.329	3.714	41.0	0
15	16	50	3.912	0.198	0.228	3.942	51.5	[−1.5]0
16	17	58	4.060	0.148	0.127	4.069	58.5	[−0.5]0
17	18	61	4.111	0.050	0.026	4.095	60.0	1.0
18	19	59	4.078	−0.033	−0.074	4.021	55.7	3.3
19	20	55	4.007	−0.070	−0.175	3.846	46.8	8.2
20	21	50	3.912	−0.095	−0.276	3.570	35.5	14.5
21	22	44	3.784	−0.128	−0.377	3.193	24.4	19.6
22	23	41	3.714	−0.071	−0.478	2.715	15.1	25.9
23	24	41	3.714	0.000	−0.578	2.137	8.5	32.5
24	25	42	3.738	0.024	−0.679	1.458	4.3	37.7
25	26	44	3.784	0.047	−0.780	0.678	2.0	42.0
26	27	44	3.784	0.000	−0.881	−0.203	0.8	43.2
27	28	44	3.784	0.000	−0.982	−1.185	0.3	43.7
28	29	42	3.738	−0.047	−1.082	−2.267	0.1	41.9
29	30	41	3.714	−0.024	−1.183	−3.450	0.0	41.0
30	31	39	3.664	−0.050				39.0
31	32	37	3.611	−0.053				37.0
32	33	36	3.584	−0.027				36.0
—	—	—	—	—				—
. . . (Table abbreviated) . . .								
—	—	—	—	—	—	—	—	—
48	49	1	0.000	−0.693				1.0
49	50	1	0.000	0.000				1.0

regression to estimate the numbers of fish in the first cohort, N_1 (Table A6.2). Column 6 in Table A6.2 contains the values of the differences in natural logarithms *calculated* from the regression line – hence the subscript in $dt(\ln[N_c])$. For example, for length interval 9 cm to 10 cm, the value is estimated as:

$$dt(\ln[N_c]) = (\text{intercept} + \text{slope}) \times 9$$
$$= (1.74 - 0.1008) \times 9 = 0.833$$

Column 7 contains the natural logarithms, $\ln[N_1]$, of the estimated numbers in the first

cohort, N_1. One of the last 'clean' length classes (on the rising left-hand side of the first cohort) which is uncontaminated by other cohorts is selected – for example the length class of 14 cm to 15 cm which contains 41 individuals. As this length class is uncontaminated, the value of $\ln[N_1]$ in column 7 will be the same as the value of $\ln[N_{1+}]$ in column 3 – i.e. 3.714. Values of $\ln[N_1]$ for subsequent rows are estimated as the following example for 15 cm to 16 cm:

$$\ln[N_1]_{15-16} = \ln[N_1]_{14-15} + dt(\ln[N_c]_{15-16})$$
$$= 3.714 + 0.228 = 3.942$$

The numbers, N_1, in the first cohort (column 8) for the clean start point (14–15 cm) and smaller size classes are taken to be the same as those for the total frequency, N_{1+}, in column 3. After the start length class, the calculated number, N_1, is estimated as that for 15 cm to 16 cm:

$$N_1 = \exp[\ln(N_1)] = 51.5$$

Calculations in column 8 are continued until the numbers in the first cohort approach zero. Column 9, which is estimated as N_{1+} minus N_1 (or column 3 minus column 8), contains the numbers of individuals in the remaining cohorts. Negative numbers [in square brackets] are assumed to be due to random fluctuations, and are listed as zeros.

The remaining distribution, N_{2+}, has a new uncontaminated left-hand side, which may be treated in a similar manner to the initial distribution, N_{1+}, to separate out the second cohort. The above process is repeated to separate out the remaining cohorts. For each cohort separated out from the overall length–frequency distribution, the mean is calculated, and these means may be used to estimate growth. As long as the pseudo-cohorts can be assumed to represent year classes, a Ford–Walford plot may be used, for example, to estimate the von Bertalanffy growth parameters K and L_∞.

Appendix 7: Statistical tables

Table A7.1 Degrees of freedom (df), values of t for 95% confidence limits (df = $n − 1$ for one parameter, or $n − 2$ for two parameters), and critical values of the correlation coefficient, r, above which the relationship is significant at the 95% level (df = $n − 2$).

df	t	r	df	t	r
1	12.71	0.997	11	2.201	0.553
2	4.303	0.950	12	2.179	0.532
3	3.182	0.878	13	2.160	0.514
4	2.776	0.811	14	2.145	0.497
5	2.571	0.754	15	2.131	0.482
6	2.447	0.707	20	2.086	0.423
7	2.365	0.666	40	2.021	0.304
8	2.306	0.632	60	2.000	0.250
9	2.262	0.602	100	1.990	0.195
10	2.228	0.576	∞	1.960	0.062

Table A7.2 Random numbers (from 1 to 20) and random letters (A to M).

Numbers (1 to 20)				Letters (A to M)			
18	07	11	02	I	K	M	A
17	00	06	08	J	I	I	L
06	03	18	08	E	D	C	J
02	05	05	10	L	F	H	B
17	09	20	13	H	K	D	K
01	10	04	16	D	G	D	K
20	13	11	10	C	I	B	B
11	06	01	19	F	E	F	I

Glossary

Adaptive management A resource management programme in which management actions are deliberately used as experimental manipulations of the managed system to test predictions of alternative models.

Amnesic shellfish poisoning (ASP) Poisoning resulting from the consumption of shellfish contaminated with toxins produced by diatoms including *Pseudonitzschia* spp.

Anthropogenic effects In environmental science, the effects, usually negative, of human activities on the marine environment.

Antifouling paint A paint that leaches out poisonous material at a low rate to prevent the growth of marine species on ship's hulls.

Artisanal fishery A small-scale, low-cost, and labour intensive fishery in which the catch, which often consists of a great variety of different species, is sold and consumed locally.

Autotroph An organism that is capable of producing organic compounds from inorganic raw materials; plants are autotrophs that use sunlight and nutrients to produce tissue (compare heterotroph).

Bilateral and multilateral agreements An arrangement whereby foreign fishers pay a fee to access fish stocks not fully utilized by national fishers.

Biodiversity The variability among living organisms from all sources and the ecological complexes of which they are part. This includes diversity within species and ecosystems.

Bioerosion The breaking down of substrates, usually coral, by the actions of various living organisms referred to as bioeroders.

Biological (biochemical) oxygen demand (BOD) A measure of the amount of pollution in the form of organic material in water: the greater the number of bacteria, the greater the demand for oxygen, and the more polluted the water.

Biomass The sum of weights (total mass) of individuals in a stock or population.

Brackish water A mixture of sea water and fresh water (as occurs near the mouths of rivers).

Bycatch The part of the catch which is taken incidentally to the target species and which is discarded.

Cartilage (in sharks) A type of connective tissue, consisting of living cells embedded in a rubbery matrix stiffened by elastic fibres of collagen.

Catch quota The maximum catch permitted to be taken from a fishery; such a limit applied to the total catch from a fishery is often referred to as a global quota (as distinct from an individual quota).

Catchability coefficient (q) The proportion of the total stock caught by one unit of fishing effort.

Chemosynthetic (chemoautotrophic) organism One that can gain energy directly from inorganic chemical compounds instead of from sunlight; in deep, hot-water vents, for example, certain bacteria use sulphur expelled from beneath the earth's crust.

Ciguatera Fish poisoning resulting from the consumption of fish that have accumulated toxins produced by particular dinoflagellates, including the benthic species *Gambierdiscus toxicus*, which is found in association with coral reefs.

Closures (in fisheries) Banning of fishing at times or seasons (temporal closures) or in particular areas (spatial closures), or a combination of both.

Colony (in biology) a community of animals or plants of the same species forming a physically connected structure or living close together.

Co-management (cooperative management) Either informal or legal arrangements between government representatives, community groups and other user groups, to take responsibility for, and manage, a fishery resource and its environment on a cooperative basis.

Commensalism A form of symbiosis in which two species are closely associated and one gains advantages while not harming the other – see symbiosis.

Community A collection of populations of plants and animals inhabiting a particular area.

Community-based fisheries management (CBFM) Arrangements under which a community takes responsibility for managing its adjacent marine environment and species.

Competitive exclusion principle The principle that two species cannot simultaneously occupy the same niche in the same place.

Convention on International Trade in Endangered Species of Wild Fauna and Flora (CITES) A convention under which commercial trade in threatened species is either banned or regulated.

Coriolis effect The deflection of moving objects including water currents (to the right in the northern hemisphere and to the left in the southern hemisphere) as a result of the earth's rotation towards the east and the difference in the speed of the earth's surface at different latitudes.

Critical habitats (or key habitats) Habitats that are crucial in the life cycle of marine species; typically, nursery and spawning areas such as estuaries, mangroves, seagrass meadows and reefs.

Customary marine tenure (CMT) Legal, traditional or *de facto* control of areas of water and resources by indigenous people.

Demersal Living on, or near, the sea floor.

Detritus Material resulting from the disintegration of dead plants and animals.

Diarrhetic shellfish poisoning (DSP) Poisoning resulting from the consumption of shellfish such as mussels contaminated with toxins produced by particular species of dinoflagellates.

Dispersal envelope The probability distribution of dispersal distances around a source location, such as a no-take fish reserve.

Dissolved organic matter (DOM) Organic matter that is dissolved in water rather than existing as particles (compare detritus).

Ecological resilience Capacity of a natural ecosystem to recover from disturbance.

Ecological succession Successive changes in the structure of a community resulting from ecological modification.

Ecologically sustainable development (ESD) Development based on the sustainable use both of species and ecosystems, the maintenance of essential ecological processes, and the preservation of biological diversity. Use that benefits the present generation without diminishing those to future generations.

Ecology The study of the interactions between groups of organisms and their environment.

Ecosystem A relatively self-contained, dynamic system composed of a natural community along with its physical environment.

Ecotourism Tourism directed towards and intended to support conservation efforts for natural environments that may be threatened.

Environmental impact assessment (EIA) A formal process for evaluating the likely and possible risks or impacts of a proposed activity or development on the environment; also referred to as an environmental impact statement (EIS).

Escape gap A gap built into a trap or net to allow the escapement of small individuals.

Escherichia coli (E. coli) Bacteria commonly found in human intestines. Its presence in water is used as an indicator of contamination by sewage.

Estuary A semi-enclosed area at the mouth of a creek or river where freshwater and seawater meet and mix.

Eutrophic (of a body of water) Water so rich in nutrients that it encourages a dense growth of plants, the decomposition of which uses up available oxygen and therefore kills animal life.

Exclusive economic zones An area of sea out to 200 nautical miles from coastlines or outer reefs, in which an adjacent country has control and responsibilities.

Extinction The total disappearance of a species.

Extirpation The local extinction of stocks or sub-stocks from an area.

Fisheries management Management to ensure that catches from a fish stock are sustainable in the long-term and that benefits to fishers and communities are optimized.

Fisheries regulations Controls designed to either restrict the amount, or efficiency, of fishing (input controls), or to restrict the total catch (output controls) to predefined limits in a fishery.

Fishery management plan A document that identifies stakeholders, details objectives, and specifies the management regulations that apply to a fishery.

Fishing mortality The death of fish caused by a fishing operation.

Global warming A term used to denote the accelerated warming of the earth's surface due to anthropogenic (human activity-related) releases of greenhouse gases generated from industrial activity and deforestation.

Greenhouse effect The trapping of the sun's warmth at the earth's surface caused by high levels of carbon dioxide and other gases that allow the passage of incoming solar radiation but reduce the amount of reflected radiation.

Groundfish A broad array of demersal species that are captured by towed gear, such as trawls.

Growth overfishing A level of fishing in which many small individuals are caught before they grow to a size at which the stock biomass is maximized. A level of fishing greater than that required to maximize yield (or value) per recruit.

Guyot (or tablemount) A single flat-topped volcanic peak rising from the sea floor.

Gyre A current that flows in an elliptical path around 30 degrees of latitude in both hemispheres.

Habitat The place where an organism lives.

Herbivore An animal that eats plant material.

Heterotroph An organism that is incapable of producing organic compounds from inorganic raw materials and must, therefore, consume other organisms to obtain nourishment and energy (compare autotroph).

Indicator A measurable variable whose values and fluctuations reflect the state of a fishery in relation to some objective and its reference point. If the objective is a particular level of spawning stock biomass, the obvious indicator is the spawning stock biomass itself or some measurable index of it (such as catch rate of sexually mature individuals).

Individual transferable quota (ITQ) A catch limit or quota allocated to an individual fisher, who then has a guaranteed share (which may be either harvested or traded) of the total allowable catch of a particular resource species.

Input controls (on fisheries) Limitations on the amount of fishing effort or restrictions on the number, type and size of fishing vessels or fishing gear in a fishery.

Integrated coastal management (ICM) Coastal management which takes into account the interdependence of ecosystems, with the involvement of many different agencies and stakeholders.

Joint venture A partnership between foreign and local fishers.

Key habitat See critical habitat.

Keystone species A predator that maintains the structure of a community by removing a competitively dominant species and making it possible for many other species to live in a particular habitat.

Large marine ecosystems (LMEs) Large regions of ocean that share similar hydrographic characteristics, contain trophically-dependent populations of aquatic species and have been identified as such for monitoring and management purposes.

Licence limitation The restriction of fishing to those people, fishing units or vessels holding licences in a fishery.

Marine protected area (MPA) Any area of intertidal or subtidal terrain, together with its overlying waters and associated flora, fauna, historical and cultural features, which has been reserved by legislation or other effective means to protect part or all of the enclosed environment (from IUCN).

Maximum economic yield (MEY) The theoretical greatest difference between total revenues and total costs of exploiting a fish stock under existing environmental conditions.

Maximum legal size A regulation in which captured individuals larger than a prescribed maximum size must be returned to the sea; usually justified on the grounds that larger individuals produce a greater number of eggs, and are often less marketable, than smaller individuals.

Maximum sustainable yield (MSY) The highest theoretical equilibrium yield that can be continuously taken (on average) from a stock under existing environmental conditions without significantly affecting the reproduction process. Its use as an objective is dangerous, but it is commonly used as a limit.

Minimum legal size A regulation in which captured individuals smaller than a prescribed minimum size must be returned to the sea; usually justified on the grounds that growth of smaller individuals produces a greater harvestable biomass, and that the size of the spawning stock is increased.

Minimum mesh size The smallest size of mesh permitted in nets and traps; imposed on the basis that smaller individuals will escape unharmed.

Minimum stock size maintenance A management objective which aims to ensure (at least with some level of probability) that the stock does not decrease below some minimum 'buffer' level – often the lowest historical level which is not associated with observed decreases in recruitment or stock stability.

Mutualism A form of symbiosis in which two species are closely associated and both gain advantages – see symbiosis.

Natural mortality The death of fish by natural causes, including predation.

Neurotoxic shellfish poisoning (NSP) Poisoning resulting from the consumption of shellfish contaminated with toxins from the dinoflagellate *Gymnodinium breve* and related species.

Niche The position or role taken by an organism within an ecosystem.

No-take fishery reserve A marine protected area within which extractive fishing activities are not permitted.

Nutrients In the context of the marine environment, dissolved food material (mainly nitrates and phosphates) required by autotrophs to produce organic matter.

Omnivore An organism that obtains food from more than one trophic level.

Open-access fishery A fishery with no restriction on the number of fishers or fishing units; an unmanaged fishery.

Output controls (in fisheries management) Limitations on the weight of the catch (a quota), or the allowable size, sex, or reproductive condition of individuals in the catch.

Overexploitation The situation where so many fish are removed from a stock that reproduction cannot replace the numbers lost.

Ozone A gas (O_3) that forms a layer, at altitudes from 20–50 km, in the stratosphere and acts to protect the earth from ultraviolet radiation.

Paralytic shellfish poisoning (PSP) Poisoning resulting from the consumption of shellfish contaminated with toxins from certain dinoflagellates including *Gymnodinium catenatum*.

Parasitism An association between two species in which one (the parasite) draws nourishment from the other species (the host) to the latter's detriment.

Pelagic Living in the surface layers of the sea.

Photosynthesis The process by which organic material is formed from water, nutrients and carbon dioxide using energy absorbed from sunlight.

Phytoplankton Small plants, which drift in the sunlit surface layers of the sea.

Pollution (marine) The introduction by humans, either directly or indirectly, of any substance (or energy such as heat) into the marine environment which results, or is likely to result, in harm to marine life and hazards to human health.

Population A group of individuals belonging to the same species; in fisheries biology, the word 'stock' is sometimes used interchangeably (and loosely) with 'population'.

Precautionary principle The principle that the lack of full scientific certainty shall be not be used as a reason for postponing cost-effective measures to prevent environmental degradation where there are threats of serious or irreversible damage.

Primary productivity The rate of formation of organic compounds from inorganic material.

Property rights A degree of resource ownership by an individual fisher, group or community.

Quota A limit on the weight of fish which may be caught in a particular stock or area; a bag limit is a quota (usually in numbers of fish caught) applied to recreational fishers.

Recreational fishery One in which fishing is for sport or entertainment with often a secondary aim of securing food for domestic consumption.

Recruitment overfishing A level of fishing in which spawning stocks are reduced to the level at which there is an obvious relationship between stock and recruitment, and the number of recruits produced are insufficient to maintain the population.

Recruitment The addition of a cohort of young animals to a population.

Reference point The state or value of some indicator (say spawning stock size) which corresponds to a desirable position (a target reference point) or an undesirable position (limit reference point) that requires urgent action.

Risk assessment (in fisheries management) Management that involves taking uncertainty into account; allows decisions on alternative management options to achieve an objective within an acceptable level of risk.

Rotational closures, periodic harvesting or pulse fishing The closing and opening of a fishery, or parts of a fishery, on a rotational basis.

Seamount A single volcanic peak with a rounded top rising from the sea floor.

Septic tank An underground tank in which the organic matter in sewage is decomposed through bacterial activity.

Sewage Waste matter, particularly human faeces and urine, conveyed in sewers which are part of a sewerage system.

Species A distinct group of animals or plants in which the individuals are able to breed among themselves and produce viable offspring, but are unable to breed with individuals of other groups.

Spillover (in relation to fish reserves) Increases in the numbers of a species in fished areas surrounding a no-take reserve resulting from the net movement of juveniles and adults spilling out of the reserve.

Stakeholders The different people, groups, communities and organizations that have an interest in a particular activity, resource or area.

Stock assessment Quantitative studies that lead to predictions of how stocks will respond under various management actions.

Subsistence fishery One in which a major part of the catch is used by the fishers and their families for food and a lesser part may be sold.

Super profits (economic rent) Profits above those needed to make a return on investment, when compared to the returns which could be generated in other sectors.

Symbiosis A close association between two species which either advantages both species (mutualism) or advantages one while not harming the other (commensalism).

Target species The resource species at which a fishing operation is directed.

Technology creep A gradual increase in the efficiency of fishing gear and methods, which results in an increase in effective fishing effort.

Thermocline An area of the vertical water column in which there are rapid changes in temperature with increasing depth.

Total allowable catch (TAC)   The maximum catch allowed from a fishery in accordance with a specified management plan.

Traditional fishery   A fishery that has existed in indigenous communities for many generations and in which customary patterns of exploitation and management have developed.

Tributyl tin (TBT)   A constituent of antifouling paint that kills marine animals or interferes with their reproduction. The use of TBT is banned in many countries.

Trophic cascade   Changes in the relative abundances of species in different trophic levels caused by the change of abundance of one species at one trophic level: e.g. if species A is removed by fishing, prey species B may increase, and B's prey may subsequently decrease.

Trophic level   A feeding level containing organisms that obtain their nourishment in a similar way and from a similar source.

Tsunami   A sea wave (or series of waves) caused by underwater earthquakes or other disturbances.

Upwelling   The vertical movement of water from the depths up into the surface layers of the sea.

Vulnerability   Refers to the proportion of fish in a fishing gear's area of influence that is retained by the gear.

Wetlands   Low-lying terrestrial areas that are flooded by tides and either contain or are saturated with water. Examples include salt marshes, coastal swamps and mangrove forests.

Zooxanthellae   Microscopic plant cells living symbiotically within coral polyps and the mantles of giant clams.

Index

Printed and bound by CPI Group (UK) Ltd, Croydon, CR0 4YY